T0323538

Analytical Heat Transfer

Analytical Heat Transfer

Second Edition

Je-Chin Han
Lesley M. Wright

CRC Press
Taylor & Francis Group
Boca Raton London New York

CRC Press is an imprint of the
Taylor & Francis Group, an **informa** business

Second edition published 2022
by CRC Press
6000 Broken Sound Parkway NW, Suite 300, Boca Raton, FL 33487-2742

and by CRC Press
4 Park Square, Milton Park, Abingdon, Oxon, OX14 4RN

CRC Press is an imprint of Taylor & Francis Group, LLC

© 2022 Je-Chin Han and Lesley M. Wright

First edition published by CRC Press in 2012

Library of Congress Cataloging-in-Publication Data
Names: Han, Je-Chin, 1946- author. | Wright, Lesley, author.
Title: Analytical heat transfer / Je-Chin Han, Lesley Wright.
Description: Second edition. | Boca Raton, FL : CRC Press, 2022. | Includes
bibliographical references and index. | Summary: "The second edition features new content on Duhamel's superposition method, Green's function method for transient heat conduction, finite-difference method for steady state and transient heat conduction in cylindrical coordinates, and laminar mixed convection. It includes three new chapters on boiling, condensation, and heat exchanger analysis, in addition to end-of-chapter problems. Provides step-by-step mathematical formula derivations, analytical solution procedures, and demonstration examples. Includes end-of-chapter problems with an accompanying Solutions Manual for instructors. This book is ideal for undergraduate and graduate students studying basic heat transfer and advanced heat transfer"—Provided by publisher.
Identifiers: LCCN 2021059841 (print) | LCCN 2021059842 (ebook) |
ISBN 9780367758974 (hbk) | ISBN 9780367759001 (pbk) | ISBN 9781003164487 (ebk)
Subjects: LCSH: Heat—Transmission.
Classification: LCC QC320 .H225 2022 (print) | LCC QC320 (ebook) |
DDC 621.402/2—dc23/eng20220408
LC record available at https://lccn.loc.gov/2021059841
LC ebook record available at https://lccn.loc.gov/2021059842

ISBN: 978-0-367-75897-4 (hbk)
ISBN: 978-0-367-75900-1 (pbk)
ISBN: 978-1-003-16448-7 (ebk)

DOI: 10.1201/9781003164487

Typeset in Times
by codeMantra

Contents

*Topics with an asterisk can be considered as advanced heat transfer.

Preface to the Second Edition

This book, *Analytical Heat Transfer*, was first published in 2012. This book bridges the gap between undergraduate-level, basic heat transfer and graduate-level, advanced heat transfer, and it serves the need of entry-level graduate students. Since it was published, we have received many useful comments and suggestions that the book should include additional topics in conduction, convection, and radiation heat transfer. In addition to combined conduction, convection, and radiation heat transfer as an intermediate-level graduate course, the advanced-level combined conduction and radiation course, and the advanced-level graduate convection course are taught as advanced-level graduate classes at many universities. To satisfy the multiple objectives, there is a need to expand the current book by including more advanced heat transfer topics.

This text is a revision of the first edition. The primary content and framework have been based on the first edition. Keeping the same format, the revised second edition adds new material at the end of each chapter. The following relevant advanced topics (Contents with an asterisk *) have been added to the second edition: steady state and transient heat conduction in cylindrical coordinates, Duhamel's superposition method and Green's function method for transient heat conduction, mass and heat transfer analogy, external laminar flow and heat transfer with a constant pressure gradient, heat transfer in high speed flows, thermally developing heat transfer in a circular tube, flow and heat transfer in a rectangular channel, laminar mixed convection, laminar-turbulent transitional heat transfer, unsteady highly turbulent flows, film cooling flows, turbulent flow heat transfer enhancement, heat transfer in rotating channels, numerical modeling for turbulent flow heat transfer, combined radiation with conduction and/or convection, and gas radiation optically thin and optically thick limits.

We hope this revised book will be useful for (a) the intermediate-level graduate heat transfer course (Chapters 1–14, only Contents without an asterisk *), as well as for (b) the advanced-level graduate combined conduction and radiation heat transfer course (Chapters 1–5 and 11–14 including Contents with an asterisk *), and for (c) the advanced-level graduate convection heat transfer course (Chapters 6–10 and 15–16 including Contents with an asterisk *). The instructors shall have the option to choose the relevant topics for their (a), (b), and (c) courses. We would be happy to receive constructive comments and suggestions on the material presented in the second edition book.

While preparing this second edition manuscript, we heavily referenced the following books and therefore are deeply appreciative to their authors:

V. Arpaci, *Conduction Heat Transfer*, Addison-Wesley Publishing Company, Reading, MA, 1966.

D.W. Hahn and M.N. Özişik, *Heat Conduction*, Third Edition, John Wiley & Sons, Hoboken, NJ, 2012.

H. Schlichting, *Boundary-Layer Theory*, Sixth Edition, McGraw-Hill, New York, 1968.

W.M. Kays and M.E. Crawford, *Convective Heat and Mass Transfer*, Second Edition, McGraw-Hill, New York, 1980.

R. Siegel and J. Howell, *Thermal Radiation Heat Transfer*, McGraw-Hill, New York, 1972.

Finally, we would like to sincerely express special thanks to former Texas A&M University PhD student, Dr. Izzet Sahin (PhD, 2021), and current Texas A&M University PhD student, Mr. I-Lun Chen. They spent significant time and effort to prepare most of the new manuscript with advanced topics from original hand-written equations and drawings. Without their diligent and persistent contributions, this second edition book would not be possible.

Je-Chin Han
Lesley M. Wright

Preface to the First Edition

Fundamental Heat Transfer is a required course for all mechanical, chemical, nuclear, and aerospace engineering undergraduate students. This senior-level undergraduate course typically covers conduction, convection, and radiation heat transfer. Advanced Heat Transfer courses are also required for most engineering graduate students. These graduate-level courses are typically taught as individual courses named Conduction, Convection, or Radiation. Many universities also offer an Intermediate Heat Transfer or Advanced Heat Transfer course to cover conduction, convection, and radiation for engineering graduate students. For these courses, however, there are not many textbooks available that cover conduction, convection, and radiation at the graduate level.

I have taught an Intermediate Heat Transfer course in the Department of Mechanical Engineering at Texas A&M University since 1980. This book has evolved from a series of my lecture notes for teaching a graduate-level intermediate heat transfer course over the past 30 years. Many MS degree students majoring in Thermal and Fluids have taken this course as their only graduate-level heat transfer course. And many PhD degree candidates have taken this course to prepare for their heat transfer qualifying examinations as well as to prepare for their advanced-level courses in conduction, convection, or radiation. This book bridges the gap between undergraduate-level basic heat transfer and graduate-level advanced heat transfer as well as serves the need of entry-level graduate students.

Analytical Heat Transfer focuses on how to analyze and solve the classic heat transfer problems in conduction, convection, and radiation in one book. This book emphasizes on how to model and how to solve the engineering heat transfer problems analytically, rather than simply applying the equations and correlations for engineering problem calculations. This book provides many well-known analytical methods and their solutions such as Bessel functions, separation of variables, similarity method, integral method, and matrix inversion method for entry-level engineering graduate students. It is unique in that it provides (1) detailed step-by-step mathematical formula derivations, (2) analytical solution procedures, and (3) many demonstration examples. This analytical knowledge will equip graduate students with the much-needed capability to read and understand the heat-transfer-related research papers in the open literature and give them a strong analytical background with which to tackle and solve the complex engineering heat transfer problems they will encounter in their professional lives.

This book is intended to cover intermediate heat transfer between the undergraduate and the advanced graduate heat transfer levels. It includes 14 chapters and an Appendix:

Chapter 1 Heat Conduction Equations
Chapter 2 1-D Steady-State Heat Conduction
Chapter 3 2-D Steady-State Heat Conduction
Chapter 4 Transient Heat Conduction
Chapter 5 Numerical Analysis in Heat Conduction

Chapter 6 Heat Convection Equations
Chapter 7 External Forced Convection
Chapter 8 Internal Forced Convection
Chapter 9 Natural Convection
Chapter 10 Turbulent Flow Heat Transfer
Chapter 11 Fundamental Radiation
Chapter 12 View Factor
Chapter 13 Radiation Exchange in a Nonparticipating Medium
Chapter 14 Radiation Transfer through Gases
Appendix A Mathematical Relations and Functions

There are many excellent undergraduate and graduate heat transfer text books available. Although I do not claim any new ideas in this book, I do attempt to present the subject in a systematic and logical manner. I hope this book is a unique compilation and is useful for graduate entry-level heat transfer study.

While preparing this manuscript, I heavily referenced the following books and therefore am deeply appreciative to their authors:

W. Rohsenow and H. Choi, *Heat, Mass, and Momentum Transfer*, Prentice-Hall, Inc., Englewood Cliffs, NJ, 1961.

A. Mills, *Heat Transfer*, Richard D. Irwin, Inc., Boston, MA, 1992.

K. Vincent Wong, *Intermediate Heat Transfer*, Marcel Dekker, Inc., New York, 2003.

F. Incropera and D. Dewitt, *Fundamentals of Heat and Mass Transfer*, Fifth Edition, John Wiley & Sons, Hoboken, NJ, 2002.

Finally, I would like to sincerely express special thanks to my former student Dr. Zhihong (Janice) Gao (PhD, 2007). Janice spent a lot of time and effort to type most of the manuscript from my original *Intermediate Heat Transfer Class Notes* in 2005–2007. Without her diligent and persistent contributions, this book would be impossible. In addition, I would like to extend appreciation to my current PhD students Mr. Jiang Lei and Ms. Shiou-Jiuan Li for their help in completing the book and drawings in 2009–2010.

Je-Chin Han

Authors

Je-Chin Han is a university distinguished professor and Marcus Easterling Endowed Chair Professor at Texas A&M University. He earned a BS degree from National Taiwan University in 1970, an MS degree from Lehigh University in 1973, and a ScD from MIT in 1976, all in mechanical engineering. He has been working on turbine blade cooling, film cooling, and rotating coolant-passage heat transfer research for the past 40 years. He is the co-author of more than 250 journal papers, and lead author of the books: *Gas Turbine Heat Transfer and Cooling Technology*, *Analytical Heat Transfer*, and *Experimental Methods in Heat Transfer and Fluid Mechanics*. He has served as editor, associate editor, and honorary board member for eight heat transfer-related journals. He received the 2002 ASME Heat Transfer Memorial Award, the 2004 International Rotating Machinery Award, the 2004 AIAA Thermophysics Award, the 2013 ASME Heat Transfer Division 75th Anniversary Medal, the 2016 ASME IGTI Aircraft Engine Technology Award, and the 2016 ASME and AICHE Max Jakob Memorial Award. He is a fellow of ASME and AIAA and an Honorary Member of ASME.

Lesley M. Wright is an associate professor and Jana and Quentin A. Baker '78 faculty fellow at Texas A&M University. Prior to joining Texas A&M, she was a member of the mechanical engineering faculty at Baylor University for 10 years. She earned a BS in engineering in 2001 from Arkansas State University and an MS and a PhD in mechanical engineering from Texas A&M University in 2003 and 2006, respectively. Currently she is investigating enhanced convective cooling technology, including heat transfer enhancement for gas turbine cooling applications. This experimental research has led to the development of innovative cooling technology for both turbine blade film cooling and internal heat transfer enhancement. In addition, Dr. Wright continues to investigate the effect of rotation on the thermal performance of rotor blade cooling passages. Her research interests have also led to the development of novel experimental methods for the acquisition of detailed surface and flow measurements in highly turbulent flows. She has co-authored the book *Experimental Methods in Heat Transfer and Fluid Mechanics* and more than 100 refereed journal and conference papers. She is a fellow of ASME.

1 Heat Conduction Equations

1.1 INTRODUCTION: CONDUCTION, CONVECTION, AND RADIATION

1.1.1 CONDUCTION

Conduction is caused by the temperature gradient through a solid material. For example, Figure 1.1 shows that heat is conducted through a wall of a building or a container from the high-temperature side to the low-temperature side. This is a one-dimensional (1-D), steady-state, heat conduction problem if T_1 and T_2 are uniform. According to Fourier's conduction law, the temperature profile is linear through the plane wall.

1.1.1.1 Fourier's Conduction Law

$$q'' = -k\frac{dT}{dx} = k\frac{T_1 - T_2}{L} \tag{1.1}$$

and

$$q'' \equiv \frac{q}{A_c} \quad \text{or} \quad q = q'' A_c$$

where q'' is the heat flux (W/m²), q is the heat rate (W or J/s), k is the thermal conductivity of solid material (W/m K), A_c is the cross-sectional area for conduction, perpendicular to heat flow (m²), and L is the conduction length (m).

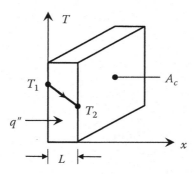

FIGURE 1.1 1-D heat conduction through a building or container wall.

DOI: 10.1201/9781003164487-1

One can predict heat rate or heat loss through the plane wall by knowing T_1, T_2, k, L, and A_c. This is the simple 1-D steady-state problem. However, in actual applications, there are many two-dimensional (2-D) or three-dimensional (3-D) steady-state heat conduction problems; there are cases where heat generation occurs in the solid material during heat conduction. Also, transient heat conduction problems take place in many engineering applications. In addition, some special applications involve heat conduction with a moving boundary. These more complicated heat conduction problems will be discussed in the following chapters.

1.1.2 CONVECTION

Convection is caused by fluid flow motion over a solid surface. For example, Figure 1.2 shows that heat is removed from a heated solid surface to cooling fluid. This is a 2-D boundary-layer flow and heat transfer problem. According to Newton, the heat removal rate from the heated surface is proportional to the temperature difference between the heated wall and the cooling fluid. The proportionality constant is known as the heat transfer coefficient; the same heat rate from the heated surface can be determined by applying Fourier's Conduction Law to the cooling fluid.

1.1.2.1 Newton's Cooling Law

$$q'' = -k_f \frac{dT}{dy}\Big|_{\text{at wall}} = h(T_s - T_\infty) \tag{1.2}$$

Also,

$$h = \frac{q''}{T_s - T_\infty} = \frac{-k_f \frac{dT}{dy}\Big|_{y=0}}{T_s - T_\infty} \tag{1.3}$$

and

$$q'' = \frac{q}{A_s} \quad \text{or} \quad q = q'' A_s$$

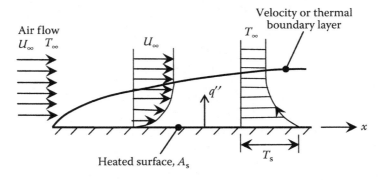

FIGURE 1.2 Velocity and thermal boundary layer.

TABLE 1.1

Typical Values of Heat Transfer Coefficient

Type of Convection	h, $W/m^2 \cdot K$
Natural convection	
Caused by ΔT: air	5
Caused by ΔT: water	25
Forced convection	
Caused by fan, blower: air	25–250
Caused by pump: water	50–20,000
Boiling or condensation	
Caused by phase change	
Water $\rightleftharpoons$ Steam	10,000–100,000
Freon $\rightleftharpoons$ Vapor	2500–50,000

---

where T_s is the surface temperature (°C or K), T_∞ is the fluid temperature (°C or K), h is the heat transfer coefficient (W/m^2 K), k_f is the thermal conductivity of fluid (W/mK), A_s is the surface area for convection, exposed to the fluid (m^2).

It is noted that the heat transfer coefficient depends on fluid properties (such as air or water as the coolant), flow conditions (i.e., laminar or turbulent flows), surface configurations (such as flat surface or circular tube), and so on. The heat transfer coefficient can be determined experimentally or analytically. This textbook focuses on analytical solutions. From Equation (1.3), the heat transfer coefficient can be determined by knowing the temperature profile in the cooling fluid during convection. With this analytical profile, the temperature gradient near the wall, dT/dy, can be used to determine the heat transfer coefficient. However, this requires solving the 2-D boundary-layer equations and will be the subject of the following chapters. Before solving 2-D boundary-layer equations, one needs the heat transfer coefficient as the convection boundary condition (BC) in order to solve the heat conduction problem. Therefore, Table 1.1 provides some typical values of heat transfer coefficient in many convection problems. As can be seen, in general, forced convection provides more heat transfer than natural convection; water as a coolant removes much more heat than air; and boiling or condensation, involving a phase change, has a much higher heat transfer coefficient than single-phase convection.

1.1.3 RADIATION

Radiation is caused by electromagnetic waves from solids, liquid surfaces, or gases. For example, Figure 1.3 shows that heat is radiated from a solid surface at a temperature greater than absolute zero. According to Stefan-Boltzmann, the radiation heat rate is proportional to the absolute temperature of the surface raised to the fourth power, the Stefan–Boltzmann constant, and the surface emissivity. The surface emissivity primarily depends on material, wavelength, and temperature. It is between 0 and 1. In general, the emissivity of metals is much less than nonmetals. Note that radiation from a surface can travel through air, as well as through a vacuum environment.

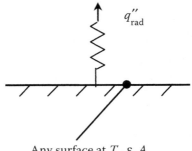

Any surface at T_s, ε, A_s

FIGURE 1.3   Radiation from a solid surface.

1.1.3.1   Stefan–Boltzmann Law

For a real surface,

$$q'' = \varepsilon \sigma T_s^4 \tag{1.4}$$

For an ideal (black) surface, $\varepsilon = 1$

$$q'' = \sigma T_s^4$$

and

$$q'' = \frac{q}{A_s} \quad \text{or} \quad q = q'' A_s$$

where ε is the emissivity of the real surface, $\varepsilon = 0 - 1$ ($\varepsilon_{\text{metal}} < \varepsilon_{\text{nonmetal}}$), T_s is the absolute temperature of the surface, K ($K = {}^\circ\text{C} + 273.15$), σ is the Stefan–Boltzmann constant, $\sigma = 5.67 \times 10^{-8}\,\text{W/m}^2\,\text{K}^4$, and A_s is the surface area for radiation (m²).

1.1.4   COMBINED MODES OF HEAT TRANSFER

In real applications, often radiation occurs concurrently as conduction or convection is present. This scenario represents combined modes of heat transfer. For example, Figure 1.4 shows heat transfer between two surfaces involving radiation and convection simultaneously.

Assume surface $A \ll$ surrounding sky or building wall surface, $T_\infty \neq T_{\text{sur}}$ (or, $T_\infty = T_{\text{sur}}$).

The total heat flux from the surface A is due to convection and radiation, and can be found as

$$q'' = q''_{\text{conv}} + q''_{\text{rad,net}} \tag{1.5}$$

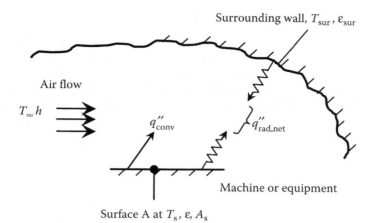

FIGURE 1.4 Heat transfer between two surfaces involving radiation and convection.

where

$$q''_{conv} = h(T_s - T_\infty)$$

$$q''_{rad,net} = \varepsilon\sigma T_s^4 - \alpha\varepsilon_{sur}\sigma T_{sur}^4$$

$$= \varepsilon\sigma\left(T_s^4 - T_{sur}^4\right), \quad \text{if } \varepsilon = \alpha, \ \varepsilon_{sur} = 1$$

$$= \varepsilon\sigma\left(T_s^2 + T_{sur}^2\right)(T_s + T_{sur})(T_s - T_{sur})$$

$$= h_r(T_s - T_{sur})$$

Therefore, from Equation (1.5)

$$q'' = h(T_s - T_\infty) + h_r(T_s - T_{sur}) \tag{1.6}$$

where

$$h_r = \varepsilon\sigma\left(T_s^2 + T_{sur}^2\right)(T_s + T_{sur}) \tag{1.7}$$

Also

$$q'' \equiv \frac{q}{A_s} \quad \text{or} \quad q = q'' A_s$$

where α is the absorptivity, T_{sur} is the surrounding wall temperature (°C or K), ε_{sur} is the emissivity of the surrounding wall, h_r is the radiation heat transfer coefficient (W/m² K), and A_s is the surface area for radiation (m²). Total heat transfer rate can be determined by knowing T_s, T_{sur}, h, ε, ε_{sur}, σ, and A_s.

1.2 GENERAL HEAT CONDUCTION EQUATIONS

If the temperature profile inside a solid material $T(x, y, z, t)$ is known, the heat rate q through the solid can be determined as shown in Figure 1.5. As far as thermal stress is concerned, it is equally important to predict the temperature profile in high-temperature applications. In this chapter, the general heat conduction equations will be derived. The general heat conduction equation can be used to solve various real problems with the appropriate BCs and initial condition. The heat conduction can be modelled as 1-D, 2-D, or 3-D depending on the nature of the problem:

1-D $T(x)$ for steady state or $T(x, t)$ for transient problems
2-D $T(x, y)$ for steady state or $T(x, y, t)$ for transient problems
3-D $T(x, y, z)$ for steady state or $T(x, y, z, t)$ for transient problems

To determine the temperature profile, the following should be given:
1. Initial condition and BCs
2. Material thermal conductivity k, density ρ, specific heat C_p, and diffusivity $\alpha = k/\rho C_p$

1.2.1 DERIVATIONS OF GENERAL HEAT CONDUCTION EQUATIONS

The general form of the conservation of energy in a small control volume of solid material is

$$E_{in} - E_{out} + E_g = E_{st} \qquad (1.8)$$

where $E_{in} - E_{out}$ is the net heat conduction, E_g is the heat generation, and E_{st} is the energy stored in the control volume.

Figure 1.6 shows the conservation of energy in a differential control volume in a 3-D Cartesian (rectangular) coordinate. If we consider energy conservation in a 1-D system (x-direction only),

$$q_x - \left(q_x + \frac{\partial q_x}{\partial x} dx \right) + E_g = E_{st} \qquad (1.9)$$

The conduction heat rates can be evaluated from Fourier's Law,

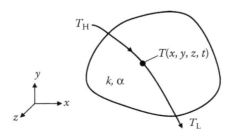

FIGURE 1.5 Heat conduction through a solid medium.

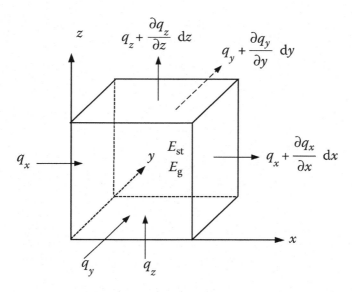

FIGURE 1.6 The volume element for deriving the heat conduction equation.

$$q_x = -kA_x \frac{\partial T}{\partial x}$$ (1.10)

With control surface area $A_x = dy\,dz$, Equation (1.9) can be written as

$$-\frac{\partial}{\partial x}\left(-k\,dy\,dz\frac{\partial T}{\partial x}\right)dx + E_g = E_{st}$$ (1.11)

The thermal energy generation can be represented by

$$E_g = \dot{q}\,dx\,dy\,dz$$ (1.12)

where $\dot{q}$ is the energy generation per unit volume, $dx\,dy\,dz$.

The energy storage can be expressed as

$$E_{st} = \frac{\partial\left(\rho\,dx\,dy\,dz \cdot C_p \cdot T\right)}{\partial t}$$ (1.13)

Substituting Equations (1.12) and (1.13) into Equation (1.11), we have

$$-\frac{\partial}{\partial x}\left(-k\frac{\partial T}{\partial x}\right)dx\,dy\,dz + \dot{q}\,dx\,dy\,dz = \frac{\partial\left(\rho\,dx\,dy\,dz \cdot C_p \cdot T\right)}{\partial t}$$ (1.14)

Dividing out the dimensions of the small control volume, $dx\,dy\,dz$, Equation (1.14) is simplified as

$$\frac{\partial}{\partial x}\left(k\frac{\partial T}{\partial x}\right)+\dot{q}=\frac{\partial(\rho C_p T)}{\partial t} \qquad (1.15)$$

If we consider energy conservation, in a 3-D system (x-direction, y-direction, z-direction), the heat equation can be written as

$$\frac{\partial}{\partial x}\left(k\frac{\partial T}{\partial x}\right)+\frac{\partial}{\partial y}\left(k\frac{\partial T}{\partial y}\right)+\frac{\partial}{\partial z}\left(k\frac{\partial T}{\partial z}\right)+\dot{q}=\frac{\partial(\rho C_p T)}{\partial t} \qquad (1.16)$$

The thermal conductivity k, which is a function of temperature, is often difficult to determine. If we assume that k, ρ, and C_p are constants, then Equation (1.16) is simplified as

$$\frac{\partial^2 T}{\partial x^2}+\frac{\partial^2 T}{\partial y^2}+\frac{\partial^2 T}{\partial z^2}+\frac{\dot{q}}{k}=\frac{1}{\alpha}\frac{\partial T}{\partial t} \qquad (1.17)$$

where $\alpha = k/\rho C_p$ is the thermal diffusivity.

In a 3-D, cylindrical coordinate system as shown in Figure 1.7a, the heat conduction equation has the form of

$$\frac{1}{r}\frac{\partial}{\partial r}\left(r\frac{\partial T}{\partial r}\right)+\frac{1}{r^2}\frac{\partial^2 T}{\partial\phi^2}+\frac{\partial^2 T}{\partial z^2}+\frac{\dot{q}}{k}=\frac{1}{\alpha}\frac{\partial T}{\partial t} \qquad (1.18)$$

In a 3-D spherical coordinate system as shown in Figure 1.7b, it has the form of

$$\frac{1}{r^2}\frac{\partial}{\partial r}\left(r^2\frac{\partial T}{\partial r}\right)+\frac{1}{r^2\sin\theta}\frac{\partial}{\partial\theta}\left(\sin\theta\frac{\partial T}{\partial\theta}\right)+\frac{1}{r^2\sin^2\theta}\frac{\partial^2 T}{\partial\phi^2}+\frac{\dot{q}}{k}=\frac{1}{\alpha}\frac{\partial T}{\partial t} \qquad (1.19)$$

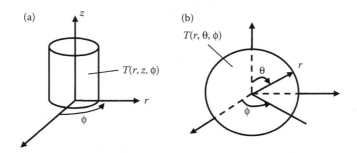

FIGURE 1.7   (a) Cylindrical coordinate system. (b) Spherical coordinate system.

1.3 BOUNDARY AND INITIAL CONDITIONS

The physical conditions existing on the boundary should be known in order to solve the heat conduction equation and determine the temperature profile within the medium. Moreover, the initial condition $T(x, 0) = T_i$ should also be known if the heat transfer is time-dependent.

1.3.1 BOUNDARY CONDITIONS

There are three types of BCs commonly found in many heat transfer applications [1].

1. Specified surface temperature $T(0, t) = T_s$, as shown in Figure 1.8
2. Specified surface heat flux, as shown in Figure 1.9a and b
   a. Finite heat flux

$$-k\frac{\partial T(0,T)}{\partial x} = q''_s \qquad (1.20)$$

   b. Adiabatic or insulated surface, which is a special case

$$-k\frac{\partial T(0,T)}{\partial x} = q''_s = 0 \qquad (1.21)$$

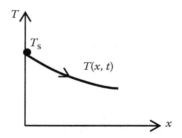

FIGURE 1.8 Boundary conditions—given surface temperature.

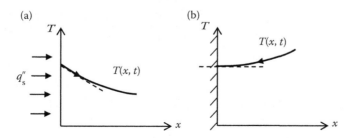

FIGURE 1.9 (a) Finite heat flux. (b) Adiabatic surface.

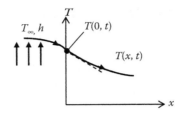

FIGURE 1.10   Convective boundary conditions.

FIGURE 1.11   Initial condition.

3. Specified surface convection, as shown in Figure 1.10

$$-k\frac{\partial T(0,T)}{\partial x} = h\big[T_\infty - T(0,t)\big] \tag{1.22}$$

1.3.2   INITIAL CONDITIONS

An initial condition, as shown in Figure 1.11, is required for the transient heat transfer problem.

$$T(x,0) = T_i \tag{1.23}$$

1.4   SIMPLIFIED HEAT CONDUCTION EQUATIONS

1. Steady state, $\partial T/\partial t = 0$, no heat generation, $\dot{q} = 0$,

$$\text{1-D} \quad \frac{\partial^2 T}{\partial x^2} = 0$$

$$\text{2-D} \quad \frac{\partial^2 T}{\partial x^2} + \frac{\partial^2 T}{\partial y^2} = 0$$

$$\text{3-D} \quad \frac{\partial^2 T}{\partial x^2} + \frac{\partial^2 T}{\partial y^2} + \frac{\partial^2 T}{\partial z^2} = 0$$

2. Steady state, 1-D with a heat source (heater application)

$$\frac{\partial^2 T}{\partial x^2} + \frac{\dot{q}}{k} = 0$$

Steady state, 1-D with heat sink–fin application

$$\frac{\partial^2 T}{\partial x^2} - \frac{\dot{q}}{k} = 0$$

3. Transient, without heat generation $\dot{q} = 0$

$$\text{1-D} \quad \frac{\partial^2 T}{\partial x^2} = \frac{1}{\alpha} \frac{\partial T}{\partial t}$$

$$\text{2-D} \quad \frac{\partial^2 T}{\partial x^2} + \frac{\partial^2 T}{\partial y^2} = \frac{1}{\alpha} \frac{\partial T}{\partial t}$$

$$\text{3-D} \quad \frac{\partial^2 T}{\partial x^2} + \frac{\partial^2 T}{\partial y^2} + \frac{\partial^2 T}{\partial z^2} = \frac{1}{\alpha} \frac{\partial T}{\partial t}$$

The above-mentioned three types of BCs can be applied to 1-D, 2-D, or 3-D heat conduction problems, respectively. For example, as shown in Figure 1.12,

$$x = 0, \qquad -k\frac{\partial T(0,y,t)}{\partial x} = 0 \quad \left(\text{adiabatic surface}\right)$$

$$x = a, \quad -k\frac{\partial T(a,y,t)}{\partial x} = h\left[T(a,y,t) - T_\infty\right] \quad \left(\text{surface convection}\right)$$

$$y = 0, \qquad -k\frac{\partial T(x,0,t)}{\partial y} = q'' \quad \left(\text{surface heat flux}\right)$$

$$y = b, \qquad T(x,b,t) = T_s \quad \left(\text{surface temperature}\right)$$

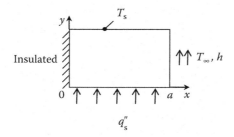

FIGURE 1.12 Heat conduction in 2-D system with various boundary conditions.

Examples

1.1 Consider 1-D steady-state heat conduction through a plane wall, without heat generation, as shown in Figure 1.13.

a. If the left side surface is at a constant temperature T_1, and the right side surface is at a lower constant temperature T_2, determine the temperature and heat flux distributions through the plane wall, $T(x)$, $q''(x)$. If the plane wall has a heat conduction cross-sectional area A_c, determine the total heat transfer rate, $q(x)$.
b. At $x = 0$, $T = T_1$, but the right side surface is exposed to a convection fluid at a lower temperature, T_∞, with a convection heat transfer coefficient h. Determine $T(x)$, T_2, $q''(x)$, and $q(x)$.
c. From case (b), in addition to a convection fluid, the right side surface also has radiation heat transfer with the surrounding surface at temperature $T_{sur} = T_\infty$. Determine $T(x)$, T_2, $q''(x)$, and $q(x)$.
d. If the left side surface receives a constant heat flux q_s'', and the right side surface is at a lower constant temperature T_2, determine $T(x)$, T_1, $q''(x)$, and $q(x)$.
e. The left side surface receives a constant heat flux, q_s'', but the right side surface is exposed to a convection fluid at a lower temperature, T_∞, with a convection heat transfer coefficient h, determine $T(x)$, T_1, T_2, $q''(x)$, and $q(x)$.
f. From the case (e), in addition to convection fluid, the right side surface also has radiation heat transfer with the surrounding surface at temperature $T_{sur} = T_\infty$, determine $T(x)$, T_1, T_2, $q''(x)$, and $q(x)$.

Solution

a. The 1-D heat conduction equation is

$$\frac{d^2T}{dx^2} = 0$$

The solution of temperature distribution is

$$\frac{dT}{dx} = C_1$$

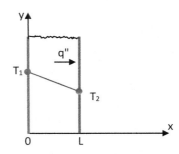

FIGURE 1.13 1-D heat conduction through a plane wall.

$$T(x) = C_1 x + C_2$$

Apply the BCs:

$$x = 0, \quad T = T_1, \quad T = T_1 = C_1 \cdot 0 + C_2$$

$$x = L, \quad T = T_2, \quad T = T_2 = C_1 \cdot L + C_2$$

Therefore, $C_2 = T_1$ and $C_1 = \dfrac{T_2 - T_1}{L}$

Temperature distribution: $T(x) = \dfrac{T_2 - T_1}{L} x + T_1$

Heat flux distribution: $q''(x) = -k \dfrac{\partial T}{\partial x}$

$$q''(x) = -k \frac{T_2 - T_1}{L}, \quad q''(x) = k \frac{T_1 - T_2}{L}$$

Heat transfer rate: $q(x) = -kA_c \dfrac{\partial T}{\partial x}$

$$q(x) = kA_c \frac{T_1 - T_2}{L}$$

b. The temperature $T(x)$, heat flux $q''(x)$, and heat transfer rate $q(x)$ distributions are the same as those shown in (a), but T_2 is unknown. Therefore, we need to determine T_2 based on the convection fluid temperature, T_∞, and heat transfer coefficient, h:

$$\text{at } x = L, -k \frac{\partial T}{\partial x} = h(T_2 - T_\infty)$$

$$\text{from (a) } \frac{\partial T}{\partial x} = \frac{T_2 - T_1}{L}$$

$$-k \frac{T_2 - T_1}{L} = h(T_2 - T_\infty)$$

$$\text{Therefore, } T_2 = \frac{T_1 + \dfrac{hL}{k} T_\infty}{1 + \dfrac{hL}{k}}$$

c. Same as part (b), but T_2 now can be determined as:

$$\text{at } x = L, -k \frac{\partial T}{\partial x} = (h + h_r)(T_2 - T_\infty)$$

$$T_2 = \frac{T_1 + \dfrac{(h+h_r)L}{k}T_\infty}{1 + \dfrac{(h+h_r)L}{k}}$$

d. Same as part (a), but T_1 is unknown. We need to determine T_1 based on a given surface heat flux BCs as:

$$\text{at } x = 0, -k\frac{\partial T}{\partial x} = q_s'', \text{given}$$

$$\text{From (a)} - k\frac{T_2 - T_1}{L} = q_s''$$

$$\text{Therefore, } T_1 = T_2 + \frac{L}{k}q_s''$$

e. Same as part (d), but T_2 is unknown, need to determine T_2 based on surface convection BC as:

$$\text{at } x = L, -k\frac{\partial T}{\partial x} = q_s'' = h(T_2 - T_\infty)$$

$$T_2 = T_\infty + \frac{q_s''}{h}$$

$$T_1 = T_\infty + \frac{q_s''}{h} + \frac{L}{k}q_s''$$

f. Same as part (e), but T_2 now can be determined as:

$$\text{at } x = L - k\frac{\partial T}{\partial x} = q_s'' = (h+h_r)(T_2 - T_\infty)$$

$$T_2 = T_\infty + \frac{q_s''}{h+h_r}$$

$$T_1 = T_\infty + \frac{q_s''}{h+h_r} + \frac{L}{k}q_s''$$

1.2 Consider 1-D heat conduction through a hollow cylindrical wall without heat generation, as shown in Figure 1.14. Determine the temperature, heat flux, and heat transfer rate distributions through the cylindrical wall with the same BCs as shown in Example 1.1 (a) to (f).

Solution

a. 1-D heat conduction equation is

$$\frac{1}{r}\frac{d}{dr}\left(r\frac{dT}{dr}\right) = 0$$

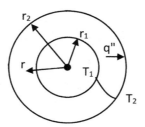

FIGURE 1.14 1-D heat conduction through a hollow cylindrical wall.

The solution of temperature distribution is

$$r\frac{dT}{dr} = C_1$$

$$\frac{dT}{dr} = \frac{C_1}{r}$$

$$T(r) = C_1 \ln r + C_2$$

$$\text{at} = r_1, T = T_1 = C_1 \ln r_1 + C_2$$

$$\text{at} = r_2, T = T_2 = C_1 \ln r_2 + C_2$$

Solve for at C_1 and C_2

$$C_1 = \frac{T_1 - T_2}{\ln \dfrac{r_1}{r_2}}$$

$$C_2 = T_1 - \frac{T_1 - T_2}{\ln \dfrac{r_1}{r_2}} \ln r_1$$

$$T = \frac{T_1 - T_2}{\ln \dfrac{r_1}{r_2}} \ln r + T_1 - \frac{T_1 - T_2}{\ln \dfrac{r_1}{r_2}} \ln r_1$$

$$T = T_1 + \frac{T_1 - T_2}{\ln \dfrac{r_1}{r_2}} \ln \frac{r}{r_1}$$

Temperature distribution: $T = T_1 - \dfrac{T_1 - T_2}{\ln \dfrac{r_2}{r_1}} \ln \dfrac{r}{r_1}$

Heat flux: $q''(r) = -k\dfrac{\partial T}{\partial r}$

$$\frac{\partial T}{\partial r} = \left[-\frac{T_1 - T_2}{\ln \frac{r_2}{r_1}} \frac{1}{r} \right]$$

where

$$q''(r) = -k \left[-\frac{T_1 - T_2}{\ln \frac{r_2}{r_1}} \frac{1}{r} \right]$$

Heat flux distribution: $q''(r) = k \dfrac{T_1 - T_2}{\ln \frac{r_2}{r_1}} \dfrac{1}{r}$

Heat transfer rate distribution: $q(r) = q''(r) A_c$

$$q(r) = k \frac{T_1 - T_2}{\ln \frac{r_2}{r_1}} \frac{1}{r} 2\pi r l$$

$$q(r) = k 2\pi l \frac{T_1 - T_2}{\ln \frac{r_2}{r_1}}$$

where l is the length of hollow cylindrical tube.

The temperature distribution is a logarithm with radius r:

$$T(r) = T_1 - \frac{T_1 - T_2}{\ln \frac{r_2}{r_1}} \ln \frac{r}{r_1}$$

Heat flux distribution depends on the radius r:

$$q''(r) = k \frac{T_1 - T_2}{\ln \frac{r_2}{r_1}} \frac{1}{r}$$

$$q''(r_1) = k \frac{T_1 - T_2}{\ln \frac{r_2}{r_1}} \frac{1}{r_1} \quad \text{at} \quad r = r_1$$

$$q''(r_2) = k \frac{T_1 - T_2}{\ln \frac{r_2}{r_1}} \frac{1}{r_2} \quad \text{at} \quad r = r_2$$

But heat transfer rate is a constant value at any radius r:

$$q(r) = k 2\pi l \frac{T_1 - T_2}{\ln \frac{r_2}{r_1}} = q(r_1) = q(r_2)$$

b. Same as part (a), but the outer cylinder surface is exposed to a convection fluid at a lower temperature, T_∞, with a convection heat transfer coefficient, h. In this case, T_2 is unknown, it can be determined as:

$$\text{at } r_2, \quad -k\frac{\partial T}{\partial r} = h(T_2 - T_\infty)$$

where $\dfrac{\partial T}{\partial r} = -\dfrac{T_1 - T_2}{\ln\dfrac{r_2}{r_1}}\dfrac{1}{r_2}$ from part (a) at $r = r_2$

solve for T_2;

$$T_2 = \frac{T_1 + \dfrac{hr_2}{k}\ln\dfrac{r_2}{r_1}T_\infty}{1 + \dfrac{hr_2}{k}\ln\dfrac{r_2}{r_1}}$$

c. Same as part (b), in this case, T_2 can be solved as:

$$T_2 = \frac{T_1 + \dfrac{(h + h_r)r_2}{k}\ln\dfrac{r_2}{r_1}T_\infty}{1 + \dfrac{(h + h_r)r_2}{k}\ln\dfrac{r_2}{r_1}}$$

d. Same as part (a), but cylinder inner wall receives a constant surface heat flux q_s''; therefore T_1 is unknown and it can be determined as

$$r = r_1, \quad q_s'' = -k\frac{\partial T}{\partial r}$$

$$q_s'' = k\frac{T_1 - T_2}{\ln\dfrac{r_2}{r_1}}\frac{1}{r_1} \text{ from part(a)}$$

Solve for T_1:

$$T_1 = T_2 + q_s''\frac{r_1}{k}\ln\frac{r_2}{r_1}$$

e. Same as part (d), but the cylinder outer wall is exposed to a convection fluid at a lower temperature, T_∞, with a convection heat transfer coefficient, h. In this case, T_2 is unknown, it can be determined as:

$$\text{at } r_2, -k\frac{\partial T}{\partial r} = h(T_2 - T_\infty)$$

where $\dfrac{\partial T}{\partial r} = -\dfrac{T_1 - T_2}{\ln\dfrac{r_2}{r_1}}\dfrac{1}{r_2}$ from part(a)

$$\text{solve for } T_2 \colon T_2 = \frac{T_1 + \dfrac{h r_2}{k} \ln \dfrac{r_2}{r_1} T_\infty}{1 + \dfrac{h r_2}{k} \ln \dfrac{r_2}{r_1}}, \text{ where } T_1 = T_2 + q''_s \frac{r_1}{k} \ln \frac{r_2}{r_1} \text{ from part (d)}$$

f. same as part (e), in this case, T_2 can be solved as:

$$T_2 = \frac{T_1 + \dfrac{(h + h_r) r_2}{k} \ln \dfrac{r_2}{r_1} T_\infty}{1 + \dfrac{(h + h_r) r_2}{k} \ln \dfrac{r_2}{r_1}}$$

where $T_1 = T_2 + q''_s \dfrac{r_1}{k} \ln \dfrac{r_2}{r_1}$ from part (d)

REMARKS

In general, heat conduction problems, regardless of 1-D, 2-D, 3-D, steady or unsteady, can be solved analytically if the thermal conductivity is a given constant and thermal BCs are known constants. The problems will be analyzed and solved in Chapters 2–4. However, in real-life applications, there are many materials whose thermal conductivities vary with temperature and location, $k\,(T) \sim k\,(x, y, z)$. In these cases, the heat conduction equation shown in Equation (1.16) becomes a nonlinear equation and is harder to solve analytically.

In addition, in real engineering applications, it is not easy to determine the precise convection BC shown in Figures 1.10 and 1.12. These require detailed knowledge of complex convection heat transfer to be discussed in Chapters 6–10.

PROBLEMS

1.1 Derive Equation 1.18.

1.2 Derive Equation 1.19.

1.3

    a. Write the differential equation that expresses transient heat conduction in 3-D (x, y, z coordinates) with constant heat generation and constant conductivity.

    b. Simplify the differential equation in (a) to show steady-state conduction in one dimension, assuming constant conductivity.

    c. If the BCs are: $T = T_1 \ldots$ at $\ldots x = x_1$, $T = T_2$ at $x = x_2$, solve the second-order differential equation to yield a temperature distribution (T). Express the answer (T) in terms of (T_1, T_2, x, x_1, x_2).

    d. Using the Fourier Law and the results from (c), develop an expression for the heat rate per unit area, assuming constant conductivity. Express your answer in terms of (k, T_1, T_2, x_1, x_2).

1.4 Consider 1-D, steady-state, heat conduction through a plane wall without heat generation, as shown in Figure 1.13. Determine the temperature, heat flux, and heat transfer rate distributions, as those shown in example 1.1 (a), (b), and (c), but the left-side wall is exposed to a high-temperature convection fluid $T_{\infty 1}$, with a convection heat transfer coefficient h_1.

a. at $x = 0, -k\dfrac{\partial T}{\partial x} = h_1\left(T_{\infty 1} - T_1\right)$

   at $x = L, T = T_2$

b. at $x = 0, -k\dfrac{\partial T}{\partial x} = h_1\left(T_{\infty 1} - T_1\right)$

   at $x = L, -k\dfrac{\partial T}{\partial x} = h_2\left(T_2 - T_{\infty 2}\right)$

c. at $x = 0, -k\dfrac{\partial T}{\partial x} = h_1\left(T_{\infty 1} - T_1\right)$

   at $x = L, -k\dfrac{\partial T}{\partial x} = \left(h_2 + h_r\right)\left(T_2 - T_{\infty 2}\right)$

1.5 Consider 1-D, steady-state, heat conduction through a hollow cylindrical tube without heat generation, as shown in Figure 1.14. Determine the temperature, heat flux, heat transfer rate distributions as those shown in example 1.2 (a), (b), and (c), but the cylinder inner wall is exposed to a high temperature convection fluid $T_{\infty 1}$, with convection heat transfer coefficient h_1.

a. at $r = r_1, -k\dfrac{\partial T}{\partial r} = h_1\left(T_{\infty 1} - T_1\right)$

   at $r = r_2, T = T_2$

b. at $r = r_1, -k\dfrac{\partial T}{\partial r} = h_1\left(T_{\infty 1} - T_1\right)$

   at $r = r_2, -k\dfrac{\partial T}{\partial r} = h_2\left(T_2 - T_{\infty 2}\right)$

c. at $r = r_1, -k\dfrac{\partial T}{\partial r} = h_1\left(T_{\infty 1} - T_1\right)$

   at $r = r_2, -k\dfrac{\partial T}{\partial r} = \left(h_2 + h_r\right)\left(T_2 - T_{\infty 2}\right)$

REFERENCE

1. F. Incropera and D. Dewitt, *Fundamentals of Heat and Mass Transfer*, Fifth Edition, John Wiley & Sons, New York, 2002.

2 1-D Steady-State Heat Conduction

2.1 CONDUCTION THROUGH PLANE WALLS

For 1-D steady-state heat conduction through the plane wall shown in Figure 2.1, without heat generation, the heat conduction Equation (1.17) can be simplified as

$$\frac{\partial^2 T}{\partial x^2} = 0$$
$$\frac{dT}{dx} = C_1 \tag{2.1}$$

Equation (2.1) has the general solution

$$T = c_1 x + c_2 \tag{2.2}$$

with boundary conditions:

$$\text{at } x = 0, \quad T = T_{s1} = c_1 \cdot 0 + c_2 = c_2$$
$$\text{at } x = L, \quad T = T_{s2} = c_1 L + c_2$$

Solving for c_1 and c_2,

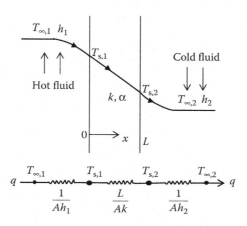

FIGURE 2.1 Conduction through a plane wall and thermal–electrical network analogy.

DOI: 10.1201/9781003164487-2

$$c_1 = \frac{T_{s2} - T_{s1}}{L}, \quad c_2 = T_{s1},$$

Substituting c_1 and c_2 into Equation (2.2), the temperature distribution is

$$T(x) = T_{s,1} - \frac{T_{s,1} - T_{s,2}}{L} x \tag{2.3}$$

Applying Fourier's Conduction Law, one obtains the heat transfer rate through the plane wall

$$q = -kA \frac{\partial T}{\partial x} = kA \frac{T_{s,1} - T_{s,2}}{L} = \frac{T_{s,1} - T_{s,2}}{(L/KA)} \tag{2.4}$$

The heat transfer rate divided by the cross-sectional area of the plane wall, the heat flux, is

$$q'' = \frac{q}{A} = -k \frac{\partial T}{\partial x} = k \frac{T_{s,1} - T_{s,2}}{L} \tag{2.5}$$

At the convective surfaces, from Newton's Cooling Law, the heat transfer rates are

$$q = Ah_1 \left(T_{\infty,1} - T_{s,1} \right) = \frac{T_{\infty,1} - T_{s,1}}{(1/Ah_1)} \tag{2.6}$$

and

$$q = Ah_2 \left(T_{s,2} - T_{\infty,2} \right) = \frac{T_{s,2} - T_{\infty,2}}{(1/Ah_2)} \tag{2.7}$$

Applying the analogy between the heat transfer and electrical network, one may define the thermal resistance based on the concept of electrical resistance. The *thermal resistance for conduction* in a plane wall is

$$R_{\text{cond}} = \frac{L}{kA} \tag{2.8}$$

The *thermal resistance for convection* is then

$$R_{\text{conv}} = \frac{1}{Ah} \tag{2.9}$$

The *total thermal resistance* may be expressed as

$$R_{\text{tot}} = \frac{1}{Ah_1} + \frac{1}{kA} + \frac{1}{Ah_2} = \frac{1}{UA} \tag{2.10}$$

where U is the overall heat transfer coefficient.

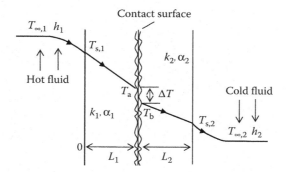

FIGURE 2.2 Temperature drop due to thermal contact resistance between surface a and surface b.

Similar to electric current, the heat rate through the wall is

$$q = \frac{T_{\infty,1} - T_{\infty,2}}{(1/Ah_1) + (L/kA) + (1/Ah_2)} = \frac{T_{\infty,1} - T_{\infty,2}}{R_{tot}} = UA(T_{\infty,1} - T_{\infty,2}) \qquad (2.11)$$

Figure 2.2 shows conduction through two plane walls with thermal contact resistance between them, R_{tc}, the total heat resistance becomes

$$R_{tot} = \frac{1}{Ah_1} + \frac{L_1}{k_1 A} + \frac{R_{tc}}{A} + \frac{L_2}{k_2 A} + \frac{1}{Ah_2} \qquad (2.12)$$

where $R_{tc} = (T_a - T_b)/q$ = pre-determined (depends on contact material surface roughness and contact pressure). $R''_{tc} = R_{tc}/A$. If thermal contact resistance is not considered, $R_{tc} = 0$.

2.1.1 Conduction through Circular Tube Walls

1-D steady-state heat conduction, without heat generation, in the radial system shown in Figure 2.3, can be simplified from Equation (1.18)

$$\frac{1}{r}\frac{d}{dr}\left(r\frac{dT}{dr}\right) = 0$$

$$r\frac{dT}{dr} = c_1 \qquad (2.13)$$

The general solution of Equation (2.13) is

$$T(r) = c_1 \ln r + c_2$$

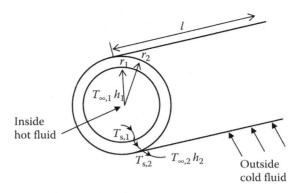

FIGURE 2.3 Conduction through circular tube wall.

with boundary conditions

$$\text{at } r = r_1, \qquad T = T_{s,1} = c_1 \ln r_1 + c_2$$
$$\text{at } r = r_2, \qquad T = T_{s,2} = c_1 \ln r_2 + c_2$$

Solving for c_1 and c_2, one obtains the temperature distribution

$$T(r) = T_{s,1} - \frac{T_{s,1} - T_{s,2}}{\ln(r_2/r_1)} \ln\frac{r}{r_1} \qquad (2.14)$$

The heat transfer rate can be determined from Fourier's Conduction Law as

$$q = -kA\frac{dT}{dr} = -k\,2\pi rl\frac{dT}{dr} = \frac{T_{s,1} - T_{s,2}}{\left(\ln\left(\dfrac{r_2}{r_1}\right)\right)/(2\pi kl)} \qquad (2.15)$$

At the convective surface, from Newton's Cooling Law, the heat transfer rates are

$$q = A_1 h_1\left(T_{\infty,1} - T_{s,1}\right) = \frac{T_{\infty,1} - T_{s,1}}{\left(1/A_1 h_1\right)}$$

and

$$q = A_2 h_2\left(T_{s,2} - T_{\infty,2}\right) = \frac{T_{s,2} - T_{\infty,2}}{\left(1/A_2 h_2\right)}$$

where the cross-sectional area for conduction is $A = 2\pi rl$, $A_1 = 2\pi r_1 l$, $A_2 = 2\pi r_2 l$.

Applying the electrical-thermal analogy, the heat transfer rate is expressed as

$$q = \frac{T_{\infty,1} - T_{\infty,2}}{(1/h_1 \, 2\pi r_1 l) + \left(\ln(r_2/r_1)/2\pi kl\right) + (1/h_2 \, 2\pi r_2 l)} = \frac{T_{\infty,1} - T_{\infty,2}}{R_{tot}}$$

(2.16)

$$= UA\left(T_{\infty,1} - T_{\infty,2}\right)$$

where U is the overall heat transfer coefficient, $UA = U_1 A_1 = U_2 A_2$.

$UA = U \cdot 2\pi rl$, area is based on radius, r
$U_1 A_1 = U_1 \cdot 2\pi r_1 l$, area is based on radius, r_1
$U_2 A_2 = U_2 \cdot 2\pi r_2 l$, area is based on radius, r_2

If we consider radiation heat loss between the tube outer surface and a surrounding wall,

$$q_{radiation} = A_2 \varepsilon \sigma \left(T_{s,2}^4 - T_{sur}^4\right)$$

$$= A_2 \varepsilon \sigma \left(T_{s,2}^2 + T_{sur}^2\right)\left(T_{s,2} + T_{sur}\right)\left(T_{s,2} - T_{sur}\right)$$

$$= A_2 h_r \left(T_{s,2} - T_{sur}\right)$$

where

$$h_r = \varepsilon \sigma \left(T_{s,2}^2 + T_{sur}^2\right)\left(T_{s,2} + T_{sur}\right)$$

and

$$q = q_{conv} + q_{rad}$$

$$= A_2 h_2 \left(T_{s,2} - T_{\infty,2}\right) + A_2 h_r \left(T_{s,2} - T_{sur}\right)$$

If $T_{sur} = T_{\infty,2}$, then

$$R_{tot} = \frac{1}{h_1 \, 2\pi r_1 l} + \frac{\ln(r_2/r_1)}{2\pi kl} + \frac{1}{(h_2 + h_r) 2\pi r_2 l}$$

(2.17)

For three concentric cylindrical walls, with radius r_1, r_2, r_3, r_4, respectively, the total heat resistance becomes

$$R_{tot} = \frac{1}{h_1 \, 2\pi r_1 l} + \frac{\ln r_2/r_1}{2\pi k_1 l} + \frac{\ln r_3/r_2}{2\pi k_2 l} + \frac{\ln r_4/r_3}{2\pi k_3 l} + \frac{1}{h_4 \, 2\pi r_4 l}$$

If thermal contact resistance is considered between the first and second concentric, cylindrical walls, add R_{tc}/A to R_{tot}, where $R_{tc} = (T_a - T_b)/q$, $R_{tc}'' = R_{tc}/A$, and $A = 2\pi r_2 l$.

2.1.2 CRITICAL RADIUS OF INSULATION

Consider a tube of insulating material with inside radius r_i at a constant temperature, T_i. At the outside radius of the insulating tube, r_o, a surface heat transfer coefficient h may be assumed for convection from the outside surface of the insulation to the atmosphere at temperature T_∞. From Equation (2.16) for this case:

$$q = \frac{T_i - T_\infty}{\left(\ln(r_o/r_i)/2\pi kl\right) + \left(1/h2\pi r_o l\right)}$$

If l, T_i, T_∞, h, k, and r_i are all assumed to remain constant while r_o varies, the rate of heat transfer, q, is a function of r_o alone. As r_o increases, the term $(1/hr_o)$ decreases but the term $(\ln r_o/r_i)/k$ increases; hence, it is possible that q might have a maximum value. Take the derivative of the above equation with respect to r_o; then set $dq/dr_o = 0$ and solve for $(r_o)_{\text{critical}}$, the critical radius for which q is a maximum [1],

$$\left(r_o\right)_{\text{critical}} = \frac{k}{h} \tag{2.18}$$

where k is the conductivity of insulation material and h is the outside convection coefficient from insulation material. If r_i is less than $(r_o)_{\text{critical}}$, q is increased as insulation is added until $r_o = (r_o)_{\text{critical}}$. Further increases in r_o cause q to decrease. If, however, r_i is greater than $(r_o)_{\text{critical}}$, any addition of insulation will decrease q (heat loss rate).

2.2 CONDUCTION WITH HEAT GENERATION

The 1-D, steady-state, heat conduction equation with heat generation in the plane wall is (from Equation 1.17)

$$\frac{\partial^2 T}{\partial x^2} + \frac{\dot{q}}{k} = 0$$

$$\frac{dT}{dx} = -\frac{\dot{q}}{k}x + c_1 \tag{2.19}$$

The general solution is

$$T = -\frac{\dot{q}}{2k}x^2 + c_1 x + c_2$$

with asymmetrical boundary conditions [2] shown in Figure 2.4,

$$at \quad x = L, \quad T = T_{s,2} = -\frac{\dot{q}}{2k}L^2 + c_1 L + c_2$$

$$at \quad x = -L, \quad T = T_{s,1} = -\frac{\dot{q}}{2k}(-L)^2 + c_1(-L) + c_2$$

Solving for c_1 and c_2, one obtains the temperature distribution

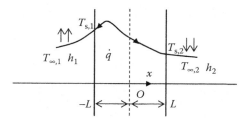

FIGURE 2.4 Plane wall heat conduction with internal heat generation and asymmetrical boundary conditions.

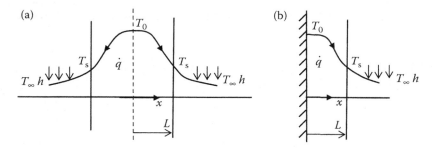

FIGURE 2.5 Plane wall heat conduction with internal heat generation. (a) Symmetric boundary conditions. (b) Adiabatic surface at midplane.

$$T(x) = \frac{\dot{q}L^2}{2k}\left(1 - \frac{x^2}{L^2}\right) + \frac{T_{s,2} - T_{s,1}}{2}\frac{x}{L} + \frac{T_{s,2} + T_{s,1}}{2}$$

Applying symmetric boundary conditions shown in Figure 2.5a, at $x = L$, $T = T_s$; $x = -L$, $T = T_s$, solving for c_1 and c_2, one obtains the temperature distribution

$$T(x) = T_s + \frac{\dot{q}L^2}{2k}\left(1 - \frac{x^2}{L^2}\right) \tag{2.20}$$

At the centerline of the plane wall, the temperature is

$$T_0 = T_s + \frac{\dot{q}L^2}{2k} \tag{2.21}$$

The heat flux to cooling fluid is

$$q'' = -k\frac{dT}{dx}\Big|_{x=L} = h(T_s - T_\infty) \tag{2.22}$$

From Equation (2.20), one obtains

$$\frac{dT}{dx}\Big|_{x=L} = -\frac{\dot{q}L}{k}$$

From Equation (2.22), $h(T_s - T_\infty) = -k\left(-(\dot{q}/k)L\right) = qL$.

Therefore, the surface temperature in Equation (2.20) can be determined as

$$T_s = T_\infty + \frac{\dot{q}L}{h} \tag{2.23}$$

In 1-D steady-state cylindrical medium with heat generation, the heat conduction equation is (from Equation 1.18)

$$\frac{1}{r}\frac{d}{dr}\left(r\frac{dT}{dr}\right) + \frac{\dot{q}}{k} = 0$$

$$\frac{d}{dr}\left(r\frac{dT}{dr}\right) = -\frac{\dot{q}}{k}r \tag{2.24}$$

The general solution is

$$r\frac{dT}{dr} = -\frac{\dot{q}}{2k}r^2 + c_1$$

$$\frac{dT}{dr} = -\frac{\dot{q}}{2k}r + \frac{c_1}{r}$$

$$T(r) = -\frac{\dot{q}}{4k}r^2 + c_1 \ln r + c_2$$

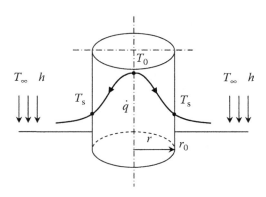

FIGURE 2.6 Cylindrical rod heat conduction with internal heat generation.

with boundary conditions as shown in Figure 2.6:

$$\text{at} \quad r = 0, \quad \frac{dT}{dr} = 0 = c_1$$

$$\text{at} \quad r = r_0, \quad T = T_s = -\frac{\dot{q}}{4k}r_0^2 + c_2$$

Solving for c_1 ($c_1 = 0$) and c_2, one obtains the temperature distribution

$$T(r) = T_s + \frac{\dot{q}r_0^2}{4k}\left(1 - \frac{r^2}{r_0^2}\right) \tag{2.25}$$

At the centerline of the cylindrical rod, the temperature is

$$T_0 = T_s + \frac{\dot{q}r_0^2}{4k} \tag{2.26}$$

The heat flux to the cooling fluid is

$$q'' = h(T_s - T_\infty) = -k\frac{dT}{dr}|_{r=r_0} \tag{2.27}$$

From Equation (2.25), one obtains

$$\frac{dT}{dr}|_{r=r_0} = -\frac{\dot{q}r_0}{2k}$$

From Equation (2.27), $h(T_s - T_\infty) = -k\left(-(\dot{q}r_0/2k)\right) = (\dot{q}r_0/2)$

Therefore, the surface temperature in Equation (2.25) can be determined as

$$T_s = T_\infty + \frac{\dot{q}r_0}{2h} \tag{2.28}$$

2.3 CONDUCTION THROUGH FINS WITH UNIFORM CROSS-SECTIONAL AREA

From Newton's Law of Cooling, the heat transfer rate can be increased by either increasing the temperature difference between the surface and fluid, the heat transfer coefficient, or the surface area. For a given problem, the temperature difference between the surface and fluid may be fixed, and increasing the heat transfer coefficient may result in more pumping power. One popular way to increase the heat transfer rate is to increase the surface area by adding fins to the heated surface. This is particularly true when the heat transfer coefficient is relatively low such as the air-side of heat exchangers (e.g., the car radiators) and air-cooled electronic components. The heat transfer rate can increase dramatically by increasing the surface area many times with the addition of numerous fins. Therefore, heat is conducted from the base surface into the fins and dissipated into the cooling fluid. However, as heat is conducted through the fins, the surface temperature decreases due to a finite thermal conductivity of the fins and the convective heat loss to the cooling fluid. This means the fin temperature is not the same as the base surface temperature and the temperature difference between the fin surface and the cooling fluid reduces along the length of the fins. It is our job to determine the fin temperature in order to calculate the heat loss from the fins to the cooling fluid.

In general, the heat transfer rate will increase with the number of fins. However, there is a limitation on the number of fins. The heat transfer coefficient will reduce if the fins are packed too close. In addition, the heat transfer rate will increase with thin fins that have a high thermal conductivity. Again, there is limitation on the thickness of thin fins due to manufacturing concerns. At this point, we are not interested in optimizing the fin dimensions but in determining the local fin temperature for a given geometry and working conditions. We assume that heat conduction through the fin is 1-D steady state because the fin is thin. The temperature gradient in the other two dimensions is neglected. We will begin with the constant cross-sectional area fins and then consider variable cross-sectional area fins. The following is the energy balance of a small control volume of the fin with heat conduction through the fin and heat dissipation into cooling fluid, as shown in Figure 2.7. The resulting temperature distributions through fins of different materials can be seen from Figure 2.8.

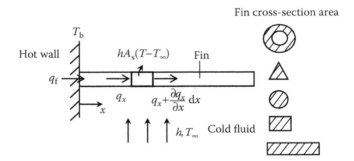

FIGURE 2.7 One-dimensional conduction through thin fins with uniform cross-sectional area.

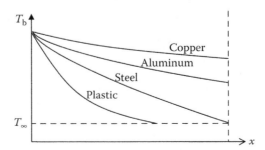

FIGURE 2.8 Temperature distributions through fins of different materials.

$$q_x - \left(q_x + \frac{dq_x}{dx} dx \right) - hA_s (T - T_\infty) = 0 \tag{2.29}$$

where $A_s = P\,dx$, P is perimeter of the fin, and q_x is from Fourier's Conduction Law shown in Equation (1.10).

$$-\frac{d}{dx} \left(-kA_c \frac{dT}{dx} \right) dx - hP\,dx(T - T_\infty) = 0 \tag{2.30}$$

If the cross-sectional area is constant, that is, $A_c = $ constant, and k is also constant, Equation (2.30) can be simplified as

$$\frac{d^2 T}{dx^2} - \frac{hP}{kA_c}(T - T_\infty) = 0$$

$$\frac{d^2 (T - T_\infty)}{dx^2} - \frac{hP}{kA_c}(T - T_\infty) = 0 \tag{2.31}$$

Letting $\theta(x) = T(x) - T_\infty$, Equation (2.31) becomes

$$\frac{d^2\theta}{dx^2} - \frac{hP}{kA_c}\theta = 0 \tag{2.32}$$

Substituting $m^2 = hP/kA_c$, Equation (2.32) becomes

$$\frac{d^2\theta}{dx^2} - m^2\theta = 0 \tag{2.33}$$

The general solution to the second-order ordinary differential equation is

$$\theta(x) = c_1 e^{mx} + c_2 e^{-mx}$$

or

$$\theta(x) = c_1 \sinh(mx) + c_2 \cosh(mx)$$

with the following boundary conditions:
   At the fin base,

$$x = 0, T = T_b, \text{ then } \theta(0) = T_b - T_\infty = \theta_b$$

At the fin tip,
   $x = L$, there are four possible cases

   1. Convection boundary condition $-k(\partial T/\partial x)|_{x=L} = h(T_L - T_\infty)$, then $(\partial\theta/\partial x)_{x=L} = (-h/k)\theta_L$

2. The fin tip is insulated $-k(\partial T/\partial x)|_{x=L} = 0$ or $(\partial\theta / \partial x)_{x=L} = 0$
3. The tip temperature is given as $T|_{x=L} = T_L$ or $\theta(L) = T_L - T\infty = \theta_L$
4. For a long fin, $L/d > 10 \sim 20$, $T|_{x=L} = T_\infty$, or $\theta(L) = T_\infty - T_\infty = 0$, which is an ideal case.

Applying the boundary conditions:

$$x = 0, \theta(0) = \theta_b = c_1 + c_2$$

$$x = L, \text{ there are four possible cases}$$

For case 1: $\dfrac{\partial\theta(L)}{\partial x} = \dfrac{h}{-k}\theta_{\mathrm{L}}$, that is, $c_1 me^{mL} + c_2(-m)e^{-mL} = -(h/k)(c_1 e^{mL} + c_2 e^{-mL})$,
solving for c_1 and c_2, one obtains the temperature distribution as follows:

$$\frac{\theta(x)}{\theta_b} = \frac{T(x) - T_\infty}{T_b - T_\infty} = \frac{\cosh m(L - x) + (h/mk)\sinh m(L - x)}{\cosh mL + (h/mk)\sinh mL} \tag{2.34}$$

The heat transfer rate through the fin base (q_f) is

$$q_f = q_x = -kA_c\frac{dT}{dx}|_{x=0} = -kA_c\frac{d\theta(0)}{dx}$$

$$= M\frac{\sinh mL + (h/mk)\cosh mL}{\cosh mL + (h/mk)\sinh mL} \tag{2.35}$$

where $M = \theta_b\sqrt{hPkA_c}$

From Appendix A.1

$$\sinh mx = \frac{e^{mx} - e^{-mx}}{2}$$

$$\cosh mx = \frac{e^{mx} + e^{-mx}}{2}$$

$$d\left(e^{mx}\right) = me^{mx}dx$$

$$d\left(e^{-mx}\right) = -me^{-mx}dx$$

$$d\left(\sinh mx\right) = \cosh mx \cdot m\,dx$$

$$d\left(\cosh mx\right) = \sinh mx \cdot m\,dx$$

For case 2: $d\theta/dx|_{x=L} = 0$, that is, $c_1 m e^{mL} - c_2 m e^{-mL} = 0$, solving for c_1 and c_2, one obtains the temperature distribution through the fin base (q_f) as

$$\frac{\theta(x)}{\theta_b} = \frac{T(x) - T_\infty}{T_b - T_\infty} = \frac{\cosh m(L - x)}{\cosh mL} \tag{2.36}$$

$$q_f = q_x = -kA_c \frac{dT}{dx}\Big|_{x=0} = -kA_c \frac{d\theta(0)}{dx} = \theta_b \left(\tanh mL\right)\sqrt{hPkA_c} \tag{2.37}$$

For case 3: $\theta(L) = \theta_L$, that is, $c_1 e^{mL} + c_2 e^{-mL} = \theta_L$, solving for c_1 and c_2, one obtains the temperature distribution and heat transfer through the fin base (q_f) as

$$\frac{\theta(x)}{\theta_b} = \frac{T(x) - T_\infty}{T_b - T_\infty} = \frac{(\theta_L/\theta_b)\sinh mx + \sinh m(L - x)}{\sinh mL} \tag{2.38}$$

$$q_f = q_x = -kA_c \frac{dT}{dx}\Big|_{x=0} = -kA_c \frac{d\theta(0)}{dx} = \theta_b \left(\frac{\cosh mL - \theta_L/\theta_b}{\sinh mL}\right)\sqrt{hPkA_c} \tag{2.39}$$

For case 4: very long fins, $\theta(L) = 0$, then $c_1 = 0$, $c_2 = \theta_b$, we obtained the following temperature distribution and heat transfer rate through the fin base (q_f) as

$$\frac{\theta}{\theta_b} = e^{-mx} \tag{2.40}$$

$$q_f = q_x = -kA_c \frac{d\theta(0)}{dx} = M = \theta_b \sqrt{hPkA_c} \tag{2.41}$$

It should be noted that the above results can be applied to any fins with a uniform cross-sectional area. This includes the fins with circular, rectangular, triangular, and other cross sections as shown in Figure 2.7. One should know how to calculate the fin cross-sectional area, A_c, and fin perimeter, P (circumferential length), for a given uniform cross-sectional area fin geometry. The characteristics of the hyperbolic functions, sinh, cosh, tanh, and their derivatives are shown in Appendix A.1 and Appendix A.2.

2.3.1   FIN PERFORMANCE

2.3.1.1   Fin Effectiveness

Most often, before adding the fins, we would like to know whether it is worthwhile to add fins to the smooth, heated surface. In this case, we define the fin effectiveness. The fin effectiveness is defined as the ratio of the heat transfer rate through the fin surface to that without the fin (i.e., convection from the fin base area).

$$\eta_\varepsilon = \frac{q_{\text{with fin}}}{q_{\text{without fin}}} \tag{2.42}$$

The fin effectiveness must be greater than unity in order to justify using the fins. Normally, it should be greater than 2 in order to include the material and manufacturing costs. In general, the fin effectiveness is greater than 5 for most of the effective fin applications. For example, for the long fins (case 4 fin tip boundary conditions), the fin effectiveness is

$$\eta_\varepsilon = \frac{q_{\text{with fin}}}{q_{\text{without fin}}} = \frac{\theta_b \sqrt{hPkA_c}}{h\theta_b A_c} = \sqrt{\frac{kP}{hA_c}} > 1 \sim 5 \tag{2.43}$$

2.3.1.2 Fin Efficiency

We would like to know the fin efficiency after we have added the fins. The fin efficiency is defined in Equation (2.44) as the ratio of heat transfer through the fin surface to that through a perfectly conducting fin (an ideal fin with infinite thermal conductivity as super conductors).

$$\eta_f = \frac{q_{\text{fin}}}{q_{\text{max}}} \tag{2.44}$$

By definition, the fin efficiency is between 0 and 1. However, the fin efficiency ranges from 0.9 to 0.95 for most fin applications. For example, for long fins (case 4 fin tip boundary condition), the efficiency is

$$\eta_f = \frac{\theta_b \sqrt{hPkA_c}}{hPL\theta_b} = \sqrt{\frac{kA_c}{hPL^2}} \geq 90\% \tag{2.45}$$

In order to achieve the higher fin effectiveness and fin efficiency, we need to have a thin fin (larger P/A_c ratio) with a relatively high thermal conductivity, k (aluminum or copper), and a low working fluid heat transfer coefficient, h (air cooling).

When calculating the total heat transfer from the system, it is necessary to include heat dissipated from the exposed base (between the fins). In a system with multiple fins, the rate of heat transfer for each fin can be calculated based on the assumed tip condition from above. Multiplying this quantity times the number of fins and adding it to the heat transfer from the base provide the total heat dissipation for the system. The rate of heat transfer from a single fin can be determined based on the fin efficiency.

$$q = q_{\text{fin}} + q_{\text{non-fin}}$$

$$q = N\eta_f q_{\text{max}} + h(T_b - T_\infty)A_{\text{non-fin}}$$

where N is the number of fins and

$$q_{max} = h\left(T_b - T_\infty\right)A_{s,fin}$$

$$q_{fin} = -kA_c\frac{dT}{dx}\Big|_{x=0}$$

It is important to point out that the temperature distribution through a fin varies depending on the aforementioned fin tip boundary conditions. In general, these temperatures decay from the fin base to the fin tip as shown in Figure 2.8. These decay curves are derived from the sinh and cosh functions shown above. In addition, the heat transfer rate through the fin depends on the temperature gradient at the fin base and the fin thermal conductivity. For example, the temperature gradient at the fin base is greater for the steel fin than the aluminum fin. However, the heat transfer rate through the fin base is higher for an aluminum fin than for a steel fin for the same fin geometry and working fluid conditions. This is because the aluminum fin has a much larger thermal conductivity than the steel fin.

2.3.2 RADIATION EFFECT

If we also consider radiation flux q_r'', the energy balance Equation (2.29) can be rewritten as

$$q_x - \left(q_x + \frac{dq_x}{dx}dx\right) - hA_s\left(T - T_\infty\right) + A_s q_r'' = 0$$

where q_r'' = constant, which is the solar radiation gain, or q_r'' = radiation loss = $-\varepsilon\sigma\left(T^4 - T_{sur}^4\right)$. If we consider q_r'' = constant, the solution of the above equation can be obtained from Equation (2.34) with

$$\theta = T - T_\infty - \frac{q_r''}{h}$$

However, if we consider $q_r'' = -\varepsilon\sigma\left(T^4 - T_{sur}^4\right)$, the above energy balance equation can be rewritten as

$$q_x - \left(q_x + \frac{dq_x}{dx}dx\right) - hA_s\left(T - T_\infty\right) - \varepsilon\sigma A_s\left(T^4 - T_{sur}^4\right) = 0$$

Furthermore, if we let $T_\infty = T_{sur}$ and $h_r = \varepsilon\sigma\left(T^2 + T_\infty^2\right)(T + T_\infty)$, the above equation can be written as

$$\frac{d^2\left(T - T_\infty\right)}{dx^2} - \frac{\left(h + h_r\right)P}{kA_c}\left(T - T_\infty\right) = 0$$

The solution of the above equation can be obtained by numerical integration.

2.4 CONDUCTION THROUGH FINS WITH A VARIABLE CROSS-SECTIONAL AREA: BESSEL FUNCTION SOLUTIONS

Most often, we would like to have a fin with a decreasing fin cross-sectional area in the heat conduction direction in order to reduce material costs. In other situations, the fin cross-sectional area may increase in the heat conduction direction, such as an annulus fin attached to a circular tube (the so-called fin tube). In these cases, we deal with steady-state, 1-D heat conduction through fins with variable cross-sectional areas. Bessel function solutions are required to solve the temperature distribution through these types of fin problems [2–4]. Figure 2.9 shows the heat conduction through variable cross-sectional area fins and heat dissipation to the working fluid.

Consider the energy balance of a small control volume in the fin,

$$q_x - \left(q_x + \frac{dq_x}{dx} dx \right) - hA_s \left(T - T_\infty \right) = 0$$

$$-\frac{dq_x}{dx} dx - hA_s \left(T - T_\infty \right) = 0$$

where $A_s = P\, dx$, P is perimeter of the fin, and q_x is from Fourier's Conduction Law shown in Equation (1.10).

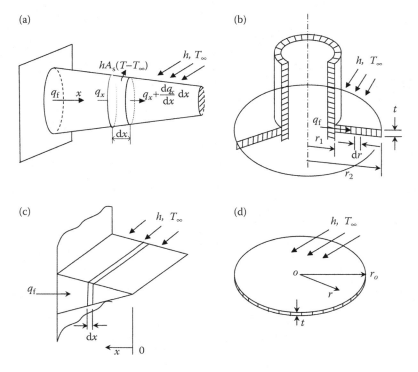

FIGURE 2.9 One-dimensional heat conduction through thin fins with variable cross-sectional area. (a) Conical fin; (b) annular fin; (c) taper fin; (d) disk fin.

$$-\frac{d}{dx}\left(-kA_c\frac{dT}{dx}\right)dx - hP\,dx(T-T_\infty) = 0$$

If k is constant, we have

$$\frac{d}{dx}\left(A_c\frac{dT}{dx}\right) - \frac{hP}{k}(T-T_\infty) = 0$$

(2.46)

$$\frac{d}{dx}\left[A_c\frac{d(T-T_\infty)}{dx}\right] - \frac{hP}{k}(T-T_\infty) = 0$$

Annulus Fin: For example, for an annulus fin with a uniform thickness, t, as shown in Figure 2.9b, the fin cross-sectional area from the centerline of the tube is $A_c = 2\pi r \cdot t$. The fin perimeter, including the top and the bottom, is $P = 2 \cdot 2\pi r$. Let $\theta = T - T_\infty$, Equation (2.46) becomes

$$\frac{d}{dr}\left[2\pi rt\frac{d\theta}{dr}\right] - \frac{h \cdot 4\pi r}{k}\theta = 0$$

$$\frac{d}{dr}\left[r\frac{d\theta}{dr}\right] - \frac{2hr}{kt}\theta = 0$$

(2.47)

$$r\frac{d}{dr}\left(r\frac{d\theta}{dr}\right) - \frac{2hr^2}{kt}\theta = 0$$

$$r\frac{d}{dr}\left(r\frac{d\theta}{dr}\right) - m^2r^2\theta = 0$$

where $m^2 = 2h/kt$.

2.4.1 BESSEL FUNCTIONS AND THEIR SOLUTIONS

1. The heat generation problem — Bessel function:
   The solution of Equation (2.47) can be a typical Bessel function with heat generation as in the following format:

$$x\frac{d}{dx}\left(x\frac{d\theta}{dx}\right) + \left(m^2x^2 - v^2\right)\theta = 0$$

(2.48)

$$\theta = a_0J_v(mx) + a_1Y_v(mx), \quad v = 0,1,2,\dots$$

(2.49)

2. The heat loss problem—modified Bessel function:
   The solution of Equation (2.47) can be the typical modified Bessel function with heat loss as in the following format:

$$x\frac{d}{dx}\left(x\frac{d\theta}{dx}\right) - \left(m^2x^2 + v^2\right)\theta = 0$$

(2.50)

$$\theta = a_0 I_v(mx) + a_1 K_v(mx), \quad v = 0,1,2,\ldots \tag{2.51}$$

Comparing Equations (2.47) and (2.50), with $r = x$, the general solution of Equation (2.47) is the same format as Equation (2.51) with $v = 0$ for the heat loss problem:

$$\theta(mr) = a_0 I_0(mr) + a_1 K_0(mr) \tag{2.52}$$

with the following boundary conditions:
At the fin base, $r = r_1$, $T = T_b$, then $\theta = T_b - T_\infty = \theta_b$.
At the fin tip, $r = r_2$, there are four possible cases as discussed before,

1. Convective boundary

$$-k \frac{\partial T}{\partial r}|_{r=r_2} = h(T_{r2} - T_\infty) \quad \text{or} \quad \frac{\partial \theta(r_2)}{\partial r} = \frac{-h}{k} \theta_{r_2}$$

2. The tip fin is insulated

$$-k \frac{\partial T}{\partial r}|_{r=r_2} = 0 \quad \text{or} \quad \frac{\partial \theta(r_2)}{\partial r} = 0$$

3. The outer tip temperature is given

$$T|_{r=r_2} = T_{r_2} \quad \text{or} \quad \theta(r_2) = T_{r_2} - T_\infty = \theta_{r_2}$$

4. For a long fin $r_2 / t > 10 \sim 20$

$$T|_{r=r_2} = T_\infty \quad \text{or} \quad \theta(r_2) = T_\infty - T_\infty = 0$$

Case 1: the tip is exposed to convection boundary condition:
From solution shown in (2.52)

$$\theta(mr) = a_0 I_0(mr) + a_1 K_0(mr)$$

At $\qquad\qquad r = r_1, \quad \theta = \theta_b = a_0 I_0(mr_1) + a_1 K_0(mr_1),$

At $\qquad\qquad r = r_2, \quad \dfrac{\partial \theta(r_2)}{\partial r} = -\dfrac{h}{k} \theta(r_2)$

$\therefore \qquad a_0 m I_1(mr_2) - a_1 m K_1(mr_2) = -\dfrac{h}{k}[a_0 I_0(mr_2) + a_1 K_0(mr_2)]$

from above, solve for a_0, a_1:

$$a_0 = \frac{\theta_b}{I_0(mr_1)} - a_1 \frac{K_0(mr_1)}{I_0(mr_1)}$$

$$a_1 = \left[\frac{\theta_b}{I_0(mr_1)} \cdot \frac{C_1}{C_2} \right] / \left[1 + \frac{K_0(mr_1)}{I_0(mr_1)} \cdot \frac{C_1}{C_2} \right]$$

where

$$C_1 = mI_1(mr_2) + \frac{h}{k} I_0(mr_2)$$

$$C_2 = mK_1(mr_2) - \frac{h}{k} K_0(mr_2)$$

Case 2: the fin tip is insulated, for example:

At $\qquad r = r_1, \quad \theta = \theta_b = a_0 I_0(mr_1) + a_1 K_0(mr_1),$

At $\quad r = r_2, \quad \dfrac{\partial \theta(r_2)}{\partial r} = a_0 \dfrac{dI_0(mr)}{dr}|_{r=r_2} + a_1 \dfrac{dK_0(mr)}{dr}|_{r=r_2}$

$$= a_0 m I_1(mr_2) - a_1 m K_1(mr_2) = 0,$$

where applying the properties of I and K,

$$\frac{d}{dr} I_0(mr) = mI_1(mr) \quad and \quad \frac{d}{dr} K_0(mr) = -mK_1(mr)$$

Solve for a_0 and a_1

$$a_0 = \frac{K_1(mr_2)}{I_0(mr_1)K_1(mr_2) + I_1(mr_2)K_0(mr_1)} \theta_b$$

$$a_1 = \frac{I_1(mr_2)}{I_0(mr_1)K_1(mr_2) + I_1(mr_2)K_0(mr_1)} \theta_b$$

One obtains the temperature distribution as

$$\frac{\theta(r)}{\theta_b} = \frac{T(r) - T_\infty}{T_b - T_\infty} = \frac{I_0(mr)K_1(mr_2) + I_1(mr_2)K_0(mr)}{I_0(mr_1)K_1(mr_2) + I_1(mr_2)K_0(mr_1)} \qquad (2.53)$$

Therefore, heat transfer through the fin base can be calculated as

$$q_f = -kA_c \frac{d\theta}{dx}\Big|_{r_1} = -k\left(2\pi r_1\right)(t)\frac{d\theta}{dx}\Big|_{r=r_1}$$

$$q_f = -2\pi k r_1 t m \frac{I_1(m r_1)K_1(m r_2) - I_1(m r_2)K_1(m r_1)}{I_0(m r_1)K_1(m r_2) + I_1(m r_2)K_0(m r_1)}\theta_b$$

(2.54)

And the fin efficiency can be determined as

$$\eta_f = \frac{q_f}{h2\pi\left(r_2^2 - r_1^2\right)\theta_b}$$

Case 3: the tip is kept at constant temperature T_2:
From solution shown in (2.52)

$$\theta(mr) = a_0 I_0(mr) + a_1 K_0(mr)$$

$$At \quad r = r_1, \quad \theta = \theta_b = a_0 I_0(m r_1) + a_1 K_0(m r_1)$$

$$At \quad r = r_2, \quad \theta = T_2 - T_\infty = \theta_2 = a_0 I_0(m r_2) + a_1 K_0(m r_2)$$

From above, solve for

$$a_0 = \frac{\theta_b K_0(m r_2) - \theta_2 K_0(m r_1)}{I_0(m r_1)K_0(m r_2) - I_0(m r_2)K_0(m r_1)}$$

$$a_1 = \frac{\theta_b I_0(m r_2) - \theta_2 I_0(m r_1)}{K_0(m r_1)I_0(m r_2) - K_0(m r_2)I_0(m r_1)}$$

Case 4: the tip is kept at same temperature as convection fluid:
The temperature distribution is same as for case 3, but $\theta(r_2) = T_\infty - T_\infty = 0$.
Disk Fin: Consider the example of a thin metal disk, as shown in Figure 2.9d, that is insulated on one side and exposed to a jet of hot air at temperature T_∞ on the other. The convection heat transfer coefficient h can be taken to be constant over the disk. The periphery at $r = R$ is maintained at a uniform temperature T_R: (a) Derive the heat conduction equation of disk; (b) determine the disk temperature distributions.

a. Energy balance (Equation 2.47):

$$\frac{d}{dr}\left(kA_c \frac{dT}{dr}\right)dr - Ph(T - T_\infty)dr = 0$$

$$A_c = 2\pi r t$$

$$P = 2\pi r$$

$$r\frac{d^2T}{dr^2} + \frac{dT}{dr} - \frac{h}{kt}r(T - T_\infty) = 0$$

Let $\theta = T - T_\infty$, $m^2 = \dfrac{h}{kt}$

$$r\frac{d}{dr}\left(r\frac{d\theta}{dr}\right) - m^2r^2\theta = 0$$

b. Modified Bessel Function (2.52):

$$\theta = a_0 I_0(mr) + a_1 K_0(mr)$$

$$\text{At} \quad r = 0, \quad \frac{d\theta}{dr} = 0 \qquad a_1 = 0, \quad \because K_1(0) \to \infty$$

$$At\ r = R, \quad \theta = T_R - T_\infty = \theta_R = a_0 I_0(mR)$$

$$\therefore \frac{\theta}{\theta_R} = \frac{I_0(mr)}{I_0(mR)}$$

The heat generation problem—Bessel Function: Heat is generated at a rate $\dot{q}$ in a long solid cylinder of radius r_o, as shown in Figure 2.6. The cylinder is immersed in a liquid at temperature T_∞. Heat transfer from the cylinder surface to the liquid can be characterized by a heat transfer coefficient h. Obtain the steady-state temperature distributions for the following cases:

a. $\dot{q} = q_o\left[1 - \left(\frac{r}{r_o}\right)^2\right]$

b. $\dot{q} = a + b(T - T_\infty)$

For case (a)

$$\frac{1}{r}\frac{d}{dr}\left(r\frac{dT}{dr}\right) + \frac{\dot{q}_o}{k}\left[1 - \left(\frac{r}{r_o}\right)^2\right] = 0$$

$$r\frac{dT}{dr} = -\frac{\dot{q}_o}{k}\left[\frac{1}{2}r^2 - \frac{1}{4}\frac{r^4}{r_o^2}\right] + C_1$$

$$\frac{dT}{dr} = -\frac{\dot{q}_o}{k}\left[\frac{1}{2}r - \frac{1}{4}\frac{r^3}{r_o^2}\right] + \frac{C_1}{r}$$

$$T = -\frac{q_o}{k}\left(\frac{1}{4}r^2 - \frac{1}{16}\frac{r^4}{r_0^2}\right) + C_1 \ln r + C_2$$

$$\text{At} \quad r = 0, \quad \frac{dT}{dr} = 0 \quad \therefore C_1 = 0$$

$$\text{At} \quad r = r_0, \quad T = T_s = -\frac{q_o}{k}\left(\frac{1}{4}r_o^2 - \frac{1}{16}r_o^2\right) + C_2$$

$$\therefore T = T_s + \frac{q_o}{k}\left(\frac{3}{16}r_o^2 - \frac{1}{4}r^2 + \frac{1}{16}\frac{r^4}{r_o^2}\right)$$

From (2.27),

$$h(T_s - T_\infty) = -k\frac{dT}{dr}\bigg|_{r=r_0} = -k\left(-\frac{1}{4}r_0\right)\frac{q_o}{k}$$

$$\text{Where} \quad T_s = T_\infty + \frac{q_o}{4}\frac{r_0}{h}$$

For case (b)

$$\frac{1}{r}\frac{d}{dr}\left(r\frac{dT}{dr}\right) + \frac{1}{k}\left[a + b(T - T_\infty)\right] = 0$$

$$r^2\frac{d}{dr}\left(r\frac{dT}{dr}\right) + \frac{b}{k}r^2\left[T - \left(T_\infty - \frac{a}{b}\right)\right] = 0$$

$$\text{Let} \quad \theta = \left[T - \left(T_\infty - \frac{a}{b}\right)\right], \quad m^2 = \frac{b}{k}$$

$$r^2\frac{d}{dr}\left(r\frac{d\theta}{dr}\right) + m^2 r^2 \theta = 0 \quad \text{Bessel Function}$$

From 2.48, with $v = 0$, $x = r$

From 2.49, $\theta = a_0 J_0(mr) + a_1 Y_0(mr)$

From A.3.1, $\dfrac{d\theta}{dr} = -ma_0 J_1(mr) - ma_1 Y_1(mr)$

$$\text{At} \quad r = 0, \quad \frac{d\theta}{dr}\bigg|_0 = 0$$

$$\text{But} \quad Y_1(0) = -\infty \quad \therefore a_1 = 0$$

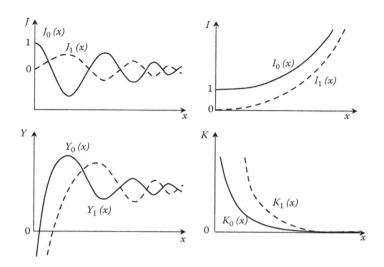

FIGURE 2.10 The characteristics of Bessel functions.

$$At \quad r = r_o, \quad \theta|_{r_0} = \theta_s = \left[T_s - \left(T_\infty - \frac{a}{b} \right) \right] = a_0 J_0 (mr_o)$$

$$\therefore \theta = \theta_s \frac{J_0 (mr)}{J_0 (mr_0)}$$

$$Where \quad h(T_s - T_\infty) = -k \frac{dT}{dr} \bigg|_{r=r_0} = -k \frac{d\theta}{dr} \bigg|_{r=r_0} = km\theta_s \frac{J_1 (mr_0)}{J_0 (mr_0)}$$

It is important to point out that temperature decreases from the fin base to the fin tip depending on the specified fin tip boundary conditions. The temperature curve again is a combination of Bessel functions I_0 and K_0 with the temperature decaying from the base to the tip. The characteristics of Bessel functions J, Y, I, and K and their derivatives are shown in Figure 2.10 and Appendix A.3.

2.4.2 RADIATION EFFECT

If we also consider the radiation effect, $q_r'' = constant = positive value$, then the above solution can be used by substituting $\theta = T - T_\infty - (q_r''/h)$. However, if we consider $q_r'' = -\varepsilon \sigma (T^4 - T_{sur}^4)$, and $T_\infty = T_{sur}$, $h_r = \varepsilon \sigma (T^2 + T_\infty^2)(T + T_\infty)$, the energy balance Equation (2.46) can be written as

$$\frac{d}{dx} \left[A_c \frac{d(T - T_\infty)}{dx} \right] - \frac{(h + h_r)P}{k} (T - T_\infty) = 0$$

The solution of the above equation can be obtained by numerical integration.

Examples

2.1 A composite cylindrical wall (Figure 2.11) is composed of two materials of thermal conductivity k_A and k_B, which are separated by a very thin, electric resistance heater for which interfacial contact resistances with material A and B are $R''_{c,A}$ and $R''_{c,B}$, respectively. Liquid pumped through the tube is at a temperature $T_{\infty,i}$ and provides a convection coefficient h_i at the inner surface of the composite. The outer surface is exposed to ambient air, which is at $T_{\infty,o}$ and provides a convection coefficient of h_o. Under steady-state conditions, a uniform heat flux of q''_h is dissipated by the heater.

   a. Sketch the equivalent thermal circuit of the system and express all resistances in terms of relevant variables.
   b. Obtain an expression that may be used to determine the heater temperature, T_h.
   c. Obtain an expression for the ratio of heat flows to the outer and inner fluids, q'_o/q'_i. How might the variables of the problem be adjusted to minimize this ratio?

Solution

   a. See the sketch shown in Figure 2.11.
   b. Performing an energy balance for the heater, $\dot{E}_{in} = \dot{E}_{out}$, it follows that

$$q''_h(2\pi r_2) = q'_i + q'_o = \frac{T_h - T_{\infty,i}}{(h_i 2\pi r_1)^{-1} + (\ln(r_2/r_1)/2\pi k_B) + R''_{tc,B}}$$

$$+ \frac{T_h - T_{\infty,o}}{(h_o 2\pi r_3)^{-1} + (\ln(r_3/r_2)/2\pi k_A) + R''_{tc,A}}$$

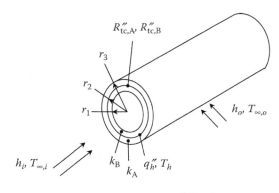

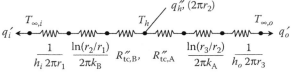

FIGURE 2.11   Thermal circuit of a composite cylindrical wall and all resistance.

c. From the circuit,

$$\frac{q_o'}{q_i'} = \frac{(T_h - T_{\infty,o})}{(T_h - T_{\infty,i})} \cdot \frac{(h_i 2\pi r_1)^{-1} + (\ln(r_2/r_1)/2\pi k_B) + R_{tc,B}''}{(h_o 2\pi r_3)^{-1} + (\ln(r_3/r_2)/2\pi k_A) + R_{tc,A}''}$$

2.2 Heat is generated at a rate q in a large slab of thickness 2L, as shown in Figure 2.5a. The side surfaces lose heat by convection to a liquid at temperature T_∞. Obtain the steady-state temperature distributions for the following cases:

a. $\dot{q} = \dot{q}_o\left[1 - (x/L)^2\right]$, with x measured from the center plane.
b. $\dot{q} = a + b(T - T_\infty)$

Solution

a. $\dot{q} = \dot{q}_o\left[1 - (x/L)^2\right]$

$$\frac{d^2T}{dx^2} = -\frac{\dot{q}_o}{k}\left[1 - \left(\frac{x}{L}\right)^2\right]$$

$$\frac{dT}{dx} = -\frac{\dot{q}_o}{k}\left[x - \frac{1}{3}\frac{x^3}{L^2}\right] + C_1$$

$$T = -\frac{\dot{q}_o}{k}\left[\frac{1}{2}x^2 - \frac{1}{12}\frac{x^4}{L^2}\right] + C_1 x + C_2$$

Boundary conditions: $x = 0$, $\dfrac{dT}{dx} = 0$ (symmetry); $x = L$,

$$-k\frac{dT}{dx} = h(T - T_\infty)$$

Substituting the boundary conditions into the above equation to replace C_1 and C_2,

$$T - T_\infty = \frac{\dot{q}_o L^2}{2k}\left[\frac{5}{6} + \frac{1}{6}\left(\frac{x}{L}\right)^4 - \left(\frac{x}{L}\right)^2 + \frac{4}{3Bi}\right]$$

b. $\dot{q} = a + b(T - T_\infty)$

$$\frac{d^2T}{dx^2} + \frac{b}{k}\left[T - \left(T_\infty - \frac{a}{b}\right)\right] = 0$$

Solving the above equation,

$$T - \left(T_\infty - \frac{a}{b}\right) = C_1 \cos(b/k)^{1/2} x + C_2 \sin(b/k)^{1/2} x$$

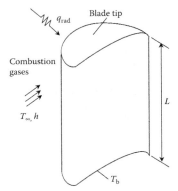

FIGURE 2.12 A turbine blade modeled as a fin with constant cross-sectional area.

Applying boundary conditions,

$$T - \left(T_\infty - \frac{a}{b}\right) = \frac{h\cos(b/k)^{1/2}(a/b)x}{h\cos(b/k)^{1/2}L + (b/k)^{1/2}\sin(b/k)^{1/2}L}$$

2.3 A long gas turbine blade (Figure 2.12) receives heat from combustion gases by convection and radiation. If re-radiation from the blade can be neglected $T_s \ll T_\infty$, determine the temperature distribution along the blade. Assume

  a. The blade tip is insulated.
  b. The heat transfer coefficient on the tip equals that on the blade sides. The cross-sectional area of the blade may be taken to be constant.

Solution

If the blade is at a much lower temperature than the combustion gases, radiation emitted by the blade will be much smaller than the absorbed radiation and can be ignored to simplify the problem. An energy balance on an element of fin Δx long gives

$$-kA_C \frac{dT}{dx}\big|_x + kA_C \frac{dT}{dx}\big|_{x+\Delta x} - hP\Delta x(T - T_\infty) + q_{rad}P\Delta x = 0$$

Dividing by Δx and letting $\Delta x \to 0$,

$$\frac{d^2T}{dx^2} - \frac{hP}{kA_c}(T - T_\infty) + \frac{q_{rad}P}{kA_c} = 0$$

Since q_{rad} is a constant, we rewrite this equation as

$$\frac{d^2T}{dx^2} - \frac{hP}{kA_c}\left[T - \left(T_\infty + \frac{q_{rad}}{h}\right)\right] = 0$$

which defines an "effective" ambient temperature $T'_\infty = T_\infty + q_{rad}/h$. Then the solutions can be obtained by replacing T_∞ with T'_∞.

a. Insulated blade tip.

$$\frac{T - (T_\infty + q_{rad}/h)}{T_b - (T_\infty + q_{rad}/h)} = \frac{\cosh m(L-x)}{\cosh mL}$$

b. Tip and side heat transfer coefficient equal.

$$\frac{T - (T_\infty + q_{rad}/h)}{T_b - (T_\infty + q_{rad}/h)} = \frac{\cosh m(L-x) + (h/mk)\sinh m(L-x)}{\cosh mL + (h/mk)\sin mL}$$

2.4 The attached figure shows a straight fin of triangular profile (Figure 2.13). Assume that this is a thin fin with $w \gg t$. Derive the heat conduction equation of fin: determine the temperature distributions in the fin analytically and determine the fin efficiency.

Solution

From Figure 2.13,

$$\frac{d}{dx}\left(A_c \frac{dT}{dx}\right) - \frac{h}{k}\frac{dA_S}{dx}(T - T_\infty) = 0$$

$$\frac{tw}{L} \times \frac{d^2T}{dx^2} + \frac{tw}{L}\frac{dT}{dx} - \frac{2hw}{k}(T - T_\infty) = 0 \tag{2.55}$$

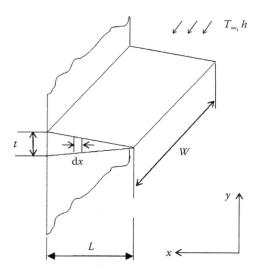

FIGURE 2.13 A triangular straight fin with variable cross-sectional area.

$$x\frac{d^2T}{dx^2}+\frac{dT}{dx}-\frac{2h}{kt}L(T-T_\infty)=0$$

$$x^2\frac{d^2\theta}{dx^2}+x\frac{d\theta}{dx}-m^2Lx\theta=0$$

(2.56)

where

$$A_c=w\frac{t}{L}x$$

$$dA_s=2w\,dx$$

$$\theta\equiv T-T_\infty$$

$$m^2\equiv\frac{2h}{kt}$$

$$A_s\simeq 2wL$$

But we need $z^2\,(d^2\theta/dz^2)+z(d\theta/dz)-z^2\theta=0$, for the modified Bessel function solution. So, $z^2\sim x, z\sim\sqrt{x}$

From the solution form, $I_0\left(2m\sqrt{xL}\right)$, imply $z\sim\sqrt{x}=2m\sqrt{xL}$

$$\frac{dz}{dx}=2m\sqrt{L}\cdot\frac{1}{2}\cdot x^{-1/2}=m\sqrt{L}\frac{1}{\sqrt{x}}=\frac{2m^2L}{z}$$

$$\frac{d\theta}{dx}=\frac{d\theta}{dz}\cdot\frac{dz}{dx};\quad\frac{d^2\theta}{dx^2}=\frac{d}{dz}\left(\frac{d\theta}{dx}\right)\cdot\frac{dz}{dx};$$

Substituting into Equation (2.56)

$$z^2\frac{d^2\theta}{dz^2}+z\frac{d\theta}{dz}-z^2\theta=0$$

(2.57)

The solution to the above equation is

$$\theta=C_1I_o(z)+C_2K_o(z)$$

Boundary conditions

at $\qquad x=0,\quad K_o\to\infty,\quad C_2=0,$

at $\qquad\qquad x=L,\quad\theta=\theta_b,$

$$\theta=\theta_b\cdot\frac{I_0\left(2m\sqrt{xL}\right)}{I_0\left(2mL\right)}$$

$$q_f=+kA_c\frac{dT}{dx}|_{x=L}=\theta_b ktwm\frac{I_1\left(2mL\right)}{I_0\left(2mL\right)}$$

$$\eta_f=\frac{q_f}{h\theta_b 2wL}=\frac{1}{mL}\frac{I_1\left(2mL\right)}{I_0\left(2mL\right)}$$

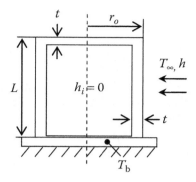

FIGURE 2.14   A hollow transistor modeled as a fin.

2.5 A hollow transistor (Figure 2.14) has a cylindrical cap of radius r_o and height L, and is attached to a base plate at temperature T_b. Show that the heat dissipated is

$$q_f = 2\pi k r_o t \left(T_b - T_\infty\right) m \left[\frac{I_0\left(mr_o\right)\sinh mL + I_1\left(mr_o\right)\cosh mL}{I_0\left(mr_o\right)\cosh mL + I_1\left(mr_o\right)\sinh mL}\right]$$

where the metal thickness is t, $m = (h/kt)^{1/2}$, and the heat transfer coefficient on the sides and top is assumed to be the same, h, and outside temperature, T_∞.

Solution

The cap is split into two fins: (1) a straight fin of width $2\pi r_o$ and length L, and (2) a disk fin of radius r_o. For the straight fin with $T = T_b$ at $x = 0$,

$$T_1 - T_\infty = C_1 \sinh mx + \left(T_b - T_\infty\right)\cosh mx; \quad m = \left(h/kt\right)^{1/2} \tag{2.58}$$

For the disk with $dT/dr = 0$ at $r = 0$,

$$T_2 - T_\infty = C_2 I_0\left(mr\right); \quad m = \left(h/kt\right)^{1/2} \tag{2.59}$$

The constants C_1 and C_2 are determined by matching the temperature and heat flow at the joining,

$$T_1(L) = T_2(r_o); \quad dT/dx|_{x=L} = -dT/dr|_{r=r_o}; \tag{2.60}$$

Since kA_c is the same for both fins at the junction. Substituting Equations (2.58) and (2.59) into Equation (2.60) gives

$$C_2 I_0\left(mr_o\right) = C_1 \sinh mL + \left(T_b - T_\infty\right)\cosh mL$$
$$-C_2 I_1\left(mr_o\right) = C_1 \cosh mL + \left(T_b - T_\infty\right)\sinh mL$$

Solving,

$$C_1 = -(T_b - T_\infty)\frac{I_0(mr_o)\sinh mL + I_1(mr_o)\cosh mL}{I_0(mr_o)\cosh mL + I_1(mr_o)\sinh mL}$$

The heat dissipation is the base heat flow of fin 1,

$$q_f = -kA_c\frac{dT_1}{dx}|_{x=0} = -kA_C mC_1$$

$$= 2\pi kr_o t(T_b - T_\infty)\frac{I_0(mr_o)\sinh mL + I_1(mr_o)\cosh mL}{I_0(mr_o)\cosh mL + I_1(mr_o)\sinh mL}$$

REMARKS

This chapter deals with 1-D steady-state heat conduction through the plane wall with and without heat generation, a cylindrical tube with and without heat generation, and fins with constant and variable cross-sectional areas. Although it is a 1-D steady-state conduction problem, there are many engineering applications. For example, heat losses through building walls, heat transfer through tubes, and heat losses through fins. In undergraduate heat transfer, we normally ask you to calculate heat transfer rates through the plane walls, circular tubes, and fins, with given dimensions, material properties, and thermal boundary conditions. Therefore, you can choose the correct formulas and use them with the given values and obtain the results.

However, at this intermediate heat transfer level, we are more focused on how to solve the heat conduction equation with various thermal boundary conditions and how to obtain the temperature distributions for a given physical problem. For example, how to solve the temperature distributions for the plane walls, circular tubes, with and without heat generation, and fins with a constant or variable cross-sectional area, with various thermal boundary conditions. In particular, we have introduced one of the very powerful mathematical tools, Bessel function, to solve the fins with variable cross-sectional area with various thermal boundary conditions. This is the only thing new compared to the undergraduate heat transfer.

PROBLEMS

2.1 The performance of gas turbine engines may be improved by increasing the tolerance of the turbine blades to hot gases emerging from the combustor. One approach to achieving high operating temperatures involves application of a *thermal barrier coating* (TBC) to the exterior surface of a blade, while passing cooling air through the blade. Typically, the blade is made from a high-temperature superalloy, such as Inconel ($k \approx 25$ W/m K) while a ceramic, such as zirconia ($k \approx 1.3$ W/m K), is used as a TBC. Consider conditions for which hot gases at $T_{\infty, o} = 1700$ K and cooling air at $T_{\infty, i} = 400$ K provide outer- and inner-surface convection coefficients of $h_o = 1000$ W/m^2 K and $h_i = 500$ W/m^2 K, respectively. If a 0.5-mm-thick zirconia TBC is attached to a 5-mm-thick Inconel blade wall by means of a metallic bonding agent, which provides an interfacial thermal resistance of $R''_{t, c} = 10^{-4}$ m^2 K/W, can the Inconel be maintained at a temperature that is below its maximum allowable value of 1250 K? Radiation effects may be neglected, and the turbine blade may be approximated as a plane wall. Plot the temperature distribution with and without the TBC. Are there any limits to the thickness of the TBC?

2.2   Consider 1-D heat conduction through a circular tube, as shown in Figure 2.3. Determine the heat loss per tube length for the following conditions:

Given:

Find: $q/L =$?

Steam Inside the Pipe	Air Outside the Pipe	Steel Pipe AISI1010
$T_{\infty 1} = 250°C$	$T_{\infty 2} = 20°C$	$2r_1 = 60\,\text{mm}$
$h_1 = 500\ \text{W/m}^2\text{K}$	$h_2 = 25\text{W/m}^2\text{K}$	$2r_2 = 75\,\text{mm}$
		$\varepsilon = 0.8$
		$k = 58.7\ \text{W/mK at 400K}$

2.3   Heat is generated at a rate $\dot{q}$ in a long solid cylinder of radius r_o, as shown in Figure 2.6. The cylinder has a thin metal sheath and is immersed in a liquid at temperature T_∞. Heat transfer from the cylinder surface to the liquid can be characterized by a heat transfer coefficient h. Obtain the steady-state temperature distributions for the following cases:

a. $\dot{q} = \dot{q}_o \left[1 - (r / r_o)^2\right]$.

b. $\dot{q} = a + b(T - T_\infty)$.

2.4   A hollow, cylindrical, copper tube is used to improve heat removal from a transistor. The copper tube is attached to the top of a round, disk-shaped transistor, as shown in Figure 2.15. The disk-shaped transistor dissipates

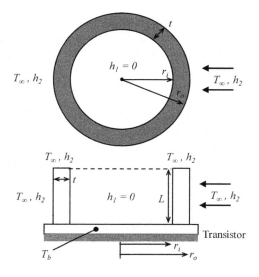

FIGURE 2.15   A hollow cylindrical copper tube fin.

0.2 W at steady-state operation. The backside of the base transistor is well insulated. At 300K, the copper has a conductivity, $k = 403$ W/mK. Other given parameters include:

$L = 15$ mm, $r_o = 7.75$ mm, $r_i = 7.5$ mm, $t = 0.25$ mm; Assume: $h_1 = 0$ (no cooling)

Air cooling: $T_\infty = 25°C$, $h_2 = 50$ W/m²K

Find: The copper tube temperature distribution and T_b with and without the tubular fin.

2.5 A thin metal disk, as shown in Figure 2.9d, is insulated on one side and exposed to a jet of hot air at temperature T_∞ on the other. The convection heat transfer coefficient, h, can be taken as constant over the disk. The periphery at $r = R$ is maintained at a uniform temperature T_R.

a. Derive the heat conduction equation of disk.

b. Determine the disk temperature distributions.

c. Do you think the disk temperature is higher at the center or the periphery? Why?

2.6 Given a relatively thin annular fin with a uniform thickness, as shown in Figure 2.9b, is affixed to a tube. The inner and outer radii of the fin are r_i and r_o, respectively, and the thickness of the fin is t. The tube surface (i.e., the base of the annular fin) is maintained at a temperature of T_b or $T(r_i) = T_b$. Both the top and the bottom surfaces are exposed to a fluid at T_∞. The convection heat transfer coefficient between the fin surfaces and the fluid is h.

a. Derive the steady-state heat conduction equation of the annular fin and propose a solution of the annular fin temperature distribution with the associate boundary conditions.

b. If during the air cooling, the annular fin has also received radiation energy from the surrounding environment, T_{sur}, sketch, compare, and comment on the fin temperature profile $T(r)$ with that in (a)?

2.7 Both sides of a very thin metal disk, as shown in Figure 2.9d, are heated by convective hot air at temperature T_∞. The convection heat transfer coefficient, h, can be taken constant over the disk. The periphery at $r = R$ is maintained at a uniform temperature T_R.

a. Derive the steady-state heat conduction equation of the disk.

b. Propose a solution method and the associated boundary conditions that can be used to determine the disk temperature distributions. Sketch the disk temperature profile $T(r)$. Do you think the disk temperature is higher at the center or the periphery? Why?

c. If during the air heating, the disk has also emitted a net uniform radiation flux q''_{rad} to the surrounding environment, derive the steady-state heat condition equation of the disk and propose a solution method to determine the disk temperature distributions. Sketch, compare, and comment on the disk temperature profile with that in (b)?

2.8 The front surface of a very thin metal disk is cooled by convective air at temperature T_∞ while the back surface is perfectly insulated, as shown in Figure 2.9d. The convection heat transfer coefficient h can be taken to be constant over the front surface of the disk. The periphery at $r = R$ is maintained at a uniform temperature, T_R, by a heat source.

  a. Derive the steady-state heat conduction equation of disk.
  b. Determine the disk temperature distributions with the associated boundary conditions. Sketch the disk temperature profile $T(r)$.
  c. If during the air cooling, the disk has also emitted a net uniform radiation flux q'_{rad} to the surrounding environment, derive the steady-state heat conduction equation of the disk and determine the disk temperature distributions. Sketch, compare, and comment on the disk temperature profile with that in (b)?

2.9   A thin, conical pin fin is shown in Figure 2.9a. Analytically determine the temperature distribution through the pin fin. Also, determine the heat flux through the pin fin base. Given: Pin fin tip temperature: $T_R > T_\infty$; Pin fin height: l; Pin fin base diameter: d; Pin fin base temperature: T_b; Cooling air at T_∞, h; Hot wall at T_b

2.10 The wall of a furnace has a height $L = 1$ m and is at a uniform temperature of 500 K. Three materials of equal thickness $t = 0.1$ m, having the properties listed in the table attached, are placed in the order shown in Figure 2.16 to insulate the furnace wall. Air at $T_\infty = 300$ K blows past the outer layer of insulation at a speed of $V_\infty = 1.5$ m/s as shown in the figure.

  a. Assuming 1-D, steady conduction through the insulation layers, calculate the heat flux from the wall to the surroundings, q. For this, choose the most appropriate of the following two correlations:

$$\frac{\overline{h}L}{k} = 0.664\,\mathrm{Re}_L^{1/2}\,\mathrm{Pr}^{1/3}\left(\mathrm{Re}_L < 10^5;\text{laminar flow}\right)$$

$$\frac{\overline{h}L}{k} = \left(0.037\,\mathrm{Re}_L^{0.8} - 850\right)\mathrm{Pr}^{1/3}\left(\mathrm{Re}_L < 10^5;\text{turbulent flow}\right)$$

where $\overline{h}$ is an average heat transfer coefficient, Re_L is the Reynolds number based on L and V_∞, and Pr is the Prandtl number of air.

  b. For the arrangement shown in the figure, calculate the temperatures at surfaces 2, 3, and 4.

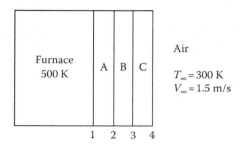

FIGURE 2.16   A furnace composite plane wall model.

c.   How should the materials A, B, C be ordered to obtain the steepest
     temperature gradient possible between surfaces 1 and 2?
d.   For this new arrangement, calculate the temperatures at surfaces 2, 3,
     and 4.

Material	k (W/mK)	ρ (kg/m³)	μ (kg/ms)	c_p (kJ/kgK)
A	100			
B	10			
C	1			
Air	0.026	1.177	1.846×10^{-5}	1.006

2.11 A very long copper rod with a relatively small diameter is moving in a vac-
      uum with a constant velocity, V (Figure 2.17). The long rod is moving from
      one constant temperature region, T_0, at $x = 0$, to another temperature region,
      T_L (at $x = L$). Solve for the steady-state temperature distribution in the rod
      between $x = 0$ and L. Neglect thermal radiation.
      a. Write down the governing equation for the axial temperature distribu-
          tion when $V = 0$.
      b. Determine the axial temperature distribution which satisfies the speci-
          fied boundary conditions when $V = 0$.
      c. Write down the governing equation for the axial temperature distribu-
          tion when $V \neq 0$.
      d. Determine the axial temperature distribution which satisfies the speci-
          fied boundary conditions when $V \neq 0$. (Make any necessary assumption.
          For example, the thermal conductivity of the copper rod is k...)
2.12 A thin long rod extends from the side of a probe that is in outer space. The
      base temperature of the rod is T_b. The rod has a diameter, D, a length, L, and
      its surface is at an emissivity, ε. State all relevant assumptions and boundary
      conditions.
      a. Starting with an energy balance on a differential cross section of the
          pin fin, perform an energy balance on the rod and derive a differential
          equation that could be used to solve this problem.
      b. Sketch, on the same plot, the temperature distribution along the rod, and
          compare the heat loss through the rod for the following four cases: (1)

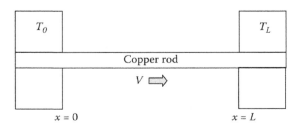

FIGURE 2.17 A moving fin model.

The rod is made of aluminum; (2) the aluminum rod is painted black; (3) the rod is made of steel; (4) the steel rod is painted black.

2.13 Given a relatively thin annular fin with uniform thickness that is affixed to a tube. The inner and outer radii of the fin are r_i and r_o, respectively, and the thickness of the fin is w. The tube surface (i.e., the base of the annular fin) is maintained at a temperature of T_b, or $T(r_i) = T_b$. Both the top and bottom surfaces of the fin are exposed to a fluid at T_∞. The convective heat transfer coefficient between the fin surfaces and the fluid is h.

a. Show that to determine the steady 1-D temperature distribution in the annular fin, $T(r)$, the governing equation may be written in the form of a modified Bessel's equation.

$$\frac{d^2 y}{dx^2} + \frac{1}{x}\frac{dy}{dx} - \left(c^2 + \frac{n^2}{x^2}\right)y = 0$$

where $y = y(x)$, and c and n are constants.

b. The general solution of the above modified Bessel's equation is $y = C_1 I_n(cx) + C_2 K_n(cx)$, where C_1 and C_2 are constants, and I_n and K_n are the modified Bessel's functions of the first and second kinds of order n, respectively. Assuming that the heat transfer on the outer surface of the fin is negligible [i.e., $dT/dr = 0$ at $r = r_o$], solve the governing equation to obtain the steady 1-D temperature distribution in the annular fin, $T(r)$.

c. Also, determine the steady 1-D temperature distribution in the annular fin numerically using the finite difference method. Give the finite-difference equations for the nodes at $r = r_i$ and $r = r_o$ and for a typical interior node. Rearrange the equations to give expressions for the temperatures at the nodes.

2.14 An engineer has suggested that a triangular fin would be more effective than a circular fin for a new natural convection heat exchanger. The fins are very long and manufactured from Al 2024T6. The triangular fin has an equilateral cross section with a base dimension of 1.0 cm and the second fin has a circular cross section with a 0.955 cm diameter. If the base temperature of the fin is maintained at 400°C, which fin will transfer more heat and which fin has a greater effectiveness?

$$q_{fin} = \left(hPkA_c\right)^{1/2} \cdot \theta_b \qquad \varepsilon_{fin} = q_{fin}/hA_c\theta_b$$

Given: $k = 25$ W/m²K; $k_{2024\text{-}T6} = 177$ W/mK; $T = 25°C$

2.15 A thin conical pin fin is attached to a hot base plate at T_b. The cooling air has temperature T_∞ and convection heat transfer coefficient h. Analytically determine the temperature profile in the pin fin. Also, determine the heat flux through the pin fin base.

2.16 Determine the temperature distribution for the annulus fin geometry as shown in Figure 2.9b. Assume that the fin tip is exposed to the same fluid and convection heat transfer coefficient as the sides of the fin.

2.17 Determine the temperature profile for the annulus fin geometry as shown in Figure 2.9b. Assume that the fin tip is fixed at a given temperature between the fin base and the convection fluid.

2.18 Determine the solutions shown in Equations (2.36) and (2.37).

2.19 Determine the solutions shown in Equations (2.38) and (2.39).

2.20 Determine the solutions shown in Equations (2.40) and (2.41).

REFERENCES

1. W. Rohsenow and H. Choi, *Heat, Mass, and Momentum Transfer*, Prentice-Hall, Inc., Englewood Cliffs, NJ, 1961.
2. F. Incropera and D. Dewitt, *Fundamentals of Heat and Mass Transfer*, Fifth Edition, John Wiley & Sons, New York, 2002.
3. A. Mills, *Heat Transfer*, Richard D. Irwin, Inc., Boston, MA, 1992.
4. V. Arpaci, *Conduction Heat Transfer*, Addison-Wesley Publishing Company, Reading, MA, 1966.

3 2-D Steady-State Heat Conduction

3.1 METHOD OF SEPARATION OF VARIABLES: GIVEN TEMPERATURE BC

Most often heat is conducted in two dimensions instead of one dimension, as discussed in Chapter 2. For example, we are interested in determining the temperature distribution in a 2-D rectangular block with appropriate boundary conditions (BCs). Once the temperature distribution is known, the associated heat transfer rate can be determined. The following are the steady-state 2-D heat conduction equations without heat generation and the typical BCs with given surface temperatures.

$$\frac{\partial^2 T}{\partial x^2} + \frac{\partial^2 T}{\partial y^2} = 0 \tag{3.1}$$

$$\frac{\partial^2 \theta}{\partial x^2} + \frac{\partial^2 \theta}{\partial y^2} = 0, \qquad \text{with } \theta = T - T_0 \tag{3.2}$$

BCs:

$x = 0$, $T = 0$ or $x = 0$, $\theta = 0$ homogeneous BC
$x = a$, $T = 0$ or $x = a$, $\theta = 0$ homogeneous BC
$y = 0$, $T = 0$ or $y = 0$, $\theta = 0$ homogeneous BC
$y = b$, $T = T_s$ or $y = b$, $\theta = T_s - T_0 = \theta_s$; nonhomogeneous BC

Here, we define a homogeneous BC as $T = 0$, or $\partial T/\partial x = 0$, $\partial T/\partial y = 0$; $\theta = 0$, or $\partial\theta/\partial x = 0$, $\partial\theta/\partial y = 0$, that is, temperature or temperature gradient at a given boundary surface (in the x- or y-direction) equals 0. In contrast, we define a nonhomogeneous BC as $T \neq 0$, $\partial T/\partial x \neq 0$, $\partial T/\partial y \neq 0$; or $\theta \neq 0$, $\partial\theta/\partial x \neq 0$, $\partial\theta/\partial y \neq 0$, that is, temperature or temperature gradient at a given boundary surface (in the x- or y-direction) does not equal 0.

Equations (3.1) and (3.2) can be solved by the method of separation of variables. With this technique, we can separate the 2-D temperature distribution, $T(x, y)$, into two independent temperature distributions, $T(x)$ and $T(y)$, respectively. The final 2-D temperature distribution is the product of the separate 1-D temperature solutions, that is, $T(x, y) = T(x) \cdot T(y)$. The following outlines the method of separation of variables [1–4]. We need four BCs, two in the x-direction and two in the y-direction, to solve the 2-D heat conduction problem. One important note is that, among four BCs, only one nonhomogeneous BC is allowed in order to apply the separation of variables method. For a given problem, we need to make sure that only one nonhomogeneous

DOI: 10.1201/9781003164487-3

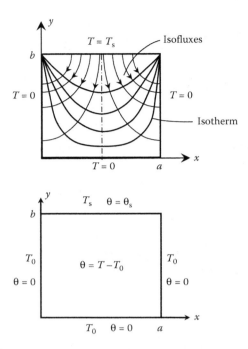

FIGURE 3.1 2-D heat conduction with three homogeneous and one nonhomogeneous boundary conditions.

BC exists either in the x- or in the y-direction. The principal of superposition will be used for the problems having two, three, or four nonhomogeneous BCs. We will begin with the simplest case, as shown in Figure 3.1, with given surface temperatures as BCs. Then we will move to more complicated cases with surface heat flux and surface convection BCs, as well as the problems requiring the principle of superposition.

Assume the form of the final solution, based on separation of variables.

$$T(x,y) = X(x)Y(y) \tag{3.3}$$

Then take the partial derivatives

$$\frac{\partial T}{\partial x} = Y\frac{\partial X}{\partial x} = Y\frac{dX}{dx}$$

$$\frac{\partial^2 T}{\partial x^2} = Y\frac{\partial^2 X}{\partial x^2} = Y\frac{d^2 X}{dx^2}$$

$$\frac{\partial T}{\partial y} = X\frac{\partial Y}{\partial y} = X\frac{dY}{dy}$$

$$\frac{\partial^2 T}{\partial y^2} = X\frac{\partial^2 Y}{\partial y^2} = X\frac{d^2 Y}{dy^2}$$

Substitute the derivatives into the 2-D conduction Equation (3.1)

$$Y\frac{d^2X}{dx^2} + X\frac{d^2Y}{dy^2} = 0$$

$$-\frac{1}{X}\frac{d^2X}{dx^2} = \frac{1}{Y}\frac{d^2Y}{dy^2}$$

(3.4)

The equality can hold only if both sides are equal to a constant as each side of Equation (3.4) is a function of an independent variable. The constant can be positive, negative, or zero. However, the positive number of the constant is the only possibility for the BCs in this case. The readers may note that zero and negative constants do not satisfy the BCs. Therefore, we express Equation (3.4) as

$$-\frac{1}{X}\frac{d^2X}{dx^2} = \frac{1}{Y}\frac{d^2Y}{dy^2} = \lambda^2$$

(3.5)

Then we have

$d^2X/dx^2 + \lambda^2X = 0 \Rightarrow X = X(x)$, the equation for two homogeneous BCs,

$d^2Y/dy^2 - \lambda^2Y = 0 \Rightarrow Y = Y(y)$, the equation for one homogeneous BC,

$X(x) = C_1\cos\lambda x + C_2\sin\lambda x$, the solution for equation with two homogeneous BCs,

$Y(y) = C_3e^{-\lambda y} + C_4e^{\lambda y}$ or $(Y(y) = C_3\sinh\lambda y + C_4\cosh\lambda y)$,

the solution for equation with one homogeneous BC.

Solve for C_1 and C_2 using the x-direction equation

at $x = 0$, $T = 0$, $\Rightarrow X = 0$, then $C_1 = 0$,

at $x = a$, $T = 0$, $\Rightarrow X = 0$, then $C_2\sin\lambda a = 0 \Rightarrow \lambda_n a = n\pi$, $n = 0, 1, 2, 3,...$ so,

$$\lambda_n = \frac{n\pi}{a}$$

Therefore,

$$X(x) = C_2\sin\frac{n\pi x}{a}$$

Solve for C_3 and C_4 using the y-direction equation at $y = 0$, $T = 0$, $Y = 0 = C_3 + C_4$ $\Rightarrow C_4 = -C_3$.

$$Y(y) = C_3e^{-\lambda_n y} - C_3e^{\lambda_n y} = C_3\left(e^{-\lambda_n y} - e^{\lambda_n y}\right)$$

With $\sinh(x) = (e^x - e^{-x})/2$, let $C_5 = C_3/2$, the above equation can be written as

$$Y(y) = C_5\sinh(\lambda_n y)$$

Then solve the product equation, and let $C_2C_5 = C_n$.

$$T(x,y) = X(x)\cdot Y(y) = C_2C_5\sin\frac{n\pi x}{a}\sinh\frac{n\pi y}{a} = \sum C_n\sin\frac{n\pi x}{a}\sinh\frac{n\pi y}{a}$$

at $y = b$, $T = T_s = \Sigma\, C_n \sin(n\pi/a)x \sinh(n\pi/a)b$.

Multiplying both sides by $\sin(m\pi x/a)\, dx$, one obtains

$$\int \sin\frac{m\pi x}{a} \cdot T_s dx = \sum \int C_n \sin\frac{n\pi x}{a} \sinh\frac{n\pi b}{a} \sin\frac{m\pi x}{a} dx$$

C_n can be determined by the integration over x,

$$C_n = \frac{\displaystyle\int T_s \sin(m\pi x/a)dx}{\displaystyle\int \sin(n\pi x/a)\sinh(n\pi b/a)\sin(m\pi x/a)dx}$$

$$C_n = 0, \qquad \text{if } m \neq n \tag{3.6}$$

$$C_n = \frac{\displaystyle\int T_s \sin(n\pi x/a)dx}{\displaystyle\int \sin^2(n\pi x/a)\sinh(n\pi b/a)dx} \qquad \text{if } m = n.$$

How does one perform integration? From the integration table, one obtains

$$1. \quad \int_0^a \sin\frac{n\pi x}{a} dx = -\frac{a}{n\pi}\cos\frac{n\pi x}{a}\Big|_0^a = \frac{a}{n\pi}\big(1 - \cos(n\pi)\big)$$

$$= \frac{a}{n\pi}\big[1 - (-1)^n\big] = \begin{cases} 0, n = \text{even} \\[2mm] \dfrac{2a}{n\pi}, n = \text{odd} \end{cases}$$

$$2. \quad \int_a^a \sin^2\frac{n\pi x}{a} dx = \frac{a}{2n\pi}\left[\frac{n\pi x}{a} - \frac{1}{2}\sin\frac{2n\pi x}{a}\right]_0^a = \frac{a}{2}$$

or

$$\int_0^a \sin^2\frac{n\pi x}{a} dx = \int_0^a \frac{1 - \cos(2n\pi x/a)}{2} dx = \frac{a}{2} - \frac{1}{2}\frac{a}{2n\pi}\sin\frac{2n\pi x}{a}\Big|_0^a = \frac{a}{2}$$

Therefore, one obtains

$$C_n = \frac{(2/n\pi)\big[1 - (-1)^n\big]T_s}{\sinh(n\pi b/a)}$$

Finally, the 2-D temperature distribution follows:

$$T(x,y) = \sum_{n=1}^{\infty} C_n \sin \frac{n\pi x}{a} \sinh \frac{n\pi y}{a} \tag{3.7}$$

$$T(x,y) = \sum_{n=1}^{\infty} \frac{(2/n\pi)\left[1-(-1)^n\right]T_s}{\sinh(n\pi b/a)} \sin \frac{n\pi x}{a} \sinh \frac{n\pi y}{a} \tag{3.8}$$

The second-order partial differential equations (PDEs) $T(x,y)$ are split into two second-order ordinary differential equations (ODEs) $T(x)$ and $T(y)$. The second-order ODE with two homogeneous BCs is the so-called eigenvalue equation. The solution of the eigenfunctions, sine and cosine, depend on the two homogeneous BCs. The eigenvalues, λ, can be determined by one of the two homogeneous BCs (either both in the x-direction, or both in the y-direction). The solution of the other second-order ODE is a decay curve of combining $e(x)$ and $e(-x)$ for an infinite-length problem, or $\sinh(x)$ and $\cosh(x)$ for a finite-length problem. The only nonhomogeneous BC will be used to solve the final unknown coefficient C_n. The integrated value C_n can be determined by performing integration of sin, sin-square or cos, cos-square, depending on the given BCs, by using the characteristics of the orthogonal functions.

If we let $\theta = T - T_0$, follow the same procedure, the 2-D temperature distribution becomes

$$\theta(x,y) = \theta(x)\theta(y) = \sum C_n \sin \frac{n\pi x}{a} \sinh \frac{n\pi y}{a} \tag{3.9}$$

$$\theta(x,y) = \sum_{n=1}^{\infty} \frac{(2/n\pi)\left[1-(-1)^n\right]\theta_s}{\sinh(n\pi b/a)} \sin \frac{n\pi x}{a} \sinh \frac{n\pi y}{a} \tag{3.10}$$

With the general solution shown in Equation (3.10), values can be assumed for T_o, T_s, a, and b. For example, if $T_o = 20°C$, $T_s = 100°C$, and $a = b = 100\,cm$, $T(x, y)$ and $q''(x, y)$ can be determined for $n = 1$ to 3. Using the given values, the two-dimensional temperature and heat flux distribution can be plotted (isotherms and isofluxes) to generate distributions similar to Figure 3.1. The lines of constant heat flow (isofluxes) are perpendicular to the isotherms. Using the temperature distribution obtained using the separation of variables technique, the heat flow per area (heat flux) in both the x- and y- directions can be calculated from Fourier's Law.

$$q''_x = -k \frac{\partial \theta}{\partial x}$$

$$q''_y = -k \frac{\partial \theta}{\partial y}$$

While the temperature within the domain is a scalar, but the heat flux is a vector as it is calculated from the temperature gradient. The magnitude of the heat flux at a specific location within the solid is a resultant of the q_x'' and q_y'' components and is directed perpendicular to the isotherm.

$$q'' = \sqrt{\left(q_x''\right)^2 + \left(q_y''\right)^2}$$

3.2 METHOD OF SEPARATION OF VARIABLES: GIVEN HEAT FLUX AND CONVECTION BCs

3.2.1 GIVEN SURFACE HEAT FLUX BC

The following shows the similar principal of using the separation of variables method to solve the 2-D heat conduction problems with one nonhomogeneous boundary specified as a surface heat flux or a surface convection condition [2]. As with the aforementioned procedure, we need to make 3 of 4 BCs homogeneous. For example, in Figure 3.2, the three temperature BCs become homogeneous by setting $\theta = T - T_1$ and the only nonhomogeneous heat flux BC becomes $q_s'' = -k\left(\partial\theta / \partial y\right)$. The solution will be the product of sine and cosine in the x-direction (two homogeneous BCs) with sinh and cosh in the y-direction (one nonhomogeneous BC). As before, the nonhomogeneous heat flux BC will be used to solve the final unknown integrated value C_n.

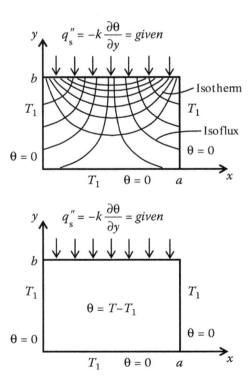

FIGURE 3.2 2-D heat conduction with three homogeneous and one heat flux nonhomogeneous boundary conditions.

Given a long rectangular bar with a constant heat flux along one edge, the other edges are isothermal. In order to obtain homogeneous BCs on the three isothermal edges, let $\theta = T - T_1$. Laplace's equation, Equation (3.2), applies to this steady 2-D conduction problem.

$$\frac{\partial^2 \theta}{\partial x^2} + \frac{\partial^2 \theta}{\partial y^2} = 0$$

BCs:

$$
\begin{aligned}
x = 0, \quad & 0 < y < b: \quad & \theta = 0, \\
y = 0, \quad & 0 < x < a: \quad & \theta = 0, \\
x = a, \quad & 0 < y < b: \quad & \theta = 0,
\end{aligned}
$$

$$y = b, \quad 0 < x < a: \quad q_s'' = -k\frac{\partial \theta}{\partial y}$$

The solution is subject to the three homogeneous BCs with θ replacing T,

$$\theta(x,y) = \sum_{n=1}^{\infty} C_n \sin\frac{n\pi x}{a} \sinh\frac{n\pi y}{a}$$

Applying the last nonhomogeneous BC, to solve for C_n,

$$\frac{\partial \theta}{\partial y}\Big|_{y=b} = -\frac{q_s''}{k} = \sum_{n=1}^{\infty} C_n\left(\sin\frac{n\pi x}{a}\right)\frac{n\pi}{a}\cosh\frac{n\pi b}{a}$$

$$
\begin{aligned}
C_n &= \frac{-(q_s''/k)\int_0^a \sin(n\pi x/a)dx}{(n\pi/a)\cosh(n\pi b/a)\int_0^a \sin^2(n\pi x/a)dx} \\[2mm]
&= \frac{-(q_s''/k)(a/n\pi)\left[1-(-1)^n\right]}{(n\pi/a)\cosh(n\pi b/a)[(a/2)]} \\[2mm]
&= \frac{-(q_s''/k)(2/n\pi)\left[1-(-1)^n\right]}{(n\pi/a)\cosh(n\pi b/a)}
\end{aligned}
$$

Therefore, we obtain

$$T(x,y) = T_1 - \sum_{n=1}^{\infty} \frac{2q_s''a\left[1-(-1)^n\right]}{kn^2\pi^2\cosh(n\pi b/a)}\sin\frac{n\pi x}{a}\sinh\frac{n\pi y}{a} \qquad (3.11)$$

From the solution of the temperature distribution across the 2-D plate with the upper surface exposed to a constant heat flux, values can be used to plot the isotherms and isofluxes through the plate. If we assume $T_1 = 20°C$, $q_s'' = 4000$ W/mK, $k = 20$ W/mK, and $a = b = 100$ cm, $T(x, y)$ and $q''(x, y)$ can be calculated for $n = 1$ to 3. Using the given values, the two-dimensional temperature and heat flux distribution can be plotted similar to what is shown in Figure 3.2.

3.2.2   GIVEN SURFACE CONVECTION BC

A similar procedure can be applied for the only nonhomogeneous convection BC problem [2] shown in Figure 3.3. Again, the solution will be a product of sine and cosine in the x-direction and sinh and cosh in the y-direction, and the nonhomogeneous convection BC will be used to solve the final unknown, integrated value of Cn.

A long rectangular bar with one side cooled by convection and the others maintained at a constant temperature, T_1. Define $\theta = T - T_1$. Laplace's equation applies to this steady 2-D conduction problem.

$$\frac{\partial^2 \theta}{\partial x^2} + \frac{\partial^2 \theta}{\partial y^2} = 0$$

which must be solved subject to BCs

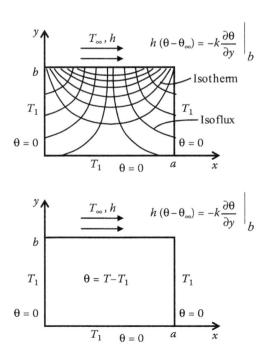

FIGURE 3.3   2-D heat conduction with one convective boundary condition.

$$x = 0, \quad 0 < y < b: \quad \theta = 0,$$
$$y = 0, \quad 0 < x < a: \quad \theta = 0,$$
$$x = a, \quad 0 < y < b: \quad \theta = 0,$$

$$y = b, \quad 0 < x < a: \begin{cases} h(T - T_\infty) = -k\dfrac{\partial T}{\partial y} \\[2mm] h\left[(T - T_1) - (T_\infty - T_1)\right] = -k\dfrac{\partial(T - T_1)}{\partial y} \\[2mm] h(\theta - \theta_\infty) = -k\dfrac{\partial \theta}{\partial y} \end{cases}$$

The solution is subject to the three homogeneous BCs with θ replacing T,

$$\theta(x,y) = \sum_{n=1}^{\infty} C_n \sin\frac{n\pi x}{a} \sinh\frac{n\pi y}{a}$$

and at $y = b$, applying nonhomogeneous BCs to solve for Cn:

$$\frac{\partial \theta}{\partial y} = \sum_{n=1}^{\infty} C_n \sin\frac{n\pi x}{a}\frac{n\pi}{a}\cosh\frac{n\pi b}{a} = -\frac{h}{k}\left[\sum_n C_n \sin\frac{n\pi x}{a}\sinh\frac{n\pi b}{a} - \theta_\infty\right]$$

$$\sum_{n=1}^{\infty} C_n \sin\frac{n\pi x}{a}\left(\frac{n\pi}{a}\cosh\frac{n\pi b}{a} + \frac{h}{k}\sinh\frac{n\pi b}{a}\right) = \frac{h}{k}\theta_\infty$$

$$C_n = \frac{(h/k)\theta_\infty \displaystyle\int_0^a \sin(n\pi x/a)dx}{\left((n\pi/a)\cosh(n\pi b/a) + (h/k)\sinh(n\pi b/a)\right)\displaystyle\int_0^a \sin^2(n\pi x/a)dx}$$

$$= \frac{(h/k)\theta_\infty(a/n\pi)\left[1 - (-1)^n\right]}{\left((n\pi/a)\cosh(n\pi b/a) + (k/h)\sinh(n\pi b/a)\right)[(a/2)]}$$

$$= \frac{\theta_\infty(2/n\pi)\left[1 - (-1)^n\right]}{(\sinh(n\pi b/a) + (n\pi/a)(k/h))\cosh(n\pi b/a)}$$

Therefore, we obtain

$$\frac{T - T_1}{T_\infty - T_1} = \sum_{n=1}^{\infty} \frac{(2/n\pi)\left[1 - (-1)^n\right]\sin(n\pi x/a)\sinh(n\pi y/a)}{\sinh(n\pi b/a) + (n\pi/a)(k/h)\cosh(n\pi b/a)} \qquad (3.12)$$

As with the previous BCs, plots of the temperature and heat flow can be generated by assuming numerical values. Taking $T_1 = 20°C$, $T_\infty = 100°C$, $k = 20$ W/mK, $h = 50$ W/m^{2}K, and $a = b = 100$ cm, $T(x, y)$ and $q''(x, y)$ can be calculated for $n = 1$ to 3 (or any value of n). Figure 3.3 shows the sample distributions.

3.3   PRINCIPLE OF SUPERPOSITION FOR NONHOMOGENEOUS BCs

In some applications, we may have all three types of surface BCs applied to a given problem. For example, Figure 3.4 shows a 2-D heat conduction problem with a given surface temperature,heat flux, and convection BCs. The problem involves three non-homogeneous BCs after letting $\theta = T - T_\infty$. We need to use the superposition principle, splitting the problem with three nonhomogeneous BCs into three individual problems [1]. For each of the problem, there is only one nonhomogeneous BC. Then the method of separation of variables can be applied. It is important to note that both the heat conduction equations and the associated BCs must satisfy the superposition principle, respectively, that is, $\theta = \theta_1 + \theta_2 + \theta_3$ for both the heat conduction equation and four BCs. From the aforementioned discussion, we know how to obtain the solution for θ_1 (two homogeneous x-BCs and one nonhomogeneous at $y = 0$), θ_2 (two homogeneous y-BCs and one nonhomogeneous at $x = 0$), and θ_3 (two homogeneous y-BCs and one nonhomogeneous at $x = b$). Applying superposition, the final temperature distributions is $\theta = \theta_1 + \theta_2 + \theta_3$. Examples 3.3 and 3.4 provide details for the superposition method.

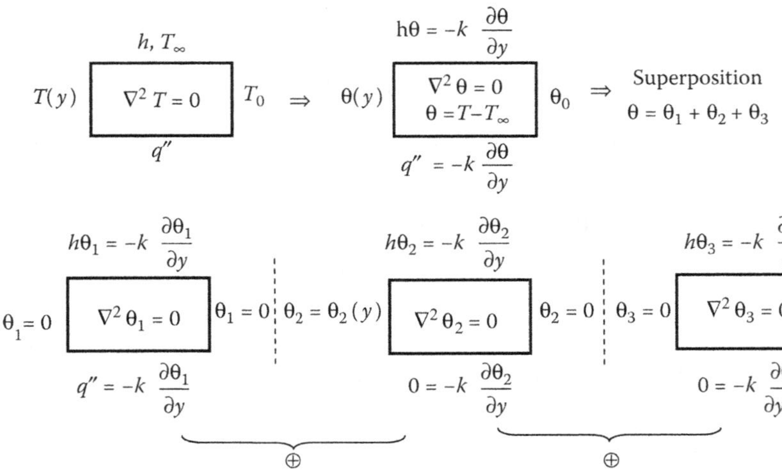

FIGURE 3.4   Principle of superposition for two-dimensional heat conduction with four non-homogeneous boundary conditions.

3.3.1 2-D HEAT CONDUCTION IN CYLINDRICAL COORDINATES

In general, heat conduction in the cylindrical coordinates, shown in Figure 3.5, is a 3D problem, i.e., $T(r,z,\phi)$. Cylindrical heat transfer can be simplified to a 2D problem if the temperature remains constant in one direction. More often, the circumferential temperature is assumed constant, so $T(r,z)$ can be determined beginning with Equation (3.13). Alternatively, if the temperature does not vary in the z-direction, the 2D temperature distribution, $T(r,\phi)$, can be determined using the separation of variables technique beginning with Equation (3.14).

$$\frac{1}{r}\frac{\partial}{\partial r}\left(r\frac{\partial T}{\partial r}\right)+\frac{\partial^2 T}{\partial z^2}=0 \Rightarrow T=R(r)Z(z) \qquad (3.13)$$

$$\frac{\partial^2 T}{\partial r^2}+\frac{1}{r}\frac{\partial T}{\partial r}+\frac{1}{r^2}\frac{\partial^2 T}{\partial \phi^2}=0 \Rightarrow T=R(r)\Theta(\phi) \qquad (3.14)$$

Case 1: Consider a short cylinder with an outer radius r_o and a height l. The bottom and outer surfaces of the cylinder are maintained at T_1 while the top surface temperature is T_2. From Equation (3.13),

$$\frac{1}{r}\frac{\partial}{\partial r}\left(r\frac{\partial T}{\partial r}\right)+\frac{\partial^2 T}{\partial z^2}=0$$

BCs,

$$T(r,0)=T_1$$

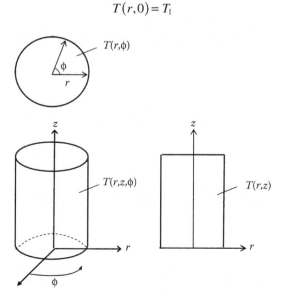

FIGURE 3.5 2-D heat conduction in cylindrical coordinates.

$$T(r,l) = T_2$$

$$T(r_o,z) = T_1$$

$$T(0,z) = \text{finite value, or } \frac{\partial T(0,z)}{\partial r} = 0$$

$$\text{Let } \theta = T - T_1, \quad \theta = \theta(r,z)$$

From the above equation and BCs,

$$\frac{1}{r}\frac{\partial}{\partial r}\left(r\frac{\partial \theta}{\partial r}\right) + \frac{\partial^2 \theta}{\partial z^2} = 0$$

$$\theta(r,0) = T_1 - T_1 = 0$$

$$\theta(r_o,z) = T_1 - T_1 = 0$$

$$\theta(r,l) = T_2 - T_1 = \theta_2$$

$$\theta(0,z) = \text{finite value, or } \frac{\partial \theta(0,z)}{\partial r} = 0$$

From the separation of variables method, assume a product solution,

$$\theta(r,z) = R(r) \cdot Z(z)$$

Take the partial derivatives and substitute the derivatives into the governing PDE,

$$Z\frac{d^2 R}{dr^2} + \frac{1}{r}Z\frac{dR}{dr} + R\frac{d^2 Z}{dz^2} = 0$$

Rearrange the above equation and group the like terms,

$$-\frac{1}{R}\left(\frac{d^2 R}{dr^2} + \frac{1}{r}\frac{dR}{dr}\right) = \frac{1}{Z}\frac{d^2 Z}{dz^2} = \lambda^2$$

The r- and z-directions should be separated,

$$\frac{d^2 R}{dr^2} + \frac{1}{r}\frac{dR}{dr} + \lambda^2 R = 0$$

$$\frac{d^2 Z}{dz^2} - \lambda^2 Z = 0$$

With focus on the r-direction equation, it can be rearranged to become,

$$r\frac{d}{dr}\left(r\frac{dR}{dr}\right) + \lambda^2 r^2 R = 0$$

This is a Bessel differential equation as shown in Equation (2.48) and the solution shown in Equation (2.49), with $x = r$, $m = \lambda, v = 0$. The solution is

$$R = a_o J_o(\lambda r) + a_1 Y_o(\lambda r)$$

at $r = 0$, from Figure 2.10, $Y_o(0) = -\infty$, but $R_o(0) = \theta(0) = $ finite temperature value, thus $a_1 = 0$, therefore,

$$R = a_o J_o(\lambda r)$$

at $r = r_o$, $\theta(r_o) = 0$, $R(r_o) = 0$, therefore,

$$J_o(\lambda r_o) = 0$$

This is an infinity of roots, as shown in Figure 2.10,

$$\lambda_n r_o, n = 1, 2, 3, \dots \infty$$

These roots have the orthogonality property,

$$\int_0^{r_o} r J_o(\lambda_n r) \cdot J_o(\lambda_m r) dr = 0 \quad \text{for } m \neq n.$$

Consider the z-direction equation, $\lambda = \lambda_n$,

$$\frac{d^2 Z}{dz^2} - \lambda^2 Z = 0$$

From Equation (2.63) with $\theta = Z$, $x = z$, $m = \lambda_n$, the solution is

$$Z = C_1 e^{\lambda_n z} + C_2 e^{-\lambda_n z}$$

at $z = 0$, $Z(0) = \theta(0) = 0$, $C_1 + C_2 = 0$, $C_2 = -C_1$

$$Z = C_1\left(e^{\lambda_n z} - e^{-\lambda_n z}\right)$$

Therefore, the product solution is

$$\theta(r,z) = R(r) \cdot Z(z)$$

$$= \sum_{n=1}^{\infty} C_n \left(e^{\lambda_n z} - e^{-\lambda_n z}\right) \cdot J_o(\lambda_n r)$$

$$= \sum_{n=1}^{\infty} C_n 2\sinh(\lambda_n z) \cdot J_o(\lambda_n r)$$

at $z = l$, $\theta(r,l) = \theta_2$, therefore,

$$\theta_2 = \sum_{n=1}^{\infty} C_n 2\sinh(\lambda_n l) \cdot J_o(\lambda_n r)$$

From orthogonality of the Bessel functions,

$$C_n = \frac{\theta_2 \int_0^{r_o} r J_o(\lambda_n r) dr}{2\sinh(\lambda_n l) \int_0^{r_o} r J_o^2(\lambda_n r) dr}$$

From Appendix A.3 the Bessel functions can be re-written to solve for C_n,

$$\int_0^{r_o} r J_o(\lambda_n r) dr = \frac{r_o}{\lambda_n} J_1(\lambda_n r_o)$$

$$\int_0^{r_o} r J_o^2(\lambda_n r) dr = \frac{r^2}{2}\left[J_o^2(\lambda_n r) + J_1^2(\lambda_n r) \right]\Big|_0^{r_o}$$

$$= \frac{r_o^2}{2} J_1^2(\lambda_n r_o)$$

$$C_n = \frac{\theta_2}{r_o \lambda_n \sinh(\lambda_n l) J_1(\lambda_n r_o)}$$

Finally, the temperature distribution is

$$\theta(r,z) = \sum_{n=1}^{\infty} \frac{2\theta_2}{r_o \lambda_n} \frac{\sinh(\lambda_n z) J_o(\lambda_n r)}{\sinh(\lambda_n l) J_1(\lambda_n r_o)}$$

$$T(r,z) = T_1 + \sum_{n=1}^{\infty} \frac{2(T_2 - T_1)}{r_o \lambda_n} \frac{\sinh(\lambda_n z) J_o(\lambda_n r)}{\sinh(\lambda_n l) J_1(\lambda_n r_o)}$$

Assuming $r_o = 10$ cm, $l = 20$ cm, $T_1 = 20°C$, $T_2 = 100°C$, $n = 1, 2,$ and 3, numerical values for the local temperature and heat flux can be determined to provide isotherms and isofluxes within the cylinder.

Case 2: Consider the short cylinder from case 1 above. The top surface is now exposed to a constant heat flux, q_s''. In this case, T_2 or θ_2 is unknown, it can be determined using the heat flux BC,

$$q_s'' = -k\frac{\partial T(r,l)}{\partial z}$$

or

$$q_s'' = -k\frac{\partial \theta(r,l)}{\partial z}$$

From case 1, $\theta(r,z)$,

$$-\frac{q_s''}{k} = \frac{\partial \theta(r,l)}{\partial z} = \sum_{n=1}^{\infty} C_n 2\lambda_n \cosh(\lambda_n l) \cdot J_o(\lambda_n r)$$

$$C_n = \frac{-\dfrac{q_s''}{k}\displaystyle\int_0^{r_o} rJ_o(\lambda_n r)\,dr}{2\lambda_n \cosh(\lambda_n l)\displaystyle\int_0^{r_o} rJ_o^2(\lambda_n r)\,dr}$$

C_n can be solved using the same procedure shown in case 1, and the temperature distribution, $\theta(r,z)$, can be determined.

Again, numerical values can be assumed and used to generate 2D plots of the temperature and heat flow through the short cylinder.

Case 3: Reconsider the short cylinder from case 1. Now the top surface is exposed to a convection fluid at temperature T_∞ with a convection heat transfer coefficient, h. In this case 3, T_2, or θ_2, is unknown, and it can be determined from the convection BC,

$$-k\frac{\partial T(r,l)}{\partial z} = h[T_\infty - T(r,l)] = h[(T_\infty - T_1) - T(r,l) + T_1]$$

or

$$-k\frac{\partial \theta(r,l)}{\partial z} = h[\theta_\infty - \theta(r,l)]$$

with

$$\theta_\infty = T_\infty - T_1$$

and

$$\theta(r,l) = T(r,l) - T_1$$

From case 1, $\theta(r,z)$,

$$-k\sum_{n=1}^{\infty} C_n 2\lambda_n \cosh(\lambda_n l) \cdot J_o(\lambda_n r)$$

$$= h\left[\theta_\infty - \sum_{n=1}^{\infty} C_n 2\sinh(\lambda_n l) \cdot J_o(\lambda_n r)\right]$$

$$C_n = \frac{\theta_\infty \int_0^{r_o} rJ_o(\lambda_n r)dr}{\left[2\sinh(\lambda_n l) - \dfrac{k}{h}2\lambda_n \cosh(\lambda_n l)\right]\int_0^{r_o} rJ_o^2(\lambda_n r)dr}$$

C_n can be solved using the same procedure shown in case 1. Thus, the temperature distribution $\theta(r,z)$ can be determined. If $r_o = 10\,\text{cm}$, $l = 20\,\text{cm}$, $T_1 = 20°\text{C}$, $T_\infty = 100°\text{C}$, $h = 50\ \text{W/m}^2\text{K}$, $k = 20\ \text{W/mK}$, $n = 1$, 2, and 3, determine $T(r,z)$. From the temperature solution, isotherms and the corresponding isofluxes can be graphically shown.

3.4 PRINCIPLE OF SUPERPOSITION FOR MULTIDIMENSIONAL HEAT CONDUCTION AND FOR NONHOMOGENEOUS EQUATIONS

3.4.1 3-D HEAT CONDUCTION PROBLEM

Sometimes we need to solve 3-D heat conduction problems in Cartesian (rectangular) coordinates as shown in Figure 3.6. Basically, we first convert the 3-D into the 2-D heat conduction problem and then solve the 2-D problem by using the separation of variables method discussed previously. The following is a brief outline on how to solve this type of problem.

The steady-state 3-D heat conduction equation without heat generation is

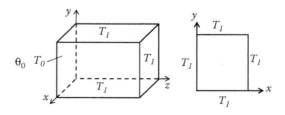

FIGURE 3.6 3-D heat conduction in Cartesian coordinates.

$$\frac{\partial^2 T}{\partial x^2} + \frac{\partial^2 T}{\partial y^2} + \frac{\partial^2 T}{\partial z^2} = 0 \tag{3.15}$$

Let $\theta = T - T_1$.

The above governing equation and the associated BCs become

$$\frac{\partial^2 \theta}{\partial x^2} + \frac{\partial^2 \theta}{\partial y^2} + \frac{\partial^2 \theta}{\partial z^2} = 0$$

$x = 0$, $\theta = 0$, or $\partial\theta/\partial x = 0$; $x = a$, $\theta = 0$, two homogeneous BCs,
$y = 0$, $\theta = 0$, or $\partial\theta/\partial y = 0$; $y = b$, $\theta = 0$, two homogeneous BCs,
$z = 0$, $\theta = \theta_0$; $z = c$, $\theta = 0$, one nonhomogeneous BC.

Let $\theta = X(x)Y(y)Z(z)$. Substitute the derivatives into the above 3-D heat conduction equation and obtain

$$-\frac{1}{X}\frac{d^2 X}{dx^2} = \frac{1}{Y}\frac{d^2 Y}{dy^2} + \frac{1}{Z}\frac{d^2 Z}{dz^2} = \lambda^2 \tag{3.16}$$

This implies

$$\frac{d^2 X}{dx^2} + \lambda^2 X = 0$$
$$-\frac{1}{Y}\frac{d^2 Y}{dy^2} = \frac{1}{Z}\frac{d^2 Z}{dz^2} - \lambda^2 = \mu^2 \tag{3.17}$$

The last equation can be further written as

$$\begin{cases} \dfrac{d^2 Y}{dy^2} + \mu^2 Y = 0 \\[2mm] \dfrac{d^2 Z}{dz^2} - \left(\lambda^2 + \mu^2\right)Z = 0 \end{cases} \tag{3.18}$$

Therefore, we need to solve two eigenvalue equations. The x-direction solution will be sine and cosine, the y-direction solution is sine and cosine, and the z-direction solution is *sinh* and *cosh*. The only nonhomogeneous BC in the z-direction will be used to determine the final unknown integrated value C_n.

From the previous discussion, the solutions for X, Y, and Z follow:

$$\begin{cases} X \cong c_1 \cos \lambda x + c_2 \sin \lambda x \\[1mm] Y \cong c_3 \cos \mu y + c_4 \sin \mu y \\[1mm] Z \cong c_5 e^{-z\sqrt{\lambda^2 + \mu^2}} + c_6 e^{z\sqrt{\lambda^2 + \mu^2}} \end{cases}$$

3.4.2 Nonhomogeneous Heat Conduction Problem

The problem of the steady-state 2-D heat conduction with uniform heat generation can be divided into two problems shown below, $\theta(x, y) = \psi(x, y) + \phi(x)$. We already know how to solve these two problems.

$$\frac{\partial^2 \theta}{\partial x^2} + \frac{\partial^2 \theta}{\partial y^2} + \frac{q}{k} = 0$$

$$\frac{\partial^2 \psi}{\partial x^2} + \frac{\partial^2 \psi}{\partial y^2} = 0 \tag{3.19}$$

$$\frac{\partial^2 \phi}{\partial x^2} + \frac{\dot{q}}{k} = 0 \tag{3.20}$$

where

$$\psi = X(x)Y(y) = \left(c_1 \cos \lambda x + c_2 \sin \lambda x\right) \cdot \left(c_3 e^{-\lambda y} + c_4 e^{\lambda y}\right)$$

Equation (3.20) can be solved using the procedure shown in Section 2.2 of Chapter 2.

Examples

3.1 A 2-D rectangular plate is subjected to the following thermal BCs:

$$
\begin{aligned}
x = 0, \quad & 0 < y < b: \quad && T = T_0 \\
x = a, \quad & 0 < y < b: \quad && T = T_0 \\
y = b, \quad & 0 < x < a: \quad && T = cx
\end{aligned}
$$

a. Derive an expression for the steady-state temperature distribution $T(x, y)$.
b. Sketch the isotherms and isofluxes.

Solution

a.

$$\frac{\partial^2 \theta}{\partial x^2} + \frac{\partial^2 \theta}{\partial y^2} = 0 \tag{3.21}$$

Let $\theta(x, y) = X(x) \cdot Y(y)$;

$$\frac{1}{X}\frac{\partial^2 X}{\partial x^2} = -\frac{\partial^2 Y}{\partial y^2} = \lambda^2$$

$$\frac{\partial^2 X}{\partial x^2} - \lambda^2 X = 0 \tag{3.22}$$

$$\frac{\partial^2 Y}{\partial y^2} + \lambda^2 Y = 0 \tag{3.23}$$

$$X = C_1 \cos(\lambda x) + C_2 \sin(\lambda x)$$
$$Y = C_3 e^{-\lambda y} + C_4 e^{\lambda y} \tag{3.24}$$
$$\theta = \left[C_1 \cos(\lambda x) + C_2 \sin(\lambda x) \right] \left[C_3 e^{-\lambda y} + C_4 e^{\lambda y} \right]$$

BCs:

i. $x = 0$: $\theta = 0$
ii. $x = a$: $\theta = 0$
iii. $y = 0$: $\theta = 0$
iv. $y = b$: $\theta = cx - T_0$

Applying BC (i) to Equation (3.24), $C_1 = 0$
Applying BC (ii) to Equation (3.24), $C_2 \sinh(\lambda a) = 0$, $\lambda_n = n\pi/a$.
Applying BC (iii) to Equation (3.24), $C_3 = -C_4$.

$$\theta(x,y) = C_2 \cdot C_4 \sin\left(\frac{n\pi x}{a}\right)\left(e^{\lambda_n y} - e^{-\lambda_n y}\right)$$

$$\theta(x,y) = \sum_{n=1}^{\infty} C_n \sin\left(\frac{n\pi x}{a}\right)\sinh\left(\frac{n\pi y}{a}\right) \tag{3.25}$$

Applying BC (iv) and using orthogonal functions to evaluate C_n,

$$C_n = \frac{\displaystyle\int_0^a (cx - T_0)\sin(n\pi x/a)dx}{\sinh(n\pi b/a)\displaystyle\int_0^a \sin^2(n\pi x/a)dx} \tag{3.26}$$

$$\int_0^a (cx - T_0)\sin\left(\frac{n\pi x}{a}\right)dx = c\left[\left(\frac{a}{n\pi}\right)^2 \sin\left(\frac{n\pi x}{a}\right) - \frac{ax}{n\pi}\cos\left(\frac{n\pi x}{a}\right)\right]_0^a$$

$$+ \frac{T_0 \cdot a}{n\pi}\cos\left(\frac{n\pi x}{a}\right)_0^a.$$

$$\int_0^a (cx - T_0)\sin\left(\frac{n\pi x}{a}\right)dx = \frac{ca^2}{n\pi}(-\cos(n\pi)) + \frac{T_0 \cdot a}{n\pi}(\cos(n\pi) - 1)$$

$$\int_0^a (cx - T_0)\sin\left(\frac{n\pi x}{a}\right)dx = \frac{ca^2}{n\pi}(-1)^{n+1} + \frac{T_0 \cdot a}{n\pi}\left[(-1)^n - 1\right]$$

$$\int_0^a (cx - T_0)\sin\left(\frac{n\pi x}{a}\right)dx = \frac{a}{n\pi}(-1)^n(-ca + T_0) - \frac{T_0 \cdot a}{n\pi}$$

$$\int_0^a \sin^2\left(\frac{n\pi x}{a}\right)dx = \left[\frac{x}{2} - \frac{1}{4n\pi}\sin\left(\frac{2n\pi x}{a}\right)\right]_0^a$$

$$= \frac{a}{2} \Rightarrow C_n = \frac{2(-1)^{n+1}(ca-T_0)-2T_0}{n\pi\sinh((n\pi b/a))}$$

Hence,

$$\theta(x,y) = \frac{2}{\pi}\sum_{n=1}^{\infty}\frac{(-1)^{n+1}(ca-T_0)-T_0}{n\sinh(n\pi b/a)}\sin\left(\frac{n\pi x}{a}\right)\sinh\left(\frac{n\pi y}{a}\right)$$

b. See the sketch in Figure 3.7.

3.2 A long rectangular bar $0 \le x \le a$, $0 \le y \le b$, shown in Figure 3.8, is heated at $x = 0$ with a uniform heat flux and is insulated at $x = a$ and $y = 0$. The side at $y = b$ loses heat by convection to a fluid at temperature T_∞.

a. Determine the temperature distribution $T(x, y)$.
b. Sketch the isotherms and isofluxes.

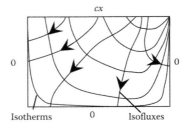

FIGURE 3.7   Sketch for the isotherms and isofluxes.

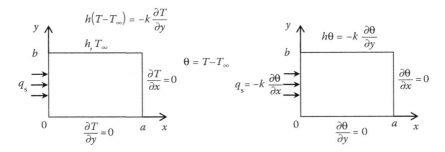

FIGURE 3.8   A long rectangular bar with heat flux as one nonhomogeneous boundary condition.

Solution

a.

$$\frac{\partial^2 \theta}{\partial x^2} + \frac{\partial^2 \theta}{\partial y^2} = 0 \tag{3.27}$$

Let $\theta(x, y) = X(x) \cdot Y(y)$:

$$\frac{1}{X}\frac{\partial^2 X}{\partial x^2} = -\frac{\partial^2 Y}{\partial y^2} = \lambda^2$$

$$\frac{\partial^2 X}{\partial x^2} - \lambda^2 X = 0 \tag{3.28}$$

$$\frac{\partial^2 Y}{\partial y^2} + \lambda^2 Y = 0 \tag{3.29}$$

$$X = C_1 \sinh(\lambda x) + C_2 \cosh(\lambda x)$$

$$Y = C_3 \sin(\lambda y) + C_4 \cos(\lambda y) \tag{3.30}$$

$$\theta = \left[C_1 \sinh(\lambda x) + C_2 \cosh(\lambda x) \right]\left[C_3 \sin(\lambda y) + C_4 \cos(\lambda y) \right]$$

BCs:
  i. $x = 0$: $q_s = -k(\partial\theta/\partial x)$
 ii. $x = a$: $\partial\theta/\partial x = 0$
iii. $y = 0$: $\partial\theta/\partial y = 0$
 iv. $y = b$: $h\theta = -k(\partial\theta/\partial y)$

Applying BC (iii) into Equation (3.30), $C_3 = 0$.
    Applying BC (ii) into Equation (3.30), $C_1\lambda \cosh(\lambda a) + C_2\lambda \sinh(\lambda a) = 0$, $C_2 = -C_1 \coth(\lambda a)$.

$$\theta = C_n \left[\sinh(\lambda x) - \coth(\lambda a)\cosh(\lambda x) \right]\cos(\lambda y) \tag{3.31}$$

Applying BC (iv) into Equation (3.31),

$$hC_n\left[\sinh(\lambda x) - \coth(\lambda a)\cosh(\lambda x)\right]\cos(\lambda b)$$

$$= +k\lambda C_n\left[\sinh(\lambda x) - \coth(\lambda a)\cosh(\lambda x)\right]\sin(\lambda b)$$

$$\lambda \tan(\lambda b) = \frac{h}{k} \tag{3.32}$$

$$\theta = \sum_{n=1}^{\infty} C_n \left[\sinh(\lambda_n x) - \coth(\lambda_n a)\cosh(\lambda_n x)\right]\cos(\lambda_n y)$$

where $\lambda_n \tan(\lambda_n b) = (h/k)$.

Applying BC (i) into Equation (3.26),

$$q_s = -k \sum_{n=1}^{\infty} C_n \left\{ \left[\lambda_n \cosh(\lambda_n x) - \coth(\lambda_n a) \lambda_n \sinh(\lambda_n x) \right] \cos(\lambda_n y) \right\}_{x=0}$$

$$q_s = -k \sum_{n=1}^{\infty} C_n \lambda_n \cos(\lambda_n y) - \frac{q_s}{k\lambda_n} \int_0^b \cos(\lambda_n y) dy = \int_0^b C_n \cos^2(\lambda_n y) dy$$

$$-\frac{q_s}{k\lambda_n} \frac{1}{\lambda_n} \sin(\lambda_n y) \Big|_0^b = C_n \left[\frac{y}{2} + \frac{2\sin(\lambda_n y)\cos(\lambda_n y)}{4\lambda_n} \right]_0^b$$

$$C_n = \frac{2q_s \sin(\lambda_n b)}{k\lambda_n \left(\sin(\lambda_n b)\cos(\lambda_n b) + b\lambda_n \right)}$$

$$\theta = \sum_{n=1}^{\infty} \frac{2q_s \sin(\lambda_n b)}{k\lambda_n \left(\sin(\lambda_n b)\cos(\lambda_n b) + b\lambda_n \right)}$$

$$\times \left[-\sinh(\lambda_n x) + \coth(\lambda_n a)\cosh(\lambda_n x) \right] \cos(\lambda_n y)$$

From the general solution above, one may assume numerical values to determine $q(x, y)$. For example, if $a = b = 100$ cm, $k = 20$ W/mK, $h = 50$ W/m²K, $T_\infty = 20°C$, $q_s = 5000$ W/m², use $n = 1$ to 3 and plot the isotherms and sketch the corresponding isofluxes.

3.3 A thin rectangular plate, $0 \le x \le a$, $0 \le y \le b$, as shown in Figure 3.9, with negligible heat loss from its sides, has the following BCs: $x = 0$, $0 < y < b$: $T = T_1$, $x = a$, $0 < y < b$: $T = T_2$, $y = 0$, $0 < x < a$: $q_s'' = 0$ (insulated) $y = b$, $0 < x < a$: $T = T_3$.

a. Determine the steady-state temperature distribution.
b. Sketch the isotherms and isofluxes.

Solution

a.

$$\frac{\partial^2 T}{\partial x^2} + \frac{\partial^2 T}{\partial y^2} = 0$$

Let $\theta = T - T_1$, then

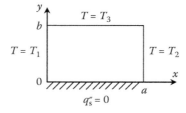

FIGURE 3.9   A thin rectangular plate with one homogeneous boundary condition.

$$\frac{\partial^2 \theta}{\partial x^2} + \frac{\partial^2 \theta}{\partial y^2} = 0$$

Let $\theta = \theta_1 + \theta_2$,

$$\frac{\partial^2 \theta_1}{\partial x^2} + \frac{\partial^2 \theta_1}{\partial y^2} = 0$$

$$\frac{\partial^2 \theta_2}{\partial x^2} + \frac{\partial^2 \theta_2}{\partial y^2} = 0$$

$$\theta_1 = \left[C_1 \cos(\lambda x) + C_2 \sin(\lambda x) \right]\left[C_3 \cosh(\lambda y) + C_4 \sinh(\lambda y) \right]$$

$$x = 0, \quad \theta_1 = 0, \qquad\qquad \Rightarrow C_1 = 0$$

$$x = a, \quad \theta_1 = 0, \quad \Rightarrow \lambda a = n\pi; \quad \lambda_n = \left(\frac{n\pi}{a} \right) \quad n = 1, 2, 3, \ldots$$

$$y = 0, \quad \frac{\partial \theta_1}{\partial y} = 0; \qquad\qquad \Rightarrow C_4 = 0$$

$$\theta_1 = \sum_{n=1}^{\infty} C_n \cosh(\lambda_n y)\sin(\lambda_n x)$$

$$\theta_1(x,b) = (T_3 - T_1) = \sum_{n=1}^{\infty} C_{n1}\cosh\left(\frac{n\pi b}{a} \right)\sin\left(\frac{n\pi x}{a} \right)$$

$$C_{n1} = \frac{(T_3 - T_1)\int_0^a \sin\left((n\pi x/a)\right)dx}{\cosh\left((n\pi b/a)\right)\int_0^a \sin^2\left((n\pi x/a)\right)dx}$$

$$= \frac{(T_3 - T_1)a \cdot \left(1 - \cos(n\pi)/n\pi\right)}{(a/2)\cosh\left((n\pi b/a)\right)} = \frac{2(T_3 - T_1)}{n\pi}\frac{\left[1 - (-1)^n\right]}{\cosh\left((n\pi b/a)\right)}$$

$$\theta_1 = \sum_{n=1}^{\infty} \frac{2(T_3 - T_1)}{\pi n}\frac{\left[1 - (-1)^n\right]}{\cosh\left((n\pi b/a)\right)} \cdot \cosh\left(\frac{n\pi y}{a} \right)\sin\left(\frac{n\pi x}{a} \right)$$

Solution for θ_2

$$\theta_2 = \left[C_1 \cosh(\lambda x) + C_2 \sinh(\lambda x) \right]\left[C_3 \cos(\lambda y) + C_4 \sin(\lambda y) \right]$$

$$y = 0, \quad \partial \theta / \partial y = 0; \qquad \Rightarrow C_4 = 0$$

$$y = b, \quad \theta_2 = 0; \quad \Rightarrow \quad \lambda_n = (n + 1/2)\pi/b$$

$$x = 0, \quad \theta_2 = 0; \quad \Rightarrow \qquad C_1 = 0$$

$$\theta_2 = \sum_{n_{odd}}^{\infty} C_{n2}\sinh(\lambda_n x)\cos(\lambda_n y)$$

$$\theta_2(a,y) = (T_2 - T_1) = \sum_{n_{odd}}^{\infty} C_{n2}\sinh(\lambda_n a)\cos(\lambda_n y)$$

$$C_{n2} = \frac{(T_2 - T_1)\int_0^b \cos((n\pi y/2b))\,dy}{\sinh((n\pi a/2b))\int_0^b \cos^2(\lambda_n y)\,dy}$$

$$= \frac{4(T_2 - T_1)\sin((n\pi/2))}{n\pi\sinh((n\pi a/2b))}$$

where $\sin((n\pi/2)) = (-1)^{(n-1/2)}$ for n odd

$$\theta_2 = \sum_{n_{odd}}^{\infty} \frac{4(T_2 - T_1)(-1)^{(n-1)/2}}{n\pi\sinh((n\pi a/2b))}\sinh\left(\frac{n\pi x}{2b}\right)\cos\left(\frac{n\pi y}{2b}\right)$$

Finally,

$$T(x,y) = T_1 + \left\{ \frac{2(T_3 - T_1)}{n\pi} \sum_{n=1}^{\infty} \frac{[1-(-1)^n]}{\cosh(n\pi b/a)}\cosh\left(\frac{n\pi y}{a}\right)\sin\left(\frac{n\pi x}{a}\right) \right.$$

$$\left. + \frac{4(T_2 - T_1)}{n\pi} \sum_{n_{odd}}^{\infty} \frac{(-1)^{(n-1)/2}}{\sinh((n\pi a/2b))}\sinh\left(\frac{n\pi x}{2b}\right)\cos\left(\frac{n\pi y}{2b}\right) \right\}$$

3.4 A long rectangular rod $0 \le x \le a$, $0 \le y \le b$, as shown in Figure 3.10, has the following thermal BCs:

$$\begin{aligned} x = 0, & \quad 0 < y < b : T = T_o \\ x = a, & \quad 0 < y < b : T = T_0 + T_1\sin(\pi y/b) \\ y = 0, & \quad 0 < x < a : T = T_0 \\ y = b, & \quad 0 < x < a : T = T_0 + T_2\sin(\pi x/a) \end{aligned}$$

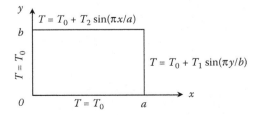

FIGURE 3.10 A long rectangular rod with two nonhomogeneous boundary conditions.

a. Determine the steady-state temperature distribution.
b. Sketch the isotherms and isofluxes.

Solution

a. Let $\theta = T - T_0$, then

$$\frac{\partial^2 \theta}{\partial x^2} + \frac{\partial^2 \theta}{\partial y^2} = 0$$

$$
\begin{aligned}
x = 0; & \quad \theta = 0 \\
x = a; & \quad \theta = T_1 \sin\left((\pi y / b)\right) \\
y = 0; & \quad \theta = 0, \\
y = b; & \quad \theta = T_2 \sin\left(\frac{\pi x}{a}\right)
\end{aligned}
$$

Let $\theta = \theta_1 + \theta_2$ with θ_1 and θ_2 satisfying the following BCs at

$$
\begin{aligned}
x = 0, & \quad \theta_1 = 0, \quad \theta_2 = 0 \\
x = a, & \quad \theta_1 = 0, \quad \theta_2 = T_1 \sin\left(\frac{\pi y}{b}\right) \\
y = 0, & \quad \theta_1 = 0, \quad \theta_2 = 0 \\
y = b, & \quad \theta_1 = T_2 \sin\left(\frac{\pi x}{a}\right) \quad \theta_2 = 0
\end{aligned}
$$

Solutions for θ_1 and θ_2 are obtained as

$$\theta_1 = \sum C_{n1} \sin\left(\frac{n\pi x}{a}\right) \cdot \sinh\left(\frac{n\pi y}{a}\right)$$

$$\theta_2 = \sum C_{n2} \sin\left(\frac{n\pi x}{b}\right) \cdot \sinh\left(\frac{n\pi y}{b}\right)$$

and $T = T_0 + \theta$

$$T = T_0 + \theta_1 + \theta_2$$

3.5 An infinitely long rod of square cross section ($L \times L$) floats in a fluid. The heat transfer coefficient between the rod and the fluid is relatively large compared with that between the rod and the ambient air, i.e., $h_f \gg h$ or $h_f \cong \infty$. Determine the steady-state temperature distribution in the rod with the associated BCs.

a. Use the analytical approach.
b. Sketch the isotherms and isoflux in the rod, if $T_f < T_\infty$ and $h = $ a constant value.

Solution

a. Assume a long rod with a square cross-section floating as a diamond-shape and treat its lower-edge point as $x = 0$, $y = 0$.

Let $\theta = T - T_\infty$,

BCs: $y = 0$, $\theta = \theta_f$

$$y = L, \frac{\partial \theta}{\partial y} = -\frac{h}{k}\theta$$

$$x = 0, \theta = \theta_f$$

$$x = L, \frac{\partial \theta}{\partial x} = -\frac{h}{k}\theta$$

Using the superposition principle:

Let $\theta = \theta_1 + \theta_2$

BCs for θ_1: $y = 0$, $\theta_1 = 0$

$$y = L, \frac{\partial \theta_1}{\partial y} = -\frac{h}{k}\theta_1$$

$$x = 0, \theta_1 = \theta_f$$

$$x = L, \frac{\partial \theta_1}{\partial x} = -\frac{h}{k}\theta_1$$

Separation of variables:

$$\theta_1 = X \cdot Y$$

$$X = C_1 \sinh \lambda_n x + C_2 \cosh \lambda_n x$$

$$Y = C_3 \sin \lambda_n y + C_4 \cos \lambda_n y$$

$$\text{at } y = 0, C_4 = 0$$

$$\text{at } y = L, \lambda_n C_3 \cos \lambda_n L = -\frac{h}{k}C_3 \sin \lambda_n L$$

$$\cot \lambda_n L = -\frac{hL}{k}\frac{1}{\lambda_n L} \quad \therefore \lambda_n L = \sqrt{}$$

at $x = L$,

$$C_1 \lambda_n \cosh \lambda_n L + C_2 \lambda_n \sinh \lambda_n L = -\frac{h}{k}\left(C_1 \sinh \lambda_n L + C_2 \cosh \lambda_n L\right)$$

$$\therefore C_1 = -C_2 \frac{\lambda_n \sinh \lambda_n L + \frac{h}{k} \cosh \lambda_n L}{\lambda_n \cosh \lambda_n L + \frac{h}{k} \sinh \lambda_n L} = -C_2 \frac{A}{B}$$

$$\therefore \theta_1 = \sum C_n \sin \lambda_n y \left(\cosh \lambda_n x - \frac{A}{B} \right)$$

at $x = 0, \theta_1 = \theta_f$, solve for C_n,

$$C_n = \frac{\theta_f \int_0^L \sin \lambda_n y \, dy}{\left(1 - \frac{A}{B}\right) \int_0^L \sin^2 \lambda_n y \, dy} = \frac{\theta_f}{\left(1 - \frac{A}{B}\right)} \cdot \frac{\frac{L}{n\pi}}{\frac{L}{2}} \left[1 - (-1)^n\right]$$

Where (A/B) should be replaced with (A/B) sinh 0 = 0 in the above C_n equation.
Let $x \Rightarrow y$, $y \Rightarrow x$, solve for θ_2
   Then $\theta = \theta_1 + \theta_2$

b. If $L = 50$ cm, $T_f = 20°C$, $T_\infty = 100°C$, $h = 50$ w/m²K, $k = 20$ W/mK, $n = 1, 2,$ and 3, determine $T(x,y)$, plot isotherms, and sketch isofluxes.

REMARKS

In this chapter, we have introduced a very powerful mathematical tool, the method of separation of variables, to solve typical 2-D heat conduction problems with various thermal BCs. In the undergraduate-level heat transfer, we normally employ the finite-difference energy balance method to solve the 2-D heat conduction problems with various thermal BCs. The finite-difference numerical methods and solutions will be discussed in Chapter 5. Here we are more focused on the analytical methods and solutions for various 2-D heat conduction problems. In general, all kinds of 2-D heat conduction problems with various BCs can be solved analytically by using superposition of separation of variables.

The most important thing for applying separation of variables is that you have to set up your problem with only one nonhomogeneous BC. If you have more than one nonhomogeneous BC, you have to employ the superposition principle and divide into two or three subproblems in order to use separation of variables. The problems become more complicated if you work with 3-D heat conduction including heat generation and complex BCs, but they are still solvable. However, if you are interested in solving for the 2-D and 3-D cylindrical coordinate systems and for the 2-D and 3-D spherical coordinate systems with complex BCs, they are beyond the intermediate-level heat transfer, you need to look at the advanced heat conduction textbook for solutions.

Another popular method is using the finite-difference method to be discussed in Chapter 5. It is particularly true when you deal with complicated BCs, such as convection. In real-life engineering applications, the convection heat transfer coefficients normally are varied along the solid surface. This will introduce additional complexity for the separation of variables because we normally assume the uniform convection BCs to simplify the problem. This will not cause any additional complexities when using the finite-difference numerical method.

PROBLEMS

3.1 A long rectangular bar $0 \leq x \leq a$, $0 \leq y \leq b$, and a, $b \ll L$, the bar length, is heated at $y = 0$ and $y = b$, respectively, to a uniform temperature T_o and is insulated at $x = 0$. The side of $x = a$ loses heat by convection to a fluid at temperature T_∞ with a convection coefficient h.

    a. Write down, step by step, a solution method and associated BCs, which can be used to determine the steady-state temperature distributions.

    b. Sketch the heat flows and the isothermal profiles in the rectangular bar.

3.2 An infinitely long rod of square cross section ($L \times L$) floats in a fluid. The heat transfer coefficient between the rod and the fluid is relatively large compared to that between the rod and the ambient air, that is, $h_f \gg h$ or $h_f \cong \infty$. Determine the steady-state temperature distributions in the rod with the associated BCs.

    a. Use the analytical approach.

    b. Sketch the isotherms and isoflux in the rod, if $T_f < T_\infty$ and $h =$ a constant value.

3.3 A long fin of rectangular cross section ($2L \times L$) with a thermal conductivity, k, is subjected to the following BCs: the left side is kept at T_o, the right side is perfectly insulated, the upper side is exposed to a constant flux, and the lower side is exposed to a convection air flow. $q'' =$ constant; steam T_o, $h = \infty$; air $h =$ constant, T_∞.

    a. Determine the temperature distribution in the fin.

    b. Approximately plot the temperature and heat flow profiles in the fin, if $T_o > T_\infty$.

3.4 Refer to Figure 3.1 and determine the temperature distributions for 2-D heat conduction with the following BCs:

(1)		(2)	
	$x = 0, T = T_o$		$x = 0, T = T_o$
	$x = a, T = T_o$		$x = a, T = T_o$
	$y = 0, T = T_o$		$y = 0, T = T_s$
	$y = b, T = T_s$		$y = b, T = T_o$
(3)	$x = 0, T = T_o$	(4)	$x = 0, T = T_s$
	$x = a, T = T_s$		$x = a, T = T_o$
	$y = 0, T = T_o$		$y = 0, T = T_o$
	$y = b, T = T_o$		$y = b, T = T_o$

3.5 Refer to Figure 3.2 and determine the temperature distributions for 2-D heat conduction with the following BCs:

(1)	$x = 0, T = T_1$	(2)	$x = 0, T = T_1$
	$x = a, T = T_1$		$x = a, T = T_1$
	$y = 0, T = T_1$		$y = 0, q_s'' = -k(\partial T / \partial y)$
	$y = b, q_s'' = -k(\partial T / \partial y)$		$y = b, T = T_1$
(3)	$x = 0, T = T_1$	(4)	$x = 0, q_s'' = -k(\partial T / \partial x)$
	$x = a, q_s'' = -k(\partial T / \partial x)$		$x = a, T = T_1$
	$y = 0, T = T_1$		$y = 0, T = T_1$
	$y = b, T = T_1$		$y = b, T = T_1$

3.6 Refer to Figure 3.3 and determine the temperature distributions for 2-D heat conduction with the following BCs:

(1)	$x = 0, T = T_1$	(2)	$x = 0, T = T_1$
	$x = a, T = T_1$		$x = a, T = T_1$
	$y = 0, T = T_1$		$y = 0, -k(\partial T/\partial y) = h(T - T_\infty)$
	$y = b, -k(\partial T/\partial y) = h(T - T_\infty)$		$y = b, T = T_1$
(3)	$x = 0, T = T_1$	(4)	$x = 0, -k(\partial T/\partial x) = h(T - T_\infty)$
	$x = a, -k(\partial T/\partial x) = h(T - T_\infty)$		$x = a, T = T_1$
	$y = 0, T = T_1$		$y = 0, T = T_1$
	$y = b, T = T_1$		$y = b, T = T_1$

3.7 Use the principle of superposition to determine the temperature distribution for 2-D heat conduction with four nonhomogeneous BCs shown in Figure 3.4.

3.8 Refer to Figure 3.6 and determine the temperature distributions for 3-D heat conduction with the following BCs:

(1)	$x = 0, T = T_1$	(2)	$x = 0, T = T_1$
	$x = a, T = T_1$		$x = a, T = T_1$
	$y = 0, T = T_1$		$y = 0, T = T_1$
	$y = b, T = T_1$		$y = b, T = T_1$
	$z = 0, T = T_0$		$z = 0, T = T_1$
	$z = c, T = T_1$		$z = c, T = T_0$

3.9 Refer to Equation (3.19) and determine the temperature distributions for 2-D heat conduction with uniform heat generation with the following BCs:

(1)	$x = 0, -k(\partial T/\partial x) = 0$	(2)	$x = 0, T = T_o$
	$x = a, -k(\partial T/\partial x) = h(T - T_\infty)$		$x = a, T = T_o$
	$y = 0, T = T_o$		$y = 0, -k(\partial T/\partial y) = 0$
	$y = b, T = T_o$		$y = b, -k(\partial T/\partial y) = h(T - T_\infty)$

3.10   Obtain an expression for the steady-state temperature distribution $T(x, y)$ in a long square bar of side a. The bar has its two sides and bottom maintained at temperature T_1, while the top side is losing heat by convection. Consider the surrounding temperature to be T_∞, the heat transfer coefficient at the top wall be denoted by h, and let the thermal conductivity of the bar be equal to k.

  a.   Sketch the domain and write the governing equation and BCs for this problem.

  b.   Define the temperature $\theta(x, y) = T(x, y) - T_1$ and find a series solution using separation of variables.

  c.   Using the BCs write the expression for the coefficients used in the series solution for $\theta(x, y)$, and write an expression.

  d.   Write an expression for $(T(x, y) - T_1)/(T_\infty - T_1)$.

3.11   You are given a very long and wide fin with a height of 2L. The base of the fin is maintained at a uniform temperature of T_b. The top and bottom surfaces of the fin are exposed to a fluid whose temperature is T_∞ ($T_\infty < T_b$). The convective heat transfer coefficient between the fin surfaces and the fluid is h.

  a.   Sketch the steady 2-D temperature distribution in the fin.

  b.   If you were to determine the steady 2-D temperature distribution in the fin using a finite-difference numerical method, you would solve a set of algebraic nodal equations simultaneously for the temperatures at a 2-D array of nodes. Derive the equation for a typical node on one of the surfaces of the fin. Please do not simplify the equation.

  c.   Using the method of separation of variables, *derive* an expression for the steady local temperature in the fin, in terms of the thermal conductivity of the fin, k, the convective heat transfer coefficient, h, the half-height of the fin, L, and the base and fluid temperatures, T_b and T_∞.

3.12   A long rectangular rubber pad of width $a = W$ and height $b = 2W$ is a component of a spacecraft structure. Its sides and bottom are bonded to a metal channel at constant temperature T_0, and the temperature distribution along the top of the pad can be approximated as a simple sine curve $T = T_0 + T_m \sin(\pi x/W)$.

  a.   Write the differential equation and BCs needed to solve for the temperature distribution in the pad.

  b.   Find the solution for the temperature distribution from the differential equation and BCs.

3.13   A long rod with a right triangular cross section has the horizontal length "a" at temperature T_1, the vertical length "b" at temperature T_2, and the inclined length perfectly insulated. Obtain an expression for the steady-state temperature distribution $T(x, y)$ in the long rod of triangular cross section as stated. Assume that the thermal conductivity of the material of the rod is constant.

3.14 A long rectangular bar $0 \leq x \leq a$, $0 \leq y \leq b$, and a, $b \ll L$, the bar length, is heated at $y = 0$ and $y = b$, respectively, to a uniform temperature T_o and is insulated at $x = 0$. The side of $x = a$ loses heat by convection to a fluid at temperature T_∞ with a convection coefficient of h_∞.

   a. Write down, step by step, a solution method and the associated BCs, which can be used to determine the bar steady-state temperature distributions. You do not need to obtain the final solution of the steady-state temperature distributions.

   b. Sketch the heat flows and the isothermal profiles in the rectangular bar.

3.15 Given a very long and wide fin with a height of $2H$. The base of the fin is maintained at a uniform temperature of T_b. The top and bottom surfaces of the fin are exposed to a fluid whose temperature is T_∞ ($T_\infty < T_b$). The convective heat transfer coefficient between the fin surfaces and the fluid is h.

   a. Derive an expression for the steady 2-D local temperature in the fin, in terms of the thermal conductivity of the fin, k, the convective heat transfer coefficient, h, the half-height of the fin, H, and the base and fluid temperature, T_b and T_∞.

   b. Sketch the steady 2-D temperature and heat flux distribution in the fin.

Note that

$$\int_0^x \left[\cos^2(ax)\right]dx = \frac{1}{4}a[2ax + \sin(2ax)] \quad \text{and}$$

$$\int_0^x \left[\cos(ax) \cdot \cos(bx)\right] = 0 \qquad \text{when } a \neq b.$$

REFERENCES

1. V. Arpaci, *Conduction Heat Transfer*, Addison-Wesley Publishing Company, Reading, MA, 1966.

2. A. Mills, *Heat Transfer*, Richard D. Irwin, Inc., Boston, MA, 1992.

3. F. Incropera and D. Dewitt, *Fundamentals of Heat and Mass Transfer*, Fifth Edition, John Wiley & Sons, New York, 2002.

4. W. Rohsenow and H. Choi, *Heat, Mass, and Momentum Transfer*, Prentice-Hall, Inc., Englewood Cliffs, NJ, 1961.

4 Transient Heat Conduction

The temperature in a solid material changes with location as well as with time, and this is the so-called transient heat conduction problem. We may have 1-D, 2-D, or 3-D transient heat conduction depending on the specific application. However, some problems can be modeled as zero-dimensional (0-D) because the temperature in a solid material uniformly changes only with time and does not depend on location. This is a special case for transient heat conduction. The finite-length solid material of 1-D, 2-D, or 3-D transient problems can be solved by the separation of variables method, and the 0-D transient problem can be solved by the lumped capacitance method. Figure 4.1 shows typical finite-length solid materials for the 1-D transient problem for the slab (or the plane wall), cylinder, and sphere. The semiinfinite solid material of the 1-D transient problem can be solved by the similarity method, the Laplace transform method, or the approximate integral method.

The unsteady 3-D heat conduction equation with heat generation is

$$\frac{\partial^2 T}{\partial x^2} + \frac{\partial^2 T}{\partial y^2} + \frac{\partial^2 T}{\partial z^2} + \frac{\dot{q}}{k} = \frac{1}{\alpha}\frac{\partial T}{\partial t} \tag{4.1}$$

The simplified case of the unsteady 1-D heat conduction equation without heat generation becomes

$$\frac{\partial^2 T}{\partial x^2} = \frac{1}{\alpha}\frac{\partial T}{\partial t} \tag{4.2}$$

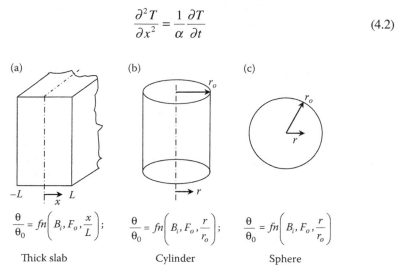

(a) $\dfrac{\theta}{\theta_0} = fn\left(B_i, F_o, \dfrac{x}{L}\right);$ Thick slab

(b) $\dfrac{\theta}{\theta_0} = fn\left(B_i, F_o, \dfrac{r}{r_o}\right);$ Cylinder

(c) $\dfrac{\theta}{\theta_0} = fn\left(B_i, F_o, \dfrac{r}{r_o}\right)$ Sphere

FIGURE 4.1   1-D transient heat conduction and the characteristic length.

DOI: 10.1201/9781003164487-4

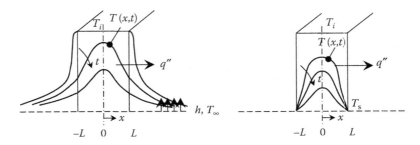

FIGURE 4.2   1-D transient problems for a slab or a plane wall.

We need both the initial and two BCs in order to solve the temperature depending on (x, t) as sketched in Figure 4.2. The solution from the separation of variables method is

$$T = T(x,t)$$

Initial condition:

$$t = 0, \quad T(x,0) = T_i$$

BCs:

$$x = 0, \quad \frac{dT}{dx} = 0$$

$$x = L, \quad T(L,t) = T_s \quad \text{or} \quad -k\frac{\partial T(L,t)}{\partial x} = h\left[T(L,t) - T_\infty\right]$$

4.1   METHOD OF LUMPED CAPACITANCE FOR 0-D PROBLEMS

In real applications, transient heat conduction in a solid material can be modeled as a 0-D problem. The important assumption is that the entire material temperature changes only with time. If that is the case, those solid geometries shown in Figure 4.1 can be solved by the following lumped capacitance method. Note that the lumped method can be applied to any irregular geometry as long as the assumption of the temperature of the entire volume uniformly changing with time is valid during the transient. Therefore, we do not need to solve the 1-D, 2-D, or 3-D transient conduction equations.

Consider the energy balance on the solid material during the cooling (or heating) process as shown in Figure 4.3:

$$\frac{d(\rho VCT)}{dt} = -hA_s(T - T_\infty) \tag{4.3}$$

Let $\theta = T - T_\infty$ then

$$\rho VC \frac{d\theta}{dt} = -hA_s\theta$$

$$\frac{d\theta}{dt} = -\frac{hA_s}{\rho VC}\theta$$

$$\int_{\theta_i}^{\theta} \frac{d\theta}{\theta} = \int_0^t -\frac{hA_s}{\rho VC}dt$$

$$\ln\frac{\theta}{\theta_i} = -\frac{hA_s}{\rho VC}t$$

$$\frac{\theta}{\theta_i} = \frac{T - T_\infty}{T_i - T_\infty} = e^{-(hA_s/\rho VC)t} = e^{-(hL_c/k)\cdot(\alpha t/L_c^2)} = e^{-Bi\cdot Fo} = f(Bi, Fo) \qquad (4.4)$$

where $Bi = (hL_c/k)$, $Fo = (\alpha t/L_c^2)$, $L_c = (V/A_s) =$ (volume/surface area).
   The above temperature decay solution is plotted in Figure 4.3. Therefore, the solid temperature can be predicted with time for a given material with a certain geometry

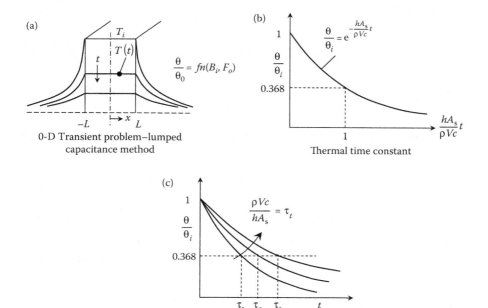

FIGURE 4.3   Method of lumped capacitance.

under the cooling or heating condition. The object with a smaller thermal time constant ($\tau_t = (\rho Vc/hA_s)$) can quickly reach the environment temperature.

The question remains, "under what condition can the lumped capacitance solution be used?" The answer is that the Biot (Bi) number must be less than 0.1. Therefore, to use the 0-D solution, the condition $Bi = (hL_c/k) < 0.1$ must be satisfied. The Biot number is defined as the ratio of surface convection (h, the convection heat transfer coefficient from the solid surface; L_c, the characteristic length of the solid material) to solid conduction (k, the thermal conductivity of the solid material). The smaller Bi number implies a small-sized object with a high-conductivity is exposed to a low convection cooling or heating fluid. For the case of a smaller Bi, the temperature inside the solid material changes uniformly (independent of location) with the environmental cooling or heating during the transient. Of course, $Bi = 0.1$ implies that we may have 10% error by using the lumped solution. The smaller Bi is better for using the lumped solution. Another point is that the solution can be applied to any geometry if the condition of $Bi < 0.1$ is valid. For a given solid geometry, the characteristic length is $L_c = $ (volume/surface area) $= (V/A_s)$. For example, as shown in Figure 4.1, the characteristic length $L_c = L$ is for a 2L-thick slab (plane wall),

$L_c = \dfrac{1}{2} R_o$ for the cylinder, and $L_c = \dfrac{1}{3} R_o$ for the spherical coordinates.

4.1.1 RADIATION EFFECT

If we also consider radiation flux q_r'' and internal heat generation, $\dot{q}$, the energy balance Equation (4.3) can be rewritten as

$$\frac{d(\rho VCT)}{dt} = -hA_s(T - T_\infty) + \dot{q}V + A_s q_r''$$

where $q_r'' = $ radiation gain from solar flux $=$ constant, or $q_r'' = $ radiation loss $= \varepsilon\sigma(T^4 - T_{sur}^4); \dot{q} = $ heat generation $= I^2 R$ due to electric current and resistance heating $=$ constant.

If we consider $q_r'' = $ constant and $\dot{q} = $ constant, the solution of the above equation can be obtained from Equation (4.4) by setting

$$\theta = \left[(T - T_\infty) - \left(\frac{\dot{q}V + A_s q_r''}{hA_s} \right) \right]$$

However, if we consider $\dot{q} = 0$, but $q_r'' = -\varepsilon\sigma(T^4 - T_{sur}^4)$, the energy balance equation can be rewritten as

$$\frac{d(\rho VCT)}{dt} = -hA_s(T - T_\infty) - \varepsilon\sigma A_s(T^4 - T_{sur}^4)$$

If we let $T_\infty = T_{sur}, h_r = \varepsilon\sigma(T^2 + T_\infty^2)(T + T_\infty)$, the above equation can be written as

$$\frac{d(T - T_\infty)}{dt} + \frac{(h + h_r)A_s}{\rho VC}(T - T_\infty) = 0$$

The solution of the above equation can be obtained by numerical integration.

4.2 METHOD OF SEPARATION OF VARIABLES FOR 1-D AND MULTIDIMENSIONAL TRANSIENT CONDUCTION PROBLEMS

4.2.1 1-D TRANSIENT HEAT CONDUCTION IN A SLAB

The solution of the 1-D transient conduction problem for a slab (plane wall) is expected as $T(x, t)$. The separation of variables method used for the 2-D steady-state heat conduction problem can be applied if we consider $T(x, t)$ similar to $T(x, y)$. In other words, we separate the temperature $T(x, y)$ into the product of $T(x) \cdot T(y)$ for the 2-D steady state and $T(x, t)$ into $T(x) \cdot T(t)$ for the 1-D transient. Then we can follow the similar procedure as before in order to solve the 1-D transient problem [1].

For a 1-D plane wall transient problem, as shown in Figure 4.4a, with the convection BC,

$$\frac{\partial^2 T}{\partial x^2} = \frac{1}{\alpha}\frac{\partial T}{\partial t}$$

Let $\theta = T - T_\infty$ and $\theta(x, t) = X(x)\tau(t)$, then,

$$\frac{\partial^2 \theta}{\partial x^2} = \frac{1}{\alpha}\frac{\partial \theta}{\partial t} \tag{4.5}$$

$$\frac{d^2 X}{dx^2} + \lambda^2 X = 0$$

(a)

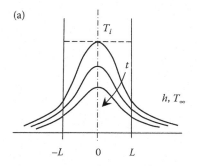

Convective boundary condition

(b)

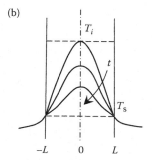

Constant surface
temperature boundary condition

FIGURE 4.4 1-D transient heat conduction.

$$\frac{d\tau}{dt} + \lambda^2 \alpha \tau = 0$$

$$X = c_1 \sin \lambda x + c_2 \cos \lambda x$$

$$\tau = c_3 e^{-\lambda^2 \alpha t}$$

Initial condition:

$$\theta(x,0) = \theta_i = T_i - T_\infty$$

BCs:

$$\begin{cases} \dfrac{\partial \theta(0,t)}{\partial x} = 0 \\ -k\dfrac{\partial \theta(L,t)}{\partial x} = h\theta(L,t) \end{cases} \Rightarrow \begin{cases} c_1 = 0 \\ -kc_2(-\sin \lambda L)\lambda = hc_2 \cos \lambda L \end{cases}$$

$$k \sin(\lambda L) \cdot \lambda = h \cos(\lambda L)$$

$$\lambda_n = \frac{h}{k} \cot(\lambda_n L)$$

$$\lambda_n L = \frac{hL}{k} \cot(\lambda_n L) = Bi \cdot \cot(\lambda_n L)$$

λ_n is determined by the convection BC.

$$\theta = c_2 \cos \lambda_n x \cdot c_3 e^{-\lambda_n^2 \alpha t} = c_n \cos(\lambda_n x) e^{-\lambda_n^2 \alpha t}$$

where $c_n = c_2 c_3$.

Applying the initial condition

$$\theta_i = c_n \cos(\lambda_n x) e^0$$

$$\theta_i \cos(\lambda_n x) = c_n \cos^2(\lambda_n x)$$

$$c_n = \frac{\theta_i \int_0^L \cos(\lambda_n x) dx}{\int_0^L \cos^2(\lambda_n x) dx} = \frac{\theta_i 2 \sin(\lambda_n L)}{\lambda_n L + \sin(\lambda_n L)\cos(\lambda_n L)}$$

$$\Rightarrow \theta = T - T_\infty = \sum c_n \cos(\lambda_n x) e^{-(\lambda_n)^2 \alpha t}$$

$$\Rightarrow \frac{\theta}{\theta_i} = \frac{T - T_\infty}{T_i - T_\infty} = \sum \frac{2\sin(\lambda_n L)\cos(\lambda_n x)}{\lambda_n L + \sin(\lambda_n L)\cos(\lambda_n L)} \cdot e^{-(\lambda_n L)^2 Fo} \qquad (4.6)$$

For $Fo = \dfrac{\alpha t}{L^2} \geq 0.2$, the first term $(n = 1)$ in the infinite series is much larger than all the other terms, thus,

$$\frac{\theta}{\theta_i} = \frac{2\sin(\lambda_1 L) \cdot \cos(\lambda_1 x)}{\lambda_1 L + \sin(\lambda_1 L) \cdot \cos(\lambda_1 L)} \cdot e^{-(\lambda_1 L)^2 Fo}$$

where $\lambda_1 L$ is the root of $\lambda_1 L \cdot \tan(\lambda_1 L) = Bi = \dfrac{hL}{k}$. At $x = 0$, the dimensionless mid-plane temperature is

$$\theta = T_0 - T_\infty = \theta_0,$$

$$\frac{\theta_0}{\theta_i} = \frac{2\sin(\lambda_1 L)}{\lambda_1 L + \sin(\lambda_1 L) \cdot \cos(\lambda_1 L)} \cdot e^{-(\lambda_1 L)^2 Fo}$$

$\sim e^{-Fo}$ for any given Bi.
  and

$$\frac{\theta}{\theta_i} \sim \cos(\lambda_1 x)$$

For any given $Bi = \dfrac{hL}{k}$, $\dfrac{\theta}{\theta_0}$ is dependent on x only and is not a function of time.

Given the general solution for transient conduction through the plane wall, numerical values can be used to generate the temperature distributions shown in Figure 4.4. For example, if $L = 10\,\text{cm}$, $K = 20\,\text{W/m} \cdot \text{K}$, $\alpha = 4 \times 10^{-6}\,\text{m}^2/\text{s}$, $T_i = 400°\text{C}$, $T_\infty = 20°\text{C}$, and $h = 10^3\,\text{W/m}^2 \cdot \text{K}$, $T(x,t)$, with $n = 1, 2, 3$ can be plotted.

In addition to $T(x, t)$, the ratio of the total energy transferred from the wall over the time t is

$$\frac{Q}{Q_o} = \frac{\int \rho C_p [T_i - T(x,t)] dV}{\rho C_p V (T_i - T_\infty)} \qquad (4.7)$$

$$= \frac{1}{V} \int \left(1 - \frac{\theta}{\theta_i}\right) dV$$

$$= \frac{1}{A_S L} \int_0^L \left(1 - \frac{\theta}{\theta_i}\right) A_S dx$$

$$= 1 - \sum \frac{2\sin(\lambda_n L)}{\lambda_n L + \sin(\lambda_n L)\cos(\lambda_n L)} \cdot \frac{\sin(\lambda_n L)}{\lambda_n L} \cdot e^{-(\lambda_n L)^2 Fo} \qquad (4.8)$$

where $Fo = (\alpha t / L^2)$.

Similarly, Figure 4.4b also includes the case for a given surface temperature. The constant surface temperature BC can also be assumed for very high convective heat transfer coefficients. In other words, if $h \to \infty$, $T_\infty \to T_s$, with $Bi \to \infty$. For the given surface temperature, or very high heat transfer coefficient, the BCs become:

$$\text{at } x = 0, \quad \frac{\partial T(0,t)}{\partial x} = 0$$

$$\text{at } x = L, \quad T(L,t) = T_s$$

Using the separation of variables technique,

$$\theta = T - T_s = X(t) \cdot \tau(t)$$

$$\frac{\partial^2 \theta}{\partial x^2} = \frac{1}{\alpha} \frac{\partial \theta}{\partial t}$$

$$X = C_1 \sin(\lambda x) + C_2 \cos(\lambda x)$$

$$\tau = C_3 e^{-\lambda^2 \alpha t}$$

at

$$x = 0, \frac{\partial \theta}{\partial x} = 0, C_1 = 0$$

$$x = L, \theta = 0, 0 = C_2 \cos(\lambda L)$$

Therefore,

$$\lambda_n L = \frac{2n+1}{2}\pi, n = 0,1,2,\ldots$$

Thus,

$$\theta = \sum C_n \cos(\lambda_n x) \cdot e^{-\lambda_n^2 \alpha t}$$

at

$$t = 0, \theta = T_i - T_s = \theta_i = \sum C_n \cos(\lambda_n x)$$

$$C_n = \frac{\theta_i \int_0^L \cos(\lambda_n x)\, dx}{\int_0^L \cos^2(\lambda_n x)\, dx} = (-1)^n \frac{2\theta_i}{\lambda_n L}$$

The general solution can then be obtained as

$$\frac{\theta}{\theta_i} = \frac{T - T_s}{T_i - T_s} = \sum \frac{2(-1)^n}{(\lambda_n L)} \cos(\lambda_n x) e^{-(\lambda_n L)^2 Fo} \tag{4.9}$$

Consider a 1-D plane wall with a thickness L, initially a spatial temperature variation within the wall is present, $F(x)$. Suddenly, the left side surface (at $x = 0$) is exposed to a convection fluid at T_∞ with a heat transfer coefficient h_1; the right-side surface (at $x = L$) is exposed to a convection fluid at T_∞ with a heat transfer coefficient h_2. Determine $T(x,t)$.

Let

$$\theta = T - T_\infty$$

$$\frac{\partial \theta}{\partial t} = \alpha \frac{\partial^2 \theta}{\partial x^2}$$

at

$$x = 0, t > 0, \quad h_1(T_\infty - T) = -k \frac{\partial T}{\partial x}$$

$$h_1 \theta = k \frac{\partial \theta}{\partial x}$$

at

$$x = L, t > 0, \quad h_2(T - T_\infty) = -k \frac{\partial T}{\partial x}$$

$$h_2 \theta = -k \frac{\partial \theta}{\partial x}$$

at

$$t = 0, \theta = T - T_\infty = F(x) - T_\infty$$

Let

$$\theta = X(x) \cdot \tau(t)$$

$$\frac{d^2 X}{dx^2} + \lambda^2 X = 0$$

$$\frac{d\tau}{dt} + \lambda^2 \tau = 0$$

The general solutions to the two ordinary differential equations are:

$$X = c_1 \cos \lambda x + c_2 \sin \lambda x$$

$$\tau = c_3 e^{-\lambda^2 t}$$

From the BC, $x = 0$

$$k\left(-c_1 \lambda \sin \lambda x + c_2 \lambda \cos \lambda x\right) = h_1 \left(c_1 \cos \lambda x + c_2 \sin \lambda x\right)$$

$$kc_2 \lambda = h_1 c_1 \text{ or } c_1 = \frac{k\lambda}{h_1} c_2$$

From the BC, $x = L$

$$-k\left(-c_1 \lambda \sin \lambda x + c_2 \lambda \cos \lambda x\right) = h_2 \left(c_1 \cos \lambda x + c_2 \sin \lambda x\right)$$

$$c_1 k\lambda \sin \lambda L - c_2 k\lambda \cos \lambda L = h_2 c_1 \cos \lambda L + h_2 c_2 \sin \lambda L$$

Solving for c_1 and c_2, and re-arranging:

$$\tan \lambda L = \frac{\sin \lambda L}{\cos \lambda L} = \frac{\lambda\left(\dfrac{h_1}{k} + \dfrac{h_2}{k}\right)}{\lambda^2 - \dfrac{h_1}{k} \cdot \dfrac{h_2}{k}}$$

or

$$\tan \lambda_n L = \frac{\lambda_n \left(\dfrac{h_1}{k} + \dfrac{h_2}{k} \right)}{\lambda_n^2 - \dfrac{h_1}{k} \cdot \dfrac{h_2}{k}}, \quad n = 1,2,3,\ldots$$

Therefore,

$$X = \lambda_n \cos \lambda_n x + \frac{h_1}{k} \sin \lambda_n x$$

$$\theta = \sum_{n=1}^{\infty} C_n \left(\lambda_n \cos \lambda_n x + \frac{h_1}{k} \sin \lambda_n x \right) \cdot e^{-\alpha \lambda_n^2 t}$$

where C_n can be determined for $t = 0$, $\theta = F(x) - T_\infty$,

$$C_n = \frac{\displaystyle\int_0^L [T(x) - T_\infty] \cdot \left(\lambda_n \cos \lambda_n x + \frac{h_1}{k} \sin \lambda_n x \right) dx}{\displaystyle\int_0^L \left(\lambda_n \cos \lambda_n x + \frac{h_1}{k} \sin \lambda_n x \right)^2 dx}$$

where the denominator of C_n can be integrated to become:

$$\frac{1}{2} \left\{ \left[\frac{h_1}{k} + \left[\lambda_n^2 + \left(\frac{h_1}{k} \right)^2 \right] \right] \cdot \left[L + \frac{\dfrac{h_2}{k}}{\lambda_n^2 + \left(\dfrac{h_2}{k} \right)^2} \right] \right\}$$

If the initial plane wall temperature is $F(x) = T_i = \text{constant}$, $F(x) - T_\infty = T_i - T_\infty = \theta_i$, and if $h_1 = h_2$, the temperature distribution $T(x, t)$ will be the same as shown in Equation (4.6).

4.2.2 MULTIDIMENSIONAL TRANSIENT HEAT CONDUCTION IN A SLAB (2-D OR 3-D)

The governing equation for multidimensional heat conduction, as shown in Figure 4.5, is

$$\frac{\partial^2 \theta}{\partial x^2} + \frac{\partial^2 \theta}{\partial y^2} + \frac{\partial^2 \theta}{\partial z^2} = \frac{1}{\alpha} \frac{\partial \theta}{\partial t} \tag{4.10}$$

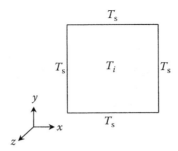

FIGURE 4.5 Multidimensional transient heat conduction.

The solution of the 3-D transient conduction problem is given as the product solution of the above-mentioned 1-D transient conduction problem:

$$\theta(x,y,z,t) = \theta_x(x,t) \cdot \theta_y(y,t) \cdot \theta_z(z,t)$$

$$\begin{cases} \dfrac{\partial^2 \theta_x}{\partial x^2} = \dfrac{1}{\alpha} \dfrac{\partial \theta_x}{\partial t} \\[2mm] \dfrac{\partial^2 \theta_y}{\partial y^2} = \dfrac{1}{\alpha} \dfrac{\partial \theta_y}{\partial t} \\[2mm] \dfrac{\partial^2 \theta_z}{\partial z^2} = \dfrac{1}{\alpha} \dfrac{\partial \theta_z}{\partial t} \end{cases} \tag{4.11}$$

where $\theta_x(x,\ t)$, $\theta_y(y,\ t)$, and $\theta_z(z,\ t)$ can be solved by the aforementioned method of separation of variables with a given surface temperature or convection BCs.

Presented below is an alternative way to determine the 3-D transient heat conduction temperature distribution.

$$\frac{\partial^2 T}{\partial x^2} + \frac{\partial^2 T}{\partial y^2} + \frac{\partial^2 T}{\partial z^2} = \frac{1}{\alpha} \frac{\partial T}{\partial t}$$

Let $T(x,\ y,\ z,\ t) = X(x) \cdot Y(y) \cdot Z(z) \cdot \tau(t)$

$$YZ\tau \frac{d^2 X}{dx^2} + XZ\tau \frac{d^2 Y}{dx^2} + XY\tau \frac{d^2 Z}{dx^2} = \frac{XYZ}{\alpha} \frac{d\tau}{dt}$$

$$\frac{1}{X} \frac{d^2 X}{dx^2} + \frac{1}{Y} \frac{d^2 Y}{dy^2} + \frac{1}{Z} \frac{d^2 Z}{dz^2} = \frac{1}{\alpha \tau} \frac{d^2 \tau}{dt}$$

Each function must be equal to a constant:

$$\frac{d^2 X}{dx^2} + \beta^2 X = 0, \quad X = c_1 \cos \beta x + c_2 \sin \beta x$$

$$\frac{d^2Y}{dy^2} + \gamma^2 Y = 0, \quad Y = c_3 \cos \gamma y + c_4 \sin \gamma y$$

$$\frac{d^2Z}{dz^2} + \eta^2 Z = 0, \quad Z = c_5 \cos \eta z + c_6 \sin \eta z$$

$$\frac{d\tau}{\partial t} + \left(\beta^2 + \gamma^2 + \eta^2\right)\tau = 0, \quad \tau = e^{-\left(\beta^2 + \gamma^2 + \eta^2\right)t}$$

Thus

$$T\left(x, y, z, t\right) = \sum_{m=1}^{\infty}\sum_{n=1}^{\infty}\sum_{p=1}^{\infty} C_{mnp} X_m Y_n Z_p e^{-\left(\beta^2 + \gamma^2 + \eta^2\right)t}$$

where β_m, γ_n, η_p, X_m, Y_n, and Z_p depend on the BCs.
At $t = 0$,

$$T\left(x, y, z, 0\right) = F\left(x, y, z\right) = \sum_{m=1}^{\infty}\sum_{n=1}^{\infty}\sum_{p=1}^{\infty} C_{mnp} X_m Y_n Z_p$$

where

$$C_{mnp} = \frac{\int_{x=0}^{a}\int_{y=0}^{b}\int_{z=0}^{c} F\left(x, y, z\right) X_m Y_n Z_p dx dy dz}{\int_{0}^{a} X_m^2 dx \cdot \int_{0}^{b} Y_n^2 dy \cdot \int_{0}^{c} Z_p^2 dz \cdot}$$

A special case, at $t = 0$, $F\left(x, y, z\right) = T_i = $ constant.

4.2.3   1-D TRANSIENT HEAT CONDUCTION IN A RECTANGLE WITH HEAT GENERATION

The governing equation for 1-D transient heat conduction with heat generation is

$$\frac{\partial^2 \theta}{\partial x^2} + \frac{\dot{q}}{k} = \frac{1}{\alpha}\frac{\partial \theta}{\partial t} \tag{4.12}$$

The temperature profile can be solved by separation of variables. For example, to use the separation of variables method, we define

$$\theta = \theta_1\left(x\right) + \theta_2\left(x,t\right)$$

$$\frac{d^2\theta_1}{dx^2} + \frac{\dot{q}}{k} = 0 \tag{4.13}$$

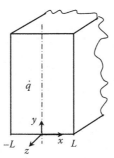

FIGURE 4.6 Multidimensional transient heat conduction with heat generation.

where θ_1 can be solved by 1-D heat conduction with internal heat generation, as shown in Chapter 2 (Section 2.2), and θ_2 can be solved by the aforementioned method of separation of variables with a given convection or surface temperature BCs.

For 3-D transient heat conduction with heat generation, as shown in Figure 4.6, the following equations can be used to solve temperature profiles $T(x, y, z, t)$ or $\theta(x, y, z, t)$:

$$\frac{\partial^2 \theta}{\partial x^2} + \frac{\partial^2 \theta}{\partial y^2} + \frac{\partial^2 \theta}{\partial z^2} + \frac{\dot{q}}{k} = \frac{1}{\alpha}\frac{\partial \theta}{\partial t} \tag{4.14}$$

$$\theta = \theta_1(x) + \theta_{2x}(x, t)\cdot\theta_{2y}(y, t)\cdot\theta_{2z}(z, t) \tag{4.15}$$

where

$$\frac{\partial^2 \theta_1}{\partial x^2} + \frac{\dot{q}}{k} = 0$$

$$\frac{\partial^2 \theta_{2x}}{\partial x^2} = \frac{1}{\alpha}\frac{\partial \theta_{2x}}{\partial t}$$

$$\frac{\partial^2 \theta_{2y}}{\partial y^2} = \frac{1}{\alpha}\frac{\partial \theta_{2y}}{\partial t}$$

$$\frac{\partial^2 \theta_{2z}}{\partial z^2} = \frac{1}{\alpha}\frac{\partial \theta_{2z}}{\partial t}$$

A general form of the solution of 1-D transient heat conduction with the convection BC applicable to a slab, a cylinder (Example 4.1), and a sphere (Example 4.2), as shown in Figure 4.1, can be written [2] as

$$\frac{\theta}{\theta_i} = \sum_{n=1}^{\infty} C_n f(\lambda_n \eta) e^{-\lambda_n^2 Fo} \tag{4.16}$$

$$\frac{Q}{Q_o} = 1 - \sum_{n=1}^{\infty} C_n f\left(\lambda_n\right) e^{-\lambda_n^2 Fo} \qquad (4.17)$$

where

$\lambda_n = \lambda_n L$ for the slab
$\lambda_n = \lambda_n r_o$ for the cylinder
$\lambda_n = \lambda_n r_o$ for the sphere

Geometry	$\theta(x, t)$	C_n	$f(\lambda_n, \eta)$	$F(\lambda_n)$
Slab	$(\partial^2\theta/\partial x^2) = (1/\alpha)$ $(\partial\theta/\partial t)$	$(2 \sin \lambda_n/(\lambda_n + \sin \lambda_n \cos \lambda_n))$	$\cos(\lambda_n(x/L))$	$\sin \lambda_n/\lambda_n$
Cylinder	$(1/r)(\partial/\partial r)(r(\partial\theta/\partial r))$ $= (1/r)(\partial\theta/\partial t)$	$(2J_1(\lambda_n)/(\lambda_n[J_0^2(\lambda_n)$ $+ J_1^2(\lambda_n)]))$	$J_o(\lambda_n(r/r_o))$	$(2J_1(\lambda_n)/\lambda_n)$
Sphere	$(1/r^2)(\partial/\partial r)(r^2(\partial\theta/\partial r))$ $= (1/r)(\partial\theta/\partial t)$	$(2[\sin \lambda_n - \lambda_n \cos \lambda_n])/$ $(\lambda_n - \sin \lambda_n \cos \lambda_n)$	$(\sin(\lambda_n(r/r_o))/$ $(\lambda_n(r/r_o))$	$\left(\left(3\left(\sin \lambda_n - \lambda_n \cos \lambda_n\right)\right)/\lambda_n^3\right)$

With the eigenvalues as

$$Bi \cos \lambda_n - \lambda_n \sin \lambda_n = 0 \text{ for the slab}, \eta = \frac{x}{L}, \quad Bi = \frac{hL}{k}, \quad Fo = \frac{\alpha t}{L^2}$$

$$\lambda_n J_1(\lambda_n) - BiJ_o(\lambda_n) = 0 \text{ for the cylinder}, \eta = \frac{r}{r_o}, \quad Bi = \frac{hr_o}{k}, \quad Fo = \frac{\alpha t}{r_o^2}$$

$$\lambda_n \cos \lambda_n + (Bi - 1) \sin \lambda_n = 0 \text{ for the sphere}, \eta = \frac{r}{r_o}, \quad Bi = \frac{hr_o}{k}, \quad Fo = \frac{\alpha t}{r_o^2}$$

4.2.4 TRANSIENT HEAT CONDUCTION IN THE CYLINDRICAL COORDINATE SYSTEM

Case 1: Consider 1-D transient heat condition in a long cylindrical rod with initial temperature T_i that is suddenly put into a convective fluid at a low temperature, T_∞, with a very high convection heat transfer coefficient, h, i.e., the surface temperature is close to fluid temperature. This is a constant surface temperature problem, $T_s = T_\infty$ = constant. Determine the temperature distribution.

$$\frac{1}{r}\frac{\partial}{\partial r}\left(r\frac{\partial T}{\partial r}\right) = \frac{1}{\alpha}\frac{\partial T}{\partial t}$$

Let

$$\theta = T - T_s$$

at $r = 0$

$$\frac{\partial \theta}{\partial r} = 0$$

$$r = r_0, \theta = T_s - T_s = 0$$

$$t = 0, \theta = T_i - T_s = \theta_i$$

$$\theta(r, t) = R(r) \cdot \tau(t)$$

$$\frac{d}{dr}\left(r\frac{dR}{dr}\right) + \lambda^2 rR = 0$$

$$R(r) = c_1 J_0(\lambda r) + c_2 Y_0(\lambda r)$$

$$\frac{d\tau}{dt} + \alpha\lambda^2\tau = 0$$

$$\tau(t) = c_3 e^{-\alpha\lambda^2 t}$$

From the BC at $r = 0$

$$\frac{\partial \theta}{\partial r} = \frac{dR}{dr} = 0$$

$$0 = -c_1\lambda J_1(0) - c_2\lambda Y_1(0)$$

$$c_2 = 0 \text{ because } Y_1(0) = -\infty$$

at $r = r_0$,

$$\theta = R = 0$$

$$0 = c_1 J_0(\lambda r_0)$$

Therefore, $\lambda_n r_0$ is the root of $J_0(\lambda_n r_0) = 0$, $n = 1, 2, 3, \ldots$
Thus

$$\theta = \sum C_n J_0(\lambda_n r) \cdot e^{-\lambda_n^2 \alpha t}$$

At $t = 0$,

$$\theta_i = \sum C_n J_0(\lambda_n r)$$

multiply both sides by $rJ_0(\lambda_m r)dr$ and integrate,

$$C_n = \frac{\theta_i \int_0^{r_0} rJ_0(\lambda_m r)dr}{\int_0^{r_0} rJ_0(\lambda_n r) \cdot J_0(\lambda_m r)dr}$$

where

$$\int_0^{r_0} rJ_0(\lambda_n r)dr = \frac{r_0}{\lambda_n} J_1(\lambda_n r_0)$$

$$\int_0^{r_0} rJ_0(\lambda_n r) \cdot J_0(\lambda_m r)dr = \frac{r_0^2}{2} J_1(\lambda_n r_0) \text{ if } m = n$$

$$\int_0^{r_0} rJ_0(\lambda_n r) \cdot J_0(\lambda_m r)dr = 0 \text{ if } m \neq n$$

Then

$$C_n = \frac{\theta_i r_0 J_1(\lambda_n r_0) / \lambda_n}{r_0^2 J_1(\lambda_n r_0) / 2} = \frac{2\theta_i}{\lambda_n r_0 J_1(\lambda_n r_0)}$$

The temperature distribution is

$$\frac{\theta(r,t)}{\theta_i} = \sum_{n=1}^{\infty} \frac{2J_0(\lambda_n r) \cdot e^{-\lambda_n^2 \alpha t}}{\lambda_n r_0 J_1(\lambda_n r_0)}$$

where $\lambda_n r_0$ is the root of $J_0(\lambda_n r_0) = 0$, $n = 1, 2, 3, \ldots$

Case 2: In this case, we consider 1-D transient conduction with a convection BC, h, T_∞, at the outer radius of a cylinder. The procedure to determine the temperature is outlined below (see Example 4.1).

Let

$$\theta = T - T_\infty, \theta_i = T_i - T_\infty$$

at $r = 0$

$$\frac{\partial \theta}{\partial r} = 0$$

At $r = r_0$

$$-k\frac{\partial \theta}{\partial r} = h\theta$$

$$R(r) = c_1 J_0(\lambda r) + c_2 Y_0(\lambda r)$$

From the BCs,

at $r = 0$

$$\frac{\partial \theta}{\partial r} = \frac{dR}{dr} = 0, c_2 = 0$$

at $r = r_0$

$$c_1 k\lambda J_1(\lambda r_0) = hc_1 J_0(\lambda r_0)$$

Therefore,

$$\lambda_n r_0 J_1(\lambda_n r_0) = \frac{hr_0}{k} J_0(\lambda_n r_0)$$

$$\theta = \sum C_n J_0(\lambda_n r) \cdot e^{-\lambda_n^2 \alpha t}$$

$$C_n = \frac{\theta_i \int_0^{r_0} r J_0(\lambda_n r)dr}{\int_0^{r_0} J_0(\lambda_n r) \cdot r J_0(\lambda_n r)dr} = \frac{2\theta_i}{\lambda_n r_0 J_1(\lambda_n r_0)}$$

The temperature distribution is

$$\frac{\theta(r,t)}{\theta_i} = \frac{T - T_\infty}{T_i - T_\infty} = \sum \frac{2Bi e^{-(\lambda_n r_0)^2 Fo} \cdot J_0(\lambda_n r)}{\left[\lambda_n^2 r_0^2 + Bi^2\right] \cdot J_0(\lambda_n r_0)}$$

where $\lambda_n r_0$ is the root of $(\lambda_n r_0) J_1(\lambda_n r_0) = Bi J_0(\lambda_n r_0)$ for $n = 1, 2, 3, \ldots$

$$Fo = \frac{\alpha t}{r_0^2}, \quad Bi = \frac{h r_0}{k}$$

If $Bi = 0.1$, $\lambda_1 r_0 = 0.4417$, $\lambda_2 r_0 = 3.8577$, $\lambda_3 r_0 = 7.0298$.

For $Fo \geq 0.2$, the first term in the infinite series is much larger than all the other terms. Thus,

$$\frac{\theta}{\theta_i} \cong \frac{2Bi \cdot e^{-(\lambda_1 r_0)^2 Fo} \cdot J_0(\lambda_1 r)}{\left[\lambda_1^2 r_0^2 + Bi^2\right] \cdot J_0(\lambda_1 r_0)}$$

where $\lambda_1 r_0$ is the root of $(\lambda_1 r_0) J_1(\lambda_1 r_0) = Bi \cdot J_1(\lambda_1 r_0)$.
at $r = 0$, the centerline, $\theta = T_0 - T_\infty = \theta_0$ and

$$\theta_0 \cong \frac{2Bi \cdot e^{-(\lambda_1 r_0)^2 Fo}}{\left[\lambda_1^2 r_0^2 + Bi^2\right] \cdot J_0(\lambda_1 r_0)}$$

Therefore, the dimensionless centerline temperature,

$$\frac{\theta_0}{\theta_i} \sim e^{-Fo} \text{ for any given } Bi, \text{ and}$$

$$\frac{\theta}{\theta_i} \sim J_0(\lambda_1 r) \text{ function of} \left(Bi, \frac{r}{r_0} \right)$$

Note for any given Bi, $\frac{\theta}{\theta_0}$ is dependent on r only and is not a function of time.

If $r_0 = 10\,\text{cm}$, $k = 20\,\text{W/m} \cdot \text{K}$, $\alpha = 4 \times 10^{-6}\,\text{m}^2/\text{s}$, $Bi = 0.1$, $T_i = 400°\text{C}$, $T_\infty = 20°\text{C}$, $Fo = 0.3$, the centerline temperature, T_0, can be determined.

Case 3: Finally, transient heat conduction in multiple dimensions within a cylindrical coordinate system is considered. The general solution is outlined below [3].

$$\frac{1}{r}\frac{\partial}{\partial r}\left(r \frac{\partial T}{\partial r} \right) + \frac{1}{r^2}\frac{\partial^2 T}{\partial \phi^2} + \frac{\partial^2 T}{\partial z^2} = \frac{1}{\alpha}\frac{\partial T}{\partial t}$$

Let

$$T(r,\phi,z,t) = R(r) \cdot \Phi(\phi) \cdot Z(z) \cdot \tau(t)$$

and substitute into the conduction equation,

$$\Phi Z \tau \left(\frac{d^2 R}{dr^2} + \frac{1}{r} \frac{dR}{dr} \right) + RZ\tau \left(\frac{1}{r^2} \frac{d^2 \Phi}{d\phi^2} \right) + R\Phi \tau \frac{d^2 Z}{dz^2} = R\Phi Z \left(\frac{1}{\alpha} \frac{\partial \tau}{\partial t} \right)$$

Now divide by $R\Phi Z \tau$,

$$\frac{1}{R} \left(\frac{d^2 R}{dr^2} + \frac{1}{r} \frac{dR}{dr} \right) + \frac{1}{\Phi} \frac{1}{r^2} \frac{d^2 \Phi}{d\phi^2} + \frac{1}{Z} \frac{d^2 Z}{dz^2} = \frac{1}{\tau} \left(\frac{1}{\alpha} \frac{\partial \tau}{\partial t} \right)$$

Let

$$\frac{r^2}{R} \left(\frac{d^2 R}{dr^2} + \frac{1}{r} \frac{dR}{dr} \right) + \frac{1}{\Phi} \frac{d^2 \Phi}{d\phi^2} = \pm (r\beta)^2$$

$$\frac{1}{Z} \frac{d^2 Z}{dz^2} = \pm \eta^2$$

$$\frac{1}{\tau} \frac{\partial \tau}{\partial t} = -\lambda^2$$

and

$$\frac{1}{\Phi} \frac{d^2 \Phi}{d\phi^2} = -\upsilon^2$$

$$\frac{r^2}{R} \left(\frac{d^2 R}{dr^2} + \frac{1}{r} \frac{dR}{dr} \right) = \upsilon^2 \pm (r\beta)^2$$

Thus

$$\frac{d^2 R}{dr^2} + \frac{1}{r} \frac{dR}{dr} + \left(\pm \beta^2 - \frac{\upsilon^2}{r^2} \right) R = 0$$

From the Bessel functions and their solutions (see Equations (2.48)–(2.51))

$$R = \text{function of } J_\upsilon (\beta r), Y_\upsilon (\beta r) \text{ or}$$

$$R = \text{function of } I_\upsilon (\beta r), K_\upsilon (\beta r).$$

with

$$\frac{d^2 \Phi}{d\phi^2} + \Phi \upsilon^2 = 0$$

$$\Phi = \text{function of } \sin(\upsilon\phi), \cos(\upsilon\phi)$$

$$\frac{d^2Z}{dz^2} \pm \eta^2 Z = 0$$

$$Z = \text{function of } \sin(\eta z), \cos(\eta z)$$

or

$$Z = \text{function of } \sinh(\eta z), \cosh(\eta z)$$

$$\frac{\partial \tau}{\partial t} + \alpha\lambda^2 \tau = 0$$

$$\tau = \text{function of } e^{-\lambda^2 \alpha t}$$

For 2-D transient, $T(r,z,t)$, i.e., $\upsilon^2 = 0$

$$\frac{d^2R}{dr^2} + \frac{1}{r}\frac{dR}{dr} \pm \beta^2 R = 0$$

$$R = \text{function of } J_0(\beta r), Y_0(\beta r) \text{ or } I_0(\beta r), K_0(\beta r)$$

For 2-D transient, $T(r,\phi,t)$, i.e., $\eta^2 = 0$

$$\frac{d^2R}{dr^2} + \frac{1}{r}\frac{dR}{dr} + \left(\beta^2 - \frac{\upsilon^2}{r^2}\right)R = 0$$

$$R = \text{function of } J_\upsilon(\beta r), Y_\upsilon(\beta r)$$

For 3-D steady state, $T(r,\phi,z)$, i.e., $\lambda^2 = 0$

$$\frac{d^2R}{dr^2} + \frac{1}{r}\frac{dR}{dr} + \left(\pm\beta^2 - \frac{\upsilon^2}{r^2}\right)R = 0$$

$$R = \text{function of } J_\upsilon(\beta r), Y_\upsilon(\beta r) \text{ or } I_\upsilon(\beta r), K_\upsilon(\beta r)$$

For general 3-D transient conduction, the temperature distribution can be determined as shown below.

$$T(r,\phi,z,t) = \sum_{m=1}^{\infty}\sum_{n=1}^{\infty}\sum_{p=1}^{\infty} C_{mnp} R(\beta_m, \upsilon_n, r) \Phi(\upsilon_n, \phi) Z(\eta_p, z) \tau(\beta_m, \eta_p, t).$$

where $\tau = e^{-\left(\beta_m^2 + \eta_p^2\right)\alpha t}$

The initial condition:

$$T(r,\phi,z,t) = F(r, \phi, z) \text{ at } t = 0,$$

$$F(r, \phi, z) = \sum_{n=1}^{\infty} C_{mnp} \Phi(\upsilon_n, \phi) \cdot \int_0^{r_0} R(\beta_m, \upsilon_n, r) dr \cdot \int_0^c Z(\eta_p, z) dz$$

where $\Phi(\upsilon_n, \phi) = a_{mnp} \sin(\upsilon_n, \phi) + b_{mnp} \cos(\upsilon_n, \phi)$

Thus a_{mnp} and b_{mnp} can be determined by multiplying by $\sin(\upsilon_n, \phi)$ and $\cos(\upsilon_n, \phi)$. Finally, we integrate over $\phi = 0$ to 2π. Additional details are provided by Özişik [3].

C_{mnp} can be obtained by multiplying by $rR(\beta_m, \upsilon_n, r) \cdot Z(\eta_p, z)$ and integrating over $dzdr$,

$$C_{mnp} = \frac{\int_{r=0}^{r_0}\int_{z=0}^c F(r,\phi,z) \cdot rR(\beta_m, \upsilon_n, r) \cdot Z(\eta_p, z) dzdr}{\displaystyle\sum_{n=1}^{\infty} \Phi(\upsilon_n, \phi) \int_0^{r_0} rR^2(\beta_m, \upsilon_n, r) dr \cdot \int_0^c Z^2(\eta_p, z) dz}$$

For a special initial condition, $T(r,\phi,z,t) = F(r,\phi,z) = T_i = $ constant value, at $t = 0$.

The 3-D transient temperature distribution $T(r,\phi, z,t)$ can be reduced to the 2-D transient temperature distribution $T(r,z,t)$, $T(r,\phi,t)$ or to 1-D transient temperature distribution $T(r,t)$.

Based on the three cases shown above, we can consider 2-D steady-state heat conduction in the cylindrical coordinate system.

$$\frac{\partial^2 T}{\partial r^2} + \frac{1}{r}\frac{\partial T}{\partial r} + \frac{1}{r^2}\frac{\partial^2 T}{\partial \phi^2} = 0$$

Let $T = R(r) \cdot \Phi(\phi)$ and insert into the above equation,

$$\Phi\left(\frac{d^2 R}{dr^2} + \frac{1}{r}\frac{dR}{dr}\right) + \frac{R}{r^2}\frac{d^2\Phi}{d\phi^2} = 0$$

$$\frac{r^2}{R}\left(\frac{d^2 R}{dr^2} + \frac{1}{r}\frac{dR}{dr}\right) = -\frac{1}{\Phi}\frac{d^2\Phi}{d\phi^2} = \lambda^2$$

Thus,

$$\frac{d^2\Phi}{d\phi^2} + \lambda^2\Phi = 0, \ \Phi = c_1\sin\lambda\phi + c_2\cos\lambda\phi$$

$$r^2\frac{d^2R}{dr^2} + r\frac{dR}{dr} - \lambda^2 R = 0, \ R = c_3 r^\lambda + c_4 r^{-\lambda}$$

BCs are needed to solve for the constants:

$$\Phi(\phi) = \Phi(\phi + 2\pi)$$

$$\left.\frac{d\Phi}{d\phi}\right|_\phi = \left.\frac{d\Phi}{d\phi}\right|_{\phi+2\pi}$$

Now choose $\lambda_n = n$, $n = 0, 1, 2, \ldots$

$$\Phi = c_1\sin n\phi + c_2\cos n\phi$$

The BCs in the radial direction:

$$r = 0, \ R \text{ is finite } c_4 = 0$$

$$r = r_0, \qquad T = f(\phi)$$

Therefore,

$$T = R(r)\Phi(\phi) = c_3 r^n \left(c_1\sin n\phi + c_2\cos n\phi\right)$$

$$= a_0 + \sum_{n=r}^{\infty} r^n \left(a_n\cos n\phi + b_n\sin n\phi\right)$$

The nonhomogeneous BC at $r = r_0$

$$f(\phi) = a_0 + \sum_{n=1}^{\infty} r_0^n \left(a_n\cos n\phi + b_n\sin n\phi\right)$$

$$a_0 = \frac{1}{2\pi}\int_0^{2\pi} f(\phi)\,d\phi, \qquad 0 \le \phi \le 2\pi$$

$$a_n = \frac{1}{\pi r_0^n} \int_0^{2\pi} f(\phi) \cdot \cos n\phi d\phi$$

$$b_n = \frac{1}{\pi r_0^n} \int_0^{2\pi} f(\phi) \cdot \sin n\phi d\phi$$

For a special case,

$$T = T_0, \qquad 0 < \phi < \pi$$

$$\text{at } r = r_0 \quad T = 0, \qquad \pi < \phi < 2\pi$$

$$a_0 = \frac{1}{2\pi} \int_0^\pi T_0 d\phi = \frac{T_0}{2}$$

$$a_n = \frac{1}{\pi r_0^n} \int_0^\pi T_0 \cdot \cos n\phi d\phi = \frac{T_0}{n\pi r_0^n} [\sin n\phi]_0^\pi = 0$$

$$b_n = \frac{1}{\pi r_0^n} \int_0^\pi T_0 \cdot \sin n\phi d\phi = \frac{-T_0}{n\pi r_0^n} [\cos n\phi]_0^\pi = \frac{2T_0}{n\pi r_0^n}$$

The temperature distribution is

$$\frac{T}{T_0} = \frac{1}{2} + \sum_{n=1}^{\infty} \frac{2}{n\pi} \left(\frac{r}{r_0}\right)^n \cdot \sin(n\phi)$$

Similarly, consider the 2-D, $T(r,z)$ case, and if $J_0(\lambda_n r_0) = 0$, $\lambda_1 r_0 = 2.4048$, $\lambda_2 r_0 = 5.5201$.

4.3 1-D TRANSIENT HEAT CONDUCTION IN A SEMIINFINITE SOLID MATERIAL

4.3.1 SIMILARITY METHOD FOR SEMIINFINITE SOLID MATERIAL

The semiinfinite solid is characterized by a single identifiable surface. If a sudden change in temperature occurs at this surface as shown in Figure 4.7, then transient 1-D conduction will take place within the solid. The similarity method can be employed to solve this kind of problem [4]. The 1-D transient heat conduction equation in a semiinfinite solid without heat generation is given by

$$\frac{\partial^2 T}{\partial x^2} = \frac{1}{\alpha} \frac{\partial T}{\partial t}$$

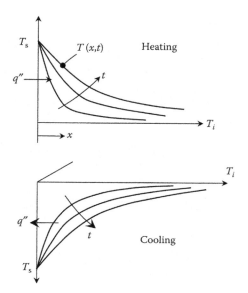

FIGURE 4.7 1-D transient heat conduction in a semiinfinite solid material.

The surface BC is

$$T(0,t) = T_s$$

With the interior BC prescribed by

$$T(\infty,t) = T_i$$

The initial condition is

$$T(x,0) = T_i$$

By applying the similarity method, we may transform the PDE, which involves two independent variables (x and t), to an ODE expressed in terms of a single similarity variable (η).

As $x \sim \sqrt{\alpha t}$ ($\sqrt{\alpha t}$ is the diffusion length), we define the similarity variable $\eta = \left(x/\sqrt{4\alpha t} \right)$; therefore $T(x, t) = T(\eta)$,

$$x = \eta\sqrt{4\alpha t} \tag{4.18}$$

Let $\theta = ((T - T_i)/(T_s - T_i))$.

The 1-D transient heat equation can be expressed as

$$\frac{\partial^2 \theta}{\partial x^2} = \frac{1}{\alpha}\frac{\partial \theta}{\partial t}$$

Transform the similarity variable into θ, that is, $\theta(x, t) = \theta(\eta)$. Therefore,

$$\frac{\partial \theta}{\partial t} = \frac{d\theta}{d\eta} \frac{\partial \eta}{\partial t} = \frac{d\theta}{d\eta} \left[\frac{-x}{2t\sqrt{4\alpha t}} \right]$$

$$\frac{\partial \theta}{\partial x} = \frac{d\theta}{d\eta} \frac{\partial \eta}{\partial x} = \frac{d\theta}{d\eta} \left(\frac{1}{\sqrt{4\alpha t}} \right)$$

$$\frac{\partial^2 \theta}{\partial x^2} = \frac{d}{d\eta} \left(\frac{\partial \theta}{\partial x} \right) \frac{\partial \eta}{\partial x} = \frac{d}{d\eta} \left[\frac{\partial \theta}{\partial \eta} \left(\frac{1}{\sqrt{4\alpha t}} \right) \right] \left(\frac{1}{\sqrt{4\alpha t}} \right) = \frac{1}{4\alpha t} \frac{\partial^2 \theta}{\partial \eta^2}$$

Inserting the above terms into the heat equation, we obtain

$$\frac{1}{4\alpha t} \frac{d^2 \theta}{d\eta^2} = \frac{1}{\alpha} \frac{-x}{2t(4\alpha t)^{1/2}} \frac{d\theta}{d\eta}$$

Rearranging, we obtain

$$\frac{\alpha}{4\alpha t} \frac{d^2 \theta}{d\eta^2} = \frac{-x}{2t(4\alpha t)^{1/2}} \frac{d\theta}{d\eta}$$

$$\frac{d^2 \theta}{d\eta^2} = \frac{-2x}{\sqrt{4\alpha t}} \frac{d\theta}{d\eta} = -2\eta \frac{d\theta}{d\eta}$$

(4.19)

For the case of given constant surface temperature,

$$x = 0, \qquad \eta = 0, \qquad \theta(0) = 1$$
$$x = \infty, \qquad \eta = \infty, \qquad \theta(\infty) = 0$$

Let $P \equiv (d\theta/d\eta)$; rearranging the equation and integrating it, we obtain

$$\frac{dP}{d\eta} = -2\eta P \Rightarrow P \equiv \frac{d\theta}{d\eta} = c_1 e^{-\eta^2}$$

$$\left[\begin{array}{l} \dfrac{dP}{P} = -2\eta d\eta \\[2ex] \displaystyle\int \dfrac{dP}{P} = \int -2\eta d\eta \\[2ex] \ln P = -\eta^2 + c \end{array} \right.$$

$$\theta = c_1 \int_0^\eta e^{-\eta^2} d\eta + c_2$$

From the BC,

$$\eta = 0, \quad \theta = 1 \quad \Rightarrow \quad c_2 = 1$$

From the BC,

$$\eta = \infty, \quad \theta = 0$$

$$0 = c_1 \int_0^\infty e^{-u^2} du + 1 = c_1 \frac{\sqrt{\pi}}{2} + 1 \quad \Rightarrow c_1 = -\frac{2}{\sqrt{\pi}}$$

Inserting c_1 and c_2 into the integral, we obtain the following results as sketched in Figure 4.8

$$\theta = \frac{T - T_i}{T_s - T_i} = 1 - \frac{2}{\sqrt{\pi}} \int_0^\eta e^{-u^2} du = 1 - \mathrm{erf}(\eta) = \mathrm{erfc}\left(\frac{x}{\sqrt{4\alpha t}}\right) = 1 - \mathrm{erf}\left(\frac{x}{\sqrt{4\alpha t}}\right) \quad (4.20)$$

and

$$q_s = -k \frac{\partial T}{\partial x} \big|_{x=0} = -k(T_s - T_i) \frac{\partial \theta}{\partial x} \big|_{x=0}$$

$$= -k(T_s - T_i) \frac{\partial}{\partial x} \left[1 - \frac{2}{\sqrt{\pi}} \int_0^\eta e^{-u^2} du \right]_{x=0}$$

$$= -k(T_s - T_i) \left[-\frac{2}{\sqrt{\pi}} e^{-\eta^2} \frac{1}{\sqrt{4\alpha t}} \right]_{\eta=0}$$

$$= \frac{k(T_s - T_i)}{\sqrt{\pi \alpha t}}$$

$$(4.21)$$

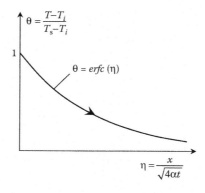

FIGURE 4.8   Solution of 1-D transient heat.

4.3.2 LAPLACE TRANSFORM METHOD FOR SEMIINFINITE SOLID MATERIAL

The 1-D transient conduction problem in a semiinfinite solid material can be solved by the Laplace transform method [1]. The heat diffusion equation for 1-D transient conduction is of the form

$$\frac{\partial^2 T}{\partial x^2} = \frac{1}{\alpha}\frac{\partial T}{\partial t}$$

Case 1: We begin with the constant surface temperature BC.

$$T(0,t) = T_s$$
$$T(\infty,t) = T_i$$

With the initial condition:

$$T(x,0) = T_i$$

Let $\theta = T - T_i$, therefore,

$$\frac{\partial^2 \theta}{\partial x^2} = \frac{1}{\alpha}\frac{\partial \theta}{\partial t}$$

$$\theta(0,t) = \theta_s = T_s - T_i$$

$$\theta(x,0) = 0$$

$$\theta(\infty,t) = 0$$

Defining the Laplace transform of a given temperature,

$$L(T) = \overline{T} = \int\limits_0^\infty T e^{-st}\,dt$$

$$L(T_i) = \frac{T_i}{s}$$

$$L(0) = 0$$

$$L\left(\frac{\partial T}{\partial t}\right) = s\overline{T} - \overline{T_i}$$

$$L\left(\frac{\partial^n T}{\partial x^n}\right) = \frac{\partial^n \overline{T}}{\partial x^n}$$

$$T(x,t) = \overline{T}(x,s)$$

(4.22)

Applying the Laplace transform to 1-D transient heat conduction,

$$L\left(\frac{\partial^2 T}{\partial x^2}\right) = L\left(\frac{1}{\alpha}\frac{\partial T}{\partial t}\right) \tag{4.23}$$

$$\frac{\partial^2 \bar{\theta}}{\partial x^2} = \frac{1}{\alpha}\left[s\bar{\theta} - \bar{\theta}(x,0)\right] \tag{4.24}$$

From the initial condition, $\bar{\theta}(x,0) = 0$.

$$\frac{d^2\bar{\theta}}{dx^2} - \frac{s}{\alpha}\bar{\theta} = 0 \Rightarrow \bar{\theta} = c_1 e^{-x\sqrt{s/\alpha}} + c_2 e^{x\sqrt{s/\alpha}}$$

Applying the Laplace transform to the BCs,

$$\begin{cases} \bar{\theta}(\infty,s) = 0, & c_2 = 0 \\ \bar{\theta}(0,s) = \dfrac{\theta_s}{s}, & c_1 = \dfrac{\theta_s}{s} \end{cases}$$

So

$$\bar{\theta} = \frac{\theta_s}{s}e^{-x\sqrt{s/\alpha}}$$

Applying the inverse Laplace transform from a given table, we obtain the following results as sketched in Figure 4.9.

$$\frac{\theta}{\theta_s} = \frac{T - T_i}{T_s - T_i} = \operatorname{erfc}\frac{x}{\sqrt{4\alpha t}} \tag{4.25}$$

Case 2: Now we consider the constant surface heat flux BC (see Example 4.3).

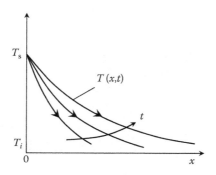

FIGURE 4.9 Solution of 1-D transient heat conduction with given surface temperature BC.

$$T(x,0) = T_i$$

$$T(\infty,t) = T_i$$

$$-k\frac{\partial T(0,t)}{\partial x} = q_s''$$

Let $\theta = T - T_i$, then

$$\theta(x,0) = 0$$

$$\theta(\infty,t) = 0$$

$$-k\frac{\partial\theta(0,t)}{\partial x} = q_s''$$

The Laplace transform solution is

$$\bar{\theta} = \frac{q_s''}{k}\frac{1}{s\sqrt{s/\alpha}}\cdot e^{-x\sqrt{s/\alpha}}$$

Applying the inverse Laplace transform from a given table, we obtain the following results as sketched in Figure 4.10.

$$T - T_i = \frac{q_s''}{k}\left[\frac{\sqrt{4\alpha t}}{\sqrt{\pi}}e^{-(x^2/4\alpha t)} - x\cdot\mathrm{erfc}\left(\frac{x}{\sqrt{4\alpha t}}\right)\right] \tag{4.26}$$

$$\text{at}\quad x = 0,\quad T_s - T_i = \frac{q_s''}{k}\frac{\sqrt{4\alpha t}}{\sqrt{\pi}} \tag{4.27}$$

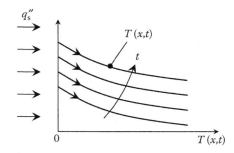

FIGURE 4.10 Solution of 1-D transient heat conduction with constant surface heat flux BC.

Case 3: Next, we impose the convective surface BC (See Example 4.4).

$$T(x,0) = T_i$$

$$T(\infty,t) = T_i$$

$$-k\frac{\partial T(0,t)}{\partial x} = h\big(T_\infty - T(0,t)\big)$$

Let $\theta = T - T_i$

$$\theta(x,0) = 0$$

$$\theta(\infty,0) = 0$$

$$-k\frac{\partial \theta(0,t)}{\partial x} = h\big(\theta_\infty - \theta(0,t)\big)$$

Applying the inverse Laplace transform from a given table, we obtain the following results as sketched in Figure 4.11.

$$\frac{\theta}{\theta_\infty} = \frac{T - T_i}{T_\infty - T_i} = \left[\mathrm{erfc}\left(\frac{x}{\sqrt{4\alpha t}}\right) - e^{\left(x(h/k) + \alpha(h^2/k^2)t\right)}\mathrm{erfc}\left(\frac{x}{\sqrt{4\alpha t}} + \frac{h}{k}\sqrt{\alpha t}\right) \right] \quad (4.28)$$

$$q'' = -k\frac{\partial T(0,t)}{\partial x} = h\big[T_\infty - T(0,t)\big] \quad (4.29)$$

4.4   HEAT CONDUCTION WITH MOVING BOUNDARIES

There are many engineering applications involving heat conduction with moving boundaries such as freezing or melting for solar storage systems. Other examples are related to high-temperature droplet evaporation and ablation applications. The problems can be solved by using the similarity method [5].

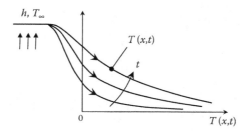

FIGURE 4.11   Solution of 1-D transient heat conduction convective surface BC.

4.4.1 Freezing and Solidification Problems Using the Similarity Method

Freezing—The Neumann solution (exact solution) [5], as sketched in Figure 4.12:
Governing equation:

$$\theta = T - T_m$$

$$\frac{\partial^2 \theta_1}{\partial x^2} = \frac{1}{\alpha_1}\frac{\partial \theta_1}{\partial t} \tag{4.30}$$

$$\frac{\partial^2 \theta_2}{\partial x^2} = \frac{1}{\alpha_2}\frac{\partial \theta_2}{\partial t}$$

BCs (4.30):

$$x = 0, \quad \theta_1 = \theta_s \tag{4.30a}$$

$$x = \infty, \quad \theta_2 = \theta_\infty \tag{4.30b}$$

$$x = \delta(t), \quad \theta_1 = 0 \tag{4.30c}$$

$$\theta_2 = 0 \tag{4.30d}$$

$$k_1 \frac{\partial \theta_1}{\partial x} - k_2 \frac{\partial \theta_2}{\partial x} = L\rho_1 \frac{d\delta}{dt} \tag{4.30e}$$

where L represents the latent heat of melting. Equation (4.30e) represents the energy balance at the interface. In other words, *(conduction heat flux from water) – (conduction heat flux to ice) = (heat flux released to make ice from water)*.
Initial conditions:

$$t = 0, \quad \delta = 0, \quad T = T_\infty$$

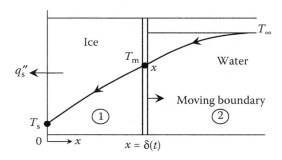

FIGURE 4.12 Heat conduction with moving boundary problem—freezing.

Assume $\rho_1 = \rho_2$.

Neumann applied the similarity method:

$$\theta_1 = c_1 + c_2 \text{erf} \frac{x}{\sqrt{4\alpha_1 t}} \tag{4.30f}$$

$$\theta_2 = c_1' + c_2' \text{erf} \frac{x}{\sqrt{4\alpha_2 t}} \tag{4.30g}$$

At $x = \delta$, $\delta \sim \sqrt{t}$ to obtain $\theta_1 = 0$, $\delta = b\sqrt{t}$.

Insert Equations (4.30a)–(4.30d) into Equations (4.30f) and (4.30g)

$$c_1 = \theta_s$$

$$c_2 = \frac{-\theta_s}{\text{erf}\left(b/\sqrt{4\alpha_1}\right)}$$

$$c_2' = \frac{\theta_\infty}{\text{erfc}\left(b/\sqrt{4\alpha_2}\right)}$$

$$c_1' = \theta_\infty - \frac{\theta_\infty}{\text{erfc}\left(b/\sqrt{4\alpha_2}\right)}$$

From Equation (4.30e) with $x = \delta(t) = b\sqrt{t}$:

$$\frac{-k_1\theta_s \exp\left(-b^2/4\alpha_1\right)}{\sqrt{\pi\alpha_1}\,\text{erf}\left(b/\sqrt{4\alpha_1}\right)} - \frac{+k_2\theta_\infty \exp\left(-b^2/4\alpha_2\right)}{\sqrt{\pi\alpha_2}\,\text{erfc}\left(b/\sqrt{4\alpha_2}\right)} = L\rho_1 \frac{b}{2} \tag{4.30h}$$

$$\because \frac{\partial}{\partial x}\left[\text{erf}\frac{x}{\sqrt{4\alpha_1 t}}\right] = \frac{2}{\sqrt{\pi}} \frac{\partial}{\partial x} \int_0^\eta e^{-\eta^2} d\eta$$

$$= \frac{2}{\sqrt{\pi}} \frac{\partial \eta}{\partial x} \frac{d}{d\eta} \int_0^\eta e^{-\eta^2} d\eta$$

$$= \frac{2}{\sqrt{\pi}} \frac{1}{\sqrt{4\alpha_1 t}} \exp\left(\frac{-x^2}{4\alpha_1 t}\right)$$

Therefore, b can be obtained from Equation (4.30h) and:

$$\delta = b\sqrt{t}$$

θ_1 can be obtained from Equation (4.30f), θ_2 can be obtained from Equation (4.30g), and q_1'' can be determined:

$$q_s'' = q_1'' = k_1 \frac{\partial \theta_1}{\partial x}\bigg|_0 = -k\theta_s \left[1 - \frac{1}{\text{erf}\left(b/\sqrt{4\alpha_1}\right) \cdot \sqrt{\pi\alpha_1 t}}\right]$$

A special case exists when "slow" freezing takes place. As sketched in Figure 4.13, the temperature distribution through the ice is approximately linear ($\theta_\infty = 0$) and can be obtained from Equation (4.33f).

$$\frac{\theta_1}{\theta_s} = 1 - \frac{\text{erf}\left(x/\sqrt{4\alpha_1 t}\right)}{\text{erf}\left(b/\sqrt{4\alpha_1}\right)} \tag{4.31}$$

b can be obtained from Equation (4.33h) with $\theta_\infty = 0$ and $\delta = b\sqrt{t}$.

In another special case, the heat transfer from the liquid phase to the solid-liquid interface is controlled by convection. Therefore, the interface energy balance is written as

$$k_1 \frac{\partial \theta_1}{\partial x} - h\left(T_\infty - T_m\right) = L\rho_1 \frac{d\delta}{dt}$$

where $\theta_\infty = T_\infty - T_m = constant$. Again, solve for b with $\delta = b\sqrt{t}$.

4.4.2 MELTING AND LIQUIFICATION PROBLEMS USING THE SIMILARITY METHOD

From Figure 4.14, we allow θ_1 = liquid phase and θ_2 = solid phase.

Let $\theta = T - T_m$,

$$\text{at } x = 0, \quad \theta_1 = T - T_m = \theta_s$$

$$\text{at } x = \infty, \quad \theta_2 = T_i - T_m = \theta_i$$

$$\text{at } x = \delta(t), \quad \theta_1 = T_m - T_m = 0$$

$$\theta_2 = T_m - T_m = 0$$

$$-k_1 \frac{\partial \theta_1}{\partial x} + k_2 \frac{\partial \theta_2}{\partial x} = L\rho_1 \frac{d\delta}{dt} \tag{4.32}$$

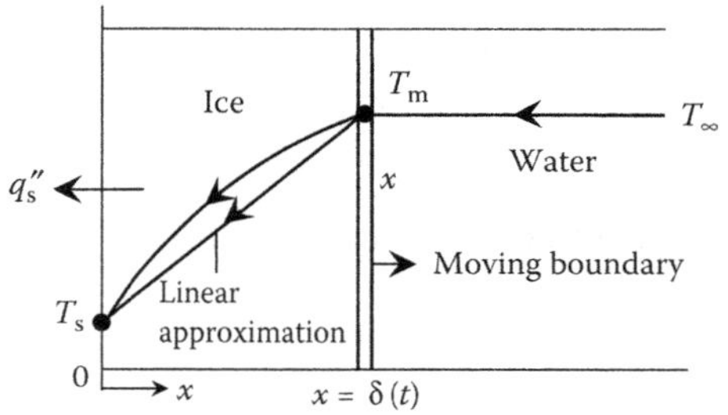

FIGURE 4.13 Slow freezing: $T_\infty \cong T_m$.

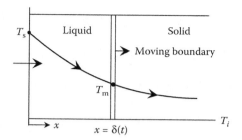

FIGURE 4.14 Heat conduction with moving boundary problems.

L is latent heat of melting. This is the energy balance at the interface: (conduction heat flux from liquid) – (conduction heat flux to solid) = (heat flux required from liquid to melt solid).

For the initial condition:

$$\text{at } t = 0, \quad \delta(t) = 0$$

$$T = T_i$$

$$\theta = T_i - T_m = \theta_i$$

From the similarity solution shown in 4.4.1,

$$\theta_1 = c_1 + c_2 \text{erf} \frac{x}{\sqrt{4\alpha_1 t}}$$

$$\theta_2 = c_1' + c_2' \text{erf} \frac{x}{\sqrt{4\alpha_2 t}}$$

From the BC:

$$c_1 = \theta_s, \quad c_2 = -\frac{\theta_s}{\text{erf}\left(b/\sqrt{4\alpha_1}\right)}$$

$$c_2' = \frac{\theta_i}{\text{erfc}\left(b/\sqrt{4\alpha_2}\right)}, \quad c_1' = \theta_i - \frac{\theta_i}{\text{erfc}\left(b/\sqrt{4\alpha_2}\right)}$$

At the interface:

$$x = \delta(t) = b\sqrt{t},$$

$$\frac{k_1 \theta_s \exp\left(-b^2/4\alpha_1\right)}{\sqrt{\pi\alpha_1} \cdot \text{erf}\left(b/\sqrt{4\alpha_1}\right)} + \frac{k_2 \theta_i \exp\left(-b^2/4\alpha_2\right)}{\sqrt{\pi\alpha_2} \cdot \text{erfc}\left(b/\sqrt{4\alpha_2}\right)} = L\rho_1 \frac{b}{2} \tag{4.33}$$

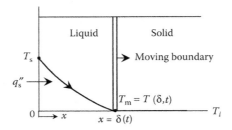

FIGURE 4.15 Slow melting: $T_m \equiv T_i$.

Thus, b can be obtained from the above implicit solution, allowing for the solution of θ_1 and θ_2.

Special case: Again, in the unique case of "slow" melting, as sketched in Figure 4.15, $\theta_2 = 0$ and θ_1 can be obtained from the above solution.

$$\frac{\theta_1}{\theta_s} = 1 - \frac{\mathrm{erf}\left(x/\sqrt{4\alpha_1 t}\right)}{\mathrm{erf}\left(b/\sqrt{4\alpha_1}\right)} \tag{4.34}$$

where b can be obtained from the above with $\theta_i = 0$, $\quad \dfrac{k_1 \theta_s \exp\left(-b^2/4\alpha_1\right)}{\sqrt{\pi\alpha_1} \cdot \mathrm{erf}\left(b/\sqrt{4\alpha_1}\right)} = L\rho_1 \dfrac{b}{2}$
q_s'' can be determined at $x = 0$ as

$$q_s'' = q_1'' = -k\frac{\partial\theta_1}{\partial x} = k\theta_s\left[1 - \frac{1}{\mathrm{erf}\left(b/\sqrt{4\alpha_1}\right) \cdot \sqrt{\pi\alpha_1 t}}\right]$$

If convection dominates at the liquid–solid interface, the energy balance at $x = \delta(t)$ is

$$h(T_s - T_m) + k_2\frac{\partial\theta_2}{\partial x} = L\rho_1\frac{d\delta}{dt}$$

where $T_s - T_m = \theta_s = $ constant.

4.4.3 ABLATION

Ablating heat shields have been successful in satellite and missile re-entry to the earth's atmosphere as a means of protecting the surface from aerodynamic heating. In this application, the high heat flux generated at the surface first causes an initial transient temperature rise until the surface reaches the melting temperature, T_m. Ablation (melting of the surface) begins and follows a second short transient period, and then a steady-state ablation velocity is reached. The melted material is assumed to run off immediately. The problem can be simplified to 1-D transient heat conduction with a moving boundary due to ablation. Figure 4.16 shows ablation at the

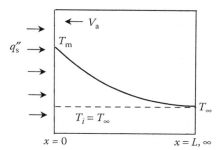

FIGURE 4.16 Ablation at surface of flat wall.

surface of the flat plate with a moving ablation velocity, V_a, moving to the left [6]. The heat conduction equation for this reference frame is Equation (4.38) with an added enthalpy flux term associated with the moving velocity V_a:

$$\frac{\partial}{\partial x}\left(k\frac{\partial T}{\partial x}\right) + \rho c V_a \frac{\partial T}{\partial x} = \rho C \frac{\partial T}{\partial t} \tag{4.35}$$

The problem can be solved in three stages: (1) the initial transient before the surfaces reach T_m, (2) the second transient period during ablation, and (3) after the steady-state ablation velocity has been reached, the temperature distribution in the material is steady. The initial transient problem has been solved previously (i.e., $V_a = 0$, given the surface heat flux BC). The second transient problem, Equation (4.35), can be solved by the finite-difference method. For the steady-state problem with constant ablation velocity, Equation (4.35) becomes

$$\frac{\partial}{\partial x}\left(k\frac{\partial T}{\partial x}\right) = -\rho C V_a \frac{\partial T}{\partial x} \tag{4.36}$$

With the accompanying low thermal conductivity material (such as glasses and plastics) the temperature gradient at the surface is very steep so that $x = L$ may be considered as $x = \infty$. Proper BCs are as follows:

$$x = 0, \quad T = T_m$$
$$x = \infty, \quad T = T_\infty = T_i \tag{4.37}$$
$$x = \infty, \quad \frac{\partial T}{\partial x} = 0$$

For constant properties k, ρ, c, and for the case of constant surface heat flux q_s'', Equation (4.36) is solved by integrating twice and evaluating the integration constants with Equation (4.40). Let $\theta = (dT/dx)$, then

$$\frac{d}{dx}\theta = -\frac{V_a}{\alpha}\theta$$

$$\frac{d\theta}{\theta} = -\frac{V_a}{\alpha}dx$$

$$\int\frac{d\theta}{\theta} = \int -\frac{V_a}{\alpha}dx$$

$$\ln\theta = -\frac{V_a}{\alpha}x + C$$

$$\theta = \frac{dT}{dx} = C_1 e^{-(V_a/\alpha)x} + C_2$$

where at $x = \infty$, $(dT/dx) = 0 = \theta$, $\therefore C_2 = 0$.

Then $(dT/dx) = C_1 e^{-(V_a/\alpha)x}$,

$$\int dT = \int C_1 e^{-(V_a/\alpha)x}dx$$

$$T = -C_1\frac{\alpha}{V_a}e^{-(V_a/\alpha)x} + C_3$$

$$\text{at } x = \infty, \quad T = T_\infty = C_3$$

$$\text{at } x = 0, \quad T = T_m = -C_1\frac{\alpha}{V_a} + C_3$$

$$-C_1 = (T_m - C_3)\frac{V_a}{\alpha} = (T_m - T_\infty)\frac{V_a}{\alpha}$$

Therefore,

$$T = (T_m - T_\infty)e^{-(V_a/\alpha)x} + T_\infty$$

$$\frac{T - T_\infty}{T_m - T_\infty} = e^{-(V_a/\alpha)x} \tag{4.38}$$

Applying the energy balance on the surface in order to determine the ablation velocity, V_a (assuming L as the heat of ablation of the material):

$$q_s'' - \rho L V_a = -k\frac{\partial T}{\partial x}\Big|_{x=0} = \rho C V_a(T_m - T_\infty) \tag{4.39}$$

$$V_a = \frac{q_s''}{\rho L + \rho C(T_m - T_\infty)} \tag{4.40}$$

The total heat conducted into the solid material evaluated with the temperature distribution, Equation (4.38), is

$$q_c'' = \rho C \int_0^\infty (T - T_\infty) dx = \frac{k(T_m - T_\infty)}{V_a} \tag{4.41}$$

The total heat transferred to the surface in time t is $q_s'' \cdot t$. Then for this period of time t, the fraction of the total heat transferred which was conducted into the solid material is obtained by substituting Equation (4.40) into Equation (4.41):

$$\frac{q_c''}{q_s''} = \frac{k(T_m - T_\infty)[\rho L + \rho C(T_m - T_\infty)]}{q_s'' \cdot q_s'' \cdot t} \tag{4.42}$$

Comparing Equations (4.40) and (4.42), a large magnitude of $[\rho L + \rho C(T_m - T_\infty)]$ is desirable to reduce the amount of material ablated, but a small magnitude is desirable to reduce the fraction of q_s'' which is conducted into the solid material. A compromise is necessary.

4.5 DUHAMEL'S THEOREM FOR TIME-DEPENDENT BOUNDARY CONDITION PROBLEMS

Duhamel's theorem can be applied for transient heat conduction problems with a time-dependent BC. The solution is based on that of the same transient problem with time-independent BCs that has been presented in the previous sections. Duhamel's theorem can be applied for 1-D, 2-D, and 3-D transient heat conduction problems with time-dependent BCs in rectangular and cylindrical coordinate systems. A few selective cases are discussed in the below sections. Hahn and Özişik provide additional details for the application Duhamel's theorem with various time-dependent BCs [3].

4.5.1 DUHAMEL'S THEOREM FOR THE 1-D PLANE WALL PROBLEM

Case 1: Consider a 1-D plane wall with thickness L. Initially the plane wall temperature is zero at $t = 0$. Suddenly, when $t > 0$, the right-side surface at $x = L$ is exposed to a time-dependent BC, $f(t)$, while the left-side surface at $x = 0$ keeps remains at the zero temperature.

$$\frac{\partial^2 T}{\partial x^2} = \frac{1}{\alpha} \frac{\partial T}{\partial t}$$

$$
\begin{aligned}
T(x,0) &= 0 &&\text{at} &&t = 0 \\
T(0, t) &= 0 &&\text{at} &&x = 0, t > 0 \\
T(L, t) &= f(t) &&\text{at} &&x = L, t > 0
\end{aligned}
$$

For this time-independent problem, $\phi(x,t)$,

Let $T(L,t) = \text{constant} = 1$ for example. The solution is

$$\phi(x,t) = \frac{x}{L} + \frac{2}{L}\sum_{n=1}^{\infty} e^{-\alpha\lambda_n^2 t} \cdot \frac{(-1)^n}{\lambda_n} \cdot \sin\lambda_n x$$

where $\lambda_n = \dfrac{n\pi}{L}$.

From Duhamel's theorem, the solution of the time-dependent BC is the solution of time-independent BC with $f(t)$ over a time interval.

$$T(x,t) = \int_{\tau=0}^{t} \phi(x, t-\tau) \cdot \frac{df(\tau)}{d\tau} d\tau$$

For the time-dependent BC:

$$\phi(x,t-\tau) = \frac{x}{L} + \frac{2}{L}\sum_{n=1}^{\infty} e^{-\alpha\lambda_n^2 (t-\tau)} \cdot \frac{(-1)^n}{\lambda_n} \cdot \sin\lambda_n x$$

Assume $f(t) = bt$, a linear increase with t, for $t = 0$ to $t = \tau_1$ and then $f(t) = 0$, returns to zero, for $t > \tau_1$. Thus

$$\frac{df(\tau)}{d\tau} = b \quad \text{for} \quad 0 < t < \tau_1$$

$$T(x,t) = \int_{\tau=0}^{t} \left[\frac{x}{L} + \frac{2}{L}\sum_{n=1}^{\infty} e^{-\alpha\lambda_n^2 (t-\tau)} \cdot \frac{(-1)^n}{\lambda_n} \cdot \sin\lambda_n x \right] \cdot b \cdot d\tau$$

$$= b\frac{x}{L}t + b\frac{2}{L}\sum_{n=1}^{\infty} \frac{(-1)^n}{\lambda_n^3}\left(1 - e^{-\alpha\lambda_n^2 t}\right) \cdot \sin\lambda_n x$$

for $t > \tau_1$, the BC has a discontinuity at time $t = \tau$, i.e., $\Delta f_1 = b\tau_1$

$$T(x,t) = \int_{\tau=0}^{\tau_1} \phi(x, t-\tau) \cdot \frac{df(\tau)}{d\tau} d\tau + \phi(x, t-\tau_1) \cdot \Delta f_1$$

$$= \int_{\tau=0}^{\tau_1} \phi(x, t-\tau) \cdot b \cdot d\tau - \phi(x, t-\tau) \cdot b\tau_1$$

The temperature distribution is

$$T(x,t) = \int_{\tau=0}^{\tau_1} \left[\frac{x}{L} + \frac{2}{L} \sum_{n=1}^{\infty} e^{-\alpha\lambda_n^2(t-\tau)} \cdot \frac{(-1)^n}{\lambda_n} \cdot \sin\lambda_n x \right] b \, d\tau$$

$$- b\tau_1 \left[\frac{x}{L} + \frac{2}{L} \sum_{n=1}^{\infty} e^{-\alpha\lambda_n^2(t-\tau_1)} \frac{(-1)^n}{\lambda_n} \cdot \sin\lambda_n x \right]$$

where the first term integral equals

$$b\frac{x}{L}\tau_1 + \frac{2b}{L}\sum_{n=1}^{\infty}\frac{(-1)^n}{\alpha\lambda_n^3} \cdot \left(1 - e^{-\alpha\lambda_n^2\tau_1}\right) \cdot \sin\lambda_n x$$

Case 2: Consider a time-dependent BC having continuous changes with time and many step changes with time.

$$T(x,t) = \int_{\tau=0}^{t} \phi(x,\, t-\tau)\frac{df(\tau)}{d\tau} d\tau + \sum_{n=1}^{N-1} \phi(x,\, t-\tau_n)\Delta f_n$$

For example, the plane wall left side surface at $x = 0$ is exposed to a time-dependent heat flux or convection boundary while the right side surface at $x = L$ is insulated. Initially, the temperature of the entire wall is zero.

$$\frac{\partial^2 T}{\partial x^2} = \frac{1}{\alpha}\frac{\partial T}{\partial t}$$

$$T(x,0) = 0 \qquad \text{for} \qquad t = 0$$

$$-k\frac{\partial T}{\partial x} = f(t) \qquad \text{at } x = 0, \qquad t > 0$$

$$\frac{\partial T}{\partial x} = 0 \qquad \text{at } x = L, \qquad t > 0$$

For the case of the time-independent BC problem:

$$\frac{\partial^2 \phi(x,t)}{\partial x^2} = \frac{1}{\alpha}\frac{\partial \phi(x,t)}{\partial t}$$

$$-k\frac{\partial \phi}{\partial x} = \text{constant} = 1 \text{ at } x = 0, t > 0$$

$$\frac{\partial \phi}{\partial x} = 0 \text{ at } x = L, t > 0$$

$$\phi(x,0) = 0 \text{ for } t = 0$$

The solution is

$$\phi(x,0) = \frac{\alpha t}{Lk} + \frac{2}{Lk} \sum_{n=1}^{\infty} \frac{\cos \lambda_n x}{\lambda_n^2} \left(1 - e^{-\alpha \lambda_n^2 t}\right)$$

where

$$\lambda_n = \frac{n\pi}{L}$$

Using Duhamel's theorem for $0 < t < \tau_1$, assume $f(\tau) = t$ for example,

$$T(x,t) = \int_{\tau=0}^{t} \phi(x, t-\tau) \frac{df(\tau)}{d\tau} d\tau$$

where

$$\frac{df(\tau)}{d\tau} = 1$$

The time-dependent solution $\phi(x, t-\tau)$ is obtained from the equation above by replacing t with $(t-\tau)$. Now we have:

$$T(x,t) = \int_{\tau=0}^{t} \frac{\alpha}{Lk}(t-\tau) + \frac{2}{Lk} \sum_{n=1}^{\infty} \frac{\cos \lambda_n x}{\lambda_n^2} \left(1 - e^{\alpha \lambda_n^2 (t-\tau)}\right) d\tau$$

$$= \frac{\alpha t^2}{2Lk} + \frac{2}{Lk} \sum \frac{\cos \lambda_n x}{\lambda_n^2} t - \frac{2}{Lk} \sum \frac{\cos \lambda_n x}{\alpha \lambda_n^4} \left(1 - e^{-\alpha \lambda_n^2 t}\right)$$

For $t > \tau_1$, the boundary surface $f(t)$ has only one discontinuity (one step change) at time $t = \tau_1$ and the resulting change in $f(t)$ is a decrease in the amount $\Delta f_1 = -\tau_1$. Then Duhamel's theorem reduces to

$$T(x,t) = \int_{\tau=0}^{t} \phi(x, t-\tau) \frac{df(\tau)}{d\tau} d\tau + \phi(x, t-\tau_1) \Delta f_1$$

Assume

$$\frac{df(\tau)}{d\tau} = 1, \quad \Delta f_1 = -\tau_1$$

and from the above time-independent solution of ϕ,

$$T(x,t) = \int\limits_{t=0}^{t_1} \left[\frac{\alpha}{Lk}(t-\tau) + \frac{2}{Lk}\sum_{n=1}^{\infty}\frac{\cos\lambda_n x}{\lambda_n^2} - \frac{2}{Lk}\sum_{n=1}^{\infty}\frac{\cos\lambda_n x}{\lambda_n^2}e^{-\alpha\lambda_n^2(t-\tau)} \right] d\tau$$

$$= \tau_1 \left[\frac{\alpha}{Lk}(t-\tau_1) + \frac{2}{Lk}\sum_{n=1}^{\infty}\frac{\cos\lambda_n x}{\lambda_n^2} - \frac{2}{Lk}\sum_{n=1}^{\infty}\frac{\cos\lambda_n x}{\lambda_n^2}e^{-\alpha\lambda_n^2(t-\tau_1)} \right]$$

Note that

$$f(\tau) = -k\frac{\partial T}{\partial x}\bigg|_{x=0} \quad \text{can be heat flux or convection,}$$

$$f(\tau) = q_s''(\tau), \qquad \text{at} \quad x = 0, \text{or}$$
$$f(\tau) = h(T_\infty - T) \qquad \text{at} \quad x = 0,$$

where either h or T_∞ changes with time.

4.5.2 DUHAMEL'S THEOREM FOR THE 1-D SEMIINFINITE SOLID PROBLEM

Case 1: Consider a 1-D semiinfinite solid that is initially at zero temperature. For time $t > 0$, the boundary surface at $x = 0$ is exposed to a temperature variation as $f(t)$. Determine the temperature distribution $T(x,t)$ at time $t > 0$. The 1-D transient heat conduction equation and BCs are:

$$\frac{\partial^2 T}{\partial x^2} = \frac{1}{\alpha}\frac{\partial T}{\partial t}$$

$$T(0,t) = f(t) \quad \text{at} \quad x = 0, \quad t > 0$$
$$T(\infty,t) = 0 \quad \text{at} \quad x = \infty, \quad t > 0$$
$$T(x,0) = 0 \quad \text{for} \quad t = 0$$

Consider the solution of the same problem with a time-independent BC, i.e.,

$$\frac{\partial^2 \phi}{\partial x^2} = \frac{1}{\alpha}\frac{\partial \phi}{\partial t}$$

$$\phi(0,t) = \text{constant} = 1 \quad \text{at} \quad x = 0, t > 0$$
$$\phi(\infty,t) = 0 \quad \text{at} \quad x = \infty, t > 0$$
$$\phi(x,0) = 0 \quad \text{for} \quad t = 0$$

From Duhamel's theorem, $T(x,t)$ with a time-dependent BC can be solved with $\phi(x,t)$ from the time-independent BC problem as the below equation.

$$T(x,t) = \int_{\tau=0}^{t} f(\tau) \frac{\partial \phi(x, t-\tau)}{\partial t} d\tau$$

The solution of the 1-D transient, semiinfinite solid with a constant surface BC can be found from Equation (4.20) using $T_i = 0$, $T_s = 1$, $T = \phi$:

$$\phi(x,t) = 1 - \text{erf}\left(\frac{x}{\sqrt{4\alpha t}}\right) = \text{erfc}\left(\frac{x}{\sqrt{4\alpha t}}\right) = 1 - \frac{2}{\sqrt{\pi}} \int_{0}^{\eta} e^{-\eta^2} d\eta$$

Thus,

$$\frac{\partial \phi(x, t-\tau)}{\partial t} = \frac{x}{\sqrt{4\alpha\pi} \cdot (t-\tau)^{3/2}} \cdot e^{-\left(\frac{x}{4\alpha(t-\tau)}\right)^2}$$

We now insert this into the above time-dependent solution to yield:

$$T(x,t) = \frac{x}{\sqrt{4\alpha\pi}} \int_{\tau=0}^{t} f(\tau) \frac{1}{(t-\tau)^{3/2}} e^{-\left(\frac{x}{4\alpha(t-\tau)}\right)^2} d\tau$$

Letting

$$\eta = \frac{x}{\sqrt{4\alpha(t-\tau)}}, \quad t - \tau = \frac{x^2}{4\alpha\eta}, \quad d\tau = \frac{2}{\eta}(t-\tau)d\eta$$

Therefore,

$$T(x,t) = \frac{2}{\sqrt{\pi}} \int_{\frac{x}{\sqrt{4\alpha t}}}^{\infty} f\left(t - \frac{x^2}{4\alpha\eta^2}\right) \cdot e^{-\eta^2} d\eta$$

If we consider the periodic surface temperature variation:

$$f(t) = T_s \cos(\omega t - c)$$

Then

$$T(x,t) = \frac{2}{\sqrt{\pi}} \int_{\frac{x}{\sqrt{4\alpha t}}}^{\infty} T_s \cos\left(\omega\left(t - \frac{x^2}{4\alpha\eta^2}\right) - c\right) \cdot e^{-\eta^2} d\eta$$

or

$$T(x,t) = \frac{2T_s}{\sqrt{\pi}} \int_0^\infty \cos\left(\omega\left(t - \frac{x^2}{4\alpha\eta^2}\right) - c\right) \cdot e^{-\eta^2} d\eta$$

$$-\frac{2T_s}{\sqrt{\pi}} \int_0^{\frac{x}{\sqrt{4\alpha t}}} \cos\left(\omega\left(t - \frac{x^2}{4\alpha\eta^2}\right) - c\right) \cdot e^{-\eta^2} d\eta$$

The second term on the right,

$$\int_0^{\frac{x}{\sqrt{4\alpha t}}} = \int_0^0 = 0 \qquad \text{if } t \to \infty$$

Therefore $T(x,t) =$ the first term on the right only

$$= T_s \cos\left(\omega t - x\left(\frac{\omega}{2\alpha}\right)^{1/2} - c\right) \cdot e^{-x\left(\frac{\omega}{2\alpha}\right)^{1/2}}$$

The temperature oscillates with $\cos(\omega t)$ at any location x.
    If consider surface temperature variation as

$$f(t) = T_s \sqrt{t}$$

Then

$$T(x,t) = \frac{2}{\sqrt{\pi}} \int_{\frac{x}{\sqrt{4\alpha t}}}^\infty T_s \sqrt{t} \cdot e^{-\eta^2} d\eta$$

Case 2: Now we will consider a time-dependent convective BC at $x = 0$ on the semi-infinite surface.

$$-k\frac{\partial T}{\partial x} = h(T_\infty - T) \qquad \text{at } x = 0, \quad t > 0$$

From Equation (4.28), the time-independent BC, $T(x,t)$, is

$$\frac{T(x,t)}{T_\infty - T_i} = \text{erfc}\left(\frac{x}{\sqrt{4\alpha t}}\right) - e^{\left(\frac{xh}{k} + \alpha\frac{h^2}{k^2}\right)} \cdot t \cdot \text{erfc}\left(\frac{x}{\sqrt{4\alpha t}} + \frac{h}{k}\sqrt{\alpha t}\right)$$

Consider $x = 0$ (the surface exposed to convection),

$$\frac{T(0,t)}{T_\infty - T_i} = 1 - e^{\left(\frac{h^2 \alpha t}{k^2}\right)} \cdot \text{erfc}\left(\frac{h}{k}\sqrt{\alpha t}\right)$$

If the convective fluid temperature increases with time, $\sim \sqrt{t}$, the time history of the $T_\infty(t) \sim \sqrt{t}$ can be simulated by a series of time steps, which are incorporated into the above equation using Duhamel's superposition theorem. The solution of the local surface temperature can be expressed as, at $x = 0$,

$$T(0,t) - T_i = \sum_{n=1}^{N}\left\{1 - e^{\frac{h^2 \alpha(t-t_n)}{k^2}} \cdot \text{erfc}\left(\frac{h\sqrt{\alpha(t-t_n)}}{k}\right)\right\} \cdot \left(T_{\infty,n} - T_{\infty,n-1}\right)$$

Note that the above comes from Duhamel's theorem, as mentioned before, for the case of N step changes,

$$T(x,t) = \sum \phi\left[x,(t - n\Delta t)\right] \cdot \Delta f_i$$

When t is in the time interval, $(n-1)\Delta t < t < n\Delta t$, this result can be written in a general form as

$$T(x,t) = \sum_{n=1}^{N} \phi(x, t - n\Delta t) \cdot \Delta f_i \cdot U(t - n\Delta t)$$

where

$$U(t - n\Delta t) = \text{the unit step function.}$$

Case 3: Now we will present the case of a heat flux BC which varies with time.

$$-k\frac{\partial T}{\partial x} = q_s''(t) \quad \text{at } x = 0, \quad t > 0$$

The time-independent BC, $q_1'' = $ constant, the $T(x,t)$ can be obtained from Equation (4.26).

Following Duhamel's superposition theorem, the time-dependent solution, $T(x,t)$, can be obtained, for a given $q_s''(t) \sim \sqrt{t}$ or others, using N step changes,

$$T(x,t) = \sum_{n=1}^{N} \phi(x, t - n\Delta t) \cdot \Delta f_i$$

4.5.3 DUHAMEL'S THEOREM FOR CYLINDRICAL COORDINATE SYSTEMS

The temperature distribution in the radial direction can be obtained by solving the energy equation in cylindrical coordinates:

$$\frac{\partial^2 T(r,t)}{\partial r^2} + \frac{1}{r}\frac{\partial T(r,t)}{\partial r} = \frac{1}{\alpha}\frac{\partial T(r,t)}{\partial t}$$

$$T(r_0,t) = f(t) \quad \text{at } r = r_0, \quad t > 0$$

$$T(r,0) = 0 \quad \text{for} \quad t = 0$$

For the time-independent solution

$$\frac{\partial^2 \phi(r,t)}{\partial r^2} + \frac{1}{r}\frac{\partial \phi(r,t)}{\partial r} = \frac{1}{\alpha}\frac{\partial \phi(r,t)}{\partial t}$$

$$\phi(r_0,t) = \text{constant} = 1 \quad \text{at } r = r_0, t > 0$$

Duhamel's theorem:

$$T(r,t) = \int_{\tau=0}^{t} f(\tau) \cdot \frac{\partial \phi(r,t,-\tau)}{\partial t} \cdot d\tau$$

From the time-independent solution,

$$\phi(r,t) = 1 - \frac{2}{r_0}\sum_{n=1}^{\infty}\frac{J_0(\lambda_n r)}{J_1(\lambda_n r_0)} \cdot e^{-\alpha\lambda_n^2 t}$$

where λ_n values are the positive roots of $J_0(\lambda_n r_0) = 0$.

Thus, the time-dependent solution can be written as

$$T(r,t) = \frac{2\alpha}{r_0}\sum_{n=1}^{\infty} e^{-\alpha\lambda_n^2 t} \cdot \lambda_n \cdot \frac{J_0(\lambda_n r)}{J_1(\lambda_n r_0)} \cdot \int_0^t e^{\alpha\lambda_n^2 t} \cdot f(\tau)d\tau$$

4.6 GREEN'S FUNCTION FOR TIME-DEPENDENT, NONHOMOGENEOUS PROBLEMS

The method using Green's function can be applied to solve transient heat conduction problems with time-dependent, nonhomogeneous BCs. The solutions are based on the solutions of the same problem with time-independent, nonhomogeneous BCs that have been shown in the previous sections. The Green's function method can be applied for 1-D, 2-D, and 3-D transient heat conduction problems in rectangular and cylindrical coordinate systems. A few selected cases are outlined in the below sections. The text from Hahn and Özişik provides additional details for the application of Green's function to transient heat conduction problems [3].

4.6.1   GREEN'S FUNCTION IN A RECTANGULAR COORDINATE SYSTEM

Consider 1-D transient heat conduction in a plane wall of thickness L with time-dependent heat generation and a time-dependent BC.

$$\frac{\partial^2 T}{\partial x^2} + \frac{1}{k}\dot{q}(x,t) = \frac{1}{\alpha}\frac{\partial T}{\partial t}$$

$$
\begin{aligned}
T(x,0) &= F(x) & \text{for} && t &= 0 \\
T(0,t) &= f_1(t) & \text{at } x = 0, && t &> 0 \\
T(L,t) &= f_2(t) & \text{at } x = L, && t &> 0
\end{aligned}
$$

For the case of the homogeneous problem

$$\frac{\partial^2 \psi}{\partial x^2} = \frac{1}{\alpha}\frac{\partial \psi}{\partial t}$$

$$
\begin{aligned}
\psi(x,0) &= F(x) & \text{for} && t &= 0 \\
\psi &= 0 & \text{at } x = 0, && t &> 0 \\
\psi &= 0 & \text{at } x = L, && t &> 0
\end{aligned}
$$

The solution is obtained as

$$\psi(x,t) = \int_{x'=0}^{L}\left[\frac{2}{L}\sum_{n=1}^{\infty}\sin\lambda_n x \cdot \sin\lambda_n x' \cdot e^{-\alpha\lambda_n^2 t}\right]\cdot F(x')dx'$$

where $\lambda_n = \dfrac{n\pi}{L}$, $n = 1,2,3,\ldots$

The nonhomogeneous solution can be related to Green's function as

$$\psi(x,t) = \int_{x'=0}^{x} G(x,t|x',\tau)\Big|_{\tau=0}\cdot F(x')dx'$$

Thus

$$G(x,t|x',\tau)\Big|_{\tau=0} = \frac{2}{L}\sum_{n=1}^{\infty}\sin\lambda_n x \cdot \sin\lambda_n x' \cdot e^{-\alpha\lambda_n^2 t}$$

The desired Green's function is obtained by replacing t by $t - \tau$ in the above equation,

$$G(x,t \mid x', \tau) = \frac{2}{L} \sum_{n=1}^{\infty} \sin \lambda_n x \cdot \sin \lambda_n x' \cdot e^{-\alpha \lambda_n^2 (t-\tau)}$$

The Green's function, $G(x,t \mid x', \tau)$, for the 1-D transient problem represents the temperature at the location x, at time t, due to an instantaneous point source of unit strength, located at the point x', releasing its energy spontaneously at time $t = \tau$. The solution for $T(x,t)$ is expressed in terms of the 1-D Green's function $G(x, t \mid x', \tau)$ as

$$T(x, t) = \int_{x'=0}^{L} G(x, t \mid x', \tau) \Big|_{\tau=0} \cdot F(x')dx' + \frac{\alpha}{k} \int_{\tau=0}^{t} d\tau \cdot \int_{x'=0}^{L} G(x, t \mid x', \tau) \cdot \dot{q}(x',\tau) \cdot dx'$$

$$+ \alpha \int_{\tau=0}^{t} \frac{\partial G(x, t \mid x', \tau)}{\partial x'} \Big|_{x'=0} \cdot f_1(\tau)d\tau - \alpha \int_{\tau=0}^{t} \frac{\partial G(x, t \mid x', \tau)}{\partial x'} \Big|_{x'=L} \cdot f_2(\tau)d\tau$$

The right-hand-side, 1st term is for the homogeneous problem, $\dot{q}(x',t) = 0$, the 2nd term is for the nonhomogeneous problem, $\dot{q}(x',t)$ is given, the 3rd and 4th terms are for time-dependent BCs. Thus, the solution in terms of Green's function:

$$T(x, t) = \frac{2}{L} \sum_{n=1}^{\infty} e^{-\alpha \lambda_n^2 t} \cdot \sin \lambda_n x \cdot \int_{x'=0}^{L} \sin \lambda_n x' \cdot F(x')dx' +$$

$$\frac{\alpha}{k} \frac{2}{L} \sum_{n=1}^{\infty} e^{-\alpha \lambda_n^2 t} \cdot \sin \lambda_n x \cdot \int_{\tau=0}^{t} e^{\alpha \lambda_n^2 t} d\tau \cdot \int_{x'=0}^{L} \sin \lambda_n x' \cdot \dot{q}(x',\tau)dx' +$$

$$\alpha \frac{2}{L} \sum_{n=1}^{\infty} e^{-\alpha \lambda_n^2 t} \cdot \lambda_n \cdot \sin \lambda_n x \int_{\tau=0}^{t} e^{\alpha \lambda_n^2 \tau} \cdot f_1(\tau)d\tau$$

$$-\alpha \frac{2}{L} \sum_{n=1}^{\infty} (-1)^n e^{-\alpha \lambda_n^2 t} \cdot \lambda_n \cdot \sin \lambda_n x \int_{\tau=0}^{t} e^{\alpha \lambda_n^2 \tau} \cdot f_2(\tau)d\tau$$

Special Case 1: If $F(x) = 0$, $f_1(t) = 0$, $f_2(t) = 0$, and $\dot{q}(x,t) = \dot{q} = $ constant.

$$T(x,t) = \frac{2\alpha}{kL} \sum_{n=1}^{\infty} e^{-\alpha \lambda_n^2 t} \cdot \sin \lambda_n x \cdot \int_{x'=0}^{L} \sin \lambda_n x' \cdot \dot{q} \cdot dx'$$

Special Case 2: If $F(x) = f_1(t) = f_2(t) = 0$, and $\dot{q}(x,t) = \dot{q}(t) \cdot \delta(x - a)$ for $t > 0$, a plane surface heat source of strength $\dot{q}(t)$ at $x = a$ releases its heat continuously,

$$T(x,t) = \frac{2\alpha}{kL} \sum_{n=1}^{\infty} e^{-\alpha\lambda_n^2 t} \cdot \sin \lambda_n x \cdot \sin \lambda_n a \cdot \int_{\tau=0}^{t} e^{\alpha\lambda_n^2 \tau} \cdot \dot{q}(\tau) \cdot d\tau$$

Special Case 3: If $F(x) = f_1(t) = f_2(t) = 0$, and $\dot{q}(x,t) = \dot{q}(x) \cdot \delta(t-a)$ for $t > 0$, a distributed heat source of strength $\dot{q}(x)$ releases its heat spontaneously at time $t = 0$,

$$T(x,t) = \frac{2}{L} \sum_{n=1}^{\infty} e^{-\alpha\lambda_n^2 t} \cdot \sin \lambda_n x \cdot \int_{x'=0}^{L} \left[\frac{\alpha}{k} \dot{q}(x') \right] \sin \lambda_n x' \cdot dx'$$

where $\frac{\alpha}{k} \dot{q}(x') = F(x)$. This implies that an instantaneous volume source of strength $\frac{\alpha}{k} \dot{q}(x)$ releasing its heat spontaneously at time $t = 0$ is equivalent to an initial temperature distribution $F(x)$.

4.6.2 Green's Function in a Cylindrical Coordinate System

Consider 1-D transient heat conduction in a cylinder of radius r_0, with time-dependent heat generation and a time-dependent BC.

$$\frac{\partial^2 T}{\partial r^2} + \frac{1}{r}\frac{\partial T}{\partial r} + \frac{1}{k}\dot{q}(r,t) = \frac{1}{\alpha}\frac{\partial T}{\partial t}$$

$$T(r,0) = F(r) \quad \text{for } t = 0$$
$$T(r_0, t) = f(t) \quad \text{at } r = r_0, \quad t > 0$$

For the homogeneous case,

$$\frac{\partial^2 \psi}{\partial r^2} + \frac{1}{r}\frac{\partial \psi}{\partial r} = \frac{1}{\alpha}\frac{\partial \psi}{\partial t}$$

$$\psi(r,0) = F(r) \quad \text{for } t = 0$$
$$\psi(r_0, t) = f(t) \quad \text{at } r = r_0, \quad t > 0$$

The solution is

$$\psi(r,t) = \int_{r'=0}^{r_0} r' \left[\frac{2}{r_0^2} \sum_{n=1}^{\infty} e^{-\alpha\lambda_n^2 t} \cdot \frac{J_0(\lambda_n r)}{J_1^2(\lambda_n r_0)} \cdot J_0(\lambda_n r') \right] \cdot F(r') dr'$$

where the λ_n values are positive roots of $J_0(\lambda_n r_0) = 0$.

The solution in terms of Green's function is

$$\psi(r,t) = \int_{r'=0}^{r_0} r' G(r,t|r',\tau)\Big|_{\tau=0} \cdot F(r')dr'$$

where

$$G(r,t|r',\tau)\Big|_{\tau=0} = \frac{2}{r_0^2} \sum_{n=1}^{\infty} e^{-\alpha\lambda_n^2 t} \cdot \frac{J_0(\lambda_n r)}{J_1^2(\lambda_n r_0)} \cdot J_0(\lambda_n r')$$

The desired Green's function is obtained by replacing t by $t - \tau$,

$$G(r,t|r',\tau)\Big|_{\tau=0} = \frac{2}{r_0^2} \sum_{n=1}^{\infty} e^{-\alpha\lambda_n^2(t-\tau)} \cdot \frac{J_0(\lambda_n r)}{J_1^2(\lambda_n r_0)} \cdot J_0(\lambda_n r')$$

Thus, the nonhomogeneous solution $T(r,t)$ based on Green's function:

$$T(r,t) = \int_{r'=0}^{r_0} r' G(r,t|r',\tau)\Big|_{\tau=0} \cdot F(r')dr' + \frac{\alpha}{k} \int_{\tau=0}^{\tau} d\tau \cdot \int_{r'=0}^{r_0} r' G(r,t|r',\tau) \cdot \dot{q}(r',\tau)dr'$$

$$-\alpha \int_{\tau=0}^{t} \left[r' \frac{\partial G}{\partial r'} \right]_{r'=r_0} \cdot f(\tau) \cdot d\tau$$

where

$$\left[r' \frac{\partial G}{\partial r'} \right]_{r'=r_0} = -\frac{2}{r_0} \sum_{n=1}^{\infty} e^{-\alpha\lambda_n^2(t-\tau)} \cdot \lambda_n \cdot \frac{J_0(\lambda_n r)}{J_1(\lambda_n r_0)}$$

Thus, the solution in terms of Green's function:

$$T(x,t) = \frac{2}{r_0} \sum_{n=1}^{\infty} e^{-\alpha\lambda_n^2 t} \frac{J_0(\lambda_n r)}{J_1^2(\lambda_n r_0)} \int_{r'=0}^{r_0} r' J_0(\lambda_n r') \cdot F(r')dr'$$

$$+ \frac{2\alpha}{kr_0^2} \sum_{n=1}^{\infty} e^{-\alpha\lambda_n^2 t} \frac{J_0(\lambda_n r)}{J_1^2(\lambda_n r_0)} \cdot \int_{\tau=0}^{t} e^{-\alpha\lambda_n^2 t} d\tau + \int_{r'=0}^{r_0} r' J_0(\lambda_n r') \cdot \dot{q}(r',\tau)dr'$$

$$+ \frac{2\alpha}{r_0} \sum_{n=1}^{\infty} e^{-\alpha\lambda_n^2 t} \cdot \lambda_n \cdot \frac{J_0(\lambda_n r)}{J_1^2(\lambda_n r_0)} \cdot \int_{\tau=0}^{t} e^{\alpha\lambda_n^2 \tau} \cdot f(\tau) \cdot d\tau$$

Special Case 1: If $F(r) = 0$, $f(r_0, t) = 0$, and $\dot{q}(r, t) = \dot{q} =$ constant.

$$T(r,t) = \frac{2\dot{q}}{kr_0} \sum_{n=1}^{\infty} \frac{J_0(\lambda_n r)}{\lambda_n^3 J_1^2(\lambda_n r_1)} - \frac{2\dot{q}}{kr_0} \sum_{n=1}^{\infty} e^{-\alpha\lambda_n^2 t} \cdot \frac{J_0(\lambda_n r)}{\lambda_n^3 J_1^2(\lambda_n r_0)}$$

If $t \to \infty$, steady state, $e^{-\alpha\lambda_n^2 t} = 0$, thus

$$T(r,\infty) = \frac{2\dot{q}}{kr_0} \sum_{n=1}^{\infty} \frac{J_0(\lambda_n r)}{\lambda_n^3 J_1(\lambda_n r_1)} = \frac{\dot{q}(r_0^2 - r^2)}{4k}$$

therefore, the solution is

$$T(r,t) = \frac{\dot{q}(r_0^2 - r^2)}{4k} - \frac{2\dot{q}}{kr_0} \sum_{n=1}^{\infty} e^{-\alpha\lambda_n^2 t} \cdot \frac{J_0(\lambda_n r)}{\lambda_n^3 J_1(\lambda_n r_0)}$$

Special Case 2: If $F(r) = 0$, $f(r_0, t) = 0$, and $\dot{q}(r, \tau) = \dot{q}(\tau)\frac{1}{2\pi r'} \cdot \delta(r' = 0)$,

$$T(r,t) = \frac{\alpha}{k\pi r_0^2} \sum_{n=1}^{\infty} e^{-\alpha\lambda_n^2 t} \cdot \frac{J_0(\lambda_n r)}{J_1^2(\lambda_n r_0)} \cdot \int_{\tau=0}^{t} e^{-\alpha\lambda_n^2 \tau} \dot{q}(\tau) d\tau$$

Special Case 3: If $F(r) = 0$, $f(r_0, t) = 0$, and $\dot{q}(r, \tau) = \dot{q}(r')\delta(\tau - 0)$,

$$T(r,t) = \frac{2}{r_0^2} \sum_{n=1}^{\infty} e^{-\alpha\lambda_n^2 t} \cdot \frac{J_0(\lambda_n r)}{J_1^2(\lambda_n r_0)} \cdot \int_{r'=0}^{r_0} r' J_0(\lambda_n r') \cdot \frac{\alpha\dot{q}(r')}{k} dr'$$

where $\frac{\alpha}{k}\dot{q}(r') = F(r)$, this implies an instantaneous volume heat source of strength $\frac{\alpha}{k}\dot{q}(r')$ releasing its heat spontaneously at time $t = 0$ is equivalent to an initial temperature distribution $F(r)$.

Examples

4.1 Solve transient temperature profiles of a convectively cooled cylinder, as shown in Figure 4.1b, by separation of variables.

Solution

Let $\theta = \dfrac{T - T_\infty}{T_i - T_\infty}$

$$\frac{1}{r}\frac{\partial}{\partial r}\left(r\frac{\partial \theta}{\partial r}\right) = \frac{1}{\alpha}\frac{\partial \theta}{\partial t} \quad \text{or} \quad \frac{\partial^2 \theta}{\partial r^2} + \frac{1}{r}\frac{\partial \theta}{\partial r} = \frac{1}{\alpha}\frac{\partial \theta}{\partial t}$$

BCs:

i. $r = 0,\quad \frac{\partial \theta}{\partial r}\Big|_{r=0} = 0$

ii. $r = r_o,\quad -k\frac{\partial \theta}{\partial r}\Big|_{r=r_o} = h\theta_{r_O}$

Separation of variables:

$$\theta = R(r)\tau(t)$$

$$\frac{\partial \tau}{\partial t} + \lambda^2 \tau = 0;\quad \tau = C_3 e^{-\alpha\lambda^2 t}$$

$$\frac{\partial^2 R}{\partial r^2} + \frac{1}{r}\frac{\partial R}{\partial r} + \lambda^2 R = 0$$

$$R(r) = C_1 J_0(\lambda r) + C_2 Y_0(\lambda r)$$

Applying BCs

$$\frac{\partial R}{\partial r}\Big|_{r=0} = -C_1\lambda J_1(0) + C_2\lambda Y_1(0) = 0,\quad J_1(0) = 0,\quad \Rightarrow C_2 = 0$$

$$-k\frac{\partial R}{\partial r}\Big|_{r=r_O} = kC_1\lambda J_1(\lambda r_O) = h\theta_R,\quad \lambda J_1(\lambda r_O) = \frac{h}{k}J_0(\lambda r_O),\quad \lambda_n = \lambda r_O$$

$$\lambda_n J_1(\lambda_n) - Bi J_0(\lambda_n) = 0$$

where $Bi = (hr_o/k)$,

$$\theta = \sum_{n=1}^{\infty} C_n e^{-(\alpha/r_O^2)\lambda_n^2 t} J_0\left(\lambda_n \frac{r}{r_O}\right)$$

$$\text{at } t = 0,\quad \theta = 1$$

$$1 = \sum_{n=1}^{\infty} C_n J_0\left(\lambda_n \frac{r}{r_O}\right)$$

$$C_n = \frac{\displaystyle\int_0^{r_O} rJ_0(\lambda_n(r/r_O))\,dr}{\displaystyle\int_0^{r_O} rJ_0^2(\lambda_n(r/r_O))\,dr} = \frac{(r_O^2/\lambda_n)J_1(\lambda_n)}{(r_O^2/2)[J_0^2(\lambda_n) + J_1^2(\lambda_n)]}$$

$$\theta = \sum_{n=1}^{\infty} \frac{2J_1(\lambda_n)J_0(\lambda_n(r/r_O))}{\lambda_n[J_0^2(\lambda_n) + J_1^2(\lambda_n)]}\cdot e^{-\lambda_n^2 F_O}$$

where $F_0 = \left(\alpha t / r_o^2 \right)$

4.2 Solve transient temperature profiles of a convectively cooled sphere, as shown in Figure 4.1c, by separation of variables.

Solution

Let $\theta = \dfrac{T - T_\infty}{T_i - T_\infty}$

The 1-D transient conduction equation is:

$$\frac{1}{r^2} \frac{\partial}{\partial r} \left(r^2 \frac{\partial \theta}{\partial r} \right) = \frac{1}{\alpha} \frac{\partial \theta}{\partial t}$$

This equation can be re-written as:

$$\frac{1}{r} \frac{\partial^2}{\partial r^2} (r\theta) = \frac{1}{\alpha} \frac{\partial \theta}{\partial t}$$

BCs:

    i. $r = r_o - k \dfrac{\partial \theta}{\partial r} = h \theta_{r_o}$

    ii. $t = 0 \ \theta = 1$

Let $U(r, t) = r\theta(r, t)$,

$$\frac{\partial^2 U}{\partial r^2} = \frac{1}{\alpha} \frac{\partial U}{\partial t}$$

$$\text{BCs} \left\{ \begin{array}{ll} U = 0 & \text{at } r = 0 \\[2ex] \dfrac{\partial U}{\partial r} + \left(\dfrac{h}{k} - \dfrac{1}{r_o} \right) U = 0 & \text{at } r = r_o \end{array} \right.$$

$$U = r \quad \text{for } t = 0$$

$$U = R(r)\tau(t)$$

$$\frac{\partial \tau}{\partial t} + \alpha \lambda^2 \tau = 0; \quad \Rightarrow \tau(t) = C_3 e^{-\alpha \lambda^2 t}$$

$$\frac{\partial^2 R}{\partial r^2} + \lambda^2 R(r) = 0$$

$$R(r) = C_1 \sin(\lambda r) + C_2 \cos(\lambda r)$$

$$\text{at } r = 0, \quad U = 0, \Rightarrow C_2 = 0$$

$$C_1\lambda\cos(\lambda r_o)+\left(\frac{h}{k}-\frac{1}{r_o}\right)C_1\sin(\lambda r_o)=0$$

$$\lambda_n = \lambda r_o$$

$$\lambda_n \cos(\lambda_n)+(B_i-1)\sin(\lambda_n)=0$$

where $B_i = (hr_o/k)$,

$$U = \sum_{n=1}^{\infty} C_n \sin\left(\lambda_n \frac{r}{r_o}\right) e^{-\lambda_n^2 F_0}$$

where $F_0 = \left(\alpha t/r_o^2\right)$

At $t = 0$, $\theta = 1$, $U = r$,

$$C_n = \frac{\displaystyle\int_0^{r_o} r\sin(\lambda_n(r/r_o))\,dr}{\displaystyle\int_0^{r_o} \sin^2(\lambda_n(r/r_o))\,dr} \cdot \frac{\lambda_n/r_o}{\lambda_n/r_o}$$

$$C_n = \frac{2(\sin(\lambda_n)-\lambda_n\cos(\lambda_n))}{\lambda_n - \sin(\lambda_n)\cos(\lambda_n)}$$

$$\theta = \sum_{n=1}^{\infty} \frac{2[\sin(\lambda_n)-\lambda_n\cos(\lambda_n)]}{\lambda_n - \sin(\lambda_n)\cos(\lambda_n)} \cdot \frac{\sin\left(\lambda_n\left(\dfrac{r}{r_o}\right)\right)}{\lambda_n\left(\dfrac{r}{r_o}\right)} \cdot e^{-\lambda_n^2 F_0}$$

$$= \frac{U}{r}$$

4.3 Solve transient temperature profiles in a semiinfinite solid body using the Laplace transform for the following BCs of *constant surface heat flux*. If at time $t = 0$ the surface is suddenly exposed to a constant heat flux q_s''—for example, by radiation from a high-temperature source.

Solution

1-D transient:

$$\frac{\partial^2 T}{\partial x^2} = \frac{1}{\alpha}\frac{\partial T}{\partial t}$$

Initial condition:

$t = 0; \quad T = T_i$

BCs:

  i. $x = 0; -k\dfrac{\partial T}{\partial x} = q_s''$

  ii. $x \to \infty; T = T_i$

  Let $\theta = T - T_i$,

$$\frac{\partial^2 \theta}{\partial x^2} = \frac{1}{\alpha}\frac{\partial \theta}{\partial t}$$

Initial condition:

$$t = 0; \quad \theta = 0$$

BCs:

  i. $x = 0; -k\dfrac{\partial \theta}{\partial x} = q_s''$

  ii. $x \to \infty; \theta = 0$

Applying the Laplace transform,

$$\frac{\partial^2 \tilde{\theta}}{\partial x^2} - \frac{s}{\alpha}\tilde{\theta} = 0 \tag{1}$$

BCs:

  i. $x = 0; -k\dfrac{\partial \tilde{\theta}}{\partial x} = \dfrac{q_s''}{s}$

  ii. $x \to \infty; \tilde{\theta} = 0$

Solving,

$$\tilde{\theta} = C_1 e^{-x\sqrt{s/\alpha}} + C_2 e^{x\sqrt{s/\alpha}} \tag{2}$$

Applying BC (ii), $C_2 = 0$.

  Applying BC (i), $C_1 = \left(q_s''/s\right)\left(1/k\sqrt{s/\alpha}\right)$

  Substituting this into above (2),

$$\tilde{\theta} = \frac{q_s''}{s}\frac{1}{k\sqrt{s/\alpha}}e^{-x\sqrt{s/\alpha}} \tag{3}$$

Rearranging,

$$\tilde{\theta} = \frac{q''/_{k\sqrt{\alpha}}}{s^{3/2}} e^{-(x/\sqrt{\alpha})\sqrt{s}}$$

Applying the Laplace inverse,

$$\theta = \frac{q_s''}{k}\sqrt{\alpha}\left[2\sqrt{\frac{t}{\pi}}\exp\left(-\frac{x^2}{4\alpha t}\right) - \frac{x}{\sqrt{\alpha}}\,\mathrm{erfc}\left(\frac{x}{2\sqrt{\alpha t}}\right)\right]$$

$$T - T_i = \frac{q_s''}{k}\left[\left(\frac{4\alpha t}{\pi}\right)^{1/2}\exp\left(-\frac{x^2}{4\alpha t}\right) - x\,\mathrm{erfc}\,\frac{x}{(4\alpha t)^{1/2}}\right]$$

4.4 Solve transient temperature profiles in a semiinfinite solid body using the Laplace transform for the following BCs of *convective heat transfer to the surface.* If at time $t = 0$ the surface is suddenly exposed to a fluid at temperature T_∞, with a convective heat transfer coefficient h.

Solution

1-D transient

$$\frac{\partial^2 T}{\partial x^2} = \frac{1}{\alpha}\frac{\partial T}{\partial t}$$

Initial condition:

$$t = 0; \quad T = T_i$$

BCs:

i. $x = 0; -k\dfrac{\partial\theta}{\partial x} = h(\theta_\infty - \theta_0)$

ii. $x \rightarrow \infty; T = T_i$

Let $\theta = T - T_i$,

$$\frac{\partial^2\theta}{\partial x^2} = \frac{1}{\alpha}\frac{\partial\theta}{\partial t}$$

Initial condition:

$$t = 0; \quad \theta = 0$$

BCs:

i. $x = 0; -k\dfrac{\partial\theta}{\partial x} = h(\theta_\infty - \theta_0)$

ii. $x \rightarrow \infty; \theta = 0$

Applying the Laplace transform,

$$\frac{\partial^2 \tilde{\theta}}{\partial x^2} - \frac{s}{\alpha}\tilde{\theta} = 0 \tag{1}$$

BCs:

i. $x = 0$; $-k\dfrac{d\tilde{\theta}(0,t)}{dx} = h\left(\dfrac{\theta_\infty}{s} - \tilde{\theta}(0,t)\right)$

ii. $x \to \infty$; $\tilde{\theta} = 0$

By solving,

$$\tilde{\theta} = C_1 e^{-x\sqrt{s/\alpha}} + C_2 e^{x\sqrt{s/\alpha}} \tag{2}$$

Applying BC (ii), $C_2 = 0$.
  Applying BC (i),

$$C_1 = \frac{(h/k)\theta_\infty}{\left((h/k) + \left(\sqrt{s/\alpha}\right)\right)s}$$

Substituting this into above (2),

$$\tilde{\theta} = \frac{(h/k)\theta_\infty}{\left((h/k) + \left(\sqrt{s/\alpha}\right)\right)s} e^{-x\sqrt{s/\alpha}} \tag{3}$$

By re-arranging,

$$\tilde{\theta} = \theta_\infty \frac{(h/k)\sqrt{\alpha}}{\left((h/k)\sqrt{\alpha} + \sqrt{s}\right)s} e^{-\left(x/\sqrt{\alpha}\right)\sqrt{s}}$$

Applying the Laplace inverse,

$$\frac{\theta}{\theta_\infty} = -e^{(h/k)x}e^{(h^2\alpha/k^2)t}erfc\left(\frac{h}{k}\sqrt{\alpha t} + \frac{x}{\sqrt{4\alpha t}}\right) + erfc\left(\frac{x}{\sqrt{4\alpha t}}\right)$$

$$\frac{T - T_i}{T_\infty - T_i} = erfc\frac{x}{(4\alpha t)^{1/2}} - e^{hx/k + (h/k)^2\alpha t}erfc\left(\frac{x}{(4\alpha t)^{1/2}} + \frac{h}{k}(\alpha t)^{1/2}\right)$$

REMARKS

There are many engineering problems involving 0-D, 1-D, 2-D, or 3-D transient heat conduction with various thermal BCs. In the undergraduate-level heat transfer, we have focused primarily on how to apply the lumped capacitance solutions to solve relatively simple engineering problems. For 1-D and multidimensional transient conduction problems, we normally do not go through the detailed mathematical equations and solutions. Instead, students are expected to apply these formulas to solve many engineering problems given the solid material geometry with appropriate thermal properties and thermal BCs.

In the intermediate-level heat transfer, with Chapter 4, we have introduced several very powerful mathematical tools such as similarity method, Laplace transform method, and integral approximate method, in addition to the separation of variables method already mentioned in Chapter 3. Specifically, the separation of variables method is convenient to solve the transient conduction problems with finite-length dimensions such as the plate, cylinder, and sphere with various thermal BCs. However, the similarity method, Laplace transform method, or integral approximate method are more appropriate to solve the transient conduction problems with semi-infinite solid material for various thermal BCs.

1-D transient heat conduction with moving boundaries belongs to advanced heat conduction material. Duhamel's theorem can be applied for transient heat conduction problems with a time-dependent BC. The method using Green's function can be applied to solve transient heat conduction problems with time-dependent, nonhomogeneous BCs. Both Duhamel's theorem and Green's function method belong to advanced heat conduction material.

PROBLEMS

4.1 The wall of a rocket nozzle is of thickness $L = 25\,\text{mm}$ and is made from a high-alloy steel for which $\rho = 8000\,\text{kg/m}^3$, $c = 500\,\text{J/kg K}$, and $k = 25$ W/m K. During a test firing, the wall is initially at $T_i = 25°C$ and its inner surface is exposed to hot combustion gases for which $h = 500$ W/m^{2}K and $T_\infty = 1750°C$. The firing time is limited by the nozzle inner-wall temperature when it reaches 1500°C. The outer surface is well insulated.

   a. Write the transient heat conduction equation, the associated BCs, and determine the nozzle wall temperature distributions. (*Note*: The diameter of the nozzle is much larger than its thickness. No need to perform integration of orthogonal functions.)

   b. Sketch several nozzle wall temperature profiles during transient heating.

   c. To increase the firing time, changing the wall thickness L is considered. Should L be increased or decreased? Why? The value of the firing time could also be increased by selecting a wall material with different thermophysical properties. Should materials of larger or smaller values of ρ, c, and k be chosen?

4.2 One side of a metal plane wall is insulated. The wall has a thickness of L and is initially at temperature T_i when suddenly the other side is heated by forced convection from water at temperature T_∞ with a convection heat transfer coefficient h.

   a. Outline, step by step, the procedures and the associated initial and BCs that may be used to solve the temperature distributions in the plane wall. You do not need to solve the transient temperature distribution.

    b. Sketch the temperature profiles in the plane wall during the heating process. Also, estimate the surface temperature at the final steady-state condition.

4.3 A long metal plane wall with a thickness of 2L is initially at temperature T_i and suddenly both sides are heated by convection fluid flow at temperature T_∞ with a convection heat transfer coefficient h. Outline the procedures that may be used to solve the temperature distributions in the plane wall and sketch the temperature profiles in the plane wall during the heating process for two different cases.

    a. Fluid flow is natural convection air.

    b. Fluid flow is forced convection water.

    c. Also, estimate the surface temperature at the final steady-state condition. Which fluid flow will reach the steady temperature faster? Why? Make appropriate assumptions in order to justify your answers.

4.4 A large flat plate (with a thickness of 2L) initially at T_i is suddenly plunged into a liquid bath at T_∞. Derive an expression for the instantaneous temperature distribution in the plate, if:

    a. The heat transfer coefficient between the two surfaces of the plate and the liquid, h, is given as constant and finite.

    b. The heat transfer coefficient h between the two surfaces of the plate and the liquid is very large so that the temperatures on the two surfaces of the plate may be assumed to change abruptly to the temperature of the liquid (i.e., $T(L, t) = T_\infty$, for $t > 0$).

    c. Sketch the instantaneous temperature distribution in the plate for both (a) and (b) if $T_\infty > T_i$.

4.5 A semiinfinite solid is initially at a uniform temperature T_i and suddenly exposed at its surface to a constant heat flux q''.

    a. Determine the temperature history in the solid.

    b. Approximately plot the temperature profiles for the following BCs:

       i. Constant q'' at the surface.

      ii. Constant temperature at the surface, $T_s(0,t) > T_i$.

     iii. Constant fluid temperature T_f and h, $T_f > T_i$.

4.6 A semiinfinite carbon steel block is initially at a uniform temperature T_i and its surfaces are suddenly exposed simultaneously to a constant irradiation flux q'' and ambient air of h, T_∞.

    a. Determine the temperature history in the solid and sketch the temperature profiles in the solid assuming $T_i = T_\infty$. Analytical method.

    b. If a semiinfinite pure copper block will be operated at the same condition, state that the time required for the surface of the block to reach T_s will be longer or shorter than that of the steel block. Why?

4.7 A liquid confined in a half-space $x > 0$ is initially at a temperature T_i which is higher than its freezing temperature T_m. For times $t > 0$, the surface at $x = 0$ is subjected to the following BCs. Plot the temperature history both in the liquid and the solid.

    a. Constant heat flux is q''; this q'' is removed away from the surface.

    b. Constant surface temperature T_o, $T_o < T_m$.

    c. If natural convection takes place in the liquid region with a constant h, plot the temperature profiles and freezing distance δ with time in both case (a) and case (b). Explain the difference.

4.8   A solid simulated as a half-space, $x > 0$, is initially at a temperature T_i which is equal to its melting temperature T_m. For time $t > 0$, the surface at $x = 0$ is subjected to the following BCs. Plot the temperature history both in liquid and solid:

    a. Constant heat flux is q''; this q'' is applied to the surface.

    b. Constant surface temperature T_s, $T_s > T_m$.

    c. Convection to the surface with heating fluid at T_∞, h.

4.9 Refer to Equation (4.14) and solve the temperature distributions for the 3-D block with the following BCs:

$$x = \pm a, T = T_s$$

a. $y = \pm b, T = T_s$

$$z = \pm c, T = T_s$$

b.
$$
\begin{aligned}
x &= \pm a, & \left(-K \partial T(\pm a,t)/\partial x\right) &= h\left[T(\pm a,t) - T_\infty\right] \\
y &= \pm b, & \left(-K \partial T(\pm b,t)/\partial y\right) &= h\left[T(\pm b,t) - T_\infty\right] \\
z &= \pm c, & \left(-K \partial T(\pm c,t)/\partial z\right) &= h\left[T(\pm c,t) - T_\infty\right]
\end{aligned}
$$

4.10 A semiinfinite solid is initially at a uniform temperature T_s. The surface of the semiinfinite solid is suddenly exposed to a constant temperature T_w.

    a. Write the governing equations and the BCs.

    b. Nondimensionalize the governing equations and BCs by appropriate choice of temperature variable, length scale, and timescale.

4.11 An infinite body of cold liquid initially at uniform temperature T_s is brought in contact with a heated horizontal wall of infinite length maintained at a constant temperature (T_w). It is expected that after infinite time the liquid temperature profile will be linear within a thermal boundary layer of thickness δ. Neglect gravity or body forces and liquid convection and assume that heat transfer in the liquid is by conduction only.

    a. Write the governing equations and the BCs.

    b. Nondimensionalize the governing equations and BCs by appropriate choice of temperature variable, length scale, and timescale.

    c. Solve the governing equations to obtain the transient temperature profile. Verify if the solution for the transient temperature profile satisfies the linear temperature profile when steady-state conditions are reached (at infinite time).

4.12 A plate is initially at temperature T_i when laid on an insulated surface and cooled by air flow at temperature T_∞, with the heat transfer coefficient h. The length of the plate is l, width w, and thickness b ($l \gg w$, $l \gg b$, $w \gg b$). The thermal diffusivity of the plate is α and the thermal conductivity of the plate material is k. The viscosity of air is μ and density is ρ. Estimate the time required for the bottom surface of the plate to cool to T_b, when

    a. The plate material is made of copper (the conduction resistance in the slab is negligible).

    b. The plate material is made of plastic (the convection resistance on the slab is negligible).

4.13 A long metal plane wall with a thickness of $2L$ is initially at temperature T_i and suddenly both sides are heated by a convection fluid flow at temperature T_∞. Outline the procedures and solve the temperature distributions in the plane wall and sketch the temperature profiles in the plane wall during the heating process for two different cases.

    a. Fluid flow is natural convection air.

    b. Fluid flow is forced convection water.

4.14 Consider a large wall, separating two fluids at $T_{\infty 1}$ and $T_{\infty 2}$ ($T_{\infty 1} < T_{\infty 1}$). To prevent heat transfer from the hot fluid at $T_{\infty 1}$ to the wall, a thin foil guard heater (of negligible thickness) on the surface of the wall exposed to the hot fluid at $T_{\infty 1}$ is used to raise the surface temperature to $T_{\infty 1}$. Sketch the instantaneous temperature distributions at several different times in the fluids near the wall and in the wall, before and after the surface temperature is raised with the thin foil guard heater, until a steady state is reached.

    (b) Consider a large wall, separating two fluids at $T_{\infty 1}$ and $T_{\infty 2}$ ($T_{\infty 2} < T_{\infty 1}$). Instead of the thin foil guard heater in (a), heat is generated uniformly in the wall to prevent heat transfer from the fluid at $T_{\infty 1}$ to the wall. Sketch the instantaneous temperature distributions in the fluids at several different times near the wall and in the wall before and after heat is generated uniformly in the wall to prevent heat transfer from the hot fluid to the wall, until a steady state is reached.

    (c) Consider a large wall with a thickness of 2L its surfaces maintained at T_1 and T_2 ($T_2 \neq T_1$). Heat is generated uniformly in the wall. Beginning with the steady-state, 1-D, heat conduction equation and appropriate BCs ($T = T_1$ at $x = -L$ and $T = T_2$ at $x = +L$), derive an expression for the steady-state 1-D temperature distribution in the wall, $T(x)$. Using the expression, *show* that the rate of heat generation is equal to the sum of the rates of heat transfer from the two surfaces.

4.15 A plane wall has constant properties and no internal generation and is initially at a uniform temperature T_i. Suddenly, the surface $x = L$ is exposed to a heating process with a fluid at T_∞ having a convection coefficient h. At the same instant, the electrical heater is energized providing a constant heat flux q_0'' at $x = 0$.

    a. On $T-x$ coordinates, sketch the temperature distributions for the following conditions: initial condition ($t < 0$), steady-state condition ($t \to \infty$), and for two intermediate times.

    b. On $q_x'' - x$ coordinates, sketch the heat flux corresponding to the four temperature distributions of (a).

    c. On $q_x'' - t$ coordinates, sketch the heat flux at the locations $x = 0$ and $x = L$. That is, show qualitatively how $q_x''(0,t)$ and $q_x''(L,t)$ vary with time.

    d. Derive an expression for the steady-state temperature at the heater surface, $T(0, \infty)$, in terms of q_0'', T_∞, k, h, and L.

4.16 The plane wall has constant properties and a uniform internal generation of $\dot{q}\left(W/m^3\right)$ that activates only when the electric heater is energized. The wall is initially at a uniform temperature T_i. Suddenly, the surface $x = L$ is exposed to a cooling process with a fluid at T_∞ having a convection coefficient h. At the same instant, the electrical heater is energized providing a constant heat flux q_0'' at $x = 0$.

   a. On $T–x$ coordinates, sketch the temperature distributions for the following conditions: initial condition $(t \leq 0)$, steady-state condition $(t \to \infty)$, and for two intermediate times.

   b. On $q_x'' - x$ coordinate, sketch the heat flux corresponding to the four temperature distributions of (a).

   c. On $q_x'' - t$ coordinates, sketch the heat flux at the locations $x = 0$ and $x = L$. That is, show qualitatively how $q_x''(0,t)$ and $q_x''(L,t)$ vary with time.

   d. Derive an expression for the steady-state temperature at the heater surface, $T(0,\infty)$, in terms of, q_0'', $\dot{q}$, T_∞, k, h, and L.

4.17 A 10 m-long, 2 cm-diameter copper rod is immersed in a heating bath at a uniform temperature of 100°C. This rod is suddenly exposed to an air stream at 20°C with a heat transfer coefficient of 200 W/m² K. Find the time required for the copper rod to cool to an average temperature of 25°C. Write all assumptions if necessary. The thermal conductivity, specific heat, and density of copper are 401 W/m K, 385 J/kg K, and 8933 kg/m³, respectively.

4.18 Determine the temperature profile for 1-D transient heat conduction problem in a cylinder at constant surface temperature as shown in Figure 4.1b.

4.19 Determine the temperature profile for 1-D transient heat conduction problem in a sphere at constant surface temperature as shown in Figure 4.1c.

4.20 Determine the temperature profile for 1-D transient heat conduction problem in a vertical plate at constant surface temperature as shown in Figure 4.4b.

4.21 A solid material confined in a half-space $x > 0$ is initially at a temperature T_i, which is the same as its melting temperature, T_m. For times $t > 0$, the surface at $x = 0$ is kept at a constant temperature, T_s. If T_s is higher than its melting temperature T_m, we need to understand the temperature history both in the liquid and solid.

   a. Write down the heat conduction equation and the associated BCs which can be used to solve the time-dependent temperature profiles in both the liquid and solid. Determine the time-dependent temperature profiles in both the liquid and solid. You do not need to obtain the final solutions.

   b. Sketch the temperature history and melting distance δ with time both in the liquid and solid in the same diagram, i.e., sketch $T(x, t)$ with $\delta(t)$ at $t = 0, t = t_1, t = t_2, t = t_3$, etc., respectively, in the same diagram.

4.22 Consider a flat plate with a thickness of L initially at T_i. The left-hand-side of the plate (at $x = 0$) is insulated and right-hand-side of the plate (at $x = L$) is suddenly heated by air flow at T_∞ $(T_i < T_\infty)$. The heat transfer coefficient between the surface of the plate (at $x = L$) and hot air is given as h.

   a. Write down heat conduction equation and the associated BCs which can be used to solve the time-dependent temperature profiles in the flat plate.

Determine the time-dependent temperature profiles in the flat plate. You do not need to perform the integration of orthogonal function for C_n.

b. If the flat plate is a stainless material, sketch the temperature history of the plate in the same diagram, i.e., sketch $T(x, t)$ at $t = 0$, $t = t_1$, $t = t_2$, $t = t_3$, and steady state in the same diagram. If the flat plate is an aluminum material, sketch the similar temperature history of the plate in the same diagram for comparison. Comment on which material's surface temperature (at $x = L$) will increase faster during the same heating process and why.

REFERENCES

1. V. Arpaci, *Conduction Heat Transfer*, Addison-Wesley Publishing Company, Reading, MA, 1966.
2. A. Mills, *Heat Transfer*, Richard D. Irwin, Inc., Boston, MA, 1992.
3. D.W. Hahn and M.N. Özişik, *Heat Conduction*, Third Edition, John Wiley & Sons, Hoboken, NJ, 2012.
4. F. Incropera and D. Dewitt, *Fundamentals of Heat and Mass Transfer*, Fifth Edition, John Wiley & Sons, New York, 2002.
5. B. Mikic, *Conduction Heat Transfer, Class Notes*, MIT, Cambridge, MA, 1974.
6. W. Rohsenow and H. Choi, *Heat, Mass, and Momentum Transfer*, Prentice-Hall, Inc., Englewood Cliffs, NJ, 1961.

5 Numerical Analysis in Heat Conduction

5.1 FINITE-DIFFERENCE ENERGY BALANCE METHOD FOR 2-D STEADY-STATE HEAT CONDUCTION

The previously discussed principle of separation of variables is a powerful method to solve the 2-D heat conduction problem. However, the solutions become tedious for various nonhomogeneous boundary conditions (BCs). With the help of modern computers, the heat conduction problem can be easily solved by using the finite-difference energy balance method for complex BCs. For example, Figure 5.1 shows the typical numerical grid distribution for 2-D heat conduction with given surface temperatures as BCs. It requires detailed mathematical procedures if we choose the separation of variables method to solve this problem. Of course, the accuracy of the numerical solutions depends on the number of finite-difference nodes used for the energy balance calculations. In general, the accuracy improves with an increasing number of grid points. It should be noted that each grid point actually represents the temperature of a small area, $\Delta x \Delta y$. Therefore, we obtain the discrete temperature distribution using the finite-difference method. However, when Δx and Δy become very small (approaching zero), the temperature distribution predicted by the finite-difference method will be the same as that calculated using the separation of variables method.

In general, the grid size in the x-direction is not necessarily the same as that in the y-direction. We need to use a smaller grid size (more grid points) in the direction of a high-temperature gradient. The energy balance can be applied to each

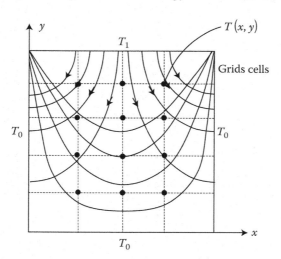

FIGURE 5.1 Finite difference method to solve 2-D heat conduction problem.

DOI: 10.1201/9781003164487-5

grid point, shown in Figure 5.1. The number of unknown temperatures is the same as the number of energy balance equations (the number of grid points). Therefore, the unknown temperatures can be solved using a system of equations. Note that we do not need to perform the energy balance on the boundary points if the boundary temperatures are already given. However, we need to perform the energy balance on the boundary points if the boundary is exposed to a heat flux or convection, in which their boundary temperatures are unknown and must be determined using the finite-difference method. The following outlines the finite-difference method to solve the 2-D heat conduction problem shown in Figure 5.2. We can begin the energy balance at the interior points and then extend the energy balance to the boundary nodes with various BCs.

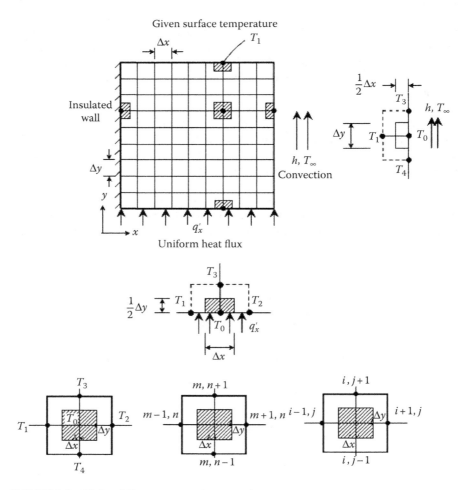

FIGURE 5.2 Finite difference method to solve 2-D heat conduction problem with variable boundary conditions.

The energy balance at the interior nodes:

$$\sum_{i=1}^{4} q_{i \to 0} + \dot{q}\left(\Delta x \cdot \Delta y \cdot 1\right) = 0 \tag{5.1}$$

$$k\Delta y \cdot 1 \cdot \frac{T_1 - T_0}{\Delta x} + k\Delta y \cdot 1 \cdot \frac{T_2 - T_0}{\Delta x} + k\Delta x \cdot 1 \cdot \frac{T_3 - T_0}{\Delta y}$$
$$+ k\Delta x \cdot 1 \cdot \frac{T_4 - T_0}{\Delta y} + \dot{q}\left(\Delta x \cdot \Delta y\right) = 0 \tag{5.2}$$

If $\Delta x = \Delta y$,

$$T_0 = \frac{1}{4}\left(T_1 + T_2 + T_3 + T_4 + \frac{\dot{q}\Delta x \Delta x}{k}\right) \tag{5.3}$$

The energy balance can also be applied to the boundary nodes (not needed if the surface temperature is given):

Convection boundary on the surface nodes:

$$k\Delta y \cdot 1 \cdot \frac{T_1 - T_0}{\Delta x} + k\frac{\Delta x}{2} \cdot 1 \cdot \frac{T_4 - T_0}{\Delta y} + k\frac{\Delta x}{2} \cdot 1 \cdot \frac{T_3 - T_0}{\Delta y}$$
$$+ h\Delta y\left(T_\infty - T_0\right) + \dot{q}\frac{\Delta x}{2} \cdot \Delta y = 0 \tag{5.4}$$

Uniform heat flux at the surface nodes:

$$k\frac{\Delta y}{2} \cdot 1 \cdot \frac{T_1 - T_0}{\Delta x} + k\Delta x \cdot 1 \cdot \frac{T_3 - T_0}{\Delta y} + k\frac{\Delta y}{2} \cdot 1 \cdot \frac{T_2 - T_0}{\Delta x}$$
$$+ q_s''\Delta x \cdot 1 + \dot{q}\frac{\Delta y}{2} \cdot \Delta x = 0 \tag{5.5}$$

If insulation is applied to the surface nodes, then $q_s'' = 0$.

In general, T_0, T_1, T_2, T_3, and T_4 can be replaced by $T_{m,\,n}$, $T_{m-1,n}$, $T_{m+1,n}$, $T_{m,\,n-1}$, and $T_{m,\,n+1}$ or by $T_{i,\,j}$, $T_{i-1,j}$, $T_{i+1,j}$, $T_{i,\,j-1}$, and $T_{i,\,j+1}$, where $m = 1, 2, 3, \ldots n = 1, 2, 3, \ldots$ or $i = 1, 2, 3, \ldots, j = 1, 2, 3, \ldots$ to obtain $T_{1,1}$, $T_{1,2}$, $T_{1,3}$, $\ldots$, $T_{2,1}$, $T_{2,2}$, $T_{2,3}$, $\ldots$, and $T_{3,1}$, $T_{3,2}$, $T_{3,3}$, $\ldots$.

Another approach is to make grid nodes directly from the heat conduction equation. The 2-D steady-state heat conduction equation with heat generation is

$$\frac{\partial^2 T}{\partial x^2} + \frac{\partial^2 T}{\partial y^2} + \frac{\dot{q}}{k} = 0$$

The finite-differential format of the steady-state 2-D heat conduction equation with heat generation can be written as

$$\frac{\dfrac{T_{m-1,n} - T_{m,n}}{\Delta x} + \dfrac{T_{m+1,n} - T_{m,n}}{\Delta x}}{\Delta x} + \frac{\dfrac{T_{m,n-1} - T_{m,n}}{\Delta y} + \dfrac{T_{m,n+1} - T_{m,n}}{\Delta y}}{\Delta y} + \frac{\dot{q}}{k} = 0 \qquad (5.6)$$

Allowing $\Delta x = \Delta y$, one obtains

$$\left(T_{m-1,n} - T_{m+1,n} + T_{m,n-1} + T_{m,n+1}\right) + \frac{\dot{q}}{k}(\Delta x)^2 = 4T_{m,n} \qquad (5.7)$$

where $m = 1, 2, 3,\ldots, n = 1, 2, 3,\ldots$.

The above linear equation can be applied to all interior nodes. Theoretically, one would obtain $m \times n$ linear equations, and therefore, the temperature distribution, $T(x, y) = T_{m,\,n}$, can be solved using the matrix method [1,2]. For example, let $T_{m,\,n} = T_1, T_2, T_3, \ldots, T_N$, and the above linear equations can be applied $T_1, T_1, T_3, \ldots, T_N$. Rearranging the equation, one obtains

$$a_{11}T_1 + a_{12}T_2 + a_{13}T_3 + \cdots + a_{1N}T_N = C_1$$
$$a_{21}T_1 + a_{22}T_2 + a_{23}T_3 + \cdots + a_{2N}T_N = C_2$$
$$\vdots \qquad\qquad\qquad\qquad\qquad\qquad (5.8)$$
$$a_{N1}T_1 + a_{N2}T_2 + a_{N3}T_3 + \cdots + a_{NN}T_N = C_N$$

Using the matrix notation, these equations can be expressed as

$$[A][T] = [C] \qquad (5.9)$$

where

$$[A] = \begin{bmatrix} a_{11} & a_{12} & \cdots & a_{1N} \\ a_{21} & a_{22} & \cdots & a_{2N} \\ \vdots & & & \\ a_{N1} & a_{N2} & \cdots & a_{NN} \end{bmatrix}, \quad [T] = \begin{bmatrix} T_1 \\ T_2 \\ \vdots \\ T_N \end{bmatrix}, \quad [C] = \begin{bmatrix} C_1 \\ C_2 \\ \vdots \\ C_N \end{bmatrix}$$

The solution may be expressed as

$$[T] = [A]^{-1}[C] \qquad (5.10)$$

where $[A]^{-1}$ is the inverse of $[A]$ that is defined as

$$[A]^{-1} = \begin{bmatrix} b_{11} & b_{12} & \cdots & b_{1N} \\ b_{21} & b_{22} & \cdots & b_{2N} \\ \vdots & & & \\ b_{N1} & b_{N2} & \cdots & b_{NN} \end{bmatrix}$$

Therefore, the temperature can be determined by

$$
\begin{aligned}
T_1 &= b_{11}C_1 + b_{12}C_2 + \cdots + b_{1N}C_N \\
T_2 &= b_{21}C_1 + b_{22}C_2 + \cdots + b_{2N}C_N \\
&\quad\vdots \\
T_N &= b_{N1}C_1 + b_{N2}C_2 + \cdots + b_{NN}C_N
\end{aligned}
\tag{5.11}
$$

Example 5.1

We use Figure 5.3 as an example to demonstrate how to solve the 2-D heat conduction problem using the finite-difference method. Let $\Delta x = \Delta y$ and $\dot{q} = 0$. Rearrange the temperatures from the energy balance on nodes 1, 2, 3,....

$$
T_1 = \frac{1}{4}(T_s + T_s + T_2 + T_3) \Rightarrow
$$

$$
-4T_1 + T_2 + T_3 + 0 + 0 + 0 + 0 + 0 = -2T_s
$$

$$
T_2 = \frac{1}{4}(T_1 + T_4 + T_1 + T_s)
$$

$$
2T_1 - 4T_2 + 0 + T_4 + 0 + 0 + 0 + 0 = -T_s
$$

$$
T_3 = \frac{1}{4}(T_s + T_5 + T_4 + T_1)
$$

$$
T_4 = \frac{1}{4}(T_3 + T_6 + T_3 + T_2)
$$

$$
T_5 = \frac{1}{4}(T_s + T_7 + T_6 + T_3)
$$

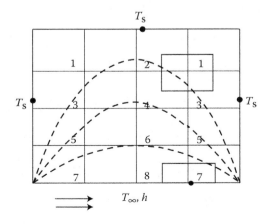

FIGURE 5.3 Example of using finite difference method to solve 2-D heat conduction problem.

$$T_6 = \frac{1}{4}\left(T_5 + T_8 + T_5 + T_4\right)$$

$$T_7 = \frac{1}{\left(4 + 2(h\Delta x/k)\right)}\left(2T_5 + T_8 + T_s + 2\frac{h\Delta x}{k}T_\infty\right)$$

$$T_8 = \frac{1}{\left(2 + (h\Delta x/k)\right)}\left(T_6 + T_7 + \frac{h\Delta x}{k}T_\infty\right)$$

Place the temperatures on the left-hand side of the equations, and the constants on the right-hand side. We can form a coefficient matrix [A], temperature matrix [T], and column matrix [C]. The linear equations of the finite-difference energy balance on each grid point can be represented by the product of $[A][T] = [C]$. Therefore, the temperature distribution can be obtained if we know how to solve [T] from [A] and [C]. Therefore, the primary task for this method is how to obtain [A] and [C]. The matrix can be solved numerically using a programming code such as MATLAB®. The solution for [T] is

$$[T] = [A]^{-1}[C]$$

$$[A] = \begin{bmatrix} -4 & 1 & 1 & 0 & 0 & 0 & 0 & 0 \\ 2 & -4 & 0 & 1 & 0 & 0 & 0 & 0 \\ 1 & 0 & -4 & 1 & 1 & 0 & 0 & 0 \\ 0 & 1 & 2 & -4 & 0 & 1 & 0 & 0 \\ 0 & 0 & 1 & 0 & -4 & 1 & 1 & 0 \\ 0 & 0 & 0 & 1 & 2 & -4 & 0 & 1 \\ 0 & 0 & 0 & 0 & 2 & 0 & -\left(4 + \frac{2h}{k}\Delta x\right) & 1 \\ 0 & 0 & 0 & 0 & 0 & 1 & 1 & -\left(2 + \frac{h}{x}\Delta x\right) \end{bmatrix}$$

$$[C] = \begin{bmatrix} -T_s \\ -T_s \\ -T_s \\ 0 \\ -T_s \\ 0 \\ -T_s - \frac{2h}{k}\Delta x T_\infty \\ -\frac{h}{k}\Delta x T_\infty \end{bmatrix} \qquad [T] = \begin{bmatrix} T_1 \\ T_2 \\ T_3 \\ T_4 \\ T_5 \\ T_6 \\ T_7 \\ T_8 \end{bmatrix}$$

Example 5.2

We use Figure 5.4 with $T(x, y) = T_{m, n}$, where $m = 1, 2, 3, 4, 5$ and $n = 1, 2, 3, 4, 5$. We need to solve the temperature column matrix $[T]$ with $\Delta x = \Delta y$ and $\dot{q} = 0$. Provide the linear equations of the finite-difference, energy balance method on each mode, and obtain a coefficient matrix, $[A]$, and a column matrix $[C]$. Therefore, the temperature matrix can be solved by $[A][T] = [C]$, $[T] = [A]^{-1}[C]$, where all $[T]$, $[A]$, and $[C]$ are equal:

$$[T] = \begin{bmatrix} T_{1,1} & T_{2,1} & T_{3,1} & T_{4,1} & T_{5,1} \\ T_{1,2} & T_{2,2} & T_{3,2} & T_{4,2} & T_{5,2} \\ T_{1,3} & T_{2,3} & T_{3,3} & T_{4,3} & T_{5,3} \\ T_{1,4} & T_{2,4} & T_{3,4} & T_{4,4} & T_{5,5} \\ T_{1,5} & T_{2,5} & T_{3,5} & T_{4,5} & T_{5,5} \end{bmatrix} = \begin{bmatrix} T_{1,1} \\ T_{2,1} \\ \vdots \\ T_{1,1} \\ T_{2,2} \\ \vdots \\ T_{1,5} \\ T_{2,5} \\ \vdots \\ T_{5,5} \end{bmatrix}$$

$$[C] = \begin{bmatrix} C_{1,1} & C_{2,1} & C_{3,1} & C_{4,1} & C_{5,1} \\ C_{1,2} & C_{2,2} & C_{3,2} & C_{4,2} & C_{5,2} \\ C_{1,3} & C_{2,3} & C_{3,3} & C_{4,3} & C_{5,3} \\ C_{1,4} & C_{2,4} & C_{3,4} & C_{4,4} & C_{4,5} \\ C_{1,5} & C_{2,5} & C_{3,5} & C_{4,5} & C_{5,5} \end{bmatrix} = \begin{bmatrix} C_{1,1} \\ C_{2,1} \\ \vdots \\ C_{1,1} \\ C_{2,2} \\ \vdots \\ C_{1,5} \\ C_{2,5} \\ \vdots \\ C_{5,5} \end{bmatrix}$$

$$[A] = \begin{bmatrix} a_{11} & a_{21} & a_{31} & a_{41} & a_{51} \\ a_{12} & a_{22} & a_{32} & a_{42} & a_{52} \\ a_{13} & a_{23} & a_{33} & a_{43} & a_{53} \\ a_{14} & a_{24} & a_{34} & a_{44} & a_{54} \\ a_{15} & a_{25} & a_{35} & a_{45} & a_{55} \end{bmatrix}$$

The above-mentioned finite-difference energy balance method can be used for solving 2-D and 3-D heat conduction problems in Cartesian, cylindrical, or spherical coordinates with various BCs. Figure 5.5 provides an example of a 3-D problem. In this problem, let $T(x, y, z) = T_{i, j, k}$, with $i = 1, 2, 3, \ldots i, j = 1, 2, 3, \ldots j$, and $k = 1, 2, 3, \ldots k$.

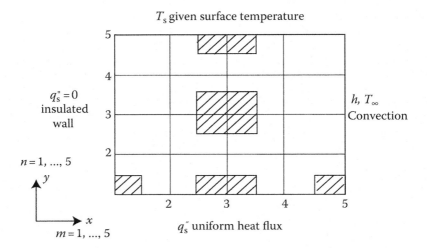

FIGURE 5.4 Example of using finite difference method to solve 2-D heat conduction problem with various boundary conditions.

For the case of a curved boundary [3], the interior nodes can be determined as shown in Figure 5.6.

As shown in Figure 5.6, given $T_{i+1,j}$, $T_{i-1,j}$, $T_{i,j+1}$, and $T_{i,j-1}$, the unknown node, $T_{i,j}$, can be determined using the appropriate energy balance equation.

$$k\Delta y \frac{T_{i-1,j} - T_{i,j}}{\Delta x} + k\Delta y \frac{T_{i+1,j} - T_{i,j}}{a\Delta x} + k\Delta x \frac{T_{i,j+1} - T_{i,j}}{b\Delta y} + k\Delta x \frac{T_{i,j-1} - T_{i,j}}{c\Delta y} = 0 \quad (5.12)$$

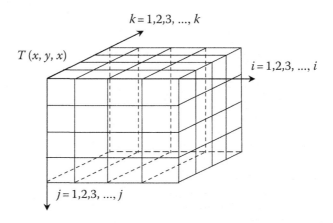

FIGURE 5.5 Finite difference method to solve 3-D heat conduction problem.

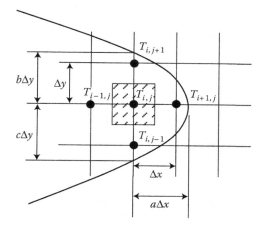

FIGURE 5.6 Finite difference method to solve the interior nodes next to the curved boundary.

5.2 FINITE-DIFFERENCE ENERGY BALANCE METHOD FOR 1-D TRANSIENT HEAT CONDUCTION

The heat equation for 1-D transient heat conduction with heat generation is

$$\frac{\partial^2 T}{\partial x^2} + \frac{\dot{q}}{k} = \frac{1}{\alpha} \frac{\partial T}{\partial t}$$

This equation is a parabolic equation. As discussed in Chapter 4, this equation can be solved analytically using the separation of variables method, similarity method, Laplace transform method, or integral method. In this chapter, we would like to solve 1-D and 2-D transient conduction problems with various BCs by using the explicit finite-difference method and the implicit finite-difference method [1].

5.2.1 Explicit Finite-Difference Method

This is a finite difference in space with an explicit form (lower bond). At a given interior point, heat conduction from the neighboring points is based on the previous time-step temperatures in order to increase that point temperature during the incremental time-step change. While this method is limited by the instability problem, it is a straightforward method that is easily used.

The finite-difference format of the above 1-D transient heat conduction equation can be written as

$$\frac{\left(T_1^p - T_2^p\right)/\Delta x + \left(T_3^p - T_2^p\right)/\Delta x}{\Delta x} + \frac{\dot{q}}{k} = \frac{1}{\alpha} \frac{T_2^{p+1} - T_2^p}{\Delta t} \tag{5.13}$$

Example 5.3

Figure 5.7 shows transient heat transfer in 1-D. The temperature at each node is changing in time and denoted with the time steps, p.

The energy balance at an interior node, for example node 2, can be determined using the energy balance.

$$\Sigma q = \text{energy storage}$$

Lower bond:

$$k \cdot y \cdot 1 \frac{T_1^p - T_2^p}{\Delta x} + k \cdot y \cdot 1 \frac{T_3^p - T_2^p}{\Delta x} + \dot{q} \cdot \Delta x \cdot y \cdot 1 = \frac{\rho C_p \Delta x \cdot y \cdot 1 \cdot \left(T_2^{p+1} - T_2^p\right)}{\Delta t} \quad (5.14)$$

Let $\dot{q} = 0$, $\alpha = \left(k/\rho C_p\right)$, one obtains

$$T_1^p + T_3^p - 2T_2^p = \frac{1}{Fo}\left(T_2^{p+1} - T_2^p\right)$$
$$T_2^{p+1} = Fo\left(T_1^p + T_3^p\right) + \left(1 - 2Fo\right)T_2^p \quad (5.15)$$

where the time increment is Δt with the time step p, that is, $t = p\Delta t$, with $p = 0$, 1, 2,.... Fo is a finite-difference form of the Fourier number $Fo = (\alpha\Delta t/\Delta x^2)$.

With a traditional hand calculation, the temperature at the $(p + 1)$ step is determined by the preceding time and temperature at the p step, as sketched in Figure 5.7. For example, at $p = 0$, $t = 0$, so the initial condition is $T_S^0 = T_1^0 = T_2^0 = T_3^0 = T_i$.

The stability requirement is $(1 - 2Fo) \geq 0$, or $Fo \leq 0.5$. This can be achieved with a combination of Δt as a very small value and Δx as a very large value.

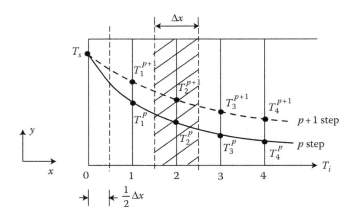

FIGURE 5.7 Finite difference energy balance method for one-dimensional transient heat conduction with given surface temperature boundary condition.

Example 5.4

Consider Figure 5.8, with a surface convection BC (node 0).

The energy balance can be applied to node 0 to solve for the first time step $(p + 1)$:

$$h\left(T_\infty - T_0^p\right) + k\frac{T_1^p - T_0^p}{\Delta x} = \frac{T_0^{p+1} - T_0^p}{\Delta t}\rho C_p\left(\frac{\Delta x}{2}\right) \tag{5.16}$$

$$T_0^{p+1} = T_0^p + \frac{2h}{\rho C_p}\frac{\Delta t}{\Delta x}\left(T_\infty - T_0^p\right) + \frac{2\alpha\Delta t}{\Delta x^2}\left(T_1^p - T_0^p\right)$$

$$= 2\mathrm{Fo}\left(T_1^p + \mathrm{Bi}\,T_\infty\right) + \left(1 - 2\mathrm{Fo} - 2\,\mathrm{Bi}\,\mathrm{Fo}\right)T_0^p \tag{5.17}$$

where

$$\mathrm{Fo} = \frac{\alpha\Delta t}{\Delta x^2} = \text{Fourier number}$$

$$\mathrm{Bi} = \frac{h\Delta x}{k} = \text{Biot number}$$

For stability, $(1 - 2\mathrm{Fo} - 2\mathrm{Bi}\,\mathrm{Fo}) \geq 0$ or $[\mathrm{Fo}\,(1 + \mathrm{Bi})] \leq 0.5$.

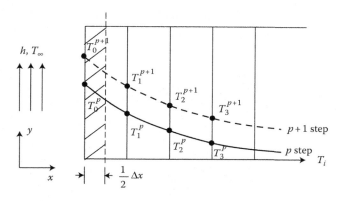

FIGURE 5.8 Finite difference energy balance method for one-dimensional transient heat conduction with convection boundary conditions.

Example 5.5

Figure 5.9 provides a surface heat flux BC.

As with the previous examples, the energy balance can be applied at node 0 to determine the boundary temperature at time step, $p + 1$.

$$q_S'' + k\frac{T_1^p - T_0^p}{\Delta x} = \frac{T_0^{p+1} - T_0^p}{\Delta t}\rho C_p\left(\frac{\Delta x}{2}\right) \tag{5.18}$$

$$T_0^{p+1} = \frac{\Delta t}{(\Delta x/2)\rho C_p}q_S'' + \frac{k\Delta t}{(\Delta x^2/2)\rho C_p}T_1^p + (1 - 2\mathrm{Fo})T_0^p \tag{5.19}$$

In this case, stability is achieved for Fo ≤ 0.5 or $(1 - 2\mathrm{Fo}) \geq 0$.

5.2.2 IMPLICIT FINITE-DIFFERENCE METHOD

This is the finite difference in space in the implicit form (upper bond). At a given interior point, heat conduction from the neighboring nodes is based on the new time-step temperatures in order to increase that point temperature during the incremental time-step change. This method does not have a stability limitation. However, the inverse matrix problem must be solved numerically. The accuracy of this method improves for large Δt and small Δx.

The energy balance at the interior nodes (e.g., node 2, as shown in Figures 5.7–5.9):

$$\Sigma\, q = \text{energy storage}$$

Upper bond:

$$k\frac{T_1^{p+1} - T_2^{p+1}}{\Delta x} + k\frac{T_3^{p+1} - T_2^{p+1}}{\Delta x} + \dot{q}\cdot\Delta x\cdot 1 = \rho C_p\Delta x\frac{T_1^{p+1} - T_2^p}{\Delta t} \tag{5.20}$$

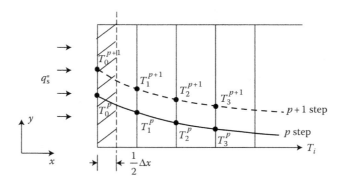

FIGURE 5.9 Finite difference energy balance method for 1-D transient heat conduction with surface heat flux boundary condition.

Allowing $\dot{q} = 0, \alpha = (k/\rho C_p), \text{Fo} = (\alpha \Delta t/\Delta x^2)$, one obtains

$$T_1^{p+1} - T_2^{p+1} + T_3^{p+1} - T_2^{p+1} = \frac{\Delta x^2}{\alpha \Delta t}\left(T_2^{p+1} - T_2^p\right)$$

$$(1 + 2\text{Fo})T_2^{p+1} - \text{Fo}\left(T_1^{p+1} + T_2^{p+1}\right) = T_2^p$$

(5.21)

As noted previously, there is no stability issue. In general, $T_0, T_1, T_2, T_3,\ldots$ can be replaced by T_m or T_i, when $m = 1, 2, 3,\ldots$ or $i = 1, 2, 3,\ldots$.

5.3 2-D TRANSIENT HEAT CONDUCTION

The above-mentioned finite-difference energy balance method can also be used for solving 2-D transient heat conduction problems [1]. Let $T(x, y, t) = T(m, n, t)$ or $T(i, j, t)$, with $m = i = 1, 2, 3,\ldots$, and $n = j = 1, 2, 3,\ldots$.

$$\frac{\partial^2 T}{\partial x^2} + \frac{\partial^2 T}{\partial y^2} + \frac{\dot{q}}{k} = \frac{1}{\alpha}\frac{\partial T}{\partial t}$$

The x-direction net heat conduction + the y-direction net heat conduction = temperature change of a small element ($\Delta x\,\Delta y \cdot 1$).

For the 2-D problem, the explicit finite-difference method (lower bond) for an interior node is:

$$\frac{\left(T_{m+1,n}^p - T_{m,n}^p\right)/\Delta x + \left(T_{m-1,n}^p - T_{m,n}^p\right)/\Delta x}{\Delta x} + \frac{\left(T_{m,n+1}^p - T_{m,n}^p\right)/\Delta y + \left(T_{m,n-1}^p - T_{m,n}^p\right)/\Delta y}{\Delta y} + \frac{\dot{q}}{k}$$

$$= \frac{1}{\alpha}\frac{T_{m,n}^{p+1} - T_{m,n}^p}{\Delta t}$$

(5.22)

Let $\Delta x = \Delta y$, $\dot{q} = 0$, $\text{Fo} = \alpha \Delta t/\Delta x^2$. The temperature at the $(p + 1)$ step is determined by the preceding time step, p, as

$$T_{m,n}^{p+1} = \text{Fo}\left(T_{m+1,n}^p + T_{m-1,n}^p + T_{m,n+1}^p + T_{m,n-1}^p\right) + (1 - 4\text{Fo})T_{m,n}^p$$

(5.23)

For this 2-D problem, the stability criterion of the interior node is $(1 - 4Fo) \geq 0$, that is, $Fo \leq 0.25$.

Recall, for the 1-D transient heat conduction problem, one obtains

$$T_m^{p+1} = \text{Fo}\left(T_{m+1}^p + T_{m-1}^p\right) + (1 - 2\text{Fo})T_m^p$$

The stability criterion for the 1-D problem was $(1 - 2\text{Fo}) \geq 0$, or $\text{Fo} \leq 0.5$.

The 2-D problem can also be solved using the implicit form of the finite-difference method (upper bond):

$$\frac{\left(T_{m+1,n}^{p+1} - T_{m,n}^{p+1}\right)/\Delta x + \left(T_{m-1,n}^{p+1} - T_{m,n}^{p+1}\right)/\Delta x}{\Delta x} + \frac{\left(T_{m,n+1}^{p+1} - T_{m,n}^{p+1}\right)/\Delta y + \left(T_{m,n-1}^{p+1} - T_{m,n}^{p+1}\right)/\Delta y}{\Delta y}$$

$$+ \frac{\dot{q}}{k} = \frac{1}{\alpha}\frac{T_{m,n}^{p+1} - T_{m,n}^{p}}{\Delta t} \tag{5.24}$$

Again, let $\Delta x = \Delta y$, $\dot{q} = 0$, $\mathrm{Fo} = \alpha\Delta t/\Delta x^2$, and temperature at $(p + 1)$ step is determined by:

$$T_{m+1,n}^{p+1} - T_{m,n}^{p+1} + T_{m-1,n}^{p+1} - T_{m,n}^{p+1} + T_{m,n+1}^{p+1} - T_{m,n}^{p+1} + T_{m,n-1}^{p+1} - T_{m,n}^{p+1} = \frac{1}{\mathrm{Fo}}\left(T_{m,n}^{p+1} - T_{m,n}^{p}\right)$$

Rearranging:

$$(1 + 4\mathrm{Fo})T_{m,n}^{p+1} - \mathrm{Fo}\left(T_{m+1,n}^{p+1} + T_{m-1,n}^{p+1} + T_{m,n+1}^{p+1} + T_{m,n-1}^{p+1}\right) = T_{m,n}^{p} \tag{5.25}$$

For a 1-D transient heat conduction problem, one obtains

$$(1 + 2\mathrm{Fo})T_{m}^{p+1} - \mathrm{Fo}\left(T_{m+1}^{p} + T_{m-1}^{p}\right) = T_{m}^{p}$$

With the implicit method there is no issue with stability, but numerical (computer) methods are required to solve the temperature matrix.

In general, the finite-difference energy balance method (explicit or implicit methods) can be used to solve 1-D, 2-D, and 3-D transient heat conduction problems in Cartesian, cylindrical, or spherical coordinates with various BCs.

Example 5.6

Figure 5.10 shows a long, square bar with opposite sides maintained at T_a and T_b, and the other two sides lose heat by convection to a fluid at T_∞. The material

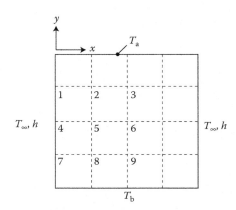

FIGURE 5.10 Finite difference method to solve a 2-D conduction problem.

conductivity of the bar is k and the convective heat transfer coefficient is h. For the given mesh, use the finite-difference energy balance method to obtain a coefficient matrix $[A]$, temperature matrix $[T]$, and a column matrix $[C]$. The temperature matrix can be solved numerically using a programming language such as MATLAB.

Figure 5.10 shows the prescribed surface conditions and the nodes. Symmetry allows us to consider just nine nodes. The interior, nodal temperatures are

$$T_2 = \frac{1}{4}(T_a + T_1 + T_3 + T_5)$$

$$T_3 = \frac{1}{4}(T_a + T_2 + T_2 + T_6)$$

$$T_5 = \frac{1}{4}(T_2 + T_4 + T_6 + T_8)$$

$$T_6 = \frac{1}{4}(T_3 + T_5 + T_5 + T_9)$$

$$T_8 = \frac{1}{4}(T_5 + T_7 + T_9 + T_b)$$

$$T_9 = \frac{1}{4}(T_6 + T_8 + T_8 + T_b)$$

For the nodes along the left boundary (nodes 1, 4, and 7, respectively), the energy balance provides:

$$-\left(\frac{k\Delta y}{\Delta x} + \frac{k\Delta x}{\Delta y} + h\Delta y\right)T_1 + \frac{k\Delta y}{\Delta x}T_2 + \frac{k\Delta x}{2\Delta y}T_4 = -\frac{k\Delta x}{2\Delta y}T_a - h\Delta y T_\infty$$

$$\frac{k\Delta x}{2\Delta y}T_1 - \left(\frac{k\Delta y}{\Delta x} + \frac{k\Delta x}{\Delta y} + h\Delta y\right)T_4 + \frac{k\Delta y}{\Delta x}T_5 + \frac{k\Delta x}{2\Delta y}T_7 = -h\Delta y T_\infty$$

$$\frac{k\Delta x}{2\Delta y}T_4 - \left(\frac{k\Delta y}{\Delta x} + \frac{k\Delta x}{\Delta y} + h\Delta y\right)T_7 + \frac{k\Delta y}{\Delta x}T_8 = -\frac{k\Delta x}{2\Delta y}T_b - h\Delta y T_\infty$$

The nodal equations should now be combined in matrix form $[A][T] = [C]$:

$$[A]=\begin{bmatrix}
-\left(\dfrac{k\Delta y}{\Delta x}+\dfrac{k\Delta x}{\Delta y}+h\Delta y\right) & \dfrac{k\Delta y}{\Delta x} & 0 & \dfrac{k\Delta x}{2\Delta y} & 0 & 0 & 0 & 0 & 0 \\[2ex]
1 & -4 & 1 & 0 & 1 & 0 & 0 & 0 & 0 \\[1ex]
0 & 2 & -4 & 0 & 0 & 1 & 0 & 0 & 0 \\[1ex]
\dfrac{k\Delta x}{2\Delta y} & 0 & 0 & -\left(\dfrac{k\Delta y}{\Delta x}+\dfrac{k\Delta x}{\Delta y}+h\Delta y\right) & \dfrac{k\Delta y}{\Delta x} & 0 & \dfrac{k\Delta x}{2\Delta y} & 0 & 0 \\[2ex]
0 & 1 & 0 & 1 & -4 & 1 & 0 & 1 & 0 \\[1ex]
0 & 0 & 1 & 0 & 2 & -4 & 0 & 0 & 1 \\[1ex]
0 & 0 & 0 & \dfrac{k\Delta x}{2\Delta y} & 0 & 0 & -\left(\dfrac{k\Delta y}{\Delta x}+\dfrac{k\Delta x}{\Delta y}+h\Delta y\right) & \dfrac{k\Delta y}{\Delta x} & 0 \\[2ex]
0 & 0 & 0 & 0 & 1 & 0 & 1 & -4 & 1 \\[1ex]
0 & 0 & 0 & 0 & 0 & 1 & 0 & 2 & -4
\end{bmatrix}$$

$$[T]=\begin{Bmatrix} T_1 \\ T_2 \\ T_3 \\ T_4 \\ T_5 \\ T_6 \\ T_7 \\ T_8 \\ T_9 \end{Bmatrix}
\qquad
[C]=\begin{bmatrix} -\dfrac{k\Delta x}{2\Delta y}T_a - h\Delta y T_\infty \\[2ex] -T_a \\[1ex] -T_a \\[1ex] -h\Delta y T_\infty \\[1ex] 0 \\[1ex] 0 \\[1ex] -\dfrac{k\Delta x}{2\Delta y}T_b - h\Delta y T_\infty \\[2ex] -T_b \\[1ex] -T_b \end{bmatrix}$$

5.4 FINITE-DIFFERENCE METHOD FOR CYLINDRICAL COORDINATES

5.4.1 STEADY HEAT CONDUCTION IN CYLINDRICAL COORDINATES

Case 1: 1-D Steady Heat Conduction in a Solid Rod with Internal Heat Generation ($T = T(r)$ only, $T \neq T(\phi, z)$).

Figure 5.11 shows the finite difference energy balance method for 1-D transient heat conduction in a cylindrical rod with internal heat generation per volume. The outer surface is exposed to a flow at temperature, T_∞, with a convection heat transfer coefficient h, at $r = r_{outer}$.

From the energy balance in each node:

$$q_{in} - q_{out} + \dot{q} = 0$$

For an interior node (m) per unit depth (in z-direction),

$$2\pi k r_{m-1} \frac{T_{m-1} - T_m}{\Delta r} - 2\pi k r_m \frac{T_m - T_{m+1}}{\Delta r} + \pi \left(r_m^2 - r_{m-1}^2 \right) \dot{q} = 0$$

where $m = 1, 2, 3, \ldots M$. For example, we have five nodes, $m = 1, 2, 3, 4, 5$; here the outer surface temperature at node 5 is T_5. We need to determine the center node temperature, T_0, from the above energy balance equation:

$$Node\ 0: 0 - 2\pi k r_0 \frac{T_0 - T_1}{\Delta r} + \pi r_0^2 \cdot \dot{q} = 0$$

Nodes 1, 2, 3, 4, and 5 can be determined as:

$$Node\ 1: 2\pi k r_0 \frac{T_0 - T_1}{\Delta r} - 2\pi k r_1 \frac{T_1 - T_2}{\Delta r} + \pi \left(r_1^2 - r_0^2 \right) \dot{q} = 0$$

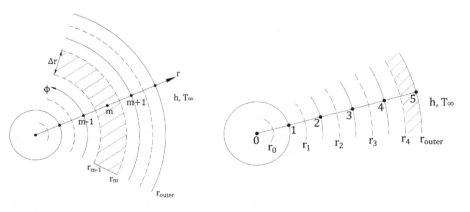

FIGURE 5.11 Finite difference method to solve the 1-D cylindrical heat conduction problem.

Node 2: $2\pi k r_1 \dfrac{T_1 - T_2}{\Delta r} - 2\pi k r_2 \dfrac{T_2 - T_3}{\Delta r} + \pi \left(r_2^2 - r_1^2 \right) \dot{q} = 0$

Node 3: $2\pi k r_2 \dfrac{T_2 - T_3}{\Delta r} - 2\pi k r_3 \dfrac{T_3 - T_4}{\Delta r} + \pi \left(r_3^2 - r_2^2 \right) \dot{q} = 0$

Node 4: $2\pi k r_3 \dfrac{T_3 - T_4}{\Delta r} - 2\pi k r_4 \dfrac{T_4 - T_5}{\Delta r} + \pi \left(r_4^2 - r_3^2 \right) \dot{q} = 0$

Node 5: $2\pi k r_4 \dfrac{T_4 - T_5}{\Delta r} - 2\pi h r_{\text{outer}} \left(T_5 - T_\infty \right) + \pi \left(r_{\text{outer}}^2 - r_4^2 \right) \dot{q} = 0$

The above 6 equations (nodes 0–6) can be re-arranged to become:

Node 0: $a_{01} T_0 + a_{02} T_1 + 0 + 0 + 0 + 0 = C_0$

Node 1: $a_{11} T_0 + a_{12} T_1 + a_{13} T_2 + 0 + 0 + 0 = C_1$

Node 2: $0 + a_{22} T_1 + a_{23} T_2 + a_{24} T_3 + 0 + 0 = C_2$

Node 3: $0 + 0 + a_{33} T_2 + a_{34} T_3 + a_{35} T_4 + 0 = C_3$

Node 4: $0 + 0 + 0 + a_{44} T_3 + a_{45} T_4 + a_{46} T_5 = C_4$

Node 5: $0 + 0 + 0 + 0 + a_{54} T_4 + a_{55} T_5 = C_5$

Therefore, the temperature matrix can be solved by

$$[A][T] = [C]$$

$$[T] = [A]^{-1}[C]$$

where temperature matrix $[T]$, coefficient matrix $[A]$, and column matrix $[C]$ are as follows:

$$[A] = \begin{bmatrix} a_{01} & a_{02} & 0 & 0 & 0 & 0 \\ a_{11} & a_{12} & a_{13} & 0 & 0 & 0 \\ 0 & a_{22} & a_{23} & a_{24} & 0 & 0 \\ 0 & 0 & a_{33} & a_{34} & a_{35} & 0 \\ 0 & 0 & 0 & a_{44} & a_{45} & a_{46} \\ 0 & 0 & 0 & 0 & a_{54} & a_{55} \end{bmatrix} \quad [T] = \begin{bmatrix} T_0 \\ T_1 \\ T_2 \\ T_3 \\ T_4 \\ T_5 \end{bmatrix} \quad [C] = \begin{bmatrix} C_0 \\ C_1 \\ C_2 \\ C_3 \\ C_4 \\ C_5 \end{bmatrix}$$

If the outer surface temperature is kept at a constant value, T_5, i.e., $T_5 = T_s =$ given, there is no need to perform the energy balance equation for node 5.

Case 2: 1-D Steady Heat Conduction in a Hollow Cylinder with Internal Heat Generation $(T = T(r)$ only, $T \neq T(\phi, z))$.

From Figure 5.12, for a hollow cylinder, there is no need to solve T_0 in this case. If $T = T_1 = $ given at $r = $ inner radius $= r_{inner}$, and $T = T_5 = $ given at $r = $ outer radius $= r_{outer}$. Therefore, T_2, T_3, and T_4 can be obtained by solving the energy balance at nodes 2, 3, and 4, respectively, as indicated in Case 1. However, if the inner surface and outer surface are exposed to working fluids at $T_{\infty i}$, h_i, and $T_{\infty o}$, h_o, as shown in Figure 5.12, the energy balance needs to be applied to node 1 and node 5 (as shown below) in order to solve for T_1, T_2, T_3, T_4, and T_5 simultaneously, using the matrix method $[A][T] = [C]$.

$$Node\ 1:\ 2\pi h_i r_{inner} (T_{\infty i} - T_1) - 2\pi k r_1 \frac{T_1 - T_2}{\Delta r} + \pi (r_1^2 - r_{inner}^2) \dot{q} = 0$$

$$Node\ 5:\ 2\pi k r_4 \frac{T_4 - T_5}{\Delta r} - 2\pi h_o r_{outer} (T_5 - T_{\infty o}) + \pi (r_{outer}^2 - r_4^2) \dot{q} = 0$$

Case 3: 2-D Steady Heat Conduction in Cylindrical Coordinates with Internal Heat Generation, $T = T(r, \phi)$, but $T \neq T(z)$.

From Figure 5.13, Let $m = 1, 2, 3, 4, 5\ldots$ in the r-direction and $n = 1, 2, 3, 4, 5\ldots$ in the ϕ-direction. Consider the energy balance for an interior node (m, n) per unit depth.

$$kr_{m-1} \Delta\phi \frac{T_{m-1,n} - T_{m,n}}{\Delta r} + kr_m \Delta\phi \frac{T_{m+1,n} - T_{m,n}}{\Delta r} + k\Delta r \frac{T_{m,n-1} - T_{m,n}}{\frac{1}{2}(r_{m-1} + r_m)\Delta\phi}$$

$$+ k\Delta r \frac{T_{m,n+1} - T_{m,n}}{\frac{1}{2}(r_{m-1} + r_m)\Delta\phi} + \dot{q}\Delta r \frac{1}{2}(r_{m-1} + r_m)\Delta\phi = 0$$

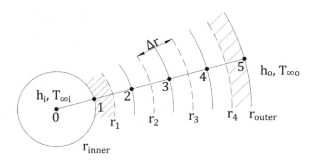

FIGURE 5.12 Finite difference method to solve the hollow cylinder heat conduction problem.

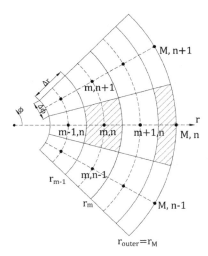

FIGURE 5.13 Finite difference method to solve the 2-D steady cylindrical heat conduction problem.

Consider a node on the outer surface (M,n) exposed to a convection boundary with h, T_∞:

$$kr_{M-1}\Delta\phi\frac{T_{M-1,n}-T_{M,n}}{\Delta r}+hr_M\cdot\Delta\phi\cdot(T_\infty-T_{M,n})+k\frac{\Delta r}{2}\frac{T_{M,n-1}-T_{M,n}}{\frac{1}{2}(r_{M-1}+r_M)\Delta\phi}$$

$$+k\frac{\Delta r}{2}\frac{T_{M,n+1}-T_{M,n}}{\frac{1}{2}(r_{M-1}+r_M)\cdot\Delta\phi}+\dot{q}\frac{\Delta r}{2}\Delta\phi\frac{1}{2}(r_{M-1}+r_M)=0$$

We can also consider an external node (M,n) that is exposed to a constant heat flux with q_s''. In the above equation, the convection term, $hr_M\Delta\phi(T_\infty-T_{M,n})$, should be replaced by heat flux term, $q_s''r_M\cdot\Delta\phi$.

For practice, determine $T(r,\phi)=T_{m,n}$ as shown in Figure 5.13, where $m=1, 2, 3, 4, 5$, and $n=1, 2, 3, 4, 5$, respectively. Write down the linear equations of the finite-difference, energy balance method for each node, and obtain a coefficient matrix $[A]$ and a column matrix $[C]$. Therefore, the temperature matrix $[T]$ can be solved by $[A][T]=[C]$ and $[T]=[A]^{-1}[C]$.

5.4.2 TRANSIENT HEAT CONDUCTION IN CYLINDRICAL COORDINATES

Case 1: 1-D Transient Heat Conduction in a Solid Rod, as shown in Figure 5.11.
Explicit Method—Interior Node per Unit Depth

$$2\pi kr_{m-1}\frac{T_{m-1}^p-T_m^p}{\Delta r}-2\pi kr_m\frac{T_m^p-T_{m+1}^p}{\Delta r}=\rho C_p\pi(r_m^2-r_{m-1}^2)\frac{T_m^{p+1}-T_m^p}{\Delta t}$$

Therefore,

$$T_m^{p+1} = 2\frac{k}{\rho C_p}\frac{\Delta t/\Delta r}{\left(r_m^2 - r_{m-1}^2\right)}\left[r_{m-1}\left(T_{m-1}^p - T_m^p\right) - r_m\left(T_m^p - T_{m+1}^p\right)\right] + T_m^p$$

where T_m is the temperature at $r = \frac{1}{2}(r_{m-1} + r_m)$.

Implicit Method—Interior Node per Unit Depth

$$2\pi k r_{m-1}\frac{T_{m-1}^{p+1} - T_m^{p+1}}{\Delta r} - 2\pi k r_m\frac{T_m^{p+1} - T_{m+1}^{p+1}}{\Delta r} = \rho C_p \pi\left(r_m^2 - r_{m-1}^2\right)\frac{T_m^{p+1} - T_m^p}{\Delta t}$$

Therefore,

$$T_m^{p+1} = \frac{2\dfrac{k}{\rho c_p}\cdot\dfrac{\Delta t/\Delta r}{r_m^2 - r_{m-1}^2}\left[r_m\cdot T_{m+1}^{p+1} + r_{m-1}\cdot T_{m-1}^{p+1}\right] + T_m^p}{1 + 2\dfrac{k}{\rho C_p}\cdot\dfrac{\Delta t/\Delta r}{\left(r_m^2 - r_{m-1}^2\right)}\cdot\left(r_{m-1} + r_m\right)}$$

Case 2: 2-D Transient Heat Conduction in a Hollow Cylinder, as shown in Figure 5.13.
Explicit Method—Interior Node per Unit Depth

$$kr_{m-1}\Delta\phi\frac{T_{m-1,n}^p - T_{m,n}^p}{\Delta r} + kr_m\Delta\phi\frac{T_{m+1,n}^p - T_{m,n}^p}{\Delta r} + k\Delta r\frac{T_{m,n-1}^p - T_{m,n}^p}{\frac{1}{2}(r_{m-1} + r_m)\Delta\phi}$$

$$+ k\Delta r\frac{T_{m,n+1}^p - T_{m,n}^p}{\frac{1}{2}(r_{m-1} + r_m)\Delta\phi} = \rho C_p \Delta r\frac{1}{2}(r_{m-1} + r_m)\Delta\phi\frac{T_{m,n}^{p+1} - T_{m,n}^p}{\Delta t}$$

Implicit Method—Interior Node per Unit Depth

$$kr_{m-1}\Delta\phi\frac{T_{m-1,n}^{p+1} - T_{m,n}^{p+1}}{\Delta r} + kr_m\Delta\phi\frac{T_{m+1,n}^{p+1} - T_{m,n}^{p+1}}{\Delta r} + k\Delta r\frac{T_{m,n-1}^{p+1} - T_{m,n}^{p+1}}{\frac{1}{2}(r_{m-1} + r_m)\Delta\phi}$$

$$+ k\Delta r\frac{T_{m,n+1}^{p+1} - T_{m,n}^{p+1}}{\frac{1}{2}(r_{m-1} + r_m)\Delta\phi} = \rho C_p \Delta r\frac{1}{2}(r_{m-1} + r_m)\Delta\phi\frac{T_{m,n}^{p+1} - T_{m,n}^p}{\Delta t}$$

If the inner or outer surfaces are exposed to a convective fluid with h and T_∞, or receive a heat flux with q_s'', the energy balance needs to be applied to these surface nodes in order to determine these temperatures.

Case 3: 1-D Transient Heat Conduction in a Hollow Cylinder.

Consider 1-D transient heat conduction in a hollow cylindrical tube, similar to Figure 5.12, but without heat generation, $T = T(r)$ only. Consider 3 nodes only, i.e., node 1 for the inner surface, node 2 for the interior, and node 3 on the outer surface. Assume the inner surface $r = r_{inner}$ is exposed to a convection fluid with $T_{\infty i}$, h_i, and the outer surface $r = r_{outer}$ receives a heat flux with q_s''. Perform the finite difference energy balance for each node.

Explicit Method

$$\text{Node 1: } 2\pi h_i r_{inner}\left(T_{\infty i} - T_1^p\right) + 2\pi k r_1 \frac{T_2^p - T_1^p}{\Delta r} = \rho C_p \pi\left(r_1^2 - r_{inner}^2\right)\frac{T_1^{p+1} - T_1^p}{\Delta t}$$

$$\text{Node 2: } 2\pi k r_1 \frac{T_1^p - T_2^p}{\Delta r} + 2\pi k r_2 \frac{T_3^p - T_2^p}{\Delta r} = \rho C_p \pi\left(r_2^2 - r_1^2\right)\frac{T_2^{p+1} - T_2^p}{\Delta t}$$

$$\text{Node 3: } 2\pi k r_2 \frac{T_2^p - T_3^p}{\Delta r} + 2\pi r_{outer} q_s'' = \rho C_p \pi\left(r_{outer}^2 - r_2^2\right)\frac{T_3^{p+1} - T_3^p}{\Delta t}$$

At a given node, heat transfer from the neighboring nodes is based on the previous time-step temperature (p) in order to increase that node temperature during the incremental time-step change. This method is limited by the instability problem; therefore, it requires a larger Δr (less accurate) and smaller Δt (more time consuming). However, the application of this method is straightforward. The time increment is Δt with a time step, p; that is, $t = p\Delta t$, with $p = 0, 1, 2, ...$. The temperature at the $p + 1$ step is determined by the temperature at the preceding time (at p step). For example, $p = 0, t = 0$, the initial condition, $T_1^0 = T_2^0 = T_3^0 = T_i = $ given.

Implicit Method

Replace all "p" in the left-hand side of the above equations on each node 1, 2, and 3 by $p + 1$.

$$\text{Node 1: } 2\pi h_i r_{inner}\left(T_{\infty i} - T_1^{p+1}\right) + 2\pi k r_1 \frac{T_2^{p+1} - T_1^{p+1}}{\Delta r} = \rho C_p \pi\left(r_1^2 - r_{inner}^2\right)\frac{T_1^{p+1} - T_1^p}{\Delta t}$$

$$\text{Node 2: } 2\pi k r_1 \frac{T_1^{p+1} - T_2^{p+1}}{\Delta r} + 2\pi k r_2 \frac{T_3^{p+1} - T_2^{p+1}}{\Delta r} = \rho C_p \pi\left(r_2^2 - r_1^2\right)\frac{T_2^{p+1} - T_2^p}{\Delta t}$$

$$\text{Node 3: } 2\pi k r_2 \frac{T_2^{p+1} - T_3^{p+1}}{\Delta r} + 2\pi r_{outer} q_s'' = \rho C_p \pi\left(r_{outer}^2 - r_2^2\right)\frac{T_3^{p+1} - T_3^p}{\Delta t}$$

At a given node, the heat transfer from the neighboring nodes is based on the new time-step temperature $(p + 1)$ in order to increase that node temperature during the incremental time-step change. This method has no instability problem. To achieve the best accuracy, larger Δt and smaller Δr increments should be used. The drawback of the implicit method is the need to develop a computer program to solve the inverse matrix problem, i.e., $[A][T] = [C]$, and $[T] = [A]^{-1}[C]$.

REMARKS

The finite-difference method is a very powerful numerical technique to solve many engineering application problems. As long as you know how to perform the basic energy balance at the interior and boundary nodes, this method can solve a variety of heat conduction problems with complex thermal BCs. In the undergraduate-level heat transfer course, students are typically required to perform the simple energy balance at any specified node inside a 2-D steady-state solid material and on the boundary.

In the intermediate-level heat transfer, we are more focused on how to perform the simple energy balance along with discretizing the heat conduction equation in order to solve the 1-D and 2-D steady-state heat conduction problems with various BCs using the inverse matrix method. We also solve the 1-D and 2-D transient heat conduction problems with various BCs by using the finite-difference implicit method and explicit method. In general, the same technique can be used to solve heat conduction problems with cylindrical and spherical coordinates.

PROBLEMS

5.1 Refer to Figure 5.4, show the matrices $[A]$, $[T]$, and $[C]$, with the following grid distributions:

(1) $m = 1, 2, 3, 4, 5$ (2) $m = 1, 2, 3, 4$ (3) $m = 1, 2, 3$
      $n = 1, 2, 3, 4, 5$ $n = 1, 2, 3, 4$ $n = 1, 2, 3$

5.2 Derive the finite-difference energy balance equations and show the matrices $[A]$, $[T]$, and $[C]$, for a 1-D hollow cylinder with the following BCs:

(1) $r = r_1, T = T_1$ (2) $r = r_1, \quad -k\left(\dfrac{\partial T}{\partial r}\right) = h_1 (T_{\infty 1} - T_1)$
      $r = r_2, T = T_2$

$r = r_2, \quad -k\left(\dfrac{\partial T}{\partial r}\right) = h_2 (T_2 - T_{\infty 2})$

(3) $r = r_1, \quad -k(\partial T / \partial r) = h_1 (T_{\infty 1} - T_1)$

$r = r_2, \quad -k(\partial T / \partial r) = q_s''$

5.3 A semiinfinite stainless-steel block is initially at $T_i = 20°C$. The surface has an emissivity $\varepsilon = 1.0$ and is placed in a large enclosure of T_{sur} of 20°C. Suddenly, the surface is exposed to a hot-air flow of $T_\infty = 600°C$ and $h = 100$ W/m²K. At a given time of t (seconds), temperatures T_1 and T_2 have been calculated as shown below. Predict the surface temperature after $(t + 100)$ s. Use the forward-difference energy balance method. Given

$\alpha = 4 \cdot 10^{-6}\,m^2/s$	$k = 15.07\ w/mK$	$\rho = 7900\,kg/m^3$
$C_p = 477\,J/kg\ K$	$\sigma = 5.67 \cdot 10^{-8} w/m^2\ K^4$	$\varepsilon = 0.1$

At time t (seconds): $T_1 = 400°C$ and $T_2 = 380°C$. Find: $T_1 =$? after $(t = \Delta t)$ seconds where $\Delta t = 100\,s$.

5.4   Given a very long and wide fin with a height of $2L$. The base of the fin is maintained at a uniform temperature of T_b. The top and bottom surfaces of the fin are exposed to a fluid whose temperature is T_∞ $(T_\infty < T_b)$. The convective heat transfer coefficient between the fin surfaces and the fluid is h.

a. Sketch the steady 2-D temperature distribution in the fin.

b. If you were to determine the steady 2-D temperature distribution in the fin using a finite-difference numerical method, you would solve a set of algebraic nodal equations simultaneously for the temperatures at a 2-D array of nodes. Derive the equation for a typical node on one of the surfaces of the fin. Please do not simplify the equation.

c. Using the method of separation of variables, *derive* an expression for the steady local temperature in the fin, in terms of the thermal conductivity of the fin, k, the convective heat transfer coefficient, h, the half-height of the fin, L, and the base and fluid temperatures, T_b and T_∞.

5.5   A 3-mm-diameter rod that is 120 mm in length is supported by two electrodes within a large vacuum enclosure. Initially, the rod is in equilibrium with the electrodes and its surroundings. Suddenly, an electrical current is passed through the rod.

a. Using a first law analysis, what is the energy balance on the rod?

b. Using $\left(A_c = \left(\pi D^2 / 4 \right), P = \pi D \right)$, derive the explicit finite-difference expression for node (n). Recall that heat generation follows the $I^2 R_e$ law and that R_e is defined as $R_e = \rho_e \Delta x / A_c$. Express your answer using the Fourier number in the explicit finite-difference form.

c. What are the stability criteria at node (n)?

5.6   Refer to Figure 5.4, use the explicit finite-difference method to derive the energy balance equations during the transient process for the following grid distributions:

(1)	$m = 1, 2, 3, 4, 5$	(2)	$m = 1, 2, 3, 4$	(3)	$m = 1, 2, 3$
	$n = 1, 2, 3, 4, 5$		$n = 1, 2, 3, 4$		$n = 1, 2, 3$

5.7   Refer to Figure 5.4, use the implicit finite-difference method to derive the energy balance equations during the transient process for the following grid distributions:

(1)	$m = 1, 2, 3, 4, 5$	(2)	$m = 1, 2, 3, 4$	(3)	$m = 1, 2, 3$
	$n = 1, 2, 3, 4, 5$		$n = 1, 2, 3, 4$		$n = 1, 2, 3$

5.8 Use the explicit finite-difference method and derive finite-difference energy balance equations for a 1-D hollow cylinder during a transient process with the following BCs:

---

(1) $\quad r = r_1,\ T = T_1$
$\quad\quad r = r_2,\ T = T_2$

(2) $\quad r = r_1,\ \ -k\left(\dfrac{\partial T}{\partial r}\right) = h_1\left(T_{\infty 1} - T_1\right)$

$\quad\quad r = r_2,\ \ -k\left(\dfrac{\partial T}{\partial r}\right) = h_2\left(T_2 - T_{\infty 2}\right)$

(3) $\quad r = r_1,\ \ -k\left(\partial T / \partial r\right) = h_1\left(T_{\infty 1} - T_1\right)$
$\quad\quad r = r_2,\ \ -k\left(\partial T / \partial r\right) = q_s''$

5.9 Use the implicit finite-difference method and derive finite-difference energy balance equations for a 1-D hollow cylinder during the transient with the following BCs:

---

(1) $\quad r = r_1,\ T = T_1$
$\quad\quad r = r_2,\ T = T_2$

(2) $\quad r = r_1,\ \ -k\left(\dfrac{\partial T}{\partial r}\right) = h_1\left(T_{\infty 1} - T_1\right)$

$\quad\quad r = r_2,\ \ -k\left(\dfrac{\partial T}{\partial r}\right) = h_2\left(T_2 - T_{\infty 2}\right)$

(3) $\quad r = r_1,\ \ -k\left(\partial T / \partial r\right) = h_1\left(T_{\infty 1} - T_1\right)$
$\quad\quad r = r_2,\ \ -k\left(\partial T / \partial r\right) = q_s''$

5.10 Derive Equations (5.23) and (5.25) for 3-D transient heat conduction problems.

REFERENCES

1. F. Incropera and D. Dewitt, *Fundamentals of Heat and Mass Transfer*, Fifth Edition, John Wiley & Sons, New York, 2002.
2. A. Mills, *Heat Transfer*, Richard D. Irwin, Inc., Boston, MA, 1992.
3. K.-F. Vincent Wong, *Intermediate Heat Transfer*, Marcel Dekker, Inc., New York, 2003.

6 Heat Convection Equations

6.1 BOUNDARY LAYER CONCEPTS

For a fluid moving over a solid body, a hydrodynamic (or velocity) boundary layer is formed near the solid surface. The hydrodynamic (or velocity) boundary layer is the region where the fluid velocity changes from its free stream value to zero at the surface. For example, considering the flow over a flat plate, as shown in Figure 6.1, due to the viscosity of the fluid, the velocity gradually decreases from its free stream, maximum value, to zero at the flat plate (assuming that the fluid particle on the surface is not moving). The fluid particle wants to move at the same velocity as the free stream, but the viscous force resists the motion as it moves over the surface. Therefore, a hydrodynamic boundary layer develops over a solid surface due to fluid viscosity. The thickness of this layer is defined based on where the velocity reaches approximately 99% of the free stream value.

The velocity profile, and the associated hydrodynamic boundary layer thickness, over a flat plate will be solved in the later sections. In general, the hydrodynamic boundary layer thickness grows with the square root of distance from the leading edge of the flat plate. Figure 6.1 shows a typical profile over a flat plate in laminar (Re < 300×10^3) and turbulent (Re > 300×10^3) boundary layer regions, respectively. Once the velocity profile over a flat plate, $u(y)$, at a given distance x is determined, the hydrodynamic boundary layer thickness, the shear stress on the surface, and the friction factor (or friction coefficient) can be obtained using the following equations.

For external flow:

$$Re = \frac{\text{Inertia force}}{\text{Viscous force}} = \frac{\rho \mathcal{U}_\infty x}{\mu} = \frac{\mathcal{U}_\infty x}{\nu}$$

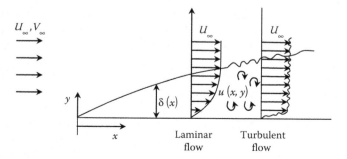

FIGURE 6.1 Hydrodynamic boundary layer over a flat plate.

DOI: 10.1201/9781003164487-6

The flow is defined as laminar if $Re < 300 \times 10^3$, and turbulent if $Re > 300 \times 10^3$.

Flow boundary-layer thickness $\delta(x) \sim \sqrt{x}$

Dynamic viscosity μ

Kinematic viscosity $\nu = \mu/\rho$

Shear stress $\tau_W = \mu \left. \dfrac{\partial u}{\partial y} \right|_{y=0} = C_f \cdot \dfrac{1}{2}\left(\rho \mathcal{U}_\infty^2 - 0\right)$

Friction coefficient

$$C_f = \frac{\tau_W}{(1/2)\left(\rho \mathcal{U}_\infty^2 - 0\right)} = \frac{\tau_W}{(1/2)\rho \mathcal{U}_\infty^2} = \frac{\mu \partial u/\partial y|_{y=0}}{(1/2)\rho \mathcal{U}_\infty^2} \sim \frac{\mu}{\rho} \frac{\left(\mathcal{U}_\infty/\delta\right)}{\mathcal{U}_\infty^2} \sim \nu \frac{1}{\delta \mathcal{U}_\infty}$$

C_f is a function of Reynolds number, that is:

$$C_f = a\,\mathrm{Re}^b; \qquad C_f = a\,\mathrm{Re}^b \sim \frac{1}{\mathrm{Re}}$$

The Reynolds number is defined as the fluid inertia force against the viscous force (i.e., the fluid particle tries to move but viscosity resists the movement) and is a combination of velocity, viscosity, and length (distance measured from the leading edge of the plate). When the Reynolds number is approximately less than 300×10^3, the motion of the fluid particles is very predictable: layer-to-layer from the free stream velocity to zero velocity on the solid surface. This velocity change creates shear stress over the solid surface. When the Reynolds number is greater than 300×10^3, the motion of the fluid particles tends to become unstable (random motion) and gradually transitions into the turbulent flow boundary layer. In the laminar boundary layer, the velocity profile gradually changes from the free stream value to zero on the surface as a parabolic shape. However, in the turbulent boundary layer, the velocity profile remains fairly uniform, approximately at the free stream value, until very near the surface and then suddenly changes to zero at the surface. This is due to turbulent mixing (the particle moves up and down, back and forth); therefore, the free stream characteristics are transported closer to the surface. From the application point of view, the shear stress (viscosity $\times$ velocity gradient at the surface, i.e., $\tau_w = \mu(\partial u/\partial y)|_{y=0}$) decreases with decreasing velocity gradient or viscosity.

Since the velocity gradient decreases (as the boundary layer grows thicker) with increasing distance from the leading edge, the shear stress (related to pressure loss) and friction factor decrease with increasing distance from the leading edge of the flat plate. However, when the flow transitions into turbulence, the pressure loss is much greater than that in the laminar flow portion. This is due to the increased pressure loss associated with overcoming the turbulent, random motion within the turbulent boundary layer.

It is noted that the boundary layer thickness decreases with increasing square root of the free stream velocity, and the friction factor decreases with increasing free stream velocity; however, the shear stress increases with increasing free stream velocity (a thinner boundary layer and larger velocity gradient), as sketched in Figure 6.2. Similarly, the friction factor decreases with increasing Reynolds number, but the shear stress increases with increasing Reynolds number.

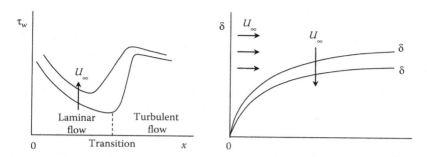

FIGURE 6.2 Hydrodynamic boundary layer, friction factor, and shear stress profile.

For a hot fluid moving over a cold surface, a thermal (temperature) boundary layer is formed around the solid surface. The thermal (or temperature) boundary layer is the region where the fluid temperature changes from its free stream value to that at the solid surface. Heat transfer can take place either from the hot fluid to the cold surface or from the heated surface to cold fluid, as shown in Figure 6.3. For example, considering the hot fluid over a cold flat plate, the temperature gradually decreases from its free stream maximum value to that at the plate, due to the conductivity of the fluid and its velocity distribution. The hot fluid particle conducts heat from the free stream into the cold surface through the velocity boundary layer. Therefore, a thermal boundary layer thickness is developed over a solid surface due to the fluid motion.

The temperature profile and associated thermal boundary layer thickness over a flat plate are solved in Chapter 7. In an ideal case (assume Pr = 1), the thermal boundary layer is identical to the hydrodynamic boundary layer, as shown comparing Figures 6.1 and 6.3. In this ideal case, the temperature profile is the same as the velocity profile through the entire boundary layer over the flat plate. Once we determine the temperature profile over a flat plate, $T(y)$ at a given distance x, the thermal boundary layer thickness, the heat flux on the surface, and the heat transfer coefficient (or Nusselt number) can be obtained as follows:

Thermal boundary-layer thickness:

$$\delta_T(x) \sim \sqrt{x}$$

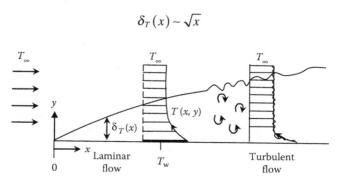

FIGURE 6.3 Thermal boundary layer over a heated flat plate.

If $Pr = 1$:

$$\delta(x) = \delta_T(x)$$

At the surface, the heat flux is:

$$q_w'' = -k \frac{\partial T}{\partial y}\bigg|_{y=0} = -k_f \frac{\partial T}{\partial y}\bigg|_{y=0} \equiv h(T_W - T_\infty)$$

The heat transfer coefficient, h, with the unit of W/m^{2}K can be expressed as

$$h = \frac{-k_f (\partial T/\partial y)\big|_{y=0}}{T_W - T_\infty} \sim \frac{-k_f \left((T_\infty - T_W)/\delta_T\right)}{T_W - T_\infty} \sim \frac{k_f}{\delta_T} \sim \frac{k_f}{\delta} \sim k_f \mathcal{U}_\infty$$

In the laminar boundary layer, the temperature profile gradually changes from the free stream value to the surface with a parabolic shape. However, in the turbulent boundary layer, the temperature profile remains relatively uniform from the free stream to near the surface and then suddenly changes to the surface value. This is due to turbulent mixing (particle moves up and down, back and forth); the hot (or cold) free stream particle can move next to the cold (or heated) surface due to random motion. From an application point of view, the heat flux (fluid conductivity · temperature gradient at the surface) and the heat transfer coefficient (heat flux/temperature difference between the free stream and the surface) decrease with decreasing temperature gradient and fluid conductivity. Because the temperature gradient decreases (the thermal boundary layer thickness increases) with increasing distance due to fluid thermal conductivity, the heat flux (related to heat transfer rate) and the heat transfer coefficient decrease with increasing distance from the leading edge of the flat plate. However, when the flow transitions into the turbulent boundary layer, the heat flux (and heat transfer coefficient) is much greater than when the flow is laminar. This is because a major portion of the heat transfer is due to turbulent, random motion in the turbulent boundary layer.

It is noted that heat flux is proportional to the heat transfer coefficient and the temperature difference between the free stream and the surface. The heat transfer coefficient increases with increasing free stream velocity (thinner hydrodynamic and thermal boundary layers) and the fluid thermal conductivity, as shown in Figure 6.4. This implies that the heat transfer coefficient, heat flux, and Nusselt number (the dimensionless heat transfer coefficient) increase with Reynolds number.

Another important parameter in heat transfer study is the role of the Prandtl number (Pr). The Prandtl number is a ratio of kinematic viscosity to thermal diffusivity, or a ratio of velocity to temperature boundary layer thickness, as sketched in Figure 6.4. For example, the thermal boundary layer thickness is identical to the hydrodynamic boundary layer thickness if $Pr = 1$, as discussed above. However, in real life, different fluids have different Prandtl numbers. In Chapters 7, 8, and 10, we will see that the Nusselt number is proportional to the Reynolds number and Prandtl number for both laminar and turbulent flows (using a different power and constants).

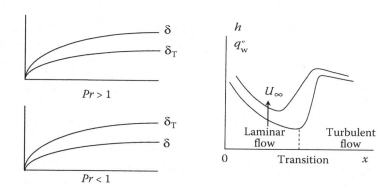

FIGURE 6.4 Thermal boundary layer, heat transfer coefficient, and heat flux profile.

Nusselt number:

$$\text{Nu} = \frac{hx}{k} = a\,\text{Re}^b\,\text{Pr}^n$$

Prandtl number:

$$\text{Pr} = \frac{\nu}{\alpha} = \frac{\mu/\rho}{k/\rho C_p} = \frac{\mu C_p}{k} \sim \frac{\delta}{\delta_T}$$

The Prandtl number is a property of fluid; it shows the ratio of momentum transfer versus heat transfer.

Air or gas $\text{Pr} = 0.7$
Water $Pr = 2 \sim 20$
Oil $Pr = 100 \sim 1000$
Liquid metal $Pr = 0.01 \sim 0.001$

When the temperature increases, the viscosity and Prandtl number for oil decrease.

6.2 GENERAL HEAT CONVECTION EQUATIONS

For a fluid moving over a heated or cooled surface, the general 3-D pressure profile $P(x, y, z, t)$, velocity profiles $u(x, y, z, t)$, $v(x, y, z, t)$, and $w(x, y, z, t)$, and temperature profile $T(x, y, z, t)$ can be obtained by solving the following continuity, momentum, and energy equations inside the hydrodynamic and thermal boundary layers over the heated or cooled solid surface [1–3].

Conservation of mass (continuity equation):

$$\frac{\partial \rho}{\partial t} + \nabla \cdot (\rho V) = 0 \tag{6.1}$$

where

$$V = iu + jv + kw$$

and

$$\nabla = i\left(\frac{\partial}{\partial x}\right) + j\left(\frac{\partial}{\partial y}\right) + k\left(\frac{\partial}{\partial z}\right)$$

are the velocity vector and the del operator for the unit vectors, i, j, and k in the x-, y-, and z-directions, respectively.

Conservation of momentum:

$$\rho\frac{DV}{Dt} = -\nabla P + \mu\nabla^2 V + \rho g \tag{6.2}$$

Conservation of energy:

$$\rho\frac{Dh}{Dt} = \frac{DP}{Dt} + \nabla \cdot k\nabla T + \mu\Phi + \dot{q} \tag{6.3}$$

where $h = e + (1/2)\ V \cdot V$, and e is the specific internal energy. Φ is often called the dissipation function with the form

$$\Phi = 2\left[\left(\frac{\partial u}{\partial x}\right)^2 + \left(\frac{\partial v}{\partial y}\right)^2 + \left(\frac{\partial w}{\partial z}\right)^2\right] + \left(\frac{\partial u}{\partial y} + \frac{\partial v}{\partial x}\right)^2 + \left(\frac{\partial u}{\partial z} + \frac{\partial w}{\partial x}\right)^2 + \left(\frac{\partial v}{\partial z} + \frac{\partial w}{\partial y}\right)^2 \tag{6.4}$$

And $\dot{q}$ is the heat generation per unit volume.

6.3   2-D HEAT CONVECTION EQUATIONS

Many real-life applications for a fluid moving over a solid body can be modeled as 2-D boundary layer flow and heat transfer problems. We need to know the 2-D velocity profiles, $u(x, y, t)$ and $v(x, y, t)$, to calculate the wall shear stress (related to pressure loss) and the friction factor along the surface for a given fluid at given flow conditions. In addition, we need the 2-D temperature profile, $T(x, y, t)$, to calculate the wall heat flux (related to heat transfer rate) and the heat transfer coefficient along the surface for a given fluid at given thermal BCs. Since the fluid is moving due to a pressure difference between the upstream and downstream fluid and the viscous boundary layer effect over the solid surface, it is necessary to perform conservation of mass (continuity equation) and momentum (momentum equation) through the boundary layer to solve for the velocity distributions over the surface. Similarly, since the heat transfer is due to the temperature difference between the free stream and the surface and the energy carried with the moving fluid, it is necessary to perform the conservation of energy (energy equation) through the thermal boundary layer to solve for temperature distributions over the heated (or cooled) surface.

Consider a small 2-D differential fluid element (dxdy) at any point within the boundary layer, the following shows a step-by-step derivation of the 2-D conservation equations for mass, momentum, and energy through hydrodynamic and thermal boundary layers over a solid surface [1–3].

Conservation of mass:

Perform a mass balance, as shown in Figure 6.5:

$$-\frac{\partial}{\partial x}(\rho u\, dy)dx - \frac{\partial}{\partial y}(\rho v\, dx)dy = \frac{\partial}{\partial t}(\rho\, dx\, dy) \tag{6.5}$$

If the flow is steady:

$$-\frac{\partial}{\partial x}(\rho u\, dy)dx - \frac{\partial}{\partial y}(\rho v\, dx)dy = 0$$

$$\frac{\partial(\rho u)}{\partial x} + \frac{\partial(\rho v)}{\partial y} = 0 \tag{6.6}$$

For incompressible flow, ρ = constant, the continuity equation can be simplified as

$$\frac{\partial u}{\partial x} + \frac{\partial v}{\partial y} = 0 \tag{6.7}$$

Conservation of momentum:

From Newton's Second Law, the net force exerted on a body equals the momentum change.

$$\sum F_x = ma_x$$

$$\sum F_y = ma_y$$

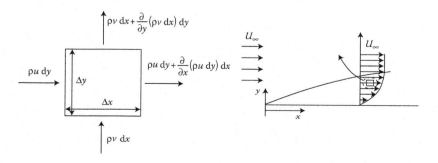

FIGURE 6.5 Conservation of mass.

where a_x is the acceleration in the x-direction, and a_y is the acceleration in the y-direction.

The force exerted on a control volume and the momentum change are shown in Figure 6.6.

$$\underbrace{\frac{\partial}{\partial x}\sigma_x}_{\substack{\text{normal}\\\text{stress}}} + \underbrace{\frac{\partial}{\partial y}\tau_{xy}}_{\substack{\text{shear}\\\text{stress}}} - \underbrace{\frac{\partial P_x}{\partial x}}_{\substack{\text{pressure}\\\text{gradient}}} = \underbrace{\frac{\partial}{\partial x}\left(\rho u^2\right) + \frac{\partial}{\partial y}\left(\rho uv\right)}_{\text{convective term}} + \underbrace{\frac{\partial}{\partial t}\left(\rho u\right)}_{\substack{\text{unsteady}\\\text{term}}} \tag{6.8}$$

From Navier–Stokes for a Newtonian, incompressible fluid:

$$\sigma_x = 2\mu\frac{\partial u}{\partial x} \tag{6.9}$$

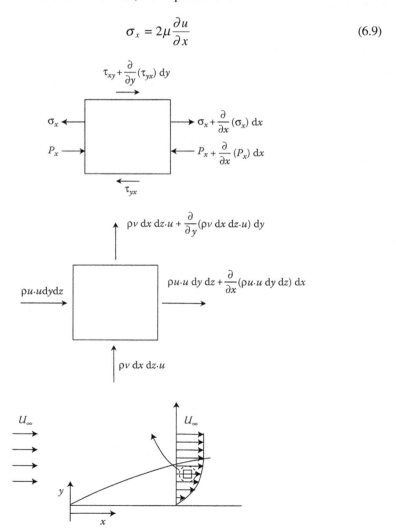

FIGURE 6.6   Conservation of momentum.

$$\tau_{xy} = \tau_{yx} = \mu\left(\frac{\partial u}{\partial y} + \frac{\partial v}{\partial x}\right) \tag{6.10}$$

Substituting Equations (6.9) and (6.10) into Equation (6.8), we obtain

$$\frac{\partial}{\partial x}\left(2\mu\frac{\partial u}{\partial x}\right) + \frac{\partial}{\partial y}\left[\mu\left(\frac{\partial u}{\partial y} + \frac{\partial v}{\partial x}\right)\right] - \frac{\partial P_x}{\partial x} = \frac{\partial}{\partial x}\left(\rho u^2\right) + \frac{\partial}{\partial y}(\rho uv) + \frac{\partial}{\partial t}(\rho u)$$

For a steady-state, constant-property flow, the left-hand side can be expressed as

$$2\mu\frac{\partial^2 u}{\partial x^2} + \mu\frac{\partial^2 u}{\partial y^2} + \mu\frac{\partial^2 v}{\partial x \partial y} = \mu\frac{\partial^2 u}{\partial x^2} + \mu\frac{\partial^2 u}{\partial y^2} + \mu\frac{\partial^2 u}{\partial x^2} + \mu\frac{\partial^2 v}{\partial x \partial y}$$

$$= \mu\frac{\partial^2 u}{\partial x^2} + \mu\frac{\partial^2 u}{\partial y^2} + \mu\frac{\partial}{\partial x}\left(\frac{\partial u}{\partial x} + \frac{\partial v}{\partial y}\right)^{\!\!\!0}$$

$$\sim \mu\frac{\partial^2 u}{\partial y^2} \quad \text{if } \mu\frac{\partial^2 u}{\partial x^2} \ll \mu\frac{\partial^2 u}{\partial y^2}$$

The right-hand side can be expanded as

$$\frac{\partial}{\partial x}\left(\rho u^2\right) + \frac{\partial}{\partial y}(\rho uv) = 2\rho u\frac{\partial u}{\partial x} + \rho v\frac{\partial u}{\partial y} + \rho u\frac{\partial v}{\partial y}$$

$$= \rho u\frac{\partial u}{\partial x} + \rho v\frac{\partial u}{\partial y} + \rho u\left(\frac{\partial u}{\partial x} + \frac{\partial v}{\partial y}\right)^{\!\!\!0}$$

Introduce the mass conservation equation, and we will obtain the x-direction momentum equation:

$$\underbrace{u\frac{\partial u}{\partial x} + v\frac{\partial u}{\partial y}}_{\text{convection}} = \underbrace{-\frac{1}{\rho}\frac{\partial P_x}{\partial x}}_{\text{pressure}} + \underbrace{v\frac{\partial^2 u}{\partial x^2} + v\frac{\partial^2 u}{\partial y^2}}_{\text{stress}} \tag{6.11}$$

Similarly, we will obtain the y-direction momentum equation from Equation (6.11) by changing u and v inside the derivatives, and considering the pressure gradient in the y-direction:

$$u\frac{\partial v}{\partial x} + v\frac{\partial v}{\partial y} = -\frac{1}{\rho}\frac{\partial P_x}{\partial y} + v\frac{\partial^2 v}{\partial x^2} + v\frac{\partial^2 v}{\partial y^2} \tag{6.12}$$

Conservation of energy:
Beginning with the more general equation for unsteady flow:
Perform energy balance as shown in Figure 6.7,

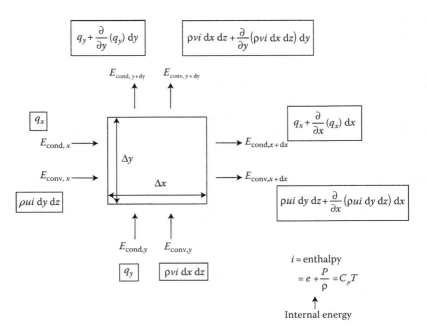

FIGURE 6.7   Conservation of energy.

$$-\frac{\partial q_x}{\partial x}dx - \frac{\partial q_y}{\partial y}dy - \frac{\partial}{\partial x}\left(\rho u \, dy \cdot C_p \cdot T\right)dx - \frac{\partial}{\partial y}\left(\rho v \, dx \cdot C_p \cdot T\right)dy = \frac{\partial\left(\rho \, dx \, dy \cdot C_p \cdot T\right)}{\partial t}$$

$$(6.13)$$

$$\underbrace{\frac{\partial\left(\rho C_p T\right)}{\partial t}}_{\text{unsteady}} + \underbrace{\frac{\partial}{\partial x}\left(\rho u C_p T\right) + \frac{\partial}{\partial y}\left(\rho v C_p T\right)}_{\text{convection}} = \underbrace{\frac{\partial}{\partial x}\left(k\frac{\partial T}{\partial x}\right) + \frac{\partial}{\partial y}\left(k\frac{\partial T}{\partial y}\right)}_{\text{heat diffusion}} + \underbrace{\frac{\dot{q}}{k}}_{\substack{\text{heat}\\\text{generation}}}$$

$$+ \underbrace{\mu\Phi}_{\substack{\text{heat dissipation}\\\text{source due to friction}}}$$

$$(6.14)$$

For steady-state and constant properties:

$$\frac{\partial(uT)}{\partial x} + \frac{\partial(vT)}{\partial y} = \alpha\left(\frac{\partial^2 T}{\partial x^2} + \frac{\partial^2 T}{\partial y^2}\right) + \frac{\mu}{\rho C_p}\Phi \qquad (6.15)$$

$$\mu\Phi = \mu\left\{\left(\frac{\partial u}{\partial y} + \frac{\partial v}{\partial x}\right)^2 + 2\left[\left(\frac{\partial u}{\partial x}\right)^2 + \left(\frac{\partial v}{\partial y}\right)^2\right]\right\} \sim \mu\left(\frac{\partial u}{\partial y}\right)^2 \qquad (6.16)$$

6.4 BOUNDARY-LAYER APPROXIMATIONS

The above-derived boundary equations are not easily solved analytically. Therefore, boundary layer approximations, as shown in Figure 6.8, can be employed to simplify the boundary equations as follows: the velocity in the x-direction is greater than that in the y-direction; the streamwise velocity change in the y-direction is greater than that in the x-direction; the temperature change in the y-direction is greater than that in the x-direction.

$$u \gg v$$

$$\frac{\partial u}{\partial y} \gg \frac{\partial u}{\partial x}, \frac{\partial v}{\partial x}, \frac{\partial v}{\partial y}$$

$$\frac{\partial T}{\partial y} \gg \frac{\partial T}{\partial x}$$

Therefore, the continuity equation remains the same. Assuming steady flow and constant fluid properties, the momentum equations in the x- and y-directions can be simplified as

$$u\frac{\partial u}{\partial x} + v\frac{\partial u}{\partial y} = -\frac{1}{\rho}\frac{\partial P}{\partial x} + v\frac{\partial^2 u}{\partial y^2} \tag{6.17}$$

$$-\frac{1}{\rho}\frac{\partial P}{\partial y} = 0 \tag{6.18}$$

For an incompressible flow, that is, $M < 0.2$, $\Phi \sim 0$, the energy equation becomes

$$u\frac{\partial T}{\partial x} + v\frac{\partial T}{\partial y} = \alpha\frac{\partial^2 T}{\partial y^2} \tag{6.19}$$

Outside of the boundary layer, we have potential flow (μ effect $\to 0$, $v \to 0$, $\partial u/\partial y \to 0$); Equation 6.17 reduces to

$$\mathcal{U}_\infty\frac{\partial \mathcal{U}_\infty}{\partial x} = -\frac{1}{\rho}\frac{\partial P}{\partial x} \tag{6.20}$$

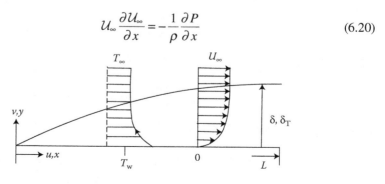

FIGURE 6.8 Boundary-layer approximations.

Note that Equation (6.18) implies that there is no pressure change in the y-direction within the boundary layer. Equation (6.20) implies that the pressure change in the x-direction within the boundary layer can be predetermined from the velocity and its velocity change in the x-direction outside of the boundary layer. Therefore, Equation (6.20) can be substituted into Equation (6.17) to solve for the velocity profiles inside the boundary layer.

6.4.1 BOUNDARY-LAYER SIMILARITY/DIMENSIONAL ANALYSIS

Here we want to generalize the application of the above-derived boundary layer approximation equations. Most often, we want to apply the boundary layer equations from a small-scale test model to a large-scale application or from a large-scale test model to a small-scale application. This is called boundary layer similarity or dimensional analysis. The following is a common way of converting dimensional parameters into nondimensional parameters [2].

Let

$$x^* = \frac{x}{L} \qquad y^* = \frac{y}{L}$$

$$u^* = \frac{u}{\mathcal{U}_\infty} \qquad v^* = \frac{v}{\mathcal{U}_\infty}$$

$$T^* = \frac{T - T_W}{T_\infty - T_W} \qquad P^* = \frac{P}{\rho \mathcal{U}_\infty^2} \qquad (6.21)$$

Then, the conservation equations can be written as

$$\frac{\partial u^*}{\partial x^*} + \frac{\partial v^*}{\partial y^*} = 0$$

$$u^* \frac{\partial u^*}{\partial x^*} + v^* \frac{\partial u^*}{\partial y^*} = -\frac{\partial P^*}{\partial x^*} + \frac{v}{\mathcal{U}_\infty L} \cdot \frac{\partial^2 u^*}{\partial y^{*2}}$$

$$\frac{\partial P^*}{\partial y^*} = 0$$

$$u^* \frac{\partial T^*}{\partial x^*} + v^* \frac{\partial T^*}{\partial y^*} = \frac{\alpha}{\mathcal{U}_\infty L} \cdot \frac{\partial^2 T^*}{\partial y^{*2}}$$

The coefficient is

$$\frac{\alpha}{\mathcal{U}_\infty L} = \frac{1}{(\mathcal{U}_\infty L/v) \cdot (v/\alpha)} = \frac{1}{\text{Re} \cdot \text{Pr}} \qquad (6.22)$$

The above similarity functional solutions can be written as

$$u^* = f_1\left(x^*, y^*, \mathrm{Re}_L, \frac{dP^*}{dx^*}\right)$$

$$\tau_W = \mu \left.\frac{\partial u}{\partial y}\right|_{y=0} = \mu \frac{\mathcal{U}_\infty}{L} \left.\frac{\partial u^*}{\partial y^*}\right|_{y^*=0}$$

where

$$\left.\frac{\partial u^*}{\partial y^*}\right|_{y^*=0} = f_2\left(x^*, \mathrm{Re}_L, \frac{dP^*}{dx^*}\right)$$

$$C_f = \frac{\tau_W}{(1/2)\rho V_\infty^2} = \frac{\mu(\mathcal{U}_\infty/L)}{(1/2)\rho V_\infty^2} f_2\left(x^*, \mathrm{Re}_L, \frac{dP}{dx^*}\right) \qquad (6.23)$$

$$C_f = \frac{2}{\mathrm{Re}_L} f_2\left(x^*, \mathrm{Re}_L, \frac{dP^*}{dx^*}\right)$$

Special case: when flow over a flat plate $dP^*/dx^* = 0$, the average friction factor can be determined from the Reynolds number as:

$$\overline{C}_f = \frac{2}{\mathrm{Re}_L} f_2(\mathrm{Re}_L) = a\,\mathrm{Re}_L^m \qquad (6.24)$$

Similarly, the temperature and heat transfer coefficient can be obtained as

$$T^* = f_3\left(x^*, y^*, \mathrm{Re}_L, \mathrm{Pr}, \frac{dP^*}{dx^*}\right)$$

$$h = \frac{-k(\partial T/\partial y)|_{y=0}}{T_W - T_\infty} \sim \frac{k}{L} \left.\frac{\partial T^*}{\partial y^*}\right|_{y^*=0} = \frac{k}{L} f_4\left(x^*, \mathrm{Re}_L, \mathrm{Pr}, \frac{dP^*}{dx^*}\right) \qquad (6.25)$$

$$\mathrm{Nu}_L \equiv \frac{hL}{k} = f_4\left(x^*, \mathrm{Re}_L, \mathrm{Pr}, \frac{dP^*}{dx^*}\right)$$

For flow over a flat plate, $dP^*/dx^* = 0$, the average Nusselt number can be determined from Reynolds number and Prandtl number as

$$\overline{Nu}_L \equiv \frac{\overline{h}L}{k} = f_5\left(Re_L, Pr\right) = aRe_L^m Pr^n \tag{6.26}$$

The above analysis concludes that for flow over a flat plate, the local friction factor (at a given location x) is a function of only the Reynolds number, and the local heat transfer coefficient, or Nusselt number (at a given location x), is a function of both the Reynolds number and the Prandtl number.

6.4.2 REYNOLDS ANALOGY

Assuming that $Pr = 1$ (approximation for air, $Pr = 0.7$), the above friction factor and the Nusselt number can be reduced to the following:

$$C_{fx} = \frac{2}{Re_L} f_2\left(x^*, Re_L\right) \tag{6.27}$$

$$Nu_x = f_4\left(x^*, Re_L, Pr\right) \tag{6.28}$$

If $f_2 = f_4$, $Pr = 1$,

$$C_f \frac{Re_L}{2} = f_2 = f_4 = Nu$$

$$\frac{1}{2}C_f = \frac{Nu}{Re \cdot Pr} = St = \frac{(hL/k)}{(\rho VL/\mu)\cdot(\mu C_p/k)} = \frac{h}{\rho C_p V} \tag{6.29}$$

Reynolds analogy:

$$\frac{1}{2}C_f = St \tag{6.30}$$

Experimentally, we obtained

$$\frac{1}{2}C_f Pr^{-2/3} = St \tag{6.31}$$

where $0.6 \leq Pr \leq 60$.

The importance of the Reynolds analogy is that one can estimate the heat transfer coefficient (or the Stanton number, St) from a given (or a predetermined) friction factor, or one can calculate the friction factor from a given (or predetermined) heat transfer coefficient (or St) for a typical 2-D boundary layer flow and heat transfer problem. The original Reynolds analogy is shown in Equation (6.30). However, Equation (6.31) is also referred to as the Reynold's analogy, including the Prandtl number effect.

6.5 MASS TRANSFER

6.5.1 THE CONCENTRATION BOUNDARY

The velocity and temperature boundary layers determine surface friction and convection heat transfer, respectively. In a similar way, the concentration boundary layer determines convective mass transfer. Consider a mixture of chemical species A and B flowing over a flat surface, the concertation of species A at the surface is $C_{A,w}$, kmol/m³, in the free stream the concentration is $C_{A,\infty}$, kmol/m³. Between the surface and the free stream, a concentration boundary layer is developed, as shown in Figure 6.9.

For steady, 2-D flow of an incompressible fluid with constant properties, the conservation of chemical species is:

$$u\frac{\partial C_A}{\partial x} + v\frac{\partial C_A}{\partial y} = D_{AB}\left(\frac{\partial^2 C_A}{\partial x^2} + \frac{\partial^2 C_A}{\partial y^2}\right)$$

Based on the boundary layer approximation, $\dfrac{\partial C_A}{\partial y} \gg \dfrac{\partial C_A}{\partial x}$, the above equation is simplified:

$$u\frac{\partial C_A}{\partial x} + v\frac{\partial C_A}{\partial y} = D_{AB}\frac{\partial^2 C_A}{\partial y^2}$$

Based on the heat and mass transfer analogy, the molar flux can be written as:

$$M_A'' = -D_{AB}\frac{\partial C_A}{\partial y}$$

where D_{AB} is the binary diffusion coefficient from A to B. The species molar flux and mass transfer coefficient at the surface can be combined:

$$M_A'' = -D_{AB}\frac{\partial C_A}{\partial y}\bigg|_{y=0}$$

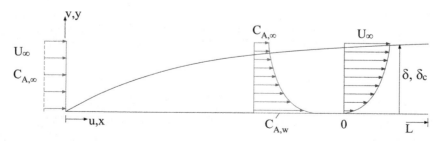

FIGURE 6.9 Concentration boundary layer over a flat plate.

$$h_m = -\frac{D_{AB}\left.\dfrac{\partial C_A}{\partial y}\right|_{y=0}}{C_{A,w} - C_{A,\infty}}$$

Based on boundary layer similarity parameters, the above concentration equation becomes:

$$C_A^* = \frac{C_A - C_{A,w}}{C_{A,\infty} - C_{A,w}}$$

$$u^* \frac{\partial C_A^*}{\partial x^*} + v^* \frac{\partial C_A^*}{\partial y^*} = \frac{D_{AB}}{\mathcal{U}_\infty L} \frac{\partial^2 C_A^*}{\partial y^{*2}} = \frac{1}{\mathrm{Re} \cdot \mathrm{Sc}} \cdot \frac{\partial^2 C_A^*}{\partial y^{*2}}$$

where

$$\frac{D_{AB}}{\mathcal{U}_\infty L} = \frac{v}{\mathcal{U}_\infty L} \cdot \frac{D_{AB}}{v} = \frac{D_{AB}/v}{\mathrm{Re}} = \frac{1}{\mathrm{Re} \cdot \mathrm{Sc}}$$

with

$$\mathrm{Re} = \frac{\mathcal{U}_\infty L}{v}; \quad \mathrm{Sc} = \frac{v}{D_{AB}}$$

The boundary conditions are specified as:

$$C_A^*\left(x^*, 0\right) = 0$$

$$C_A^*\left(x^*, \infty\right) = 1$$

The solution to this equation has the functional form of:

$$C_A^* = f\left(x^*, y^*, \mathrm{Re}, \mathrm{Sc}, \frac{dp^*}{dx^*}\right)$$

From the above heat transfer analysis, we obtain the mass transfer coefficient:

$$h_m = -\frac{D_{AB}}{L} \cdot \frac{C_{A,\infty} - C_{A,w}}{C_{A,\infty} - C_{A,w}} \cdot \left.\frac{\partial C_A^*}{\partial y^*}\right|_{y^*=0} = \frac{D_{AB}}{L} \left.\frac{\partial C_A^*}{\partial y^*}\right|_{y^*=0}$$

The species density, ρ_A, is the product of the species concentration, C_A, and the molecular weight, w_A. Therefore, the species mass flux, m_A'', can be written as the product of the species molar flux, M_A'', and the molecular weight, w_A.

$$m_A'' = M_A'' \cdot w_A$$

$$m_A'' = -D_{AB} \frac{\partial \rho_A}{\partial y}$$

$$h_m = \frac{-D_{AB} \left. \dfrac{\partial \rho_A}{\partial y} \right|_{y=0}}{\rho_{A,w} - \rho_{A,\infty}}$$

Define the dimensionless parameter, known as the Sherwood number.

$$Sh = \frac{h_m L}{D_{AB}} = \left. \frac{\partial C_A^*}{\partial y^*} \right|_{y^*=0}$$

Therefore, the local Sherwood number and the average Sherwood number become:

$$Sh = f\left(x^*, \mathrm{Re}, \mathrm{Sc}, \frac{dp^*}{dx^*} \right)$$

$$\overline{Sh} = \frac{\overline{h_m} L}{D_{AB}} = f\left(\mathrm{Re}, \mathrm{Sc}, \frac{dp^*}{dx^*} \right)$$

For flow over a flat plate:

$$\frac{dp^*}{dx^*} = 0$$

From Figure 6.9, the Schmidt number is a measure of the ratio of the velocity to concentration boundary later thickness as:

$$Sc \sim \left(\frac{\delta}{\delta_c} \right)^n, \; n \sim 3$$

The Lewis number relates Pr and Sc:

$$Le = \frac{\alpha}{D_{AB}} = \frac{Sc}{Pr}$$

6.5.2 HEAT, MASS, AND MOMENTUM TRANSFER ANALOGY

Momentum Transfer—Velocity Boundary Layer:

$$u^* = f_1\left(x^*, y^*, \mathrm{Re}_L, \frac{dp^*}{dx^*} \right)$$

If $\dfrac{dp^*}{dx^*} = 0$:

$$C_f = \frac{2}{\mathrm{Re}_L} \cdot \left.\frac{\partial u^*}{\partial y^*}\right|_0 = \frac{2}{\mathrm{Re}_L} f_2\left(x^*, \mathrm{Re}_L\right)$$

$$\overline{C_f} = \frac{2}{\mathrm{Re}_L}\overline{f_2}(\mathrm{Re}_L) = a\,\mathrm{Re}_L^m$$

Heat Transfer—Thermal (Temperature) Boundary Layer:

$$T^* = f_3\left(x^*, y^*, \mathrm{Re}_L, \mathrm{Pr}, \frac{dp^*}{dx^*}\right)$$

If $\dfrac{dp^*}{dx^*} = 0$:

$$\mathrm{Nu} = \frac{hL}{k} = \left.\frac{\partial T^*}{\partial y^*}\right|_0 = f_4\left(x^*, \mathrm{Re}_L, \mathrm{Pr}\right)$$

$$\overline{\mathrm{Nu}} = f_5(\mathrm{Re}_L, \mathrm{Pr}) = a\,\mathrm{Re}_L^m\,\mathrm{Pr}^n$$

Mass Transfer—Concentration (Density) Boundary Layer:

$$C_A^* = f_6\left(x^*, y^*, \mathrm{Re}_L, \mathrm{Sc}, \frac{dp^*}{dx^*}\right)$$

$$\mathrm{Sh} = \frac{h_m L}{D_{AB}} = \left.\frac{\partial C_A^*}{\partial y^*}\right|_0 = f_7\left(x^*, \mathrm{Re}_L, \mathrm{Sc}\right)$$

$$\overline{\mathrm{Sh}} = f_8\left(\mathrm{Re}_L, \mathrm{Sc}\right) = a\,\mathrm{Re}_L^m\,\mathrm{Sc}^n$$

Dimensionless Parameters:

$$\mathrm{Re} = \frac{\rho v L}{\mu} = \frac{vL}{v} \quad \text{Reynolds number} \sim \frac{\text{inertia force}}{\text{viscous force}}$$

$$\mathrm{Pr} = \frac{v}{\alpha} = \frac{\mu c_p}{k} \quad \text{Prandtl number} \sim \frac{\text{velocity boundary layer}}{\text{thermal boundary layer}} \sim \left(\frac{\delta}{\delta_t}\right)^n$$

$$\mathrm{Nu} = \frac{hL}{k} \quad \text{Nusselt number} \sim \text{dimensionless heat transfer coefficient}$$

$$Sc = \frac{v}{D_{AB}} \text{ Schmidt number} \sim \frac{\text{velocity boundary layer}}{\text{concentration boundary layer}} \sim \left(\frac{\delta}{\delta_c}\right)^n$$

$$Sh = \frac{h_m L}{D_{AB}} \text{ Sherwood number} \sim \text{ dimensionless mass transfer coefficient}$$

$$Le = \frac{\alpha}{D_{AB}} = \frac{Sc}{Pr} \text{ Lewis number} \sim \frac{\text{thermal boundary layer}}{\text{concentration boundary layer}} \sim \left(\frac{\delta_t}{\delta_c}\right)^n$$

$$St = \frac{h}{\rho c_p v} = \frac{Nu}{Re \cdot Pr} = \frac{1}{2} C_f \cdot Pr^{-2/3} \text{ heat transfer Stanton number, for } 0.6 < Pr < 60$$

$$St_m = \frac{h_m}{v} = \frac{Sh}{Re \cdot Pr} = \frac{1}{2} C_f \cdot S_c^{-2/3} \text{ mass transfer Stanton number, for } 0.6 < Sc < 3000$$

$$\frac{h}{h_m} = \rho c_p \left(\frac{Pr}{Sc}\right)^{-2/3} = \rho c_p Le^{2/3} \text{ relationship between heat and mass transfer}$$
coefficients

$St = St_m$ special case if $Pr = Sc = 1$

$h = \rho c_p h_m$, special case if $Pr = Sc = 1$, or $Le = 1$

6.5.3 EVAPORATIVE COOLING MASS TRANSFER

Air (species B) at T_∞ flows over the water surface (species A) at T_w. For a steady-state condition, the molar transfer rate of species A $(M_A, \text{kmol/s})$ can be expressed as:

$$M_A = h_m A_s \left[C_{A,s}(T_w) - C_{A,\infty}(T_\infty) \right]$$

where h_m (m/s) is the convection mass transfer coefficient, the molar concentrations $C_{A,s}(T_w)$ and $C_{A,\infty}(T_\infty)$ have units of kmol/m^3.

The species transfer rate may also be expressed as a mass transfer rate $(m_A, \text{kg/s})$ by multiplying both sides of the equation by the molecular weight w_A (kg/kmol) of species A, thus:

$$m_A = M_A \cdot w_A$$

$$= h_m A_s \left[\rho_{A,s}(T_w) - \rho_{A,\infty}(T_\infty) \cdot \phi_\infty \right]$$

$$= \text{water evaporative mass transfer rate, kg/s}$$

where

$$\rho_{A,s}(T_w) = C_{A,s}(T_w) \cdot w_A$$

$$= \text{saturated water vapor density at temperature } T_w, \text{kg/m}^3$$

$$= \frac{P_{A,s}}{RT_w} \left(\text{based on the ideal gas law}\right)$$

$$\rho_{A,\infty}(T_\infty) = C_{A,\infty}(T_\infty) \cdot w_A$$

$$= \text{saturated water vapor density at temperature } T_\infty, \text{kg/m}^3$$

$$= \frac{P_{A,s}}{RT_\infty} \left(\text{based on the ideal gas law}\right)$$

$$\phi_\infty = \text{humidity of air flow}$$

$$= 0 \text{ for dry air flow condition}$$

Therefore, the water evaporative heat rate (kW) can be expressed as:

$$q_{\text{evap}} = m_A \cdot h_{fg}$$

where
h_{fg} = latent heat of water evaporation, kj/kg

Note that the required energy to provide water evaporation is from the air flow ($i.e.$, $T_\infty > T_w$). The energy balance on the water surface is:

$$q_{\text{conv}} = hA_s(T_\infty - T_w) \text{ convection heat transfer}$$

$$= q_{\text{evap}} = m_A \cdot h_{fg} \text{ convection mass transfer}$$

If we consider radiation from the sun, the energy balance on the water surface becomes

$$q_{\text{conv}} + q_{\text{rad}} = q_{\text{evap}}$$

If a heater is immersed inside the water, the energy balance can be written as

$$q_{\text{conv}} + q_{\text{rad}} + q_{\text{heater}} = q_{\text{evap}}$$

6.5.4 NAPHTHALENE SUBLIMATION MASS TRANSFER

Sublimation is the phase change directly from the solid to the vapor phase. This process can be observed from frozen carbon dioxide (dry ice) at room temperature. The cloud surrounding the dry ice is a mixture of the carbon dioxide gas and the surrounding humid air. Naphthalene is a substance, commonly available, which undergoes this phase change from solid to vapor at atmospheric pressure. Moth balls, which can be purchased at most grocery stores, are over 99% naphthalene. Over time, the moth balls gradually go through the sublimation process, where the naphthalene diffuses through a space as a vapor. This vapor prevents moths, or other insects, from infesting the space.

When a surface coated with naphthalene is exposed to forced air convection, it sublimates at room temperature. Consider air flowing over a naphthalene-coated flat plate, the amount of mass that sublimates can be used to obtain the local or average mass transfer coefficient. By using the mass transfer analogy, the heat transfer coefficients can then be calculated. One of the primary advantages of using mass transfer over traditional heat transfer methods for the measurement of the convective heat transfer coefficients is that no heating of the surface or the fluid is required. Without direct heating, the challenge of accounting for heat conduction loss during the thermal experiment is eliminated. Refer to Chapter 10 of *Experimental Methods in Heat Transfer and Fluid Mechanics* by Han and Wright [4] for additional details.

The naphthalene sublimation rate $(m_A, \text{kg/s})$ can be determined as:

$$m_A = h_m A_s \left[\rho_A(T_w) - \rho_A(T_\infty) \right]$$

where

$$\rho_A(T_w) = \text{local density of naphthalene vapor at the surface at } T_w$$

$$= P_w / RT_w \left(\text{based on the ideal gas law} \right)$$

where

$R = $ naphthalene vapor gas constant

$P_w = $ the partial pressure of naphthalene vapor

$$= 47.8802 \left\{ 10^{\left[11.884 - \frac{6713}{T_w(°R)} \right]} \right\}$$

$\rho_A(T_\infty) = $ local density of naphthalene vapor at the free stream $T_\infty = 0$ for air flows over the naphthalene coated plate

For many practical applications, we would like to obtain the convection heat transfer coefficient, h, from the mass transfer coefficient, h_m, by using the heat and mass transfer analogy. In this case, we need to determine the mass transfer rate, m_A, first in order to find the mass transfer coefficient, h_m.

Consider air flowing over the naphthalene coated plate, the naphthalene sublimation rate $(m_A, \text{kg/s})$ over a period of time, Δt, can be expressed as:

$$m_A = \rho_{\text{solid naphthalene}} \cdot A_s \cdot \frac{\Delta z}{\Delta t}$$

where

$\Delta z = $ naphthalene sublimation thickness (m)

$\rho_{\text{solid naphthalene}} = $ the density of solid naphthalene at air temperature (kg/m^3)

$\Delta t = $ the duration of the sublimation (seconds)

After m_A is measured, h_m can be determined from the above equation:

$$m_A = h_m A_s \left[\rho_A(T_w) - \rho_A(T_\infty) \right]$$

and

$$Sh = h_m L / D_{AB}$$

with

$$D_{AB} = \frac{\nu}{Sc}$$

Based on the heat and mass transfer analogy, the convection heat transfer coefficient (and Nusselt number) can be calculated as:

$$Nu = \frac{hL}{k} = Sh \cdot \left(\frac{Pr}{Sc} \right)^{1/3}$$

where
  $Pr = 0.71$ for air
  $Sc = 2.5$ naphthalene to air

On the other hand, if the heat transfer coefficient is given, we can predict the mass transfer coefficient from the heat transfer coefficient by using heat and mass transfer analogy, i.e.,

$$h_m = h \rho c_p Le^{2/3} = h \rho c_p \left(\frac{Sc}{Pr} \right)^{2/3}$$

where
  $Pr = 0.71$ for air
  $Sc = 2.5$ naphthalene to air

Then the naphthalene sublimation mass transfer rate m_A can be predicted from the above equation.

REMARKS

This chapter provides the basic concept of boundary-layer flow and heat transfer; it focuses on how to derive 2-D boundary-layer conservations for mass, momentum, and energy; boundary-layer approximations; nondimensional analysis; and Reynolds analogy. Students have come across these equations in their undergraduate-level heat transfer. However, in the intermediate-level heat transfer, students are expected to fully understand how to obtain these equations. In addition, we have briefly touched on the mass transfer and the heat–mass transfer analogy.

PROBLEMS

6.1 For hot-gas flow (velocity V_∞, temperature T_∞) over a cooled convex surface (surface temperature T_s), answer the following questions:
   a. Sketch the "thermal boundary-layer thickness" distribution on the entire convex surface and explain the results.
   b. Sketch the possible local heat transfer coefficient distribution on the convex surface and explain the results.
   c. Define the similarity parameters (dimensionless parameters) that are important to determine the local heat transfer coefficient on the convex surface.
   d. Write down the relationship among those similarity parameters and give explanations.
   e. Write down how to determine the local heat flux from the convex surface.

6.2 For cold-gas flow (velocity V_∞, temperature T_∞) over a heated convex surface (surface temperature T_s), answer the following questions:
   a. Sketch the "thermal boundary-layer thickness" distribution on the entire convex surface and explain the results.
   b. Sketch the possible local heat transfer coefficient distribution on the convex surface and explain the results.
   c. Define the similarity parameters (dimensionless parameters) that are important to determine the local heat transfer coefficient on the convex surface.
   d. Write down the relationship among those similarity parameters and give explanations.
   e. Write down how to determine the local heat flux from the convex surface.

6.3 Derive Equations (6.11), (6.12), (6.15), and (6.16).

6.4 Derive Equations (6.23) and (6.25).

6.5 Derive Equation (6.30).

REFERENCES

1. W. Rohsenow and H. Choi, *Heat, Mass, and Momentum Transfer*, Prentice-Hall, Inc., Englewood Cliffs, NJ, 1961.
2. F. Incropera and D. Dewitt, *Fundamentals of Heat and Mass Transfer*, Fifth Edition, John Wiley & Sons, New York, 2002.
3. H. Schlichting, *Boundary-Layer Theory*, Sixth Edition, McGraw-Hill Book Company, New York, 1968.
4. J.C. Han and L.M. Wright, *Experimental Methods in Heat Transfer and Fluid Mechanics*, CRC Press, Boca Raton, FL, 2020.

7 External Forced Convection

7.1 LAMINAR FLOW AND HEAT TRANSFER OVER A FLAT SURFACE: SIMILARITY SOLUTION

External forced convection involves a fluid moving over the external surface of a solid body and forming hydrodynamic and thermal boundary layers around the surface. There are two well-known methods to solve external boundary-layer flow and heat transfer problems. One is the similarity method to obtain the exact solution. The other is the integral method to obtain the approximate solution. This section begins with the similarity method [1–6]. Figure 7.1 shows stream lines for flow over a flat plate. There is similarity of the velocity from stream ling to stream line. From location to another, the velocity (temperature) profiles look "similar" at each *x*-location. Like a rubber band being stretched, the master velocity profile can be stretched appropriately at each *x*-location to fit the actual velocity profile along the surface. Therefore, we define a stream function, ψ. The stream function is a function of the similarity variable, and the introduction of the stream function allows us to convert two nonlinear partial differential equations (PDEs) into a single, nonlinear PDE. Making use of similarity, the nonlinear PDE becomes a higher order, nonlinear ordinary differential equation (ODE). The final ODE can be solved using the Blasius series or numerically with a Runge–Kutta (RK) method.

Define the stream function, ψ, to satisfy the continuity equation:

$$u = \frac{\partial \Psi}{\partial y} \tag{7.1}$$

$$v = -\frac{\partial \Psi}{\partial x} \tag{7.2}$$

The continuity equation, as shown in Equation (6.7), is automatically satisfied.

$$\frac{\partial u}{\partial x} + \frac{\partial v}{\partial y} = 0$$

$$\frac{\partial}{\partial x}\left(\frac{\partial \Psi}{\partial y}\right) + \frac{\partial}{\partial y}\left(-\frac{\partial \Psi}{\partial x}\right) = 0 \tag{7.3}$$

The *x*-momentum equation, as shown in Equation (6.17) with $\partial P / \partial x = 0$, becomes:

DOI: 10.1201/9781003164487-7

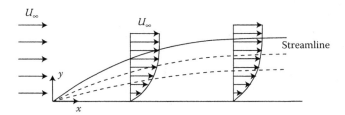

FIGURE 7.1   Stream lines for flow over a flat plate.

$$u\frac{\partial u}{\partial x} + v\frac{\partial u}{\partial y} = v\frac{\partial^2 u}{\partial y^2}$$

$$\frac{\partial \Psi}{\partial y}\left(\frac{\partial^2 \Psi}{\partial x \partial y}\right) - \frac{\partial \Psi}{\partial x}\frac{\partial^2 \Psi}{\partial y^2} = v\left(\frac{\partial^3 \Psi}{\partial y^3}\right) \tag{7.4}$$

Now, we introduce the similarity variable, η, to combine x and y into a single variable:

$$\Psi(x, y) \Rightarrow \Psi(\eta)$$

The master profile (rubber band) can be stretched from location to location. In other words, the velocity profile can be represented as:

$$\frac{u}{u_\infty} = f(\eta)$$

The similarity variable, η, is proportional to y, and the proportionality factor depends on x. We can now apply the similarity concept from Figure 7.1,

$$y \sim \sqrt{x} \Rightarrow \eta = \frac{C_2 y}{\sqrt{x}} \tag{7.5}$$

$$\Psi \sim \sqrt{x} \Rightarrow f = c_1 \frac{\Psi}{\sqrt{x}} = f(\eta) \tag{7.6}$$

η is the similarity variable and f is the similarity function.

$$\left.\begin{array}{c} u(x, y) \\ v(x, y) \end{array}\right\} \Rightarrow \Psi(x, y) \Rightarrow \Psi(\eta) \Rightarrow f(\eta)$$

In Equation (7.4), we need to replace ψ, x, and y with the new similarity definitions:

$$\eta = \frac{C_2 y}{\sqrt{x}}$$

$$f = c_1 \frac{\Psi}{\sqrt{x}} \Rightarrow \Psi = \frac{f\sqrt{x}}{c_1}$$

$$\frac{\partial \Psi}{\partial y} = \frac{\partial \Psi}{\partial \eta} \frac{\partial \eta}{\partial y} = \frac{\sqrt{x}}{c_1} f' \frac{c_2}{\sqrt{x}} = \frac{c_2}{c_1} f'$$

$$\frac{\partial \Psi}{\partial x} = \left(\frac{\partial f}{\partial x} \frac{\sqrt{x}}{C_1} + \frac{f}{C_1} \frac{1}{2} \frac{1}{\sqrt{x}} \right) = \left(\frac{\partial f}{\partial \eta} \frac{\partial \eta}{\partial x} \frac{\sqrt{x}}{C_1} + \frac{f}{C_1} \frac{1}{2} \frac{1}{\sqrt{x}} \right)$$

$$= f'\eta \left(-\frac{1}{2} \frac{1}{\sqrt{x}} \frac{1}{C_1} \right) + \frac{f}{C_1} \frac{1}{2} \frac{1}{\sqrt{x}}$$

$$= \frac{1}{2} \frac{1}{C_1 \sqrt{x}} (f - f'\eta)$$

where

$$f' = \frac{\partial f}{\partial \eta}, \quad \frac{\partial \eta}{\partial y} = \frac{C_2}{\sqrt{x}}, \quad \frac{\partial \eta}{\partial x} = -\frac{1}{2} C_2 y x^{-(3/2)} = -\frac{\eta}{2x}$$

Similarly,

$$\frac{\partial^2 \Psi}{\partial y^2} = \frac{\partial}{\partial y} \left(\frac{\partial \Psi}{\partial y} \right) = \frac{\partial}{\partial y} \left(\frac{c_2}{c_1} f' \right) = \frac{c_2}{c_1} \frac{\partial f'}{\partial \Psi} \frac{\partial \Psi}{\partial y} = \frac{c_2^2}{c_1} \frac{f''}{\sqrt{x}}$$

$$\frac{\partial^3 \Psi}{\partial y^3} = \frac{\partial}{\partial y} \left(\frac{\partial^2 \Psi}{\partial y^2} \right) = \frac{\partial}{\partial y} \left(\frac{C_2^2}{C_1} \frac{f''}{\sqrt{x}} \right) = \frac{C_2^2}{C_1} \frac{\partial}{\partial \eta} \left(\frac{f''}{\sqrt{x}} \right) \cdot \frac{\partial \eta}{\partial y}$$

$$= \frac{C_2^2}{C_1} \frac{1}{\sqrt{x}} f''' \frac{C_2}{\sqrt{x}} = \frac{C_2^3}{C_1} \frac{f'''}{x}$$

$$\frac{\partial^2 \Psi}{\partial x \partial y} = \frac{\partial}{\partial x} \left(\frac{\partial \Psi}{\partial y} \right) = \frac{\partial}{\partial x} \left(\frac{C_2}{C_1} f' \right) = \frac{\partial}{\partial \eta} \left(\frac{C_2}{C_1} f' \right) \frac{\partial \eta}{\partial x}$$

$$= \frac{C_2}{C_1} f'' \left(-\frac{\eta}{2x} \right)$$

The new definitions are now inserted into the stream function momentum Equation (7.4).

$$\frac{\partial \Psi}{\partial y} \left(\frac{\partial^2 \Psi}{\partial x \partial y} \right) - \frac{\partial \Psi}{\partial x} \frac{\partial^2 \Psi}{\partial y^2} = v \left(\frac{\partial^3 \Psi}{\partial y^3} \right)$$

$$\left(\frac{c_2}{c_1} f' \right) \left(\frac{C_2}{C_1} f'' \left(-\frac{\eta}{2x} \right) \right) - \left(\frac{1}{2} \frac{1}{C_1\sqrt{x}} (f - f'\eta) \right) \left(\frac{c_2^2}{c_1} \frac{f''}{\sqrt{x}} \right) = v \left(\frac{C_2^3}{C_1} \frac{f'''}{x} \right)$$

After simplification, we obtain:

$$ff'' + 2vc_1c_2f''' = 0 \tag{7.7}$$

This is the similarity momentum equation. The momentum equation has changed from a second-order, nonlinear PDE to a third-order, nonlinear ODE.

To solve the ODE, let us begin by assuming:

$$vc_1c_2 = 1 \tag{7.8}$$

At $y = \infty$:

$$u\big|_{y=\infty} = \mathcal{U}_\infty = \frac{\partial \Psi}{\partial y}\bigg|_{y=\infty} = f'\frac{c_2}{c_1}$$

Now, allow:

$$\frac{c_2}{c_1} = \mathcal{U}_\infty \tag{7.9}$$

From Equation (7.9), $f'(\infty) = 1$. In addition, at the wall, $u = 0$; therefore, $f'(0) = 0$. Now, with Equations (7.8) and (7.9), we obtain

$$c_2 = \sqrt{\frac{\mathcal{U}_\infty}{v}}$$

$$c_1 = \frac{1}{\sqrt{v\mathcal{U}_\infty}}$$

Therefore,

$$\eta = C_2\frac{y}{\sqrt{x}} = \frac{y}{\sqrt{x}}\sqrt{\frac{\mathcal{U}_\infty}{v}} = \frac{y}{x}\sqrt{\mathrm{Re}_x} \tag{7.10}$$

$$f = C_1\frac{\Psi}{\sqrt{x}} = \frac{\Psi}{\sqrt{v\mathcal{U}_\infty x}} \tag{7.11}$$

where f and η are similarity function and similarity variable, respectively. Finally, we obtain the following:

$$u = \frac{\partial \Psi}{\partial y} = \frac{C_2}{C_1}f' = \mathcal{U}_\infty f' \tag{7.12}$$

$$v = -\frac{\partial \Psi}{\partial x} = -\frac{1}{2}\frac{1}{C_1\sqrt{x}}(f - f'\eta) = \frac{1}{2}\sqrt{\frac{v\mathcal{U}_\infty}{x}}(f'\eta - f) \tag{7.13}$$

$$f' = \frac{df}{d\eta} = \frac{u}{\mathcal{U}_\infty} = \text{velocity profile} \tag{7.14}$$

$$f'' = \frac{d^2 f}{d\eta^2} = \frac{d(u/\mathcal{U}_\infty)}{d\eta} = \text{velocity gradient} \tag{7.15}$$

The boundary conditions (BCs) need to be transformed from x, y, u, and v into the similarity variable and function. Beginning at the surface ($y = 0$):

$$y = 0: \qquad u = v = 0$$

$$\eta = 0$$

$$f' = \frac{u}{\mathcal{U}_\infty} = 0$$

$$v = 0$$

$$f = 0$$

$$f(0) = 0 \tag{7.16}$$

Now at the edge of the boundary layer ($y \to \infty$):

$$y = \infty: \qquad u = \mathcal{U}_\infty$$

$$f'(\infty) = 1$$

The momentum equation combines with the BCs to form a third-order, ODE:

$$f''' = -\frac{1}{2} f f'' \tag{7.17}$$

Equation (7.17) is the Blasius Equation for boundary layer flow over a flat plate. It may be solved by expressing $f(\eta)$ in a power series (1908 Blasius Series Expansion) with the above-mentioned BCs as:

$$f = f_i = \frac{\alpha \eta^2}{2!} - \frac{1}{2}\frac{\alpha^2 \eta^5}{5!} + \frac{11}{4}\frac{\alpha^3 \eta^8}{8!} - \frac{375}{8}\frac{\alpha^4 \eta^{11}}{11!} + \cdots \text{for small } \eta \tag{7.18}$$

$$f - f_o = \eta - \beta - \gamma \int_\eta^\infty \left\{ \exp\left[-\left(\frac{\eta - \beta}{2}\right)^2 \right] d\eta \right\} d\eta \ldots \text{for large } \eta \tag{7.19}$$

where $\alpha = 0.332$, $\beta = 1.73$, and $\gamma = 0.231$. Thus, f' can be obtained. $f' = \alpha \eta$, $f'' = \alpha = 0.332$. Then u and v can be determined.

Equation (7.17) can also be solved by numerical integration as

$$f''' = \frac{df''}{d\eta} = -\frac{1}{2} ff''$$

$$\frac{df''}{f''} = -\frac{1}{2} f d\eta$$

Now, integrate both sides:

$$\int \frac{df''}{f''} = -\int \frac{1}{2} f \, d\eta$$

$$\ln f'' = -\int_0^\eta \frac{1}{2} f \, d\eta + C$$

$$f'' = e^{-\int_0^\eta 1/2(f \, d\eta)} \cdot C_1$$

$$f' = \int df' = \int_0^\eta \left(e^{-\int_0^\eta 1/2(f \, d\eta)} \right) d\eta \cdot C_1 + C_2$$

where

$$C_2 = 0 \, \text{at} \, \eta = 0$$

$$f' = 0$$

$$0 = \int_0^0 \left(e^{-\int_0^0 1/2(f \, d\eta)} \right) d\eta \cdot C_1 + C_2$$

$$C_2 = 0$$

C_1 can be determined with the BC $\eta \to \infty$

$$\eta = \infty \quad f' = 1$$

$$1 = \int_0^\infty \left(e^{-\int_0^\eta 1/2(f \, d\eta)} \right) d\eta \cdot C_1$$

$$C_1 = \left(1 / \int_0^\infty e^{-\int_0^\eta 1/2(f \, d\eta)} d\eta \right)$$

After grouping like terms, f can be determined by integrating f' a third time:

$$f = c_1 \int_0^\eta \int_0^\eta e^{-\int 1/2(f \, d\eta)} d\eta \, d\eta + C_3$$

The third constant of integration, C_3, is $C_3 = 0$ from $f = 0$ at $\eta = 0$.
Finally, we have:

$$f = \frac{\int_0^\eta \int_0^\eta e^{-\int_0^\eta 1/2(f \, d\eta)} d\eta \, d\eta}{\int_0^\infty e^{-\int_0^\eta 1/2(f \, d\eta)} d\eta} \qquad (7.20)$$

We can apply numerical integration (trapezoid rule) to solve for the similarity function, f. With the trapezoid rule, we will need an initial guess, and the guess is checked based on convergence of the solution. From the numerical solution, f' (u/U_∞) should converge to approximately 0.99 at the edge of boundary layer ($\eta \to 5$). For example, use of the trapezoidal rule for the numerical integration:

Choose $\eta_{max} = 5 = \Delta\eta \cdot N$, when $N = 50$ (the number of integration steps), $\Delta\eta = 0.1$
Let $\eta_{i+1} = \eta_i + \Delta\eta$ for $i = 0, 1, 2, \ldots$, with $\eta_o = 0$
Initial guess $f_i = \eta_i$
Calculate

$$\int_0^{\eta_i} f \, d\eta$$

$$e^{-\int_0^{\eta_i} f \, d\eta}$$

$$\int_0^{\eta_i} e^{-\int_0^{\eta_i} f \, d\eta} d\eta$$

$$\int_0^{\eta_i}\int_0^{\eta_i} e^{-\int_0^{\eta_i} f\,d\eta}\,d\eta\,d\eta$$

Next, calculate

$$C_1 = \frac{1}{\left(\displaystyle\int_0^{\eta_{max}} e^{-\int_0^{\eta_i} f\,d\eta}\,d\eta\right)}$$

$$f_i$$

$$f_i'$$

$$f_i''$$

Now, check the convergence. Is $\left|1 - f_i^{new}/f_i^{old}\right| < \varepsilon$, for $i = 0, 1, 2,\ldots, N$? If not, repeat the process with another initial guess.

From the tabulated data shown in Table 7.1 [2] or Figure 7.2, for a given $Re_x = \rho U_\infty x/\mu$, the velocity profile $u(x, y)$ at any location (x, y) and the shear stress f'' at the wall ($y = 0$, $\eta = 0$) can be determined.

TABLE 7.1
Flat Plate Laminar Boundary Layer Functions [2]

$\eta = y\sqrt{\dfrac{U_\infty}{vx}}$	f	$f' = \dfrac{u}{U_\infty}$	f''
0	0	0	0.332
0.4	0.027	0.133	0.331
0.8	0.106	0.265	0.327
1.2	0.238	0.394	0.317
1.6	0.420	0.517	0.297
2.0	0.650	0.630	0.267
2.4	0.922	0.729	0.228
2.8	1.231	0.812	0.184
3.2	1.569	0.876	0.139
3.6	1.930	0.923	0.098
4.0	2.306	0.956	0.064
4.4	2.692	0.976	0.039
4.8	3.085	0.988	0.022
5.0	3.284	0.991	0.017
5.2	3.482	0.994	0.011
5.6	3.880	0.997	0.005
6.0	4.280	0.999	0.002
6.4	4.679	1.000	0.001
6.8	5.079	1.000	0.000

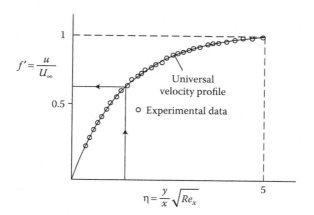

FIGURE 7.2 Graphical sketch of velocity profile from similarity.

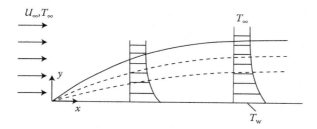

FIGURE 7.3 The thermal boundary layer concept.

The similarity function for temperature is shown in Figure 7.3. Let

$$\theta = \frac{T - T_W}{T_\infty - T_W}\left(\equiv \frac{u = 0}{U_\infty - 0} = f'\right) \tag{7.21}$$

The energy equation becomes

$$\frac{\partial \Psi}{\partial y}\frac{\partial \theta}{\partial x} - \frac{\partial \Psi}{\partial x}\frac{\partial \theta}{\partial y} = \alpha \frac{\partial^2 \theta}{\partial y^2} \tag{7.22}$$

Performing

$$\frac{\partial \theta}{\partial y} = \frac{\partial \theta}{\partial \eta}\frac{\partial \eta}{\partial y} = \theta' \frac{C_2}{\sqrt{x}}$$

$$\frac{\partial^2 \theta}{\partial y^2} = \frac{\partial}{\partial y}\left(\frac{\partial \theta}{\partial y}\right) = \frac{\partial}{\partial y}\left(\frac{C_2}{\sqrt{x}}\theta'\right) = \frac{\partial}{\partial \eta}\left(\frac{C_2}{\sqrt{x}}\theta'\right)\cdot\frac{\partial \eta}{\partial y} = \frac{C_2^2}{x}\theta''$$

$$\frac{\partial \theta}{\partial x} = \frac{\partial \theta}{\partial \eta}\cdot\frac{\partial \eta}{\partial x} = \theta'\cdot\frac{-\eta}{2x}$$

Inserting this into the energy equation, we obtain

$$\theta'' + \frac{1}{2}\mathrm{Pr}\,f\theta' = 0 \tag{7.23}$$

BCs:

$$\begin{aligned}\theta(0) &= 0 \\ \theta(\infty) &= 1\end{aligned} \tag{7.24}$$

Finally, Equation (7.23) can be solved as:

$$\int \frac{\theta''}{\theta'}\,d\eta = -\int \frac{\mathrm{Pr}}{2}f\,d\eta$$

$$\ln \theta' = -\int_0^\eta \frac{\mathrm{Pr}}{2}f\,d\eta + C$$

$$\theta' = e^{-\int_0^\eta \frac{\mathrm{Pr}}{2}f\,d\eta} \cdot C_1$$

$$\theta = C_1 \int_0^\eta e^{-\int_0^\eta \frac{\mathrm{Pr}}{2}f\,d\eta}\,d\eta + C_2$$

at $\eta = 0$, $\theta(0) = 0$, $C_2 = 0$
  at $\eta = \infty$, $\theta(\infty) = 1$,

$$C_1 = \frac{1}{\displaystyle\int_0^\infty e^{-\int_0^\eta \mathrm{Pr}/2(f\,d\eta)}\,d\eta}$$

Therefore:

$$\theta = \frac{\displaystyle\int_0^\eta e^{-\int_0^\eta \mathrm{Pr}/2(f\,d\eta)}\,d\eta}{\displaystyle\int_0^\infty e^{-\int_0^\eta \mathrm{Pr}/2(f\,d\eta)}\,d\eta} \tag{7.25}$$

$$\theta = \theta'(0)\int_0^\eta \exp\left[-\int_0^\eta \frac{\mathrm{Pr}}{2}f\,d\eta\right]d\eta \tag{7.26}$$

where

$$\theta'(0) = \frac{1}{\int_0^\infty \exp\left[-\int_0^\eta \mathrm{Pr}/2\left(f\,d\eta\right)\right] d\eta} \tag{7.27}$$

If $\mathrm{Pr} = 1$, $f' = \theta$, the thermal boundary layer is the same as the hydrodynamic boundary layer (the derivative of Equation 7.20 = Equation 7.25).

For a given Pr, θ and θ' can be determined if f'' has been solved previously. For example,

$$f = \frac{f'''}{(1/2)f''}$$

$$\int_0^\eta f\,d\eta = -2\ln f''$$

$$f'' = e^{-\int_0^\eta 1/2(f\,d\eta)}$$

$$\theta = \frac{\int_0^\eta \left(f''\right)^{\mathrm{Pr}} d\eta}{\int_0^\infty \left(f''\right)^{\mathrm{Pr}} d\eta}$$

From the tabulated data, or Figure 7.4, the temperature profile $T(x, y)$ at any location (x, y) and the heat flux at the wall $(y = 0, \eta = 0)$ can be determined. From η, we obtain θ and $T(x, y)$ for the given Prandtl number.

Note: θ can be defined as:

$$\theta = \frac{T - T_W}{T_\infty - T_W}$$

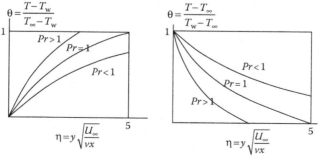

FIGURE 7.4 Graphical sketch of the temperature profile from similarity solutions.

or

$$\theta = \frac{T - T_\infty}{T_W - T_\infty}$$

as shown in Figure 7.4. The thermal boundary thickness (η) decreases as the Prandtl number increases, also shown in Figure 7.4. The thinner thermal boundary layer implies a larger θ'.

7.1.1 SUMMARY OF THE SIMILARITY SOLUTION FOR LAMINAR BOUNDARY-LAYER FLOW AND HEAT TRANSFER OVER A FLAT SURFACE

The foundational momentum equation developed in the previous section can be used to calculate boundary-layer thickness, shear stress, and the friction factor for a given Reynolds number. In addition, with the energy equation from the previous section, the heat flux, the heat transfer coefficient, and Nusselt number for a given Reynolds number and Prandtl number can also be determined.

At a given location along the plate and off the surface, x and y, the similarity variable, η, can be calculated from the freestream velocity (7.10). From Table 7.1 or Figure 7.2, the local velocity (f') can be determined from the similarity variable (η).

The boundary layer thickness can be estimated where $u/U_\infty = 0.99$. Table 7.1 shows f' reaches 0.99 at $\eta = 5$. Therefore, at a given x-location, the boundary layer thickness can be defined as $\eta = (y/x)\sqrt{\text{Re}_x} = 5$ where $y = \delta$. Based on the definition of the similarity variable, we see the boundary layer thickness is proportional to $\sqrt{x}$.

$$\frac{\delta}{x} = \frac{5}{\sqrt{\text{Re}_x}}$$

$$\delta \sim \sqrt{x}$$

The wall shear stress can be determined from f'':

$$\tau_w = \mu \frac{\partial u}{\partial y}\bigg|_{y=0} = \mu \frac{\partial}{\partial y}\left(\frac{\partial \psi}{\partial y}\right)_{y=0}$$

$$= \mu \frac{\partial^2 \psi}{\partial y^2}\bigg|_{y=0} = \mu \left[\frac{C_2^2}{C_1\sqrt{x}} f''\right]_{y=0}$$

$$= \mu \left[C_2 \frac{C_2}{C_1} \frac{1}{\sqrt{x}} f''(\eta = 0)\right]$$

$$= \mu \cdot U_\infty \cdot \sqrt{\frac{U_\infty}{xv}} \cdot (0.332)$$

From Table 7.1 (or Figure 7.2), $f''(\eta = 0) = 0.332$. Therefore:

$$\tau_w = 0.332 \cdot U_\infty^{3/2} \cdot \sqrt{\frac{\rho\mu}{x}} \qquad (7.28)$$

The shear stress decreases with increasing x because of the increasing thickness of the boundary layer (the velocity gradient decreases with increasing x). From the numerical integration and the wall shear stress, the local skin friction coefficient can also be determined:

$$C_{fx} = \frac{\tau_w}{\frac{1}{2}\rho U_\infty^2}$$

$$= \frac{0.664}{\sqrt{Re_x}}$$

$$C_{fx} \sim \frac{1}{\sqrt{x}}$$

Integrating from $x = 0 \rightarrow L$, the average skin friction coefficient for the entire plate can be determined:

$$\overline{C_{fx}} = \frac{1.328}{Re_L} \qquad (7.29)$$

where

$$Re_L = \frac{\rho U_\infty L}{\mu}$$

From the solution of the energy similarity equation, the heat transfer coefficient and Nusselt number can determined from the temperature gradient (θ') at any x-location along the surface.

$$q_w'' = -k\frac{\partial T}{\partial y}\bigg|_{y=0}$$

$$\frac{\partial T}{\partial y}\bigg|_0 = \frac{\partial T}{\partial \eta}\frac{\partial \eta}{\partial y}\bigg|_0 = (T_\infty - T_W)\cdot\theta'(\eta = 0)\cdot\sqrt{\frac{U_\infty}{\upsilon x}}$$

$$q_w'' = -k_f\left[(T_\infty - T_W)\cdot\theta'(\eta = 0)\cdot\sqrt{\frac{U_\infty}{\upsilon x}}\right]$$

$$q_w''(x) = h_x(T_W - T_\infty)$$

$$h_x = \frac{q''_w(x)}{(T_W - T_\infty)}$$

$$h_x = \frac{-k_f \left[(T_\infty - T_W) \cdot \theta'(\eta = 0) \cdot \sqrt{\dfrac{U_\infty}{vx}} \right]}{-(T_\infty - T_W)}$$

$$h_x = k_f \left[\theta'(\eta = 0) \cdot \sqrt{\frac{U_\infty}{vx}} \right] \qquad (7.30)$$

$$h_x \sim \frac{1}{\sqrt{x}}$$

From Table 7.1 (for Pr = 1) or Figure 7.4, $\theta'(\eta = 0) = 0.332 \, Pr^{1/3}$.
Now to provide the dimensionless Nusselt number:

$$Nu_x = \frac{h_x \cdot x}{k_f}$$

$$= \frac{x}{k_f} \cdot \left[k_f \cdot \theta'(\eta = 0) \cdot \sqrt{\frac{U_\infty}{vx}} \right]$$

$$= x \cdot \left[\theta'(\eta = 0) \cdot \sqrt{\frac{U_\infty}{vx}} \right]$$

$$= \theta'(\eta = 0) \cdot \sqrt{\frac{U_\infty \cdot x^2}{vx}}$$

$$Nu_x = \theta'(\eta = 0) \cdot \sqrt{Re_x} = 0.332 \sqrt{Re_x} \, Pr^{1/3}$$

We can obtain the average Nusselt number over the plate by integrating from
$x = 0 \to L$.

$$\overline{Nu} = \frac{\overline{h_x} x}{k} = 0.664 \sqrt{Re_x} \, Pr^{1/3}$$

This Nusselt number correlation is generally well accepted for flat surfaces maintained at a constant temperature and Prandtl numbers ranging between 0.5 and 15
(not applicable to oils or liquid metals). Also, as a note, the heat transfer coefficient,
and thus the Nusselt number, depends on both k_f and $\theta'(\eta = 0)$. The similarity solution method has provided an approach to determine the velocity and temperature

profiles within the boundary layer along the flat surface. The similarity variable and functions are shown in Figures 7.2 and 7.4, with Table 7.1 providing the numerical values.

REMARKS

There are many engineering applications involving external laminar flow heat transfer such as electronic component cooling and plate-type heat exchanger design. In the undergraduate-level heat transfer, there are many heat transfer correlations for Nusselt numbers as a function of Reynolds and Prandtl numbers. Students are expected to calculate heat transfer coefficients from these correlations for given Reynolds and Prandtl numbers.

For the similarity method, students are expected to know how to derive the similarity momentum and energy equations with proper velocity and thermal BCs. Students are also expected to know how to sketch and predict velocity profiles inside the boundary layer, for a given Reynolds number, from the velocity similarity solution using tables or figures; how to sketch and predict temperature profiles inside the thermal boundary layer, for a given Reynolds number and Prandtl number, from the temperature similarity solution using tables or figures. Here we focus on flow over a flat plate (zero-pressure gradient flow) with constant surface temperature BC and do not include the one at constant surface heat flux BC.

As an advanced topic, the similarity solution can be extended to include various constant pressure gradient flows (such as flow acceleration or deceleration) with variable surface temperature BCs. These can be solved using the fourth-order RK method in order to obtain the velocity and temperature profiles with the forced convection boundary-layer flow. These topics are discussed later in Section 7.3.

7.2 LAMINAR FLOW AND HEAT TRANSFER OVER A FLAT SURFACE: INTEGRAL METHOD

The other powerful method to solve boundary-layer flow and heat transfer problems is the integral approximation solution technique [1–6]. Instead of performing mass, momentum, and energy balances through a differential fluid element inside the boundary layer, as with the similarity method, the integral method performs conservation of mass, momentum, and energy across the boundary-layer thickness with a given differential element in the x-direction. It should be reiterated for fluids with a Prandtl number different from unity, such as gases, water, and oils, the hydrodynamic boundary-layer thickness is different from the thermal boundary layer.

7.2.1 MOMENTUM INTEGRAL EQUATION BY VON KARMAN

This solution method begins by employing a control volume that is infinitesimal in the x-direction but finite in the y-direction across the boundary-layer thickness, as

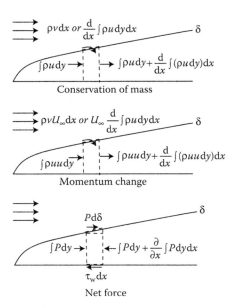

FIGURE 7.5 Integral method.

shown in Figure 7.5. We can apply mass and momentum conservation to the control volume.

From mass conservation,

$$\frac{\partial v}{\partial y} = -\frac{\partial u}{\partial x}; \quad v(\delta) = v\mid_0 - \int_0^\delta \frac{\partial u}{\partial x}\, dy = -\int_0^\delta \frac{\partial u}{\partial x}\, dy$$

$$\rho v\, dx = \frac{d}{dx} \int \left(\rho u\, dy\right) dx \qquad (7.31)$$

From momentum conservation,

$$\underbrace{-\tau_w dx + P\frac{d\delta}{dx}dx - \frac{d}{dx}\left(\int_0^\delta P\, dy\right)dx}_{\text{net force}} = \underbrace{\frac{d}{dx}\int\left(\rho uu\, dy\right)dx - U_\infty\frac{d}{dx}\int\left(\rho u\, dy\right)dx}_{\text{momentum change}} \quad (7.32)$$

where

$$-\frac{d}{dx}\int P\, dy = -\frac{d}{dx}P\int dy = -P\frac{d\delta}{dx} - \delta\frac{dP}{dx}$$

$dP/dx = -\rho U_\infty(dU_\infty/dx)$ (from the momentum differential equation, at outside of the boundary layer, $u = U_\infty$, $v = 0$, $\partial^2 U_\infty/\partial y^2 = 0$).

Therefore,

$$-\tau_w + P\frac{d\delta}{dx} - P\frac{d\delta}{dx} - \delta\left(-\rho\mathcal{U}_\infty\frac{d\mathcal{U}_\infty}{dx}\right) = \frac{d}{dx}\int\rho u u\,dy - \mathcal{U}_\infty\frac{d}{dx}\int\rho u\,dy$$

$$-\tau_w = \frac{d}{dx}\int\rho u u\,dy - \rho\mathcal{U}_\infty\frac{d\mathcal{U}_\infty}{dx}\int dy - \mathcal{U}_\infty\frac{d}{dx}\int\rho u\,dy$$

$$= \frac{d}{dx}\int\rho u u\,dy - \frac{d\mathcal{U}_\infty}{dx}\int\rho\mathcal{U}_\infty\,dy - \mathcal{U}_\infty\frac{d}{dy}\int\rho u\,dy + \frac{d\mathcal{U}_\infty}{dx}\int\rho u\,dy - \frac{d\mathcal{U}_\infty}{dx}\int\rho u\,dy$$

$$= \frac{d}{dx}\int\rho u u\,dy + \frac{d\mathcal{U}_\infty}{dx}\int\rho(u-\mathcal{U}_\infty)\,dy - \frac{d}{dx}\int\rho u\mathcal{U}_\infty\,dy$$

$$= \frac{d}{dx}\int\rho u(u-\mathcal{U}_\infty)\,dy + \frac{d\mathcal{U}_\infty}{dx}\int\rho(u-\mathcal{U}_\infty)\,dy$$

$$\tau_w = \frac{d}{dx}\int\rho u(\mathcal{U}_\infty - u)\,dy + \frac{d\mathcal{U}_\infty}{dx}\int\rho(\mathcal{U}_\infty - u)\,dy \tag{7.33}$$

For flat plate flow, $dU_\infty/dx = 0$

$$\tau_w = \mu\frac{\partial u}{\partial y}\bigg|_{y=0} = \frac{d}{dx}\int\rho u(\mathcal{U}_\infty - u)\,dy \tag{7.34a}$$

$$-v\frac{\partial u}{\partial y}\bigg|_{y=0} = \frac{d}{dx}\int u(u-\mathcal{U}_\infty)\,dy \tag{7.34b}$$

7.2.2 ENERGY INTEGRAL EQUATION BY POHLHAUSEN

The conversation of energy is applied to a similar element, as shown in Figure 7.6.

$$q_s''dx = \frac{d}{dx}\left(\int_0^{\delta_T}\rho C_p u T\,dy\right)dx$$

$$\tag{7.35}$$

$$- C_p T_\infty\frac{d}{dx}\left(\int_0^{\delta}\rho u\,dy\right)dx$$

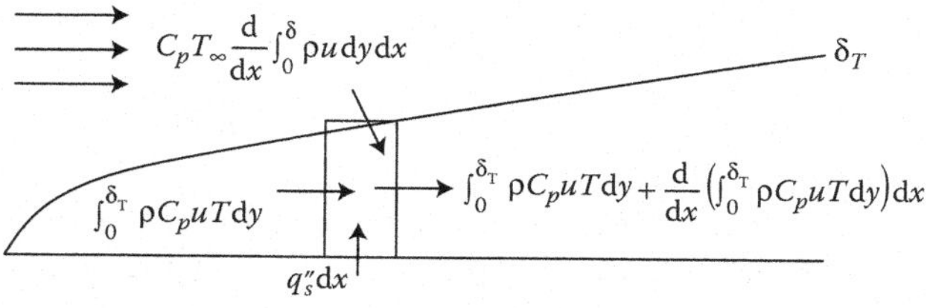

FIGURE 7.6 Conservation of energy in integral form.

$$= \frac{d}{dx} \int_0^{\delta_T} \rho C_p u (T - T_\infty) \, dy \, dx$$

$$q_s'' = -k \frac{\partial T}{\partial y}\Big|_{y=0} = \frac{d}{dx} \int_0^{\delta_T} \rho C_p u (T - T_\infty) \, dy \qquad (7.36a)$$

$$-\alpha \frac{\partial T}{\partial y}\Big|_{y=0} = \frac{d}{dx} \int_0^{\delta_T} u (T - T_\infty) \, dy \qquad (7.36b)$$

Equation (7.36b) can be compared directly to (Equation 7.34b) from the conversation of linear momentum over a flat plate ($dP/dx = 0$).

7.2.3 OUTLINE FOR THE APPLICATION OF THE INTEGRAL APPROXIMATION METHOD

This section outlines the process of using the integral momentum and energy equations to approximate the boundary layer thickness, shear stress, thermal boundary layer thickness, and heat flux along the surface. This solution technique begins by assuming a generic expression for the velocity (u) or temperature (T) profile. BCs are applied to the general (assumed) function to obtain specific profiles. From the known, approximate, velocity and temperature profiles, the integral equations are used to estimate the wall shear stress and/or heat flux. As shown in Equation (7.36), in order to solve the thermal problem, the velocity profile be known. However, this method does not require the velocity and temperature profiles are assumed in the same form.

For the hydrodynamic boundary layer (fluid mechanics problem):

Step 1: Assume the velocity profile. Below are examples of commonly assumed forms of the velocity distribution within the boundary layer.

$$u = a = \mathcal{U}_\infty$$

$$u = a + by$$

$$u = a + by + cy^2$$

$$u = a + by + cy^2 + dy^3$$

$$u = a + by + cy^2 + dy^3 + ey^4$$

$$u = a + by + cy^2 + dy^3 + ey^4 + fy^5$$

Step 2: Apply the velocity BCs to determine the coefficients a, b, c, d, e, and f. The assumed form of the velocity profile must satisfy the BCs for the given flow field.

Step 3: Insert the velocity profile into momentum integral to solve for $\delta(x)$.

Step 4: Substitute $\delta(x)$ back into the velocity profile.

Step 5: Solve for the wall shear stress, or friction factor (C_{fx}). The final solution should not be in terms of the boundary layer thickness, δ, as this is an unknown quantity.

$$C_{fx} = \frac{\tau_w}{(1/2)\rho \mathcal{U}_\infty^2} = \frac{\mu \, \partial u / \partial y|_0}{(1/2)\rho \mathcal{U}_\infty^2}$$

Similarly, for the thermal boundary layer:

Step 1: Assume a form for the temperature profile. Commonly used general equations are shown below:

$$T = a + by$$

$$T = a + by + cy^2$$

$$T = a + by + cy^2 + dy^3$$

$$T = a + by + cy^2 + dy^3 + ey^4$$

$$T = a + by + cy^2 + dy^3 + ey^4 + fy^5$$

Step 2: Apply the thermal BCs to determine the coefficients a, b, c, d, e, and f.

Step 3: Insert the temperature profile into the energy integral to solve for $\delta_T(x)$.

Step 4: Substitute $\delta_T(x)$ back into the temperature profile.

Step 5: Solve for the surface heat flux and convective heat transfer coefficient, h. Again, the expression for the heat transfer coefficient should not be a function of the unknown hydrodynamic and thermal boundary layer thicknesses, δ and δ_T.

$$h = \frac{q_w''}{T_w - T_\infty} = \frac{-k\left(\partial T / \partial y\right)_0}{T_w - T_\infty}$$

Examples

7.1 Assume second-order velocity and temperature profiles for the boundary layer flow to satisfy the following BCs. Determine the hydrodynamic and thermal boundary layer thicknesses. Also, determine the Nusselt number distribution along the surface.

$$y = 0 \qquad u = 0$$

$$y = \delta \qquad u = \mathcal{U}_\infty$$

$$y = \delta \qquad \frac{\partial u}{\partial y} = 0$$

$$y = 0 \qquad q''_w = -k \left. \frac{\partial T}{\partial y} \right|_{y=0}$$

$$y = \delta_T \qquad T = T_\infty$$

$$y = \delta_T \qquad \left. \frac{\partial T}{\partial y} \right|_{y=\delta_T} = 0$$

Solution

Based on the problem statement, according to "step 1," the assumed velocity profile is:

$$u = a + by + cy^2$$

According to "step 2," the BCs are used to determine the constants, a, b, and c. As shown above, three BCs are required to solve for the three constants; assuming a higher or lower order polynomial will require additional or fewer BCs, respectively.
From the velocity BCs:

$$y = 0, \quad u = 0, \quad a = 0$$
$$y = \delta, \quad u = \mathcal{U}_\infty, \quad \mathcal{U}_\infty = b\delta + c\delta^2$$
$$y = \delta, \quad \frac{\partial u}{\partial y} = 0, \quad 0 = b + 2c\delta$$

Solving the system of three equations with three unknowns (a, b, and c) yields:

$$a = 0 \qquad b = 2\frac{\mathcal{U}_\infty}{\delta} \qquad c = -\frac{\mathcal{U}_\infty}{\delta^2}$$

Thus, velocity profile becomes (completing "step 2"):

$$u = \frac{2\mathcal{U}_\infty}{\delta} y - \frac{\mathcal{U}_\infty}{\delta^2} y^2$$

Moving to "step 3," the velocity profile is inserted into the momentum integral equation:

$$\tau_w = \mu \left. \frac{\partial u}{\partial y} \right|_{y=0} = \frac{d}{dx} \cdot \int_0^\delta \rho u (\mathcal{U}_\infty - u) dy$$

$$\mu \frac{2\mathcal{U}_\infty}{\delta} = \frac{d}{dx} \int_0^\delta \rho \left[2\mathcal{U}_\infty \left(\frac{y}{\delta} \right) - \mathcal{U}_\infty \left(\frac{y}{\delta} \right)^2 \right] \cdot \left[\mathcal{U}_\infty - \left(2\mathcal{U}_\infty \left(\frac{y}{\delta} \right)^2 - \mathcal{U}_\infty \left(\frac{y}{\delta} \right)^2 \right) \right] dy$$

$$= \rho \mathcal{U}_\infty^2 \cdot \frac{d}{dx} \left(\frac{2}{15} \delta \right)$$

Solve for $\delta(x)$ as:

$$\delta(x) = \sqrt{\frac{30\mu x}{\rho \mathcal{U}_\infty}}$$

Now we move to the temperature profile. From the problem statement, the assumed temperature profile is a second-order polynomial. With "step 1," we have:

$$T = a + by + cy^2$$

According to "step 2," the thermal BCs are used to determine the constants, a, b, and c. With the second-order polynomial expression, three thermal BCs are used. Using these BCs, "step 2" gives us:

$$y = 0, \qquad q_w'' = -k\left.\frac{\partial T}{\partial y}\right|_{y=0} \qquad q_w'' = -kb$$

$$y = \delta_T, \qquad T = T_\infty, \qquad T_\infty = a - \frac{q''}{k}\delta_T + c\delta_T^2$$

$$y = \delta_T, \qquad \left.\frac{\partial T}{\partial y}\right|_{y=\delta_T} = 0, \qquad 0 = -\frac{q''}{k} + 2c\delta_T$$

Solving the system of equations, gives us the constants:

$$a = T_\infty + \frac{q_w''}{2k}\delta_T, \qquad b = -\frac{q_w''}{k} \qquad c = \frac{q_w''}{2k\delta_T}$$

Completing "step 2," the temperature profile becomes:

$$T = T_\infty + \frac{q_w''}{2k}\delta_T - \frac{q_w''}{k}y + \frac{q_w''}{2k\delta_T}y^2$$

With "step 3," both the velocity and temperature profiles are inserted into the energy integral equation:

$$q_w'' = \frac{d}{dx}\int_0^{\delta_T} \rho c_p u (T - T_\infty) dy$$

$$= \frac{d}{dx}\int_0^{\delta_T} \rho c_p \mathcal{U}_\infty \left[2\left(\frac{y}{\delta}\right) - \left(\frac{y}{\delta}\right)^2\right] \cdot \frac{q_w''}{k} \cdot \delta_T \left[\frac{1}{2} - \left(\frac{y}{\delta_T}\right) + \frac{1}{2}\left(\frac{y}{\delta_T}\right)^2\right] \cdot dy$$

$$= \frac{d}{dx}\left[\rho c_p \mathcal{U}_\infty \frac{q_w''}{k} \cdot \delta_T^2 \left(\frac{1}{12}r - \frac{1}{60}r^2\right)\right]$$

where

$$r = \frac{\delta_T}{\delta}$$

For the cases where:

$$\frac{\delta_T}{\delta} < 1$$

$$\left(\frac{\delta_T}{\delta}\right)^2 = r^2 \sim 0$$

The above integral equation becomes:

$$\frac{d}{dx}\left(\delta_T^2 \cdot r\right) = \frac{12k}{\rho c_p \mathcal{U}_\infty}$$

From "step 4," we can solve for $\delta_T(x)$ as:

$$\delta_T = 4.36\frac{x}{\sqrt{\mathrm{Re}_x}} \cdot \mathrm{Pr}^{-\frac{1}{3}} \sim 0.7378 \cdot \delta \cdot \mathrm{Pr}^{-\frac{1}{3}}$$

From the thermal boundary layer thickness, the heat transfer coefficient and Nusselt number can be determined ("step 5"):

$$h = \frac{q_w''}{T_w - T_\infty} = \frac{q_w''}{\frac{q_w''}{2k}\delta_T} = \frac{2k}{\delta_T}$$

$$\mathrm{Nu} = \frac{hx}{k} = \frac{2x}{\delta_T} = 0.496\sqrt{\mathrm{Re}_x} \cdot \mathrm{Pr}^{\frac{1}{3}}$$

Keeping r^2, the above energy integral equation can be written as:

$$\frac{k}{\rho c_p \mathcal{U}_\infty} = \frac{d}{d_x}\left(\frac{1}{12}r - \frac{1}{60}r^2\right)\delta_T^2$$

We can solve for $\delta_T(x)$ as:

$$\delta_T \sim \delta \mathrm{Pr}^{-\frac{1}{3}} \sim \sqrt{\frac{30\mu x}{\rho \mathcal{U}_\infty}} \cdot \mathrm{Pr}^{-\frac{1}{3}}$$

Finally, the heat transfer coefficient and Nusselt number become:

$$h = \frac{q_w''}{T_w - T_\infty} = \frac{2k}{\delta_T}$$

$$\mathrm{Nu} = \frac{hx}{k} = \frac{2x}{\delta_T} \sim \sqrt{\mathrm{Re}_x} \cdot \mathrm{Pr}^{\frac{1}{3}}$$

7.2 Building on Example 7.1, now assume third-order velocity and temperature profiles for the boundary layer flow to satisfy the BCs.

$$u = a + by + cy^2 + dy^3$$

$$T = a + by + cy^2 + dy^3$$

The BCs are given in the order they should be used to determine the respective constants, first velocity followed by temperature:

1.	$y = 0$	$u = 0$
2.	$y = \delta$	$u = \mathcal{U}_\infty$
3.	$y = \delta$	$\partial u / \partial y = 0$
4.	$y = 0$	$\partial^2 u / \partial y^2 = 0$

Based on the velocity BCs, solve for a, b, c, and d. Therefore, the velocity profile can be obtained as:

1.	$y = 0$	$T = T_w$
2.	$y = \delta_T$	$T = T_\infty$
3.	$y = \delta_T$	$\partial T / \partial y = 0$
4.	$y = 0$	$\partial^2 T / \partial y^2 = 0$

$$\frac{u - 0}{\mathcal{U}_\infty - 0} = \frac{3}{2}\frac{y}{\delta} - \frac{1}{2}\left(\frac{y}{\delta}\right)^3 = \frac{u}{\mathcal{U}_\infty}$$

Substitute the above velocity profile into the momentum integral to solve for $\delta(x)$:

$$\mu \mathcal{U}_\infty \frac{3}{2}\frac{1}{\delta} = \frac{d}{dx}\int_0^\delta \rho \mathcal{U}_\infty^2 \left[\frac{3}{2}\left(\frac{y}{\delta}\right) - \frac{1}{2}\left(\frac{y}{\delta}\right)^3\right]\left[1 - \frac{3}{2}\left(\frac{y}{\delta}\right) + \frac{1}{2}\left(\frac{y}{\delta}\right)^3\right] dy$$

$$= \frac{d}{dx}\left(\frac{39}{280}\delta\rho\mathcal{U}_\infty^2\right)$$

$$\delta \frac{d\delta}{dx} = \frac{140}{13}\frac{v}{\mathcal{U}_\infty}$$

$$\frac{1}{2}\delta^2 = \frac{140}{13}\frac{v}{\mathcal{U}_\infty}x + C$$

at $x = 0$, $\delta = 0$, $C = 0$

Therefore,

$$\delta(x) = 4.64\sqrt{(vx/\mathcal{U}_\infty)}$$

or

$$\frac{\delta(x)}{x} = \frac{4.64}{\sqrt{\mathrm{Re}_x}}$$

Substitute $\delta(x)$ back into the assumed velocity profile to obtain the final expression and the local friction factor.

$$C_{fx} = \frac{\tau_w}{(1/2)\rho \mathcal{U}_\infty^2} = \frac{\mu(\partial u/\partial y)_0}{(1/2)\rho \mathcal{U}_\infty^2} = \frac{\mu(3/2)(1/\delta)\mathcal{U}_\infty}{(1/2)\rho \mathcal{U}_\infty^2} = \frac{0.646}{\sqrt{\mathrm{Re}_x}} \qquad (7.37)$$

Based on the thermal BCs, we can solve for a, b, c, and d, and the temperature profile is obtained as:

$$\frac{T - T_w}{T_\infty - T_w} = \frac{3}{2}\left(\frac{y}{\delta_T}\right) - \frac{1}{2}\left(\frac{y}{\delta_T}\right)^3$$

Or

$$\frac{T - T_\infty}{T_w - T_\infty} = 1 - \frac{3}{2}\left(\frac{y}{\delta_T}\right) + \frac{1}{2}\left(\frac{y}{\delta_T}\right)^3$$

Put the above temperature profile into the energy integral to solve for $\delta_T(x)$. If $\mathrm{Pr} = 1$, $\delta = \delta_T$, $u = T - T_w$, this is a special case if $\mathrm{Pr} = 1$.

$$\frac{\delta_T}{x} = \frac{\delta}{x} = \frac{4.64}{\sqrt{\mathrm{Re}_x}}$$

The coefficient 4.64 is from momentum integral. This constant is 5.0 from the similarity solution. From the energy integral, we obtain

$$-k\frac{\partial T}{\partial y}\bigg|_{y=0} = k(T_w - T_\infty)\frac{3}{2}\frac{1}{\delta_T}$$

$$= \frac{d}{dx}\int_0^{\delta_T} \rho C_p \left[\frac{3}{2}\left(\frac{y}{\delta}\right) - \frac{1}{2}\left(\frac{y}{\delta}\right)^3\right]\mathcal{U}_\infty (T_w - T_\infty)\left[1 - \frac{3}{2}\left(\frac{y}{\delta_T}\right) + \frac{1}{2}\left(\frac{y}{\delta_T}\right)^3\right]dy$$

Then,

$$k\frac{3}{2}\frac{1}{\delta_T} = \rho C_p \mathcal{U}_\infty \left\{3\delta\left[\frac{\delta_T}{\delta}\frac{1}{10} - \left(\frac{\delta_T}{\delta}\right)^3\frac{1}{70}\right]\frac{d}{dx}\left(\frac{\delta_T}{\delta}\right) + 3\left[\left(\frac{\delta_T}{\delta}\right)^2\frac{1}{20} - \left(\frac{\delta_T}{\delta}\right)^4\frac{1}{280}\right]\frac{d\delta}{dx}\right\}$$

Let

$$r = \frac{\delta_T}{\delta} < 1, \quad \left(\frac{\delta_T}{\delta}\right)^3 \sim \left(\frac{\delta_T}{\delta}\right)^4 \sim 0$$

We obtain

$$2r^2\delta^2\frac{dr}{dx}+r^3\delta\frac{d\delta}{dx}=10\frac{\alpha}{\mathcal{U}_\infty}$$

From above

$$\delta^2=\frac{280}{13}\frac{vx}{\mathcal{U}_\infty},\quad \delta\frac{d\delta}{dx}=\frac{140}{13}\frac{v}{\mathcal{U}_\infty}$$

We obtain

$$2r^2=\frac{280}{13}\frac{vx}{\mathcal{U}_\infty}\frac{dr}{dx}+r^3\frac{140}{13}\frac{v}{\mathcal{U}_\infty}=10\frac{\alpha}{\mathcal{U}_\infty}$$

Next

$$4r^2x\frac{dr}{dx}+r^3=\frac{13}{14\mathrm{Pr}}$$

Let $s=r^3$

$$r^2\frac{dr}{dx}=\frac{1}{3}\frac{d}{dx}\left(r^3\right)=\frac{1}{3}\frac{ds}{dx}$$

$$s+\frac{4}{3}x\frac{ds}{dx}=\frac{13}{14\mathrm{Pr}}$$

Let

$$s'=s-\frac{13}{14\mathrm{Pr}}$$

Then

$$s'+\frac{4}{3}x\frac{ds'}{dx}=0$$

Solve for the homogenous equation as:

$$s'=C_1x^{-3/4}$$

Thus

$$s=r^3=C_1x^{-3/4}+\frac{13}{14\mathrm{Pr}}$$

At $x=0$:

$$r^3=\frac{13}{14\mathrm{Pr}}$$

$$r = \frac{\delta_T}{\delta} = \left(\frac{13}{14\,\mathrm{Pr}}\right)^{1/3} \cong 0.975\,(\mathrm{Pr})^{-1/3} \sim \mathrm{Pr}^{-1/3}$$

where

$$\frac{\delta}{x} = \frac{4.64}{\sqrt{\mathrm{Re}_x}}$$

Therefore,

$$\delta_T = \delta\,\mathrm{Pr}^{-1/3} = \frac{4.64x}{\sqrt{\mathrm{Re}_x}} \cdot \mathrm{Pr}^{-1/3}$$

$$h = \frac{-k\,(\partial T/\partial y)_{y=0}}{T_w - T_\infty} = \frac{3}{2}\frac{k}{\delta_T}$$

$$= \frac{3}{2}\frac{k}{\delta\,\mathrm{Pr}^{-1/3}} = \frac{3}{2}\frac{k}{x\left(4.64/\sqrt{\mathrm{Re}_x}\right)\mathrm{Pr}^{-1/3}}$$

For this typical case,

$$\mathrm{Nu}_x = \frac{hx}{k} = 0.323\,\mathrm{Re}_x^{1/2}\,\mathrm{Pr}^{1/3} \tag{7.38}$$

From Figure 7.7, at $x = x_o$, $\delta T = 0$, $r = 0$

$$0 = C_1\frac{1}{x_o^{3/4}} + \frac{13}{14\,\mathrm{Pr}}, \quad C_1 = -\frac{13}{14\,\mathrm{Pr}}x_o^{3/4}$$

Therefore,

$$r = \frac{\delta_T}{\delta} = 0.975\,\mathrm{Pr}^{-1/3}\left[1 - \left(\frac{x_o}{x}\right)^{3/4}\right]^{1/3}$$

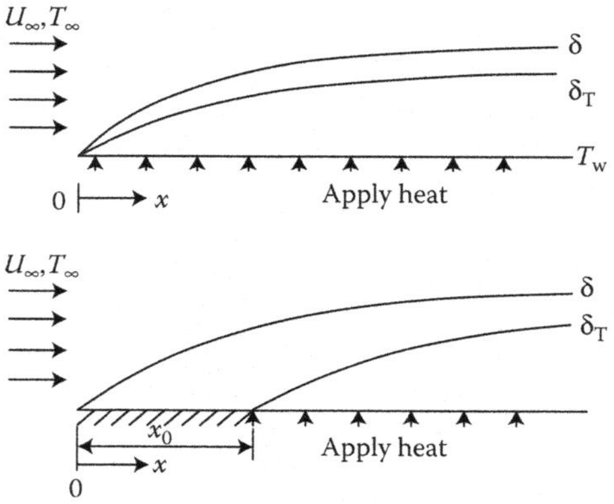

FIGURE 7.7 Unheated starting length.

For a typical case, $x_o = 0$, $\delta T = 0$,

$$\frac{\delta_T}{\delta} = 0.975\,\mathrm{Pr}^{-1/3} \simeq \mathrm{Pr}^{-1/3}$$

For the unheated, leading length problem, $x_o > 0$,

$$\mathrm{Nu}_x = \frac{hx}{k} = \frac{0.323\,\mathrm{Re}_x^{1/2}\,\mathrm{Pr}^{1/3}}{\sqrt[3]{1-\left(x_0/x\right)^{3/4}}} \tag{7.39}$$

If $x_0 = 0$, we revert to the typical case.

7.2.4 OTHER BOUNDARY LAYER PROPERTIES

The velocity and temperature gradients that develop within the viscous and thermal boundary layers are related to the shear stress and heat flux on the surface. While this is useful information, the viscous boundary layer thickness is also commonly represented in terms of other thickness definitions. These definitions are useful as the actual velocity profile merges smoothly into the free stream; therefore, the boundary layer thickness, δ, is often difficult measure. Figure 7.8 shows the actual boundary layer thickness relative to the "displacement thickness," δ^*, and the "momentum thickness," θ.

As discussed throughout this chapter, the boundary layer thickness is defined as the distance from the surface where the local velocity reaches the free stream velocity. The local boundary layer thickness is a function of the Reynolds number at a specific location on the surface.

The friction that develops along the surface creates a velocity deficit near the surface. Hence, we have the velocity profile within the boundary layer. The displacement thickness, δ^*, represents the velocity deficit created by friction within the flow. In the absence of friction, the velocity is constant down to the surface. Therefore, this thickness is the distance by which the surface would be "displaced" within a frictionless flow to provide the same mass flow rate as flow rate as the actual profile. The mass flow rate adjacent to the solid boundary is less than the mass flow rate that would pass in the same region in the absence of a boundary layer. The reduced flow rate (on a per length basis in the z-direction) can be represented as:

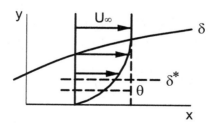

FIGURE 7.8 Conceptual representation of boundary layer, δ, displacement, δ^*, and momentum, θ, thickness.

$$\int_0^\infty \rho(\mathcal{U}_\infty - u)dy$$

Calculating the mass flow deficit and introducing the displacement thickness yields:

$$\mathcal{U}_\infty \delta^* = \int_0^\infty \mathcal{U}_\infty dy - \int_0^\infty u\,dy$$

Rearranging, the displacement thickness becomes:

$$\delta^* = \int_0^\infty \left(1 - \frac{u}{\mathcal{U}_\infty}\right)dy$$

The momentum thickness is often used to represent drag over the surface. The velocity deficit near the surface also creates a reduction in the momentum flux near the surface. The momentum deficiency, per unit length in the z-direction (compared to a frictionless flow), is:

$$\int_0^\infty \rho u(\mathcal{U}_\infty - u)dy$$

This is the thickness of the fluid layer, $\mathcal{U}_\infty$, for which the momentum flux is equal to the momentum flux deficit. Similar to the displacement thickness, the momentum thickness can be calculated by balancing the momentum flux between the actual and ideal flows:

$$\rho \mathcal{U}_\infty^2 \theta = \int_0^\infty \rho \mathcal{U}_\infty u\,dy - \int_0^\infty \rho u^2 dy$$

Rearranging to solve for the momentum thickness:

$$\theta = \int_0^\infty \frac{u}{\mathcal{U}_\infty}\left(1 - \frac{u}{\mathcal{U}_\infty}\right)dy$$

As a reference, for flow over a flat plate, the boundary layer, displacement, and momentum thicknesses can be compared (based on the similarity solution):

$$\frac{\delta}{x} = \frac{5.0}{\sqrt{\mathrm{Re}_x}}$$

$$\frac{\delta^*}{x} = \frac{1.729}{\sqrt{\mathrm{Re}_x}}$$

$$\frac{\theta}{x} = \frac{0.664}{\sqrt{Re_x}}$$

Finally, the displacement thickness and momentum thickness can be combined as the "shear factor" or "boundary layer shape factor." The shear factor, H, is the ratio of the displacement to the momentum thickness:

$$H = \frac{\delta^*}{\theta}$$

REMARKS

For the integral method, students are expected to know how to sketch and derive momentum and energy integral equations from mass, force, and heat balances across the boundary layer. Students are also expected to know how to solve for the velocity boundary-layer thickness and the friction factor from the derived momentum integral equation by assuming any velocity profile to satisfy the hydrodynamic BCs across the boundary layer. Also, you should have the tools to solve for the thermal boundary layer thickness and the heat transfer coefficient from the derived energy integral equation by assuming any temperature profile to satisfy the thermal BCs (given surface temperature or surface heat flux) across the thermal boundary layer. Note that one will get a slightly different velocity boundary-layer thickness (and friction factor) by using different velocity profiles across the boundary layer, and a slightly different thermal boundary-layer thickness (and a heat transfer coefficient) by using different temperature profiles across the thermal boundary layer. This is the nature of the integral, approximation method. Table 7.2 shows how the boundary layer thickness and friction coefficient vary with different assumed velocity profiles. Another note

TABLE 7.2
Comparison of δ and C_{fx} for Different Assumed Velocity Profiles

Assumed Profile	Δ	C_{fx}
Linear	$3.46\dfrac{x}{\sqrt{Re_x}}$	$\dfrac{0.578}{\sqrt{Re_x}}$
Second Order	$5.48\dfrac{x}{\sqrt{Re_x}}$	$\dfrac{0.729}{\sqrt{Re_x}}$
Third Order	$4.64\dfrac{x}{\sqrt{Re_x}}$	$\dfrac{0.646}{\sqrt{Re_x}}$
Fourth Order	$5.84\dfrac{x}{\sqrt{Re_x}}$	$\dfrac{0.686}{\sqrt{Re_x}}$
Blasius (Similarity) Solution	$5.0\dfrac{x}{\sqrt{Re_x}}$	$\dfrac{0.664}{\sqrt{Re_x}}$

is that velocity and thermal boundary-layer thickness is the same if Prandtl number of unity is assumed.

7.3 LAMINAR FLOW AND HEAT TRANSFER WITH A CONSTANT PRESSURE GRADIENT

The similarity method can be used to determine velocity and temperature distributions inside the laminar boundary layer formed on a surface exposed to an accelerating or decelerating mainstream flow. The acceleration or deceleration of the fluid results from a pressure gradient along the surface. With a "constant" pressure gradient, the mainstream velocity can be represented as $U = cx^m$. Similarly, with a constant temperature gradient along the surface, the temperature difference is expressed as $T_w - T_\infty = cx^n$. Figure 7.9 shows several flow configurations that can be considered as laminar boundary layer flow with a constant pressure gradient: flow over a wedge with acceleration ($m > 0$), flow in a nozzle with acceleration ($m > 0$), flow in a diffuser with deceleration ($m < 0$), flow across a cylinder with a stagnation point ($m = 1$), flow impingement on a wall with stagnation, etc. Note that flow over a flat plate with a constant velocity is a special case of $m = 0$. The boundary layer will be thinner with increased shear and heat transfer in the presence of flow acceleration and thicker with reduced shear and heat transfer with a decelerating fluid.

The similarity method presented for the zero pressure gradient flow (flat plate) can be adapted to include cases with a constant pressure gradient (m, n = constant). Recall the continuity equation with the corresponding stream function:

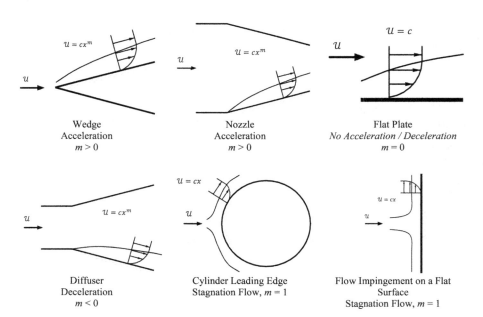

FIGURE 7.9 Sample laminar boundary flows with constant pressure gradients.

$$\frac{\partial u}{\partial x} + \frac{\partial v}{\partial y} = 0$$

$$u = \frac{\partial \psi}{\partial y}$$

$$v = -\frac{\partial \psi}{\partial x}$$

The momentum equation, in the x-direction, can be re-written to include the effect of the pressure gradient:

$$\frac{\partial u}{\partial x} + v\frac{\partial u}{\partial y} = U\frac{dU}{dx} + v\frac{\partial^2 u}{\partial y^2}$$

The mainstream velocity, U, is changing in the x-direction. Therefore, $U\dfrac{dU}{dx}$ can be re-written:

$$U\frac{dU}{dx} = -\frac{1}{\rho}\frac{dP}{dx}$$

with

$$U = cx^m$$

$$U\frac{dU}{dx} = U \cdot \left[cmx^{m-1} \right]$$

$$= Ucm\frac{x^m}{x}$$

$$cx^m = U$$

$$U\frac{dU}{dx} = Um\frac{U}{x}$$

$$U\frac{dU}{dx} = U^2\frac{m}{x}$$

Finally, we have the energy equation:

$$u\frac{\partial T}{\partial x} + v\frac{\partial T}{\partial y} = \alpha\frac{\partial^2 T}{\partial y^2}$$

7.3.1 THE RK CALCULATION TO SOLVE FOR THE SIMILARITY SOLUTION

Following the method presented in Section 7.1, the similarity variable and similarity function should be defined and substituted into the mass, momentum, and energy equations. In this case, the free stream velocity is a function of x, $U = cx^m$. Likewise, the wall to free stream temperature difference is also a function of x, $T_w - T_\infty = cx^n$. The similarity variable is:

$$\eta = y\sqrt{\frac{U(x)}{vx}} = \frac{y}{x}\sqrt{\text{Re}_x}$$

The similarity function for momentum becomes:

$$f = \frac{\Psi}{\sqrt{vxU(x)}} = \frac{\Psi}{v\sqrt{\text{Re}_x}}$$

The similarity function for energy:

$$\theta = \frac{T - T_w}{T_\infty - T_w}$$

With

$$T_w - T_\infty = cx^n$$

Including the effects of the free stream pressure and temperature gradients, m and n, the momentum and energy equations can be derived as:

$$f''' + \frac{m+1}{2}ff'' + m\left(1 - f'^2\right) = 0$$

$$\theta'' + \frac{m+1}{2}\text{Pr}f\theta' + n\text{Pr}(1 - \theta)f' = 0$$

In order to solve the higher order differential equations, it is necessary to employ numerical techniques. RK methods are commonly used to approximate solutions to ODEs. RK calculations are preferred as they provide highly accurate results without requiring derivatives, and they only require one initial point to start the calculation procedure (initial value problem). The primary disadvantage of RK calculations is the requirement to evaluate the function several times at each step of the integration. While several variations of RK methods are known, the fourth-order RK methods are the most widely used. With the application of the fourth-order RK method to the solution of boundary layer flows, one must remember, a well-defined function, $f(x, y)$ or $f'(x, y)$, is not available. However, we have a starting point (x_1, y_1), and we know the final point (x_n, y_n) at the edge of the boundary layer.

The fourth-order RK method is similar to Simpson's one-third rule. It is used to estimate multiple slopes in order to generate and improve average slope for a specific

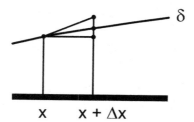

FIGURE 7.10 Boundary Layer Growth Model.

interval. A segment within the boundary layer is shown in Figure 7.10. The element shows the boundary layer growth over an element, $x + \Delta x$. Using this representation of the boundary layer growth, the Euler method with an explicit forward—difference can be coupled with the 4th order RK method to numerically approximate the velocity profile within the boundary layer.

The slope of the boundary layer growth can be written as:

$$\delta' = \frac{d\delta}{dx} = k(x, \delta)$$

From a first-order RK method, the boundary layer thickness at $x + \Delta x$ can be estimated as:

$$\delta(x + \Delta x) \cong \delta(x) + k(x, \delta) \cdot \Delta x$$

With this first-order method, the truncation error is on the order of Δx^2, and the method has been shown to "overshoot" the true point. The forth-order RK method is preferred as the truncation error reduces to the order of Δx^5. Similar to Simpson's one-third rule, the slope, k, is weighted to provide a better approximation of the slope for the boundary layer growth.

$$\delta(x + \Delta x) \cong \delta(x) + k \cdot \Delta x$$

where the weighted average slope, k, can be expressed as

$$k = \frac{1}{6}(k_1 + 2k_2 + 2k_3 + k_4)$$

with

$$k_1 = k_1(x, \delta)$$

$$k_2 = k_2\left(x + \frac{1}{2}\Delta x, \delta + \frac{1}{2}\Delta x \cdot k_1\right)$$

$$k_3 = k_3\left(x + \frac{1}{2}\Delta x, \delta + \frac{1}{2}\Delta x \cdot k_2\right)$$

$$k_4 = k_4\left(x + \Delta x, \delta + \Delta x \cdot k_3\right)$$

The fourth-order RK method can be used to solve the laminar boundary layer flow and heat transfer problem. For example, let $m = constant$ and $n = 0$, the similarity momentum and energy equations become:

$$f''' + \frac{m+1}{2}ff'' + m\left(1 - f'^2\right) = 0$$

$$\theta'' + \frac{m+1}{2}\Pr f\theta' = 0$$

The momentum equation contains a third derivative, while the energy equation is a second derivative.

To apply the fourth-order RK method to the boundary layer problem, the similarity momentum and energy equations must be linearized to become the first derivative equations. Therefore, let:

$$Z_1 = f$$

$$Z_2 = f'$$

$$Z_3 = f''$$

and

$$Z_4 = \theta$$

$$Z_5 = \theta'$$

Now the RK type similarity momentum and energy equations become five linear equations:

$$Z_1' = Z_2$$

$$Z_2' = Z_3$$

$$Z_3' = -\frac{m+1}{2}Z_1 \cdot Z_3 - m\left(1 - Z_2^2\right)$$

$$Z_4' = Z_5$$

$$Z_5' = -\frac{m+1}{2}\Pr Z_1 \cdot Z_5$$

For the boundary layer thickness, we can apply our three BCs to the momentum equation:

$$Z_1(0) = 0$$

$$Z_2(0) = 0$$

$$Z_2(\infty) = 1$$

However, Z_3 at the wall, which represents the shear stress on the wall, is unknown and must be determined.

$$Z_3(0) = f''(0) = \text{unknown}$$

Applying our thermal BCs to the energy equation:

$$Z_4(0) = 0$$

$$Z_4(\infty) = 1$$

In this case, the temperature gradient at the wall, Z_5, is unknown and must be determined using this RK method.

$$Z_5(0) = \theta'(0) = \text{unknown}$$

Recall, the similarity functions for momentum, f, and energy, θ, are only functions of the similarity variable, η. Likewise, the RK-type momentum and energy equations are also only a function of η.

$$f = f(\eta)$$

$$\theta = \theta(\eta)$$

$$Z = Z(\eta)$$

This is a typical two-point boundary value problem where we know the velocity and temperature behavior at the boundaries ($y = 0$ and $y = \delta$). As stated earlier, we know the final point (x_n, y_n) at the edge of the boundary layer, so we must make an initial guess for the velocity and temperature gradients at the wall to begin the process. The RK method can be applied working from point-to-point through the boundary layer. As $y \to$ delta, the gradients must approach zero. If the initial guess leads to incorrect estimations at the edge of the boundary layer, a new initial guess is required, and the process is repeated. If the outer BC is correctly achieved, the initial guess is correct, and the similarity functions are known. The following steps should be followed to use the fourth RK method to approximate the velocity profile, wall shear stress, temperature profile, and wall heat flux.

Step 1: Guess a value of the unknown velocity gradient at the wall, $Z_3(\eta=0)$ or $f''(\eta=0)$.

Step 2: With the known BCs ($Z_1(\eta=0)=0$ and $Z_2(\eta=0)=0$) and the guessed value of $Z_3(\eta=0)$, use the RK calculation to calculate the 4 slopes for Z_1, Z_2, and Z_3. This is k_1, k_2, k_3, and k_4 as separate values for Z_1, Z_2, and Z_3.

Step 3: Use the RK method to calculate the weighted slope, k, for each variable Z_1, Z_2, and Z_3.

Step 4: Update the values of Z_1, Z_2, and Z_3 for $\eta + \Delta\eta$ based on the weighted slope and the RK calculation.

Step 5: As $\eta \to \infty$, check if the known BCs are satisfied: $Z_2(\eta=0)=f'(\eta=0)= 0$ and $Z_2(\eta=\infty) = f'(\eta=\infty) = 1$. If the BCs are not satisfied, return to "step 1" and begin the process with a new guess for the velocity gradient at the wall.

To solve for the temperature profile through the thermal boundary layer, steps 1–5 are repeated with Z_4 and Z_5. In step 1, the unknown value of the temperature gradient, $Z_5(\eta=0)$ or $\theta'(\eta=0)$, is guessed to initiate the RK method. The four slopes (k_1, k_2, k_3, and k_4) are then calculated for both Z_4 and Z_5, and this is followed by the calculation of the weighed slopes, k, for both Z_4 and Z_5. The process is repeated in steps 1–5, from the surface to the edge of the thermal boundary layer. Upon reaching the edge of the boundary layer, the guess is checked by ensuring the BCs are satisfied, $Z_4(\eta=0)= \theta(\eta=0)=0$ and $Z_4(\eta=\infty) = \theta(\eta=\infty) = 1$. If the BCs are not satisfied, the process is repeated with a new initial guess of the temperature gradient, $Z_5(\eta=0)$.

For example, consider flow over a flat plate, $m = 0$, Pr = 1, the table below shows the guess procedure until the correct results are obtained.

η	f or Z_1	f' or Z_2	f'' or Z_3	θ or Z_4	θ' or Z_5
0	0	0	0.5 - *guess*	0	0.5 - *guess*
0.2	↓	↓	↓	↓	↓
↓					
10	▼	2 - *wrong*	▼	2 - *wrong*	▼
0	0	0	0.332 - *guess*	0	0.332 - *guess*
0.2	↓	↓	↓	↓	↓
↓					
10	▼	1 - *ok*	▼	1 - *ok*	▼

Figures 7.11–7.13 show the results of RK Fourth-Order Calculation for various boundary layer flows with constant pressure gradients: For $U = cx^m$ and $T_w - T_\infty = cx^n$. Figure 7.11 and the corresponding tabulated data, Table 7.3, are the velocity and temperature profiles for $m = 1$ to $m = -0.0904$ (flow separation) for $n = 0$ and Pr = 1. Recall the velocity profile is not a function of the Prandtl number, while the temperature profile will change if the Prandtl number deviates from one. Note that the velocity gradient at the wall is 0.332 for flow over a flat plate $m = 0$. The velocity gradient at the wall increases with $m > 1$ due to the thinner boundary layer with flow acceleration.

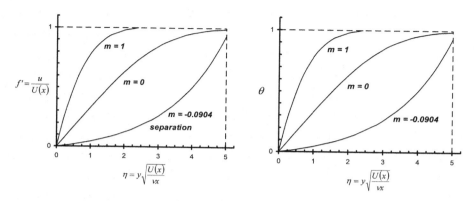

FIGURE 7.11   Runge–Kutta calculation of the velocity and temperature profiles for positive ($m = 1$), zero ($m = 0$), and adverse pressure gradients ($m = -0.904$); in all cases $Pr = 1$ with $n = 0$.

TABLE 7.3
Effect of Free Stream Acceleration (*m*) on Boundary Layer Development

m	$f''(0)$ Recall: $C_{fx} = \dfrac{2 \cdot f''(0)}{\sqrt{Re_x}}$	δ^* Displacement Thickness, δ^*	θ Momentum Thickness, θ	*H* Shear Factor or Boundary Layer Shape Factor
1 (stagnation)	1.232	0.647	0.292	2.216
0.333	0.757	0.985	0.429	2.297
0 (flat plate)	0.332	1.72	0.664	2.59
−0.05	0.213	2.117	0.751	2.818
−0.0904 (separation)	0.	3.427	0.868	3.949

Figure 7.12 shows the effect of Prandtl number on temperature profiles for $m = 0$ (solid lines) and $m = 1$ (broken lines) at $n = 0$. Note: The Reynolds analogy factor does not hold for pressure gradient flow, as shown in Table 7.4.

Figure 7.13 shows the proper correlation of the Nusselt number with the Reynolds number and Prandtl number [7]. The figure clearly shows increased heat transfer with flow acceleration.

In summary, the RK method can be used to solve a variety of laminar flow and heat transfer problems. You can apply numerical methods (using a programming language such as MATLAB®) to obtain results listed below (and partially represented in Figures 7.11 and 7.12).

a. Plot f' vs. η and f'' vs. η on the same figure for $m = -0.05, 0, 0.333$, and 1.
b. Plot θ vs. η and θ' vs. η on the same figure for $Pr = 0.72$ with $m = -0.05, 0, 0.333$, and 1 and $n = 0$.

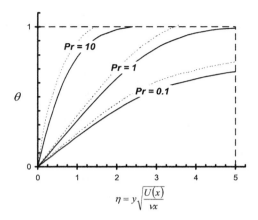

FIGURE 7.12 Runge–Kutta calculation of the temperature profiles for positive ($m = 1$, dashed lines) and zero ($m = 0$, solid lines) pressure gradients for various Prandtl numbers; in all cases $n = 0$.

TABLE 7.4

Effects of Free Stream Acceleration (m) and Prandtl Number on the Dimensionless Heat Transfer Coefficient, $\dfrac{\mathrm{Nu}_x}{\sqrt{\mathrm{Re}_x}} = \theta'_o(m, \mathrm{Pr})$, for $n = 0$.

m	Pr = 0.1	Pr = 0.72	Pr = 1	Pr = 10	$\dfrac{\mathrm{St}}{\frac{1}{2}C_{fx}} = \dfrac{\mathrm{Nu}_x}{f'(0)\sqrt{\mathrm{Re}_x}}$ (Pr = ∞)
1 (stagnation)	0.2191	0.5017	0.5708	1.3433	0.463
0 (flat plate)	0.1389	0.2957	0.3321	0.7289	1
−0.0904 (separation)	0.1085	0.2031	0.2224	0.4071	∞

---

c. Plot θ vs. η and θ' vs. η on the same figure for $m = n = 0$ with Pr = 0.1, 0.72, 10, and 50.

d. Plot θ vs. η and θ' vs. η on the same figure for $m = 0$, Pr = 0.72 with $n = -1/2$, 0, 1/2, and 1.

Nonsimilarity Condition: "Nonsimilarity" flow conditions exist in many engineering applications such as compressors and turbines in turbomachinery. Figure 7.14 shows local flow acceleration around a turbine blade, highlighting the pressure distribution on both the suction and pressure surfaces. The pressure gradient is no longer a constant value, i.e., m is not a constant value during acceleration. This type of nonsimilarity flow can be solved using the local similarity concept using the Falkner–Skan Transformation.

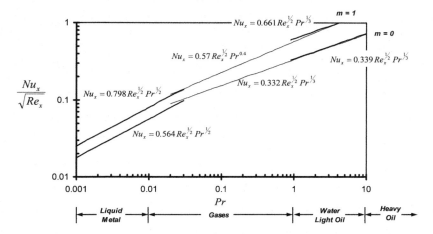

FIGURE 7.13 Correlation of the surface Nusselt numbers for various fluid and with flow acceleration.

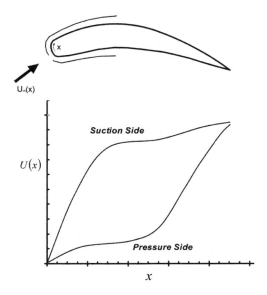

FIGURE 7.14 Nonsimilarity flow over the suction and pressure surfaces of a turbine blade

As shown previously, the free stream velocity and temperature gradients can be represented as a function of the x-location:

$$U = cx^m$$

$$T_w - T_\infty = cx^n$$

If m = constant and n = constant, the previous similarity solution can be applied. However, if $m, n \neq$ constant, the Falkner-Skan Transformation is needed for this non-similarity condition.

From the boundary layer conservation equations, the continuity equation, x-momentum equation, and energy equation are

$$u = \frac{\partial \psi}{\partial y}$$

$$v = -\frac{\partial \psi}{\partial x}$$

$$u\frac{\partial u}{\partial x} + v\frac{\partial u}{\partial y} = U\frac{dU}{dx} + v\frac{\partial^2 u}{\partial y^2}$$

$$u\frac{\partial T}{\partial x} + v\frac{\partial T}{\partial y} = \alpha\frac{\partial^2 T}{\partial y^2}$$

Again, we define the similarity variable and similarity function. The free stream velocity is a function of x, $U = cx^m$, and the wall to free stream temperature difference is also a function of x, $T_w - T_\infty = cx^n$, where m and n are not constant values. For example:

$$m = \frac{x}{U} \cdot \frac{dU}{dx}$$

$$n = \frac{x}{T_w - T_\infty} \cdot \frac{d}{dx}(T_w - T_\infty)$$

From the same procedures outlined in Section 7.1, one obtains the following similarity momentum and energy equations:

$$f = \frac{\psi}{\sqrt{vU_\infty x}} = f(\eta, x)$$

$$\eta = y\sqrt{\frac{U_\infty}{vx}}$$

$$\theta = \frac{T - T_w}{T_\infty - T_w}$$

with

$$m = \frac{x}{U} \cdot \frac{dU}{dx}$$

$$n = \frac{x}{T_w - T_\infty} \cdot \frac{d}{dx}(T_w - T_\infty)$$

Therefore,

$$\frac{u}{U} = f'(\eta, x)$$

$$\upsilon = \frac{1}{2}\sqrt{\frac{\nu U}{x}} \cdot f' - \frac{\partial}{\partial x}[(U\nu x)f]$$

The momentum and energy equations can be re-written in terms of the similarity variable and functions:

$$f''' + \frac{m+1}{2}ff'' + m(1 - f'^2) = x\left(f'\frac{\partial f'}{\partial x} - f''\frac{\partial f}{\partial x}\right)$$

$$\underbrace{\theta'' + \frac{m+1}{2}\Pr f\theta' + n\Pr(1-\theta)f' = x\Pr\left(f'\frac{\partial \theta}{\partial x} - \theta'\frac{\partial f}{\partial x}\right)}_{\text{Non-Similarity}}$$

Note that the nonsimilarity terms are in the right-hand side of the equations (non-zero). These RK type nonsimilarity momentum and energy equations can be considered as Local Similarity Equations for Local Segments Δx. For a small Δx distance, m and n can be assumed constant.

At a given x:

$$\left(\frac{\partial f'}{\partial x}\right)_i \cong \frac{f_i' - f_{i-1}'}{x_i - x_{i-1}}$$

$$\left(\frac{\partial f}{\partial x}\right)_i \cong \frac{f_i - f_{i-1}}{x_i - x_{i-1}}$$

$$\left(\frac{\partial \theta}{\partial x}\right)_i \cong \frac{\theta_i - \theta_{i-1}}{x_i - x_{i-1}}$$

where f_i', f_{i-1}', f_{i-1}, and θ_{i-1} are pre-determined at x_{i-1}. Then use RK Fourth-Order Method marching from $x = 0$ to any x by the increment $\Delta x = x_i - x_{i-1}$.

7.3.2   INTEGRAL APPROXIMATION METHODS FOR PRESSURE GRADIENT FLOWS

a. Integral Method from Pohlhausen

Recognizing the streamwise pressure gradient can be represented by the free stream acceleration, Pohlhausen extended the integral approximation method (Section 7.2) to approximate the boundary layer thickness in the presence of accelerating flows. The new form of x-momentum equation 7.33 can be obtained by defining the boundary layer displacement thickness, momentum thickness, and shape factor:

$$\tau_w = \frac{dU}{dx}\int_0^\delta \rho(U-u)\,dy + \frac{d}{dx}\int_0^\delta \rho u(U-u)\,dy$$

$$\frac{1}{2}C_{fx} = \frac{1}{U}\frac{dU}{dx}\delta^* + \frac{1}{U^2}\frac{d}{dx}\left(U^2\theta\right)$$

$$= \frac{d\theta}{dx} + (2+H)\frac{\theta}{U}\frac{dU}{dx} = f(\delta,U)$$

As defined previously, the displacement thickness is (now in terms of a variable free stream velocity):

$$\delta^* = \int_0^\delta\left(1-\frac{u}{U}\right)dy = \delta_1$$

The momentum thickness becomes:

$$\theta = \int_0^\delta \frac{u}{U}\left(1-\frac{u}{U}\right)dy = \delta_2$$

With the shape factor:

$$H \equiv \frac{\delta^*}{\theta} = \frac{\delta_1}{\delta_2}$$

Recall the wall shear stress:

$$\tau_w = C_{fx}\frac{1}{2}\rho U^2$$

Following the procedure outlined in Section 7.2.3, we can assume a dimensionless velocity profile across the boundary layer with appropriate BCs:

$$\frac{u}{U} = a + b\eta + c\eta^2 + d\eta^3 + e\eta^4 = f(\eta) = f\left(\frac{y}{\delta}\right)$$

BCs:

$\eta = 0$	$u = 0$	$\eta = 1$	$u = U$
	$v\dfrac{\partial^2 u}{\partial y^2} = -\dfrac{1}{\rho}\dfrac{\partial P}{\partial x}$		$\dfrac{\partial u}{\partial y} = 0$
	$= -U\dfrac{dU}{dx}$		$\dfrac{\partial^2 u}{\partial y^2} = 0$

Solve for the coefficients a, b, c, d, and e, thus

$$\frac{u}{U} = \left(2\eta - 2\eta^3 + \eta^4\right) + \frac{1}{6}\left(\frac{\delta^2}{v}\frac{dU}{dx}\right)\left(\eta - 3\eta^2 + 3\eta^3 - \eta^4\right)$$

In this equation, $\dfrac{\delta^2}{v}\dfrac{dU}{dx}$ varies with the local pressure gradient and is often referred to as the Pohlhausen parameter. Following "step 4" from the integral approximation method, substitute the velocity profile in the momentum integral and obtain $\delta(x)$ for a given $U(x)$. Next, with "step 5," the shear stress can be determined:

$$\tau_w = \mu\frac{\partial u}{\partial y}\bigg|_0 = \mu U\left(\frac{2}{\delta} + \frac{1}{6\delta}\cdot\frac{\delta^2}{v}\frac{dU}{dx}\right)$$

b. Walz Approximation

The integral method developed by Pohlhausen is relatively simple to implement. The process begins by assuming a velocity profile that is a function of the pressure gradient. While the resulting differential equation can be easily solved, this method lacks accuracy and is not commonly used today. While Pohlhausen assumed a fourth-order polynomial for the velocity profile, Walz assumed a profile based on local similarity. This family of profiles are solutions to the Falkner–Skan equation [7].

This integral method has been used to solve for the momentum thickness for many pressure gradient flows such as flow over an airfoil, across a cylinder, across an ellipse, in a nozzle, in a diffuser, etc., as shown in Figure 7.9. With this method, a single variable was introduced that directly correlates with the shape factor, H.

The momentum integral has the following characteristics:

$$\frac{d\left(\dfrac{\theta^2}{v}\right)}{dx} = \frac{1}{U}\cdot F\left(\frac{\theta^2}{v}\frac{dU}{dx}\right)$$

$$= \frac{a}{U} - \frac{b}{U}\frac{\theta^2}{v}\frac{dU}{dx}$$

where $a = 0.47$ and $b = 6$.

After integrating the above equation, the momentum thickness can be obtained as:

$$\frac{\theta^2}{v} = \frac{0.47}{U^6} \int_0^x U^5 dx$$

at $x = 0$, either $U = 0$ or $\theta = 0$.

Depending on the specific geometry (for instance, flow around a cylinder), an expression for the free stream velocity can be developed. For flow around a cylinder, the velocity based on the distance (or angular location) from the stagnation point:

$$U = 2\mathcal{U}_\infty \cdot \sin\theta \cong 2\mathcal{U}_\infty \theta \cong 2\mathcal{U}_\infty \frac{x}{r_o}$$

The acceleration within a nozzle can also be shown based on the linear profile of the nozzle:

$$U = a + bx^n$$

$$= U_o - ax$$

$$= 1 + x$$

$$= a\sin\left(\frac{\pi x}{2L}\right)$$

For Howarth flow,

$$a = \frac{1}{8}$$

For a given pressure gradient flow, $U(x) = $ given, the following procedure can be used to solve for the surface shear. At a given location x, calculate $U(x)$, $\theta(x)$, $\frac{dU}{dx}$, λ, $H \equiv \frac{\delta^*}{\theta}$, and $\frac{\tau_w \theta}{\mu U}$. Now τ_w can be calculated with:

$$\lambda \equiv \frac{\theta^2}{v} \frac{dU}{dx}$$

$$\text{Re}_\theta = \frac{U\theta}{v}$$

From the above empirical momentum integral relationships, we have:

$$\frac{\theta^2}{v} = \frac{0.47}{U^6} \int_0^x U^5 dx$$

$$C_f = \frac{\tau_w}{\frac{1}{2}\rho U^2}$$

For the cases of an accelerating flow without flow separation, $0 \le \lambda \le 0.1$:

$$H \equiv \frac{\delta^*}{\theta} = 2.61 - 3.75 \cdot \lambda + 5.24 \cdot \lambda^2$$

$$\frac{\tau_w \theta}{\mu U} = \text{Re}_\theta \cdot \frac{C_f}{2} = 0.22 + 1.57 \cdot \lambda - 1.8 \cdot \lambda^2$$

Conversely, for decelerating flows with an adverse pressure gradient, $-0.1 \le \lambda \le 0$, note the lower limit of λ, as separation occurs at $\lambda \cong -0.09$:

$$H \equiv \frac{\delta^*}{\theta} = \frac{0.0731}{0.14 + \lambda} + 2.088$$

$$\frac{\tau_w \theta}{\mu U} = R_\theta \cdot \frac{C_f}{2} = 0.22 + 1.402 \cdot \lambda - \frac{0.018 \cdot \lambda}{0.107 + \lambda}$$

Finally, for heat transfer:

$$q_w = \frac{d}{dx} \int_0^\delta \rho c_p u (T - T_\infty) dy$$

with

$$\frac{u}{U} = f(\eta)$$

c. Smith and Spalding's Approximation for Pressure Gradient Flow ($T_w = constant$)
Heat transfer approximations have been completed using Smith and Spalding's integral approximation method. With this approximation, the free stream velocity is variable, but it is only applicable to flows exposed to a surface at a constant temperature. This approximation of the thermal boundary layer follows a similar process as that posed by Walz. First, analogous to the momentum thickness, the conduction thickness can be expressed as:

$$\delta_c = \frac{k(T_w - T_\infty)}{q_w} = \frac{-(T_w - T_\infty)}{\left. \dfrac{\partial T}{\partial y} \right|_0}$$

Nondimensionalizing the conduction thickness:

$$\frac{d\left(\dfrac{\delta_c^2}{\nu}\right)}{dx} = \frac{1}{U} \cdot F\left(\frac{\delta_c^2}{\nu}\frac{dU}{dx},\mathrm{Pr}\right)$$

Therefore,

$$\frac{\delta_c^2}{\nu} = \frac{0.47}{U^6}\int\limits_0^x U^5 dx$$

Now the heat transfer coefficient, expressed nondimensional in terms of the Stanton number:

$$\mathrm{St} = \frac{q_w}{\rho c_p U\left(T_w - T_\infty\right)} = \frac{k}{\rho c_p U\delta_c}\cdot\frac{x}{\dfrac{x}{\mu}\mu} = \frac{\mathrm{Nu}_x}{\mathrm{Pr}\cdot\mathrm{Re}_x}$$

where the Nusselt number is:

$$\mathrm{Nu}_x = \frac{hx}{k} = \frac{q_w}{T_w - T_\infty}\cdot\frac{x}{k} = \frac{k}{\delta_c}\frac{x}{k} = \frac{x}{\delta_c}$$

Re-writing the Stanton number in terms of nondimensional lengths and velocities:

$$\mathrm{St} = \frac{c_1\left(U^*\right)^{c2}}{\left[\displaystyle\int_0^{x^*}\left(U^*\right)^{c3} dx^*\right]^{\frac{1}{2}}\cdot\mathrm{Re}_L^{\frac{1}{2}}}$$

where

$$x^* = \frac{x}{L};\; U^* = \frac{U}{U_\infty};\; \mathrm{Re}_L = \frac{U_\infty L}{\nu};\; c_1 = \mathrm{Pr}^{-1}\left(0.47\right)^{-\frac{1}{2}}$$

The constants, c_1, c_2, and c_3 in the above equation are a function of Prandtl number, and the following table shows how these constant change [7].

Pr	c_1	c_2	c_3
0.7	0.418	0.435	1.87
1.0	0.332	0.475	1.95
10	0.073	0.685	2.37

7.3.3 INTEGRAL APPROXIMATION METHOD FOR BOUNDARY-LAYER FLOW WITH A NONUNIFORM WALL TEMPERATURE

The Smith Spalding method was proposed to approximate convective heat transfer when the free stream velocity is not constant. However, this method is only applicable when the surface temperature is maintained constant. Therefore, other methods have been developed to approximate the heat transfer to or from a surface that is not held at a fixed temperature.

a. Unheated Leading Edge

Figure 7.15 shows flow over a flat plate with an unheated leading edge. The local Nusselt number for flow over a flat plate using the integral method shown in Example 2 is quoted below:

$$\mathrm{Nu}_x = \frac{hx}{k} = \frac{0.323\,\mathrm{Re}_x^{1/2}\,\mathrm{Pr}^{1/3}}{\sqrt[3]{1-\left(\dfrac{x_o}{x}\right)^{3/4}}} \text{ valid for } x > x_o$$

$$= 0.323\,\mathrm{Re}_x^{1/2}\,Pr^{1/3} \quad \text{if } x_o = 0$$

b. Nonuniform Wall Temperature—Integral Superposition

The concept with the step change of the wall temperature, shown above, can be extended to solve problems with a nonuniform wall temperature. The process involves superposition. Figure 7.16 shows the concept of N wall temperature step changes along the flat plate. Based on the above-mentioned local Nusselt number for a single step change in the wall temperature, the local Nusselt number for N step changes in the wall temperature can be obtained by summation, or integration, along the flat plate.

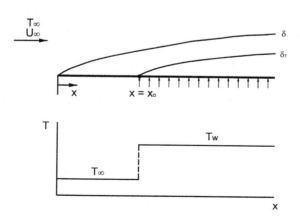

FIGURE 7.15 Flow over a flat plate with nonheated leading edge.

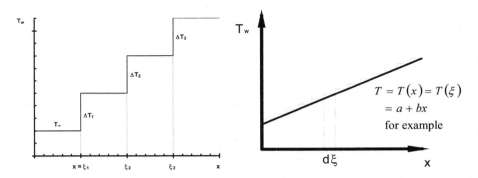

FIGURE 7.16   Nonuniform wall temperature model with superposition.

$$q_w'' = h_x (T_w - T_\infty) = \int_{\xi=0}^{x} \frac{0.323 \cdot k \cdot \mathrm{Re}_x^{1/2} \mathrm{Pr}^{1/3}}{x \cdot \left[1 - \left(\dfrac{\xi}{x}\right)^{3/4}\right]^{1/3}} \cdot \frac{dT}{d\xi} \cdot d\xi$$

$$= \sum_{n=1}^{N} \frac{0.323 \cdot k \cdot \mathrm{Re}_x^{1/2} \mathrm{Pr}^{1/3}}{x \cdot \left[1 - \left(\dfrac{\xi}{x}\right)^{3/4}\right]^{1/3}} \cdot \Delta T_n \quad \xi_n < x < \xi_{n+1}$$

For example, as shown in Figure 7.16, let $T = T(x) = a + bx$, $dT/dx = b$. The local Nusselt number becomes:

$$\mathrm{Nu}_x = 0.323 \, \mathrm{Re}_x^{1/2} \mathrm{Pr}^{1/3} \cdot \frac{a + 1.612bx}{a + bx}$$

$$= 0.323 \, \mathrm{Re}_x^{1/2} \mathrm{Pr}^{1/3} \quad \xi_n < x < \xi_{n+1}$$

If $b = 0$, $T_w = a = $ constant.

c. Nonuniform Wall Heat Flux—Integral Superposition

The superposition concept can also be applied for flow over a flat plate with a nonuniform heat flux, $q'' = q''(x)$. The wall temperature, $T(x)$, can be predicted as

$$T_w(x) - T_\infty = \int_{\xi=0}^{x} \frac{q''(\xi) \cdot d\xi}{0.323 \cdot k \left\{4.83 \cdot \mathrm{Re}_x^{1/2} \mathrm{Pr}^{1/3} \cdot \left[1 - \left(\dfrac{\xi}{x}\right)^{3/4}\right]^{2/3}\right\}}$$

For example, $q_w''(\xi) = a + bx$.

If we consider $b = 0$, then $q''_w = a = $ constant (uniform heat flux condition), the local Nusselt number is

$$\text{Nu}_x = 0.453\,\text{Re}_x^{\frac{1}{2}}\,\text{Pr}^{\frac{1}{3}}$$

Note that $\text{Nu}(x)$ for the uniform wall heat flux is greater than that for the uniform wall temperature at the same Reynolds number and Prandtl number, i.e., the coefficient $0.453 > 0.332$.

d. Integral Method for Pressure Gradient Flow with Nonuniform Wall Temperature by Lighthill (*Handbook of Heat Transfer*)

For a given pressure gradient flow $U(x)$ with a nonuniform wall temperature $T_w(x)$, the local wall heat flux and shear stress can be obtained using the following integral equations. In general, advanced numerical integrations are required to solve these integral equations.

$$q''_w(x) = 0.52 \left(\frac{k^3 \rho}{\mu^2} \cdot \text{Pr}^{\frac{1}{3}} \left[\tau_w(x) \right]^{\frac{1}{2}} \frac{\int_{\zeta=0}^{x} \dfrac{dT_w}{d\xi} \cdot d\xi}{\left(\int_0^x \tau_w^{\frac{1}{2}} dx \right)^{\frac{1}{3}}} \right)$$

After integration, the local heat transfer coefficient can be obtained:

$$h_x = \frac{q_w(x)}{T_w(x) - T_\infty}$$

where $\tau_w(x)$ could be obtained from Waltz Approximation as mentioned in the previous section as:

$$\frac{\theta^2}{v} = \frac{0.47}{U^6} \int_0^x U^5 dx$$

Given $U(x) \rightarrow \theta(x) \rightarrow \tau_w(x)$
   Given $T_w(x) \rightarrow q_w(x) \rightarrow h$

7.4   HEAT TRANSFER IN HIGH-VELOCITY FLOWS

7.4.1   ADIABATIC WALL TEMPERATURE AND RECOVERY FACTOR

a. Stagnation Temperature, T_o, or Total Temperature, T_T

When a high-speed flow approaches a sphere or thermocouple bead, as shown in Figure 7.17, dynamic energy of the flow is converted into thermal energy, which increases the fluid temperature. This is referred to as the stagnation temperature or

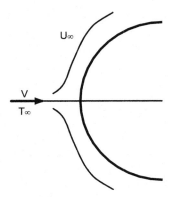

FIGURE 7.17   High-speed flow around a sphere.

the total temperature. If a gas flow is brought to rest adiabatically, the stagnation temperature can be determined as shown below. The stagnation temperature is identical to the static temperature for low-speed flows and in liquids [4]. Therefore, the total temperature, T_T, (or stagnation temperature, T_o) = static temperature T_s (or T_∞) + dynamic temperature $\left(\dfrac{u_\infty^2}{2c}\right)$:

$$T_o = T_\infty + \frac{u_\infty^2}{2c} = T_\infty\left[1+\left(\frac{\gamma-1}{2}\right)M^2\right] = T_\infty\left(1+0.2M^2\right)$$

For air:

$$\gamma = \frac{c_p}{c_v} = 1.4$$

Also:

$$M = \text{Mach number} = \frac{u_\infty}{\sqrt{\gamma RT}}$$

$\sqrt{\gamma RT}$ = speed of sound (~1000 ft/s or 313 m/s at room temperature)

If $M = 0.5$, the stagnation temperature is approximately 5% greater than the static temperature; however, if $M = 1$, the temperature deviates by 20%.

(b) Adiabatic Wall Temperature, T_{aw}, or Recovery Temperature, T_r

When a high-speed flow passes over an adiabatic surface, as shown in Figure 7.18, frictional heat $\left(\sim v\left(\dfrac{\partial u}{\partial y}\right)^2\right)$ is generated inside the boundary layer due to the viscosity and the velocity gradient. This heat increases the surface temperature, and this elevated temperature is referred to as the adiabatic wall temperature, T_{aw}, or recovery temperature, T_r. Note that the recovery temperature (T_r) = the stagnation

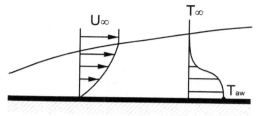

Insulated Wall

FIGURE 7.18 High-speed flow over an adiabatic surface.

temperature (T_o), if the recovery factor, $r = 1$ [4]. The adiabatic wall temperature, T_{aw} (or recovery temperature, T_r) = static temperature T_s (or T_∞) + dynamic temperature $\frac{\mathcal{U}_\infty^2}{2c}$:

If Pr = 1:

$$T_{aw} = T_\infty + \frac{\mathcal{U}_\infty^2}{2c}$$

If Pr ≠ 1:

$$T_r = T_\infty + r_c \frac{\mathcal{U}_\infty^2}{2c}, \quad \text{if} \quad Pr \neq 1$$

where

r_c = recovery factor
= $Pr^{1/2}$ for laminar flow = 0.84 (air, Pr = 0.71)
= $Pr^{1/3}$ for turbulent flow = 0.89 (air, Pr = 0.71)
= 1 if Pr = 1 (assumption, ideal)

Therefore,

$$T_{aw} = T_r = r_c(T_o - T_\infty) + T_\infty$$

$$= T_o, \text{if } r_c = 1$$

In general $T_{aw} < T_o$, $T_r < T_T$

7.4.2 HIGH-VELOCITY FLOW OVER A FLAT PLATE

The energy equation for a high-speed flow, with Pr = 1, can be written in terms of the local stagnation temperature, T^*. In the ideal case with Pr = 1, the energy equation is:

$$\rho c u \frac{\partial T^*}{\partial x} + \rho c v \frac{\partial T^*}{\partial y} = \frac{\partial}{\partial y}\left(k \frac{\partial T^*}{\partial y} \right)$$

where c = constant.

In the case of the high-speed flow, the fluid properties μ, k are now a function of temperature. The local stagnation temperate varies from the wall to the edge of the boundary layer as:

$$y = 0 \qquad T^* = T_o^* = T_0 + 0 = T_0$$
$$y = \infty \qquad T^* = T_\infty^*$$

Based on the driving temperature difference, the local, wall heat flux can be determined:

$$q_o'' = -k \left. \frac{\partial T^*}{\partial y} \right|_0 = h\left(T_o^* - T_\infty^* \right) = h\left(T_o - T_\infty^* \right)$$

Depending on the thermal BC and the level of viscous heating, the wall heat flux could be positive, negative, or zero.

If $Pr \neq 1$, the energy equation is modified to include this fluid property [4]:

$$u \frac{\partial T}{\partial x} + v \frac{\partial T}{\partial y} = \alpha \frac{\partial^2 T}{\partial y^2} + \frac{\alpha Pr}{c}\left(\frac{\partial u}{\partial y} \right)^2$$

The surface heat flux becomes:

$$q_o'' = h\left(T_o - T_{aw} \right)$$

where

$$T_{aw} = T_\infty + r_c \frac{\mathcal{U}_\infty^2}{2c}$$

Here, for laminar boundary layer heat transfer, h can be calculated from:

$$\frac{hx}{k} = Nu = 0.332 \, Re_x^{1/2} \, Pr^{1/3}$$

For high-speed flow over a flat plate, heat transfer can be negative, zero, or positive, depending on if the wall temperature is less than, equal to, or greater than the adiabatic wall temperature, as shown in Figure 7.19.

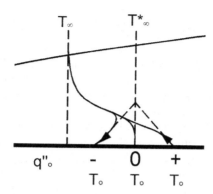

FIGURE 7.19 Heat transfer model for high-speed flow over a flat plate.

PROBLEMS

7.1   Consider a steady, incompressible, low-speed 2-D laminar boundary-layer
      flow (at U_∞, T_∞) over a flat plate at a uniform wall temperature T_w. Assume
      that there exist no body force and constant thermal and fluid properties.
      The similarity momentum and energy equations are listed here for reference:
      $f''' + (1/2)ff'' = 0$ and $\theta'' + (1/2)P_r f\theta' = 0$.

   a. From the Blasius solution of the above similarity equations, sketch the
      relations between the similarity functions $(f'\theta)$ and the similarity variable
      (η) for both water and air (i.e., sketch f' versus η for both water and air
      on the same plot; and θ versus η for both water and air on the same plot).
      Explain why they have differences, if any. You do not need to solve the
      above equations.

   b. For a given problem (i.e., U_∞, T_∞, ρ, μ, T_w, P_r are given), explain briefly
      how to determine the local velocity $u(y)$ and temperature $T(y)$ at a speci-
      fied location x, for both water and air, from the sketches in (a)?

7.2   Consider a steady, incompressible, low-speed 2-D laminar boundary-layer
      flow (at U_∞, T_∞) over a flat plate at a uniform wall temperature T_w or at a uni-
      form wall heat flux q_w'', respectively. Assume that there exist no body force
      and constant thermal and fluid properties.

   a. Based on the approximate integral method, assuming a uniform velocity
      profile inside the boundary layer, that is, $u = U_\infty$, and assuming a linear
      temperature profile inside the thermal boundary layer as $T = a + b \times y$,
      determine the local thermal boundary-layer growth along the flat plate
      (i.e., δ_T versus x) at a uniform wall temperature T_w condition. (*Note: a* and
      b are unknown constants that need to be determined.)

   b. Based on (a), determine the local Nusselt number distribution along the
      flat plate (i.e., Nu versus x) at a uniform wall temperature.

7.3   Consider a steady, incompressible, low-speed 2-D laminar boundary-layer
      flow (at U_∞, T_∞) over a flat plate at a uniform wall temperature T_w. Assume
      that there exist no body force and constant thermal and fluid properties.

The similarity momentum and energy equations are listed here for reference:
$f''' + (1/2)ff'' = 0$ and $\theta'' + (1/2)\Pr f\theta' = 0$.

a. Sketch both the velocity and the thermal boundary-layer thickness distribution for both water and liquid metal, respectively, flowing over the flat plate? (i.e., sketch δ versus x and δ_T versus x on the same plot for water, and δ versus x and δ_T versus x on the same plot for the liquid metal). Explain why they have differences, if any.

b. Define the similarity variable (η), the similarity function for temperature (θ), and the derivative of the similarity function for velocity (f')? Write the BCs which can be used for this problem?

c. From the Blasius solution of the above similarity equations, sketch the relations between the similarity functions (f', θ) and the similarity variable (η) for both water and liquid metal (i.e., sketch f' versus η for both water and liquid metal on the same plot; and θ versus η for both water and liquid metal on the same plot). Explain why they have differences, if any. You do not need to solve the above equations.

d. For a given problem (i.e., U_∞, T_∞, ρ, μ, T_w, P_r are given), explain briefly how to determine the local velocity $u(y)$ and temperature $T(y)$ at a specified location x, for both water and liquid metal, from the sketches in (c)?

e. Explain briefly how to determine the local heat transfer coefficient from the sketches in (c)? Sketch the local heat transfer coefficient over the flat plate for both water and liquid metal (i.e., sketch h versus x for both water and liquid metal on the same plot), if both are at the same free-stream velocity (U_∞)? Explain why they have differences, if any.

7.4 Consider a steady, incompressible, low-speed 2-D laminar boundary-layer flow (at U_∞, T_∞) over a flat plate at a uniform wall temperature T_w or at a uniform wall heat flux q_w'', respectively. Assume that there exist no body force and constant thermal and fluid properties.

a. Based on the integral method, sketch and write down the momentum balance as well as the energy balance across the boundary layers? (If you cannot remember, derive it).

b. Assuming a uniform velocity profile inside the boundary layer, that is, $u = U_\infty$, and assuming a linear temperature profile inside the thermal boundary layer as $T = a + b \times y$, determine the local thermal boundary-layer growth along the flat plate (i.e., δ_T versus x) at a uniform wall temperature T_w condition. (*Note:* a and b are unknown constants that need to be determined.)

c. Based on (b), determine the local Nusselt number distribution along the flat plate (i.e., Nu versus x) at a uniform wall temperature.

d. Assuming a uniform velocity profile inside the boundary layer, that is, $u = U_\infty$, and assuming a linear temperature profile inside the thermal boundary layer as $T = c + d \times y$, determine the local thermal boundary-layer growth along the flat plate (i.e., δ_T versus x) at a uniform wall heat flux q_w'' condition. (*Note:* c and d are unknown constants that need to be determined.)

e. Based on (d), determine the local Nusselt number distribution along the flat plate (i.e., Nu versus x) at a uniform wall heat flux. Explain and comment on whether the local Nusselt number distribution along the flat plate will be higher, the same, or lower than that in (c)?

7.5 Consider a steady, incompressible, low-speed 2-D laminar boundary-layer flow (at U_∞, T_∞) over a flat plate at a uniform wall temperature T_w. Assume that there exist no body force and constant thermal and fluid properties. The similarity momentum and energy equations are listed here for reference: $f''' + (1/2)ff'' = 0$ and $\theta'' + (1/2)P_f f\theta' = 0$.

a. Sketch both the velocity and the thermal boundary-layer thickness distributions, respectively, for both water and air flowing over the flat plate? (i.e., sketch δ versus x and δ_t versus x on the same plot for water and for air, respectively). Explain why they have differences, if any.

b. Define the similarity variable (η), the similarity function for temperature (θ), and the derivative of the similarity function for velocity (f')? Write the BCs that can be used for this problem?

c. From the Blasius solution of the above similarity equations, sketch the relations between the similarity functions (f', θ) and the similarity variable (η) for both water and air (i.e., sketch f' versus η for both water and air on the same plot; and θ versus η for both water and air on the same plot). Explain why they have differences, if any. You do not have to solve the above equations.

d. For a given problem (i.e., U_∞, T_∞, ρ, μ, T_w, P_r are given), explain briefly how to determine the local velocity $u(y)$ and temperature $T(y)$ at a specified location x, for both water and air, from the sketches in (c)?

e. Explain briefly how to determine the local heat transfer coefficient from the sketches in (c)? Answer whether water or air will provide a higher convective heat transfer coefficient from the surface, if both are at the same free-stream velocity (U_∞)? Why?

7.6 Consider a steady, incompressible, low-speed 2-D laminar boundary-layer flow (at U_∞, T_∞) over a flat plate at a uniform wall heat flux q_w''. Assume that there exist no body force and constant thermal and fluid properties.

a. Based on the integral method, sketch and write down the momentum balance as well as the energy balance across the boundary layers? (If you cannot remember, derive it.)

b. Assuming a uniform velocity profile inside the boundary layer, that is, $u = U_\infty$, and assuming a linear temperature profile inside the thermal boundary layer as $T = a + b \times y$, determine the local thermal boundary-layer growth along the flat plate (i.e., δ_t versus x).

c. Based on (b), determine the local Nusselt number distribution along the flat plat (i.e., Nu versus x).

d. Consider a parabolic velocity and temperature profile inside the thermal boundary layer as $u = a + b \times y + c \times y^2$, $T = a + b \times y + c \times y^2$, and comment on whether the local Nusselt number distribution along the flat plate will be higher, the same, or lower than those of a uniform velocity profile and a linear temperature profile as indicated in (b)? Explain why.

e. Consider a uniform suction through the wall (i.e., $v = -v_0$), and comment on whether the local Nusselt number distribution along the flat plate will be higher, the same, or lower than that without boundary-layer suction? Explain why.

7.7 Consider a steady, incompressible, low-speed 2-D laminar boundary-layer flow (at U_∞, T_∞) over a flat plate at a uniform wall temperature T_W. Assume that there exist no body force and constant thermal and fluid properties. The similarity momentum and energy equations are listed here for reference: $f''' + (1/2)ff'' = 0$ and $\theta'' + (1/2)\,\mathrm{Pr}f\theta' = 0$.

a. Define the similarity variable (n), the similarity function for temperature (θ), and the derivative of the similarity function for velocity (f')? Write the BCs that can be used for this problem?

b. From the Blasius solution of the above similarity equations, sketch the relations between the similarity functions (f', θ) and the similarity variable (n) for water, air, and liquid metal (i.e., sketch f' versus n for water, air, and liquid metal on the same plot; and θ versus η for water, air, and liquid metal on the same plot). Explain why they have differences, if any. Explain briefly how to determine the local velocity (u) and temperature (T) from the sketches? You do not need to solve the above equations.

c. Explain briefly how to determine the local heat transfer coefficient from the sketches in (b)? Answer whether water or liquid metal will provide a higher convective heat transfer coefficient from the surface, if both are at the same free-stream velocity (U_∞)? Explain why?

7.8 Consider a steady, incompressible, low-speed 2-D laminar boundary-layer flow (at U_∞, T_∞) over a flat plate at a uniform wall temperature T_W. Assume that there exist no body force and constant thermal and fluid properties.

a. Based on the integral method, sketch and write down the momentum balance as well as the energy balance across the boundary layers? (If you cannot remember, derive it.)

b. Assuming a uniform velocity profile inside the boundary layer that is, $u = U_\infty$, and assuming a linear temperature profile inside the thermal boundary layer as $T = a + b \times y$, determine the local thermal boundary-layer growth and the local Nusselt number distribution along the flat plate (i.e., δ_t versus X and Nu versus X).

c. Consider a uniform blowing through the wall (i.e., $v = v_0$), and comment on whether the local Nusselt number distribution along the flat plate will be higher, the same, or lower than that without boundary-layer blowing? Explain why.

7.9 Consider a steady, incompressible, low-speed 2-D laminar boundary-layer flow (at U_∞, T_∞) over a flat plate at a uniform wall temperature T_W. Assume that there exist no body force and constant thermal and fluid properties.

a. Based on the integral method, write down the final form of momentum integral equation. (If you cannot remember, please derive it.)

b. Based on the integral method, can you remember to write down the final form of the energy integral equation? (If you cannot remember, please derive it.)

c. Assuming a uniform velocity profile inside the boundary layer that is, $u = U_\infty$, and assuming a linear temperature profile inside the thermal boundary layer as $T = a + by$, determine the local Nusselt number distribution along the flat plate (i.e., Nu versus X).

d. Consider a uniform suction through the wall (i.e., $v = -v_0$) and comment on whether the local Nusselt number distribution along the flat plate will be higher, the same, or lower than those without boundary-layer suction? Explain why.

7.10 Consider a laminar air flow (at U_∞, T_∞) over a flat plate at a uniform wall heat flux q_w''.

a. Based on the integral method, can you remember to write down the momentum and energy integral equations? (If you cannot remember, please derive them).

b. Assuming a linear velocity profile inside the boundary layer as $u = a + by$, derive the velocity boundary-layer thickness distribution along the flat plate (i.e., δ versus X).

c. Assuming a linear temperature profile inside the thermal boundary layer as $T = c + dy$, derive the thermal boundary-layer thickness distribution along the flat plate (i.e., δ_T versus X).

d. Based on (b) and (c), determine the local Nusselt number distribution along the flat plate (i.e., Nu versus X).

e. Consider a uniform wall temperature (T_w) as a thermal BC at the wall and comment on whether the local Nusselt number distribution along the flat plate will be higher, the same, or lower than those of uniform wall heat flux as a thermal BC? Explain why.

f. Consider a uniform suction through the wall (i.e., $v = -v_0$) and comment on whether the local Nusselt number distribution along the flat plate will be higher, the same, or lower than those without boundary-layer suction? Explain why.

7.11 The similarity method for laminar flow over a flat plate: U_∞ is the freestream velocity, T_∞ is the free-stream temperature, and T_w is the flat plate wall temperature.

a. Write down the similarity variable, differential equations, and BCs for velocity and temperature, respectively. Then, determine velocity (u) and temperature (if $\text{Pr} = 1$) at

$$(x,y) = (2\,\text{cm}, 1/3\delta),$$

$$= (4\,\text{cm}, 1/3\delta),$$

$$= (6\,\text{cm}, 1/3\delta).$$

b. At any given x, if U_∞ increases, the friction factor will be increased or decreased. Why? How about shear stress? At any given U_∞, if x increases, the heat transfer coefficient will be increased or decreased. Why? How about heat transfer rate?

7.12 The similarity method for laminar flow over a flat plate: U_∞—free-stream velocity, T_∞—free-stream temperature, and T_w—flat plate wall temperature.

a. Write down the similarity variable, differential equations, and BCs for velocity and temperature, respectively. Then determine velocity (u) and temperature (if Pr = 1) at

$$
\begin{aligned}
(x,y) &= (1\,\text{cm},1/2\delta) \quad\text{and}\quad (x,y) = (1\,\text{cm},1/4\delta) \\
&= (3\,\text{cm},1/2\delta) \qquad\qquad = (3\,\text{cm},1/4\delta) \\
&= (9\,\text{cm},1/2\delta) \qquad\qquad = (9\,\text{cm},1/4\delta)
\end{aligned}
$$

b. At any given x, if U_∞ increases, the friction factor will be increased or decreased. Why? How about shear stress? At any given U_∞, if x increases, the heat transfer coefficient will be increased or decreased. Why? How about the heat transfer rate?

7.13 Consider the development of velocity and thermal boundary layers on a porous flat plate where air passes into the flat plate at a velocity V_o.

a. Assume that no pressure gradient exists in either the x- or the y-direction and that all fluid properties are constant. Derive the differential equation that relates boundary-layer thickness δ to distance x. A linear profile may be assumed.

b. The exact solution of the boundary-layer equations with the BCs of part (a) shows that δ approaches a constant value for large x, and that for large x, V_x and V_y are given by

$$
V_x = V_\infty \left[1 - \exp\left(\frac{V_o y}{v} \right) \right]
$$
$$
V_y = -V_o
$$

Suppose now that at some large $x = \ell$ (i.e., where δ has become constant and where V_x and V_y are given above), a step change in wall temperature occurs. Using the integral technique, derive a differential equation relating the thermal boundary-layer thickness δ_t to x ($x > \ell$). Again, assume that fluid properties are constant. A linear temperature profile may be assumed. Integrals and derivatives need not be evaluated.

7.14 Air at 1 atm and at a temperature of 30°C flows over a 0.3-m-long flat plate at 100°C with a free-stream velocity of 3 m/s. At the position $x = 0.05$ m, determine the values of the boundary-layer thickness, displacement thickness, moments thickness, wall stress, and heat transfer coefficient. Determine the values of the velocity parallel and normal to the plate surface, the values of the shear stress, and the values of temperature in the fluid at the positions ($x = 0.05$ m, $y = 0.002$ m) ($x = 0.05$ m, $y = 0.004$ m).

7.15 Using the integral method and assuming that velocity and temperature vary as

$$u = a + by + cy^2 + dy^3$$
$$T = a + by + cy^2 + dy^3$$

determine the "x" variations of the heat transfer coefficient and the wall temperature for the case of laminar flow over a flat plate with a uniform wall heat flux $(q/A)_w$. Compare the result of the heat transfer coefficient to the solution obtained in the textbook for the case of uniform wall temperature.

7.16 Laminar air flow at 1 atm pressure over a flat plate at a uniform T_w. Assume that $U_\infty = 3$ m/s, $T_\infty = 20°C$, and $T_w = 100°C$, and determine local u and T at $x = 3$ cm and $y = 0.2$ cm by using the similarity solution. Also determine the heat transfer coefficient at the same $x = 3$ cm location. Will the heat transfer coefficient be increased or decreased with increasing x for the same U_∞? Why? Will the Nusselt number be increased or decreased with increasing x for the same U_∞? Why?

7.17 Use similarity solutions: air at 300 K and 1 atm flows along a flat plate at 5 m/s. At a location 0.2 m from the leading edge, plot the u and v velocity profiles using the exact solution to the Blasius equation Also, determine the boundary-layer thickness, if it is defined as the location where $u = 0.99u_\infty$. Plot the temperature profile and determine the thermal boundary thickness if the plate temperature is 500 K.

7.18 Using integral method solutions: air at 300 K and 1 atm pressure flows along a flat plate at 5 m/s. For $x < 10$ cm, $T_s = 300$ K, whereas for 10 cm $< x < 20$ cm, $T_s = 500$ K. Calculate the heat loss from the plate and compare the result with the heat loss if the plates were isothermal at 500 K. Assume that there exists a laminar boundary layer. Compare and discuss the two cases.

7.19 Consider a boundary-layer flow over a flat plate.

a. Using the integral method, derive the continuity equation in the boundary layer for flow over a flat plate.

b. Using the integral method derive the energy conservation equation in the boundary layer for flow over a flat plate.

c. Using appropriate BCs and boundary-layer theory, show that for an inviscid fluid,

$$Nu = 0.564\,Pe^{1/2}$$

where Nu is the Nusselt number and Pe is the Peclet number. (Hint: Use a plug flow model for velocity.)

7.20 Consider a fluid approaching the leading edge of a flat plate with uniform velocity and temperature profiles U_∞ and T_∞. The flat plate is frictionless and is held at a constant heat flux of q_s''. The temperature profile at a distance x from the leading edge is given by

$$T = a_0 + a_1 y + a_2 y^2 + a_3 y^3$$

a. Using appropriate BCs evaluate a_0, a_1, a_2, and a_3.
b. Qualitatively sketch velocity and temperature profiles at a distance x_1 and x_2, respectively, from the leading edge.
c. Determine the local heat transfer coefficient (h_x) and the Nusselt number Nu_x. Express the answer in terms of the thermal boundary-layer thickness δ_t.

7.21 Consider a 2-D laminar air flow over a friction less plate. The flow approaches the leading edge of the plate with uniform velocity U_∞ and temperature T_∞. The plate is subjected to a constant wall temperature, $T_s (> T_\infty)$.
a. Qualitatively sketch hydrodynamic (δ) and thermal boundary layer (δ_t) growth as a function of distance x from the leading edge.
b. Qualitatively sketch velocity and temperature profiles at a distance x from the leading edge.
c. A third-order polynomial of the form

$$\frac{T - T_s}{T_\infty - T_s} = a_0 + a_1 \left(y/\delta_t \right) + a_2 \left(y/\delta_t \right)^2 + a_3 \left(y/\delta_t \right)^3$$

is used to describe the temperature profile. Determine the constants a_0, a_1, a_2, and a_3 using appropriate BCs.
d. Express local Nusselt numbers (Nu_x) in terms of local thermal boundary-layer thickness δ_t.
e. Set up an integral energy balance equation. Do not attempt to solve the equation.

7.22 Consider a 2-D, steady, incompressible laminar flow over a flat plate. The flow approaches the leading edge with free-stream velocity of U_∞ and temperature T_∞. The flat plate is frictionless, and it is kept at a uniform temperature of $T_s (>T_\infty)$.
a. State clearly all the boundary-layer assumptions.
b. If the temperature distribution at any axial distance x is approximated by a linear profile $(T - T_s)/(T_\infty - T_s) = y/\delta t$, derive an expression for the local Nusselt number distribution.

7.23 Consider a steady laminar viscous fluid with a free-stream velocity V_∞ and temperature T_∞ flows over a flat plate at a uniform wall temperature T_w. Assume that the thermal fluids properties are constant.
a. If the fluid has a Prandtl number of one (i.e., $Pr = 1.0$), determine the local heat transfer coefficient along the plate. You may use the method of integral approximation with the assumptions of the linear velocity and temperature profiles across the boundary layers, that is, $u = a + by$ and $T = c + dy$, where a, b, c, and d are constants.
b. If the fluid's Prandtl number is not equivalent to one (i.e., $Pr > 1$ or $Pr < 1$), outline the methods (no need to solve) in order to determine the surface heat transfer. You may use the same assumptions as in part (a). Does the heat transfer coefficient increase or decrease with the fluid Prandtl number? Explain your answers.

7.24 A flat horizontal plate has a dimension of 10 cm × 10 cm. The plate is maintained at a constant surface temperature of 300°K with a water jacket.

a. If the plate is placed in a hot air stream with a pressure of 1 atm, temperature of 400°K, and velocity of 10 m/s, sketch the local heat transfer coefficient, h_x, along the plate.

b. Determine the surface heat flux at the trailing edge of the plate.

Given: air properties at 350°K, $k = 30 \times 10^{-3}$ W/mK, $v = 20.92 \times 10^{-6}$ m²/s, Pr = 0.7:

$$Nu_x = a\,Re_x^m\,Pr^n$$

where

$a = 0.332$, $m = 1/2$, $n = 1/3$ for laminar flow

$a = 0.0296$, $m = 4/5$, $n = 1/3$ for turbulent flow

7.25 Constant properties laminar viscous fluids (ρ, Cp, K, μ = constant) with a free-stream velocity U_∞ and temperature T_∞ move over a flat plate at a uniform wall heat flux $\left(q_w'' = \text{constant}\right)$.

a. Determine the local Nusselt number ($Nu_x = h_x \cdot x/k$) along the plate by using the integral approximation method with the velocity and temperature profiles across the boundary layers as $u = a + by$ and $T = c + dy$, respectively. Assume Pr = 1 for this problem.

b. On the same plot, sketch h_x, Nu_x versus x for the uniform wall heat flux $\left(q_w'' = \text{constant}\right)$ and uniform wall temperature ($T_w = \text{constant}$) BCs, respectively. For the same flow velocity, which wall BC (q_w'' or T_w) will provide a higher heat transfer coefficient or Nusselt number? Explain why.

7.26 A viscous fluid (ρ, C_p, k, and μ are constant values) with a variable free stream velocity $U = Cx^m$ and temperature T_∞ flows over a surface at a variable wall temperature $T_w - T_\infty = Cx^n$. This is a basic constant pressure gradient flow, 2-D laminar boundary-layer convective heat transfer problem. We want to solve the velocity and temperature profiles across the boundary layers from the similarity momentum and energy equations as:

$$f''' + \frac{m+1}{2}ff'' + m\left(1 - f'^2\right) = 0$$

$$\theta'' + \frac{m+1}{2}\Pr f\theta' + n\Pr(1-\theta)f' = 0$$

a. Assume you have already solved the velocity and temperature profiles (f' vs. η, θ vs. η) from the similarity equations by the RK method, write how to determine the local heat transfer coefficient h_x along the plate.

b. Sketch θ vs. η on the same figure for $n = 0$, Pr = 0.72 with $m = 0$ and 0.5, respectively. Discuss if $m = 0$ or $m = 0.5$ will provide a high heat transfer coefficient and why?

c. Sketch θ vs. η on the same figure for $m = 0$, $n = 0$ with Pr = 0.01 (liquid metal), 0.72 (air), and 5 (water), respectively. Discuss if Pr = 0.01 or Pr = 5 will provide a high heat transfer coefficient and why?

7.27 A particular rocket ascends vertically with a velocity that increases approximately linearly with altitude. Consider a high velocity "air" flow with T_∞ and V_∞ approaching a location on the cylindrical shell of the rocket, x from the nose, at T_w. Assume T_w is greater than T_∞. This is a high-velocity boundary layer problem (the cylindrical shell of the rocket can be modeled as high-speed flow over a vertical flat plate).

a. Write down the formulas which can be used for calculating local heat transfer coefficient h_x and local wall heat flux q_w'' for a location on the cylindrical shell of the rocket, x, from the nose if the viscous dissipation (aerodynamic heating) is not negligible. Discuss whether the local heat flux at that location with viscous dissipation is a positive, zero, or negative value. Also, discuss how to determine the local heat flux at that location if the approaching flow (T_∞, V_∞) reaches supersonic condition, i.e., Mach number > 1.0.

7.28 Consider an aircraft flying at Mach 3 at an altitude of 17,500 m. Assume the aircraft has a hemispherical nose with a radius of 30 cm. If it is desired to maintain the nose at 80°C, what heat flux must be removed at the stagnation point by internal cooling? Assume that the air passes through a normal detached shock wave and then decelerates isentropically to zero at the stagnation point, then the flow near the stagnation point is approximated by low-velocity flow about a sphere.

7.29 A viscous fluid (ρ, C_p, k, and μ are constant values) with a variable free stream velocity $U = U_\infty\left(1+\dfrac{x}{L}\right)$ and temperature T_∞ flows over a curved wall (L = curved wall length) at a uniform wall heat flux, q_o''. This is a specific pressure gradient flow, 2-D laminar boundary-layer convective heat transfer problem.

a. Solve this problem using the Integral Approximation Method with the velocity and temperature profiles across the boundary layers as $u = a + by + cy^2$ and $T = a + by + cy^2$, where a, b, and c are unknown constants and need to be determined using appropriate BCs. Determine the velocity and temperature profiles for a uniform wall heat flux, q_o'', BC (i.e., determine a, b, and c).

b. Determine the hydrodynamic and thermal boundary-layer thickness along the wall by using the approximately boundary-layer momentum and energy integral equations with the velocity and temperature profiles across the boundary layers as specified in (a). You do not need to obtain the final solution, but a functional relationship such as $\delta(x, Re_L)$, $\delta_t(\delta, Pr)$ is acceptable where $Re_L = \rho U_\infty L/\mu$. Discuss the importance of the Prandtl number to affect the hydrodynamic and thermal boundary layer thickness, as well as the local heat transfer coefficient along the wall.

c. If the curved wall heat flux is now increased sinusoidally with the wall length, $q_w'' = q_o''\left[1+\sin\left(\dfrac{\pi x}{2L}\right)\right]$, outline the solution procedure to determine local heat transfer coefficient h_x and surface temperature T_w along the curved wall. Will the h_x in case (c) be higher, the same, or lower than that in case (a) and why?

d. If you consider the same viscous fluid flowing over the same curved wall made of a porous material where a small amount of fluid blows into (or, out of) the porous curved wall (boundary layer blowing or suction) at a velocity V_o, write the corresponding boundary-layer momentum and energy integral equations. Will the δ_t and h_x in case (d) be higher, the same, or lower than that in case (a) and why?

7.30 A viscous fluid (ρ, C_p, k, and μ are constant values) with a variable free stream velocity $U = U_\infty \left[1 + \sin\left(\dfrac{\pi x}{2L} \right) \right]$ and temperature T_∞ flows over a wall ($2L$ = wall length) at a uniform wall heat flux, q_o''. This is a variable pressure gradient flow, 2-D laminar boundary-layer convective heat transfer problem.

a. Solve this problem using the Integral Approximation Method with the assumed velocity and temperature profiles across the boundary layers as $u = U$ and $T = a + by$, where a and b are unknown constants and need to be determined by appropriate BCs. Determine the temperature profile for a uniform wall heat flux, q_o'', BC.

b. Determine the thermal boundary-layer thickness along the wall using the approximate boundary-layer energy integral equation with the assumed velocity and temperature profiles across the boundary layers as specified in (a). You do not need to obtain the final solution if you do not know how to perform the final integration.

REFERENCES

1. W. Rohsenow and H. Choi, *Heat, Mass, and Momentum Transfer*, Prentice-Hall, Inc., Englewood Cliffs, NJ, 1961.
2. F. Incropera and D. Dewitt, *Fundamentals of Heat and Mass Transfer*, Fifth Edition, John Wiley & Sons, New York, 2002.
3. H. Schlichting, *Boundary-Layer Theory*, Sixth Edition, McGraw-Hill, New York, 1968.
4. W.M. Kays and M.E. Crawford, *Convective Heat and Mass Transfer*, Second Edition, McGraw-Hill, New York, 1980.
5. A. Mills, *Heat Transfer*, Richard D. Irwin, Inc., Boston, MA, 1992.
6. E. Levy, *Convection Heat Transfer, Class Notes*, Lehigh University, Bethlehem, PA, 1973.
7. T. Cebeci and P. Bradshaw, *Physical and Computational Aspects of Convective Heat Transfer*, Springer-Verlag, New York, 1984.

8 Internal Forced Convection

8.1 VELOCITY AND TEMPERATURE PROFILES IN A CIRCULAR TUBE OR BETWEEN PARALLEL PLATES

Internal forced convection is flow that moves along the internal surface of a passage and forms an internal boundary layer on the surface. For example, fluid flowing through a circular tube or between two parallel plates is the most common application. Figure 8.1 shows the hydrodynamic boundary-layer development (due to viscosity) for flow entering a circular tube (or between two parallel plates). The boundary layer starts from the tube (or plate) entrance and grows along the tube (or plates) length. Unlike external flow, with internal flow, the boundary layer is not allowed to be free; boundary layer development is now confined within the internal geometry. The velocity profile changes (develops) within the entrance region of the tube (or plates). The flow becomes a "hydrodynamic, fully developed flow" when the boundary thickness is the same as the tube radius (or half the spacing between the two plates). The velocity profile no longer changes after achieving the fully developed condition; this is the primary difference between internal and external flow where the velocity profile continues to change in the flow direction for fluids flowing externally over a surface. For a laminar flow, the entrance length-to-tube diameter ratio is approximately 5%

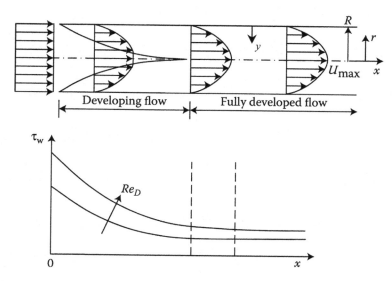

FIGURE 8.1 Velocity profile and shear stress distribution in a circular tube or between two parallel plates.

DOI: 10.1201/9781003164487-8

of the Reynolds number (based on the tube diameter). This implies that the entrance length increases with increasing Reynolds number (because a thinner boundary layer requires longer distance for the boundary layers to merge). Figure 8.1 also shows that shear stress decreases from the entrance along the tube length and becomes a constant value when the flow reaches the fully developed condition, and shear stress increases with Reynolds number (because of a thinner boundary layer from the entrance and the longer entrance length). For a turbulent flow, the entrance length is more difficult to precisely identify; the entrance length is approximately 10–20 tube diameters. Due to the constant mixing associated with turbulent flows, it is often difficult to distinguish whether the turbulent flow is fully developed or not from 10 to 20 tube diameters downstream of the tube entrance [1–4].

Figure 8.2 shows the thermal boundary-layer development (due to the thermal conductivity and velocity of the fluid) for flow entering a circular tube (or between two parallel plates). The thermal boundary layer starts from the tube (or plate) entrance and grows along the tube (or plate) length. The temperature profile continues changing from the entrance due to the addition of heat along the tube (or plate) wall. The flow becomes "thermally, fully developed flow" when the thermal boundary thickness is the same as the tube radius (or half-spacing between the two plates). The dimensionless temperature profile no longer changes after being thermally fully developed (but the temperature continues increasing). For the laminar flow, the thermal entrance length-to-tube diameter ratio is about 5% of the Reynolds number (based on the tube diameter) times Prandtl number. This implies that the thermal entrance length increases with increasing Reynolds number (because a

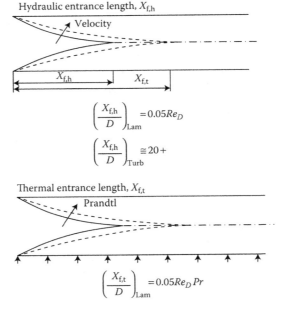

FIGURE 8.2 Hydraulic entrance length and thermal entrance length in a circular tube or between two parallel plates.

thinner boundary layer requires longer distance for the boundary layer to merge) and Prandtl number (because lower thermal conductivity requires longer distance to merge) [1–4].

Figure 8.2 also shows that the heat transfer coefficient decreases from the entrance along the tube length and becomes a constant value when the thermal boundary layer reaches the fully developed condition, and the heat transfer coefficient increases with Reynolds number (because of a thinner boundary layer from the entrance and the longer entrance length). It is noted that the thermal entrance length is identical to the hydrodynamic entrance length if Pr = 1. For turbulent flow, the thermal entrance length can only be estimated; similar to the hydrodynamic entrance length, the thermal entrance length is approximately 10–20 tube diameters. It is difficult to distinguish whether the turbulent flow is thermally and fully developed in the region from 10 to 20 tube diameters within the tube due to the mixing associated with the turbulent flow.

8.2 FULLY DEVELOPED LAMINAR FLOW AND HEAT TRANSFER IN A CIRCULAR TUBE

For fluid flow in a circular tube, the Reynolds number is defined as

$$\mathrm{Re}_D = \frac{\rho \bar{V} D}{\mu} = \frac{\bar{V} D}{\nu} = \frac{4\dot{m}}{\pi D \mu} \tag{8.1}$$

Laminar flow is observed if $\mathrm{Re}_D < 2300$.

At a certain distance from the entrance, the velocity profile, $u(r)$, no longer changes along the tube length (if the fluid properties remain constant). Correspondingly, there is no velocity component in the radial direction. Also, the axial pressure gradient required to sustain the flow against the viscous forces will be constant along the tube (no momentum change). These conditions are present when the flow is hydrodynamically and fully developed. The differential governing equations for the flow inside a circular tube are [1–4]

$$\frac{\partial u}{\partial x} + \frac{1}{r}\frac{\partial(vr)}{\partial r} = 0 \tag{8.2}$$

$$u\frac{\partial u}{\partial x} + v\frac{\partial u}{\partial r} = -\frac{1}{\rho}\frac{\partial P}{\partial x} + \nu\frac{1}{r}\frac{\partial}{\partial r}\left(r\frac{\partial u}{\partial r}\right) \tag{8.3}$$

$$u\frac{\partial T}{\partial x} + v\frac{\partial T}{\partial r} = \alpha\frac{1}{r}\frac{\partial}{\partial r}\left(r\frac{\partial T}{\partial r}\right) \tag{8.4}$$

If the flow is hydrodynamically fully developed, then

$$v = 0 \tag{8.5}$$

Therefore, the continuity Equation (8.2) becomes:

$$\frac{\partial u}{\partial x} = 0 \tag{8.6}$$

Likewise, the momentum equation simplifies to:

$$0 = -\frac{1}{\rho}\frac{\partial P}{\partial x} + v\frac{1}{r}\frac{\partial}{\partial r}\left(r\frac{\partial u}{\partial r}\right) \tag{8.7}$$

Recall, for fully developed flow $u = u(r)$ only. With the fluid velocity only changing in the radial direction and the pressure gradient $(\partial P/\partial x)$ is constant, the momentum equation becomes an ordinary differential equation. The fully developed velocity profile can be solved by integrating Equation (8.7), with respect to r, two times:

$$\frac{1}{\rho}\frac{dP}{dx} = v\frac{1}{r}\frac{d}{dr}\left(r\frac{du}{dr}\right)$$

$$\int \frac{dP}{dx}r\,dr = \int \mu d\left(r\frac{dP}{dx}\right)$$

$$\frac{dP}{dx}\frac{1}{2}r^2 = \mu r\frac{du}{dr} + C_1$$

At $r = 0$, $du/dr = 0$ (the velocity is at a maximum). Therefore, $C_1 = 0$

$$\int \frac{dP}{dx}\frac{1}{2}r\,dr = \int \mu\,du$$

$$\frac{dP}{dx}\frac{1}{4}r^2 = \mu u + C_2$$

At $r = R$, $u = 0$ (the no slip boundary condition). Therefore, $C_2 = (1/4)R^2(dP/dx)$

$$\frac{1}{4}r^2\frac{dP}{dx} = \mu u + \frac{1}{4}R^2\frac{dP}{dx}$$

$$u = \frac{1}{4\mu}\frac{dP}{dx}\left(r^2 - R^2\right)$$

Also, at $r = 0$, $u = \mathcal{U}_{max} = -(1/4\mu)R^2(dP/dx)$. The velocity profile can be re-written in terms of the centerline (or maximum velocity) $u = -\mathcal{U}_{max}\left(r^2 - R^2\right)/R^2$.

$$\frac{u}{\mathcal{U}_{max}} = 1 - \left(\frac{r}{R}\right)^2 \tag{8.8}$$

The bulk mean velocity is defined as:

$$\bar{V} = \frac{\int_{dA} \rho u \, dA}{\rho A} \equiv u_m = \bar{\mathcal{U}}$$

$$\bar{V} = \frac{1}{\pi R^2} \int_0^R u \, 2\pi r \, dr$$

(8.9)

$$= \frac{1}{\pi R^2} \int_0^R \left[1 - \left(\frac{r}{R} \right)^2 \right] \mathcal{U}_{\max} 2\pi r \, dr$$

$$= \frac{2\mathcal{U}_{\max}}{R^2} \left[\frac{1}{2} r^2 - \frac{1}{4R^2} r^4 \right]_0^R$$

$$\bar{V} = \frac{1}{2} \mathcal{U}_{\max}$$

(8.10)

Equation (8.10) is only valid for flow through a "round" tube. If the cross-section of the tube deviates from a circular tube, the relationship between the maximum (centerline) velocity and the average, bulk velocity also changes.

8.2.1 Fully Developed Flow in a Tube: Friction Factor

Knowing the fully developed velocity profile, it is possible to analytically determine the shear stress distribution through the fluid. Furthermore, based on the shear stress, we can determine the friction factor within the tube. Beginning with the shear stress definition:

$$\tau_w = \mu \frac{\partial u}{\partial y} \bigg|_{y=0} = -\mu \frac{\partial u}{\partial r} \bigg|_{r=R} = \mu \mathcal{U}_{\max} \left(\frac{2r}{R^2} \right) \bigg|_{r=R}$$

(8.11)

where $u/\mathcal{U}_{max} = 1 - (r/R)^2$, $r = R - y$, and $dr = -dy$

$$\tau_w = \mu \mathcal{U}_{\max} \frac{2}{R}$$

The friction factor can be expressed as:

$$f = \frac{\tau_w}{(1/2)\rho \bar{\mathcal{U}}^2} = \frac{\mu \mathcal{U}_{\max}(2/R)}{(1/2)\rho \left((1/2)\mathcal{U}_{\max} \right)^2} = \frac{16\mu}{\rho \bar{\mathcal{U}} D} = \frac{16}{Re_D} \sim \frac{1}{Re_D}$$

(8.12)

where $\mathrm{Re}_D = \rho \bar{U} D / \mu$

Figure 8.3 shows the force balance in the fully developed flow region. This force balance can be used to determine how the pressure changes along the length of the tube (the pressure gradient, $\partial P / \partial x$, in the flow direction):

$$\Delta P A_c + \tau_w \pi D \Delta x = 0$$

$$\Delta P = \frac{\tau_w \pi D \Delta x}{(1/4)\pi D^2}$$

$$\frac{\Delta P}{\Delta x} = \frac{-4\tau_w}{D} = -\frac{4}{D} f \frac{1}{2} \rho \bar{U}^2$$

$$= -\frac{4}{D} \frac{1}{2} \rho \bar{U}^2 \frac{16}{\mathrm{Re}_D} = -\frac{32\mu\bar{U}}{D^2} = -\frac{32\mu}{D^2} \frac{\rho}{\rho} \frac{D}{D} \frac{\mu}{\mu} \bar{U} = -\frac{32\mu^2}{D^2\rho} \frac{\rho D \bar{U}}{\mu} \qquad (8.13)$$

$$= -\frac{32\mu^2}{D^3\rho} \mathrm{Re}_D \sim \mathrm{Re}_D \qquad (8.14)$$

Also, the pumping power can be obtained as $P \cong \Delta P$ (volume flow). Figure 8.4 shows the friction factor decreases and the pressure drop increases with Reynolds number.

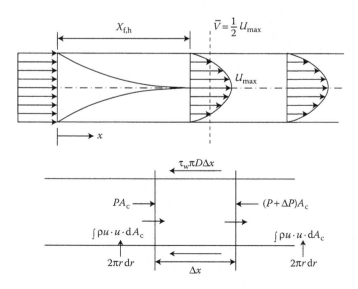

FIGURE 8.3 Force balance in the fully developed flow region.

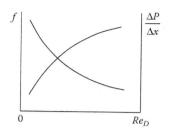

FIGURE 8.4 Friction factor and pressure drop versus Reynolds number in the fully developed flow region.

8.2.2 CASE 1: UNIFORM WALL HEAT FLUX

Laminar flow heat transfer is dependent upon the thermal boundary conditions. However, turbulent flow heat transfer is fairly independent of the thermal BCs (particularly for Prandtl number around one, such as air). Typical thermal BCs are case 1, uniform heat flux and case 2, uniform wall temperature.

Figure 8.5 shows the laminar flow in a circular tube with a uniform surface heat flux condition along with the thermal boundary layer, temperature, and the heat transfer coefficient (Nusselt number) distributions along the length of the tube. Below are the step-by-step details to obtain the results shown in Figure 8.5.

Beginning with Newton's Law of Cooling, and recognizing in this case, the surface heat flux is constant.

$$q''_w = \frac{q_w}{A_s} = h(T_w - T_b) \tag{8.15}$$

$$\underset{\text{const.}}{q''_w} = \frac{q_w}{A_s} = h\underbrace{(T_w - T_b)}_{\text{const.}} \tag{8.16}$$

From the First Law energy balance $\left(\dot{Q} = \dot{m}c_p(\Delta T)\right)$, the average, bulk mean, temperature of the fluid can be determined by integrating over the cross-sectional area of the tube $\left(\dot{m} = \rho A_c u\right)$:

$$T_b = \frac{\displaystyle\int_0^R \rho u T \cdot 2\pi r\, dr}{\displaystyle\int_0^R \rho u \cdot 2\pi r\, dr} \tag{8.17}$$

Also from the energy balance,

$$q''_w = \frac{q_w}{A_s} = \frac{\dot{m}C_p(T_{b,o} - T_{b,i})}{A_s} \tag{8.18}$$

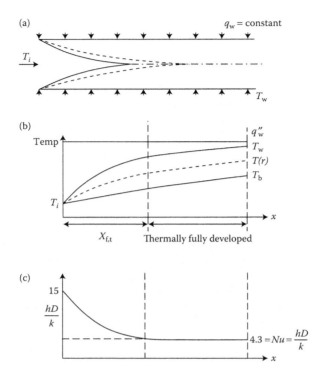

FIGURE 8.5 Laminar flow in a circular tube with the uniform surface heat flux condition. (a) Thermal boundary layer; (b) temperature; and (c) heat transfer coefficient (Nusselt number).

With q_w'' being a constant, the pipe having a fixed cross-section, and the mass flow rate being constant, from Equation (8.19), the bulk mean temperature gradient must also be constant:

$$q_w'' = \frac{\dot{m}C_p}{\pi D} \frac{dT_b}{dx} \Rightarrow \frac{dT_b}{dx} = \text{const} \qquad (8.19)$$

For thermally fully developed flow,

$$\frac{\partial}{\partial x}\left(\frac{T - T_w}{T_b - T_w}\right) = 0$$

$$\frac{T - T_w}{T_b - T_w} = f(r) \neq f(x)$$

$$\frac{\partial T}{\partial x} - \frac{\partial T_w}{\partial x} - \underbrace{\frac{T - T_w}{T_b - T_w}\left(\frac{\partial T_b}{\partial x} - \frac{\partial T_w}{\partial x}\right)}_{=0} = 0 \qquad (8.20)$$

$$\frac{\partial T}{\partial x} = \frac{\partial T_w}{\partial x} = \frac{\partial T_b}{\partial x} = \text{constant} \tag{8.21}$$

Now, Equation (8.4), the differential energy equation, can be used to determine an expression for the fully developed, temperature profile within the tube. Note, energy equation can only be solved after the velocity profile has been determined from the momentum equation.

$$\rho C_p u \frac{\partial T}{\partial x} = k \frac{1}{r} \frac{\partial}{\partial r}\left(r \frac{\partial T}{\partial r}\right) = T(r) \text{only}$$

$$\int r \underbrace{\frac{\rho C_p}{k} 2\bar{V}\left[1-\left(\frac{r}{R}\right)^2\right]}_{u} \underbrace{\frac{q''_w}{\dot{m} C_p / \pi D}}_{\partial T/\partial x} dr = \int d\left(r \frac{dT}{dr}\right) \tag{8.22}$$

The thermal boundary conditions can be expressed in terms of the centerline temperature, T_c. At the center of the tube, the temperature profile will be a minimum or maximum (depending on whether the fluid is being heated or cooled), $\partial T/\partial r = 0$. Therefore, the thermal boundary condition at the center of the tube is:

$$r = 0, \quad T = T_c \quad \text{or} \quad \frac{\partial T}{\partial r} = 0$$

$$\int \frac{\rho C_p}{k} 2\bar{V}\left(r - \frac{r^3}{R^2}\right)\frac{dT}{dx} dr = \int d\left(r \frac{dT}{dr}\right)$$

$$\frac{\rho C_p}{k} \mathcal{U}_{\max} \frac{dT}{dx}\left(\frac{1}{2}r^2 - \frac{r^4}{4R^2}\right) = r \frac{dT}{dr} + C_1$$

at $r = 0$, $\dfrac{dT}{dr} = 0$, $\therefore C_1 = 0$. Now divide r through the equation:

$$\int \frac{\rho C_p}{k} \mathcal{U}_{\max} \frac{dT}{dx}\left(\frac{1}{2}r - \frac{r^3}{4R^2}\right) dr = \int dT$$

$$\frac{\rho C_p}{k} \mathcal{U}_{\max} \frac{dT}{dx}\left(\frac{1}{2}\cdot\frac{1}{2}r^2 - \frac{1}{4}\frac{r^4}{4R^2}\right) = T + C_2$$

at $r = 0$, $T = T_C$, $\therefore C_2 = -T_C$. Therefore,

$$T - T_c = \frac{\rho C_p}{k}\frac{dT}{dx}\mathcal{U}_{\text{max}}\left(\frac{r^2}{4} - \frac{r^4}{16R^2}\right)$$

The wall temperature can be determined with $r = R$, $T = T_w$:

$$T_w - T_c = \frac{\rho C_p}{k}\frac{\partial T}{\partial x}\mathcal{U}_{\text{max}}\left(\frac{R^2}{4} - \frac{R^4}{16R^2}\right) = \frac{\rho C_p}{k}\frac{\partial T}{\partial x}\mathcal{U}_{\text{max}}\frac{3}{16}R^2$$

where

$$T_b = \frac{\int_0^R \rho u T \cdot 2\pi r\,dr}{\int_0^R \rho u \cdot 2\pi r\,dr} = T_c + \frac{7}{96}\frac{\rho C_p}{k}\frac{dT}{dx}\mathcal{U}_{\text{max}}R^2$$

$$T_w - T_b = \frac{\rho C_p}{k}\frac{dT}{dx}\mathcal{U}_{\text{max}}\left(\frac{3}{16}R^2 - \frac{7}{96}R^2\right) \tag{8.23}$$

Therefore, h can be determined by combining Equations (8.19) and (8.23):

$$h = \frac{q_w''}{T_W - T_b} = \frac{(\dot{m}C_p/\pi D)(dT_b/dx)}{\rho(C_p/k)(dT/dx)\mathcal{U}_{\text{max}}\left[(3/16)R^2 - (7/96)R^2\right]}$$

$$= \frac{(\rho \pi R^2 \bar{V} C_p/\pi 2R)(dT/dx)}{(\rho C_p/k)(dT/dx)\mathcal{U}_{\text{max}}(11/96)R^2} = \frac{(\rho C_p \pi R^2 (1/2)\mathcal{U}_{\text{max}}/\pi 2R)(dT/dx)}{(\rho C_p/k)(dT/dx)\mathcal{U}_{\text{max}}(11/96)R^2}$$

$$= \frac{k}{(44/96)R} = \frac{96}{22}\frac{k}{D}$$

Now the Nusselt number can be determined:

$$\text{Nu}_D \equiv \frac{hD}{k} = \frac{96}{22} = 4.314 \tag{8.24}$$

Alternatively, h can be determined based on the energy balance at the wall:

$$h = \frac{-k\left.\dfrac{\partial T}{\partial y}\right|_{y=0}}{T_W - T_b}$$

$$h = \frac{k \left. \frac{\partial T}{\partial r} \right|_{r=R}}{T_W - T_b}$$

As an example, if considering flow inside a circular tube with a uniform velocity (i.e., slug flow) at a uniform wall heat flux, we can determine the temperature profile and heat transfer coefficient. To start, there is no need to solve the momentum equation, as the velocity is constant across the cross-section of the tube (given in the problem statement). Therefore,

$$u = \mathcal{U}_{max} = \bar{V}$$

From the energy equation

$$\rho C_P \bar{V} \frac{\partial T}{\partial x} = k \frac{1}{r} \frac{\partial}{\partial r} \left(r \frac{\partial T}{\partial r} \right)$$

For simplicity, let c represent all the constant values:

$$c = \frac{\rho C_P}{k} \bar{V} \frac{dT}{dx}$$

Now integrate both sides of the energy equation:

$$\int c \, r \, dr = \int d \left(r \frac{dT}{dr} \right)$$

$$c \frac{1}{2} r^2 = r \frac{dT}{dr} + C_1$$

Although the velocity is constant across the tube cross-section, the temperature varies with the radius, r. As with the previous derivation, the temperature of the fluid along the centerline is a local maximum or minimum:

$$r = 0, \quad \frac{dT}{dr} = 0, \quad \therefore C_1 = 0$$

Now divide by r and integrate a second time:

$$\int c \frac{1}{2} r \, dr = \int dT$$

$$\frac{1}{2} c \cdot \frac{1}{2} r^2 = T + C_2$$

In terms of the centerline temperature:

$$r = 0, \quad T = T_C, \quad \therefore C_2 = -T_C$$

Therefore,

$$T - T_C = \frac{1}{4}cr^2 = \frac{1}{4}\frac{\rho C_P \bar{V}}{k}\frac{dT}{dx}\cdot r^2$$

In terms of the wall temperature $(r = R, \quad T = T_w)$:

$$T_w - T_c = \frac{1}{4}\frac{\rho C_P \bar{V}}{k}\frac{dT}{dx}\cdot R^2$$

Following the same procedure as outlined above, we can work our way to the heat transfer coefficient (Nusselt number):

$$T_b = \frac{\displaystyle\int_0^R \rho\bar{V}T\cdot 2\pi r dr}{\displaystyle\int_0^R \rho\bar{V}\cdot 2\pi r dr} = \text{can be obtained}$$

$$T_w - T_b = \text{can be obtained}$$

$$h = \frac{q_w''}{T_w - T_b} = \text{can be obtainted}$$

$$\frac{hD}{k} = \mathrm{Nu}_D = \text{can be obtained}$$

8.2.3 CASE 2: UNIFORM WALL TEMPERATURE

The case of a uniform wall temperature is often encountered in heat exchangers where steam might be condensing (or water evaporating) along the surface of the tube. The phase change of the water occurs at the saturation temperature, and therefore, the wall of the tube is constant at the approximate saturation temperature. This uniform wall temperature $(T_w = \text{constant})$ condition requires an iterative approach to determine the temperature profile and the Nusselt number within the tube. For fully developed flow, we have:

$$\frac{T - T_w}{T_b - T_w} \equiv f(r)$$

However, in order to attain the thermally developed condition:

$$\frac{\partial}{\partial x}\left(\frac{T-T_w}{T_b-T_w}\right)=0$$

Applying the chain rule to expand the derivative definition:

$$\frac{\partial T}{\partial x}-\frac{\partial T_w}{\partial x}-\frac{T-T_w}{T_b-T_w}\left(\frac{\partial T_b}{\partial x}-\frac{\partial T_w}{\partial x}\right)=0$$

By definition (T_w = constant), the wall temperature does not change in the x-direction. Therefore, the temperature gradient can be written as:

$$\frac{\partial T}{\partial x}=\frac{T-T_w}{T_b-T_w}\frac{dT_b}{dx} \tag{8.25}$$

Combining with the temperature gradient with the energy equation, we have:

$$\frac{1}{\alpha}\frac{\partial T}{\partial x}=\frac{1}{\alpha}\frac{T-T_w}{T_b-T_w}\frac{\partial T_b}{\partial x}=\frac{1}{ru}\frac{\partial}{\partial r}\left(r\frac{\partial T}{\partial r}\right) \tag{8.26}$$

where we know:

$$\frac{T-T_w}{T_b-T_w}\equiv f(r)$$

Let

$$\frac{\partial T}{\partial r}=\frac{\partial(T-T_w)}{\partial r}=(T_b-T_w)\frac{\partial\left(\dfrac{T-T_w}{T_b-T_w}\right)}{\partial r}$$

$$=(T_b-T_w)\cdot f'(r)$$

The energy equation now becomes:

$$\frac{1}{\alpha}f\frac{\partial T_b}{\partial x}=\frac{1}{ur}\frac{\partial}{\partial r}\left[r(T_b-T_w)f'\right]$$

or

$$\frac{1}{r}\frac{\partial}{\partial r}(rf')=\frac{1}{\alpha}u\cdot f\frac{\partial T_b}{\partial x}\frac{1}{(T_b-T_w)}$$

We also have the first law and the convective heat flux:

$$q_w'' = \frac{\dot{m}c_p \Delta T_b}{\pi D \Delta x} = h(T_w - T_b)$$

Rearranging:

$$\frac{dT_b}{dx}\left(\frac{1}{T_w - T_b}\right) = \frac{h\pi D}{\dot{m}c_p}$$

$$\frac{1}{r}\frac{\partial}{\partial r}(rf') = \frac{1}{\alpha}uf\left(\frac{-h\pi D}{\dot{m}c_p}\right)$$

Iteration is required to solve for f and f'. With this process, we first assume a temperature profile to satisfy fully developed flow:

$$f(r) = \frac{T - T_w}{T_b - T_w}$$

For example, from the temperature profile for the uniform wall heat flux result:

$$\frac{T - T_w}{T_w - T_b} = \frac{\dfrac{1}{\alpha}\dfrac{\partial T}{\partial x}\mathcal{U}_{max}\left(\dfrac{r^2}{4} - \dfrac{r^2}{16R^2} - \dfrac{R^2}{4} + \dfrac{R^2}{16R^2}\right)}{-\dfrac{1}{\alpha}\dfrac{\partial T}{\partial x}\mathcal{U}_{max}\left(\dfrac{R^2}{4} - \dfrac{R^2}{16} - \dfrac{7}{96}R^2\right)} = f(r)$$

Substituting $f(r)$ into the above energy equation (i.e., substitute into the following equation):

$$\frac{1}{r}\frac{\partial}{\partial r}(rf') = \frac{1}{\alpha}uf\left(\frac{-h\pi D}{\dot{m}c_p}\right)$$

where

$$\dot{m} = \rho A_c \overline{V} = \rho \overline{V}\frac{\pi}{4}D^2$$

$$u = 2\overline{V}\left(1 - \frac{r^2}{R^2}\right)$$

Now, check to see if the left-hand side (LHS) equals the right-hand side (RHS). If the equation is not balanced, continue guessing temperature profiles, $f(r)$, until LHS = RHS.

After the temperature profile is determined (from iteration), we can calculate the heat transfer coefficient based on the profile:

$$q_w'' = -k\frac{\partial T}{\partial y}\bigg|_0 = k\frac{\partial T}{\partial r}\bigg|_R$$

$$h = \frac{q_w''}{T_w - T_b} = 3.66\frac{k}{D}$$

$$\text{Nu} = \frac{hD}{k} = 3.66 \tag{8.27}$$

For the case of the uniform wall temperature, the temperature distribution may also be obtained by the infinite series method shown below. Again, begin with:

$$\frac{\partial T}{\partial r} = (T_b - T_w)\frac{\partial\left(\dfrac{T - T_w}{T_b - T_w}\right)}{\partial r}$$

and insert it into Equation (8.26). The solution is in the form of an infinite series:

$$\frac{T - T_w}{T_c - T_w} = \sum_{n=0}^{\infty} C_{2n}\left(\frac{r}{R}\right)^{2n}$$

where

$$C_{2n} = \frac{\lambda_0^2}{(2n)^2}(C_{2n-4} - C_{2n-2})$$

$$C_0 = 1$$

$$C_2 = -\frac{1}{4}\lambda_0^2 = -1.828397$$

$$\lambda_0 = 2.704364$$

and

$$T_c - T_w = 1.803(T_b - T_w)$$

$$\text{Nu} = \frac{1}{2}\lambda_0^2 = 3.657$$

This infinite series solution leads to the same result as shown above. Again, it must be noted, the Nusselt number shown is limited only to round tubes with a uniform wall temperature. Changing the shape of the channel will result in a different Nusselt number within the geometry.

Figure 8.6 shows the wall temperature, bulk mean fluid temperature, wall heat flux, and Nusselt number behavior in a round tube exposed to a constant surface temperature.

Laminar flow is very sensitive to the thermal boundary condition imposed along the tube wall. Figure 8.7 offers a simple comparison between the Nusselt numbers for round tubes with either the uniform surface heat flux or uniform surface temperature condition.

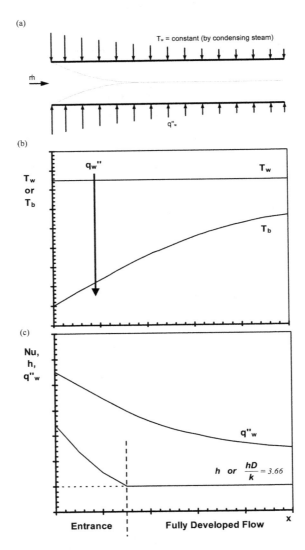

FIGURE 8.6 Laminar flow in a circular tube with the uniform surface temperature condition. (a) Thermal boundary layer; (b) temperature; and (c) wall heat flux and heat transfer coefficient (Nusselt number).

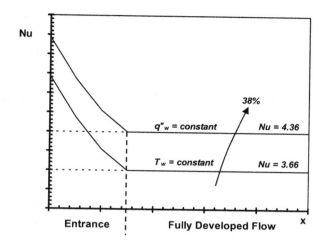

FIGURE 8.7 Round tube Nusselt number comparison between $q''_w = $ constant and $T_w = $ constant thermal boundary conditions

As noted with the derivations above, the Nusselt number is sensitive to both the thermal boundary condition and the geometry of the duct or channel. Thus far, we have focused on flow through a round tube. However, if the cross-section of the geometry changes, the process must be repeated to arrive at the Nusselt numbers for various flow configurations. Table 8.1 provides a comparison of the Nusselt numbers under various boundary conditions, velocity profiles, and flow geometries. When considering geometries that deviate from a round tube, the characteristic length used for reference changes from D (the tube diameter) to the channel "hydraulic diameter," D_h. The hydraulic diameter provides an equivalent diameter accounting for boundary

TABLE 8.1
Fully Developed, Laminar Heat Transfer

Geometry	Velocity Profile	Thermal Boundary Condition	$\mathbf{Nu}_D = \dfrac{hD}{k}$ or $\mathbf{Nu}_{D_h} = \dfrac{hD_h}{k}$
Circular tube	Parabolic	q''_w	4.36
Circular tube	Parabolic	T_w	3.66
Circular tube	Slug (liquid metal)	q''_w	8.0
Circular tube	Slug (liquid metal)	T_w	5.75
Parallel plates	Parabolic	q''_w	8.23
Parallel plates	Parabolic	T_w	7.60
Square duct	Parabolic	T_w	2.98
Triangular	Parabolic	T_w	2.35

layer growth within a specific cross section, A_c, and wetted perimeter, P. Table 8.2 provides the equivalent, hydraulic diameter for a variety of common shapes.

$$D_h = \frac{4A_c}{P}$$

Regardless of the geometry, the general definition of laminar flow does not change:

$$\text{Re}_D = \frac{\rho \bar{V} D_h}{\mu} = \frac{\bar{V} D_h}{v} \le 2300$$

TABLE 8.2

Hydraulic Diameters for Common Internal Flow Configurations

Geometry	Hydraulic Diameter Definition
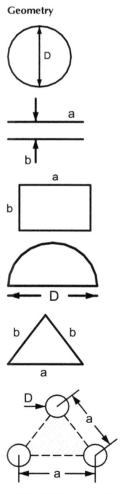	$D_h = \dfrac{4 \cdot \dfrac{\pi}{4} D^2}{\pi D} = D$
	$D_h = \dfrac{4 \cdot a \cdot b}{2(a+b)} = \dfrac{2ab}{a+b}$
	$D_h = \dfrac{2ab}{(a+b)} \cong 2b$ if $a >> b$, parallel plates
	$D_h = \dfrac{4 \cdot \dfrac{1}{2} \cdot \dfrac{\pi}{4} D^2}{\dfrac{1}{2}\pi D + D} = \dfrac{\dfrac{1}{2} \cdot \pi D}{\dfrac{1}{2}\pi + 1}$
	$D_h = \dfrac{4 \cdot \dfrac{1}{2} \cdot a \cdot \sqrt{b^2 - \left(\dfrac{1}{2}a\right)^2}}{a + 2b}$
	$D_h = \dfrac{4 \cdot \left(\dfrac{1}{2}a\sqrt{a^2 - \left(\dfrac{1}{2}a\right)^2} - 3 \cdot \dfrac{1}{6} \cdot \dfrac{\pi}{4} D^2 \right)}{3 \cdot \dfrac{1}{6} \cdot \dfrac{\pi}{4} D}$

8.3 FULLY DEVELOPED LAMINAR FLOW AND HEAT TRANSFER BETWEEN PARALLEL PLATES

8.3.1 TWO PLATES WITH SYMMETRIC, UNIFORM, HEAT FLUXES

Case 1: Fully developed, laminar flow between parallel plates

Consider a low-speed, constant-property, fully developed laminar flow between two parallel plates at $y = \pm H$, as shown in Figure 8.8. Both plates are electrically heated to give a uniform wall heat flux. Determine (a) the velocity profile, (b) the friction factor, and (c) the Nusselt number.

Based on the problem statement, the following assumptions can be made:

Low speed:	$\Phi = 0$
Constant properties:	$\rho, \mu, k, c_p = \text{constant}$
Fully developed:	$du/dx = 0$
Thermally fully developed:	$dT/dx = constant$

a. Velocity Profile

Beginning with the continuity equation:

$$v = 0$$

$$\frac{\partial u}{\partial x} + \frac{\partial v}{\partial y} = 0$$

$$\therefore \frac{\partial u}{\partial x} = 0$$

Momentum equation:

$$u\frac{\partial u}{\partial x} + v\frac{\partial u}{\partial y} = -\frac{1}{\rho}\frac{\partial P}{\partial x} + v\left(\frac{\partial^2 u}{\partial x^2} + \frac{\partial^2 u}{\partial y^2}\right)$$

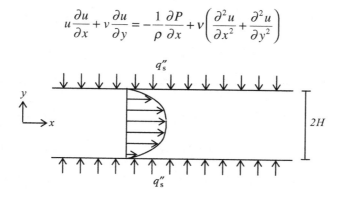

FIGURE 8.8 Two parallel plates at a uniform wall heat flux.

$$v = 0$$

$$\frac{\partial u}{\partial x} = 0$$

$$\therefore \frac{\partial^2 u}{\partial x^2} = 0$$

The momentum equation reduces to:

$$\frac{1}{\rho}\frac{\partial P}{\partial x} = v\frac{\partial^2 u}{\partial y^2} \tag{8.28}$$

with the solution (after integrating twice):

$$u = \frac{1}{2\mu}\frac{\partial P}{\partial x}y^2 + c_1 y + c_2 \tag{8.29}$$

where $v = \mu/\rho$ with the boundary conditions:
At $y = 0$, $\partial u/\partial y = 0$ (maximum velocity)
At $y = H$, $u = 0$

$$c_1 = 0$$

$$c_2 = -\frac{1}{2\mu}\frac{\partial P}{\partial x}H^2$$

Substituting the constants back in Equation (8.29) and rearranging:

$$u = -\frac{H^2}{2\mu}\frac{\partial P}{\partial x}\left[1 - \frac{y^2}{H^2}\right] \tag{8.30}$$

with

$$\mathcal{U}_{max} \text{ at } y = 0$$

Therefore:

$$\mathcal{U}_{max} = -\frac{H^2}{2\mu}\frac{\partial P}{\partial x}$$

$$u = \mathcal{U}_{max}\left[1 - \frac{y^2}{H^2}\right]$$

The bulk, mean velocity:

$$u_m = \frac{\int \rho u dA}{\rho A} = \frac{2 \int_0^H \mathcal{U}_{max} \left[1 - \frac{y^2}{H^2} \right] dy}{2H}$$

$$u_m = \frac{\mathcal{U}_{max} \left[y - (y^3/3H^2) \right]_0^H}{H} = \frac{\mathcal{U}_{max} \left((2/3)H \right)}{H}$$

$$u_m = \frac{2}{3} \mathcal{U}_{max} \tag{8.31a}$$

Thus,

$$u = \frac{3}{2} u_m \left[1 - \frac{y^2}{H^2} \right] \tag{8.31b}$$

b. Friction Factor
   By definition:

$$f \equiv \frac{\tau_W}{\frac{1}{2} \rho u_m^2}$$

From a force balance, we obtain:

$$\Delta P A_C + \tau_W (2w) \Delta x = 0$$

$$\frac{\Delta P}{\Delta X} = \frac{-2\tau_W W}{A_C} = \frac{-2\tau_W W}{2WH} = -\frac{\tau_W}{H}$$

$$\tau_W = \mu \frac{\partial u}{\partial y} \bigg|_{y=H} = \mu \left(\frac{3}{2} \right) u_m \left(\frac{2y}{H^2} \right) \bigg|_{y=H} = 3\mu \frac{u_m}{H}$$

Therefore,

$$\frac{\Delta P}{\Delta x} = \frac{-3\mu u_m}{H^2}$$

From Table 8.2, for parallel plates:

$$D_H = \frac{4A_c}{P} = \frac{4(2wH)}{2w} = 4H$$

Thus,

$$f = \frac{(3\mu u_m / H^2)(H)}{\rho u_m^2 / 2} = \frac{6\mu u_m H}{\rho u_m^2 H^2}$$

$$f = \frac{6\mu}{u_m H \rho}$$

Redefining the Reynolds number based on the hydraulic diameter:

$$\text{Re}_{D_h} = \frac{\rho u_m D_h}{\mu} = \frac{4 u_m H \rho}{\mu}$$

Finally,

$$f = \frac{24}{\text{Re}_{Dh}} \qquad (8.32)$$

c. Nusselt Number

Beginning with the energy equation:

$$\rho C_p \left(u \frac{\partial T}{\partial x} + v \frac{\partial T}{\partial y} \right) = k \left(\frac{\partial^2 T}{\partial x^2} + \frac{\partial^2 T}{\partial y^2} \right) + \dot{q} + \Phi$$

$\Phi = 0$ (low-speed flow) and $\dot{q} = 0$ (no internal heat generation)

$$\frac{dT}{dx} = \text{const}$$

$$\therefore \frac{d^2 T}{dx^2} = 0$$

The governing energy equation becomes

$$\frac{\partial^2 T}{\partial y^2} = \frac{u}{\alpha} \frac{\partial T}{\partial x} \qquad (8.33)$$

Using the velocity profile from Equation (8.31b) and integrating Equation (8.3) twice, the temperature profile becomes:

$$T = \frac{3}{2}\frac{u_m}{\alpha}\left[\frac{y^2}{2} - \frac{y^4}{12H^2}\right]\frac{dT}{dx} + c_1 y + c_2$$

Applying the thermal boundary conditions to solve for the constants of integration:
At $y = 0$, $dT/dy = 0$
At $y = H$, $T = T_s$

$$c_1 = 0$$

$$c_2 = T_s - \frac{5}{8}\frac{u_m}{\alpha}H^2\frac{dT}{dx}$$

$$T = \frac{3}{2}\frac{u_m}{\alpha}H^2\left[\frac{y^2}{2H^2} - \frac{y^4}{12H^4} - \frac{5}{12}\right]\frac{dT}{dx} + T_s \qquad (8.34)$$

Solving for the bulk mean, fluid temperature:

$$T_m \equiv \frac{\int \rho c_v u T \, dA}{\rho c_v u_m A}$$

$$T_m = \frac{2\int_0^H \rho c_v\left\{\frac{3}{2}u_m\left[1 - \frac{y^2}{H^2}\right]\right\}\left\{\frac{3}{2}\frac{u_m}{\alpha}H^2\left[\frac{y^2}{2H^2} - \frac{y^4}{12H^4} - \frac{5}{12}\right]\frac{dT}{dx} + T_s\right\}dy}{\rho c_v u_m (2H)}$$

$$T_m = \frac{9}{4}\frac{u_m H^2}{\alpha}\left\{\left[\frac{1}{6} - \frac{1}{60} - \frac{5}{12} - \frac{1}{10} + \frac{1}{84} + \frac{5}{36}\right]\frac{dT}{dx}\right\} + \frac{3}{2}\left\{T_s - \frac{T_s}{3}\right\}$$

$$T_m = -\frac{17}{35}\frac{u_m H^2}{\alpha}\frac{dT}{dx} + T_s \qquad (8.35)$$

Now we can use the surface and bulk mean temperature to determine the convective heat transfer coefficient:

$$q_s'' = h(T_s - T_m)$$

$$q_s'' = -k\frac{\partial T}{\partial y}\Big|_{y=H}$$

$$h = \frac{k\left[(3/2)(u_m/\alpha)\cdot(2/3)H\right](dT/dx)}{T_s + (17/35)(u_m/\alpha)H^2(dT/dx) - T_s}$$

$$h = \frac{35}{17}\frac{k}{H}$$

Now the Nusselt number (with $D_h = 4H$):

$$Nu_{D_h} = \frac{h(4H)}{k} = \frac{35}{17}\frac{k}{H}\left(\frac{4H}{k}\right)$$

$$Nu_{D_h} = \frac{140}{17} = 8.235$$

Case 2: Thermally and fully developed flow between parallel plates with a uniform velocity profile (slug flow).

Now consider flow between two parallel plates with a uniform velocity profile (i.e., slug flow) with a uniform wall heat flux. Determine (a) the temperature distribution and (b) the heat transfer coefficient (Nusselt number).

Based on the problem statement, $u = \bar{V}$. Therefore, there is no need to solve the momentum equation.

a. Temperature Profile

From the energy equation:

$$\frac{d^2 T}{dy^2} = \frac{\bar{V}}{\alpha}\frac{\partial T}{\partial x} = c$$

$$\frac{dT}{dy} = cy + c_1$$

$$T = \frac{1}{2}cy^2 + c_1 y + c_2$$

The thermal boundary conditions are used to solve for the constants of integration:

At $y = 0$, $dT/dy = 0$
At $y = H$, $T = T_s$

$$c_1 = 0$$

$$c_2 = T_s - C\frac{1}{2}H^2$$

Therefore,

$$T = T_s + \frac{1}{2}C\left(y^2 - H^2\right)$$

$$= T_s + \frac{1}{2}\frac{\bar{V}}{\alpha}\frac{\partial T}{\partial x}\left(y^2 - H^2\right)$$

Following the same procedure shown for Case 1:

$$T_m = \text{can be obtained}$$

$$h = \text{can be obtained}$$

$$\text{Nu}_{D_h} = \text{can be obtained}$$

8.3.2 Two Plates with Asymmetric Heat Fluxes

Case 1: Fully developed laminar flow between parallel plates with one insulated surface.

Consider low speed, constant-property, fully developed laminar flow between two parallel plates, with one plate insulated and the other uniformly heated. Determine the Nusselt number (see Figure 8.8).

For this problem, the velocity profile is fully developed. From Equation (8.31b), we have:

$$u = \frac{3}{2}u_m\left(1 - \frac{y^2}{H^2}\right)$$

where u_m = mean velocity

Now, from the energy equation:

$$\frac{\partial^2 T}{\partial y^2} = \frac{u}{\alpha}\frac{\partial T}{\partial x} = \frac{3}{2}u_m\frac{1}{\alpha}\left(1 - \frac{y^2}{H^2}\right)\frac{\partial T}{\partial x}$$

where

$$\frac{\partial T}{\partial x} = \text{constant}$$

After integration:

$$T = \frac{3}{2}\frac{u_m}{\alpha}\frac{\partial T}{\partial x}\left[\frac{1}{2}y^2 - \frac{1}{12}\frac{y^4}{H^2}\right] + c_1 y + c_2$$

Now the thermal boundary conditions are needed to solve for the constants:

At $y = -H$, $\dfrac{\partial T}{\partial y} = 0$ (insulated)

At $y = H$, $T = T_s$ (unknown surface temperature)

Solve for c_1 and c_2, the temperature profile becomes:

$$T = \frac{3}{24} \frac{u_m H^2}{\alpha} \frac{\partial T}{\partial x} \left[-\frac{y^4}{H^4} + \frac{6y^2}{H^2} + \frac{8y}{H} - 13 \right] + T_s$$

$$T_m = \frac{\displaystyle\int_{-H}^{H} uTdy}{u_m 2H} = \ldots = -\frac{52}{35} \frac{u_m H^2}{\alpha} \frac{\partial T}{\partial x} + T_s$$

$$h = \frac{q_s''}{T_s - T_m} = \frac{-k\dfrac{\partial T}{\partial y}\bigg|H}{T_s - T_m} = \frac{-2k\dfrac{u_m H}{\alpha}\dfrac{\partial T}{\partial x}}{\dfrac{52}{35}\dfrac{u_m H^2}{\alpha}\dfrac{\partial T}{\partial x}} = -\frac{70}{52}\frac{k}{H}$$

$$\mathrm{Nu}_{D_h} = \frac{hD_h}{k} = \frac{70}{52}\frac{4H}{H}\frac{k}{k} = \frac{70}{13} = 5.38$$

where

$$D_h \cong 4H$$

Case 2: Fully developed laminar flow between parallel plates with the plates unevenly heated.

The previous Case 1 is a special variation of this more general asymmetric heating. In this case the heat flux from surface 1 is twice the heat flux from the surface 2, or the heat flux from surface 1 is n times the heat flux from the surface 2. Using the momentum and energy differential equations, derive the Nusselt numbers for each of the two plates and show the Nusselt numbers can be written as:

$$\mathrm{Nu}_1 = \frac{140}{26 - 9n}$$

$$\mathrm{Nu}_2 = \frac{140n}{26n - 9}$$

If $n = 1$, $\mathrm{Nu}_1 = \mathrm{Nu}_2 = \dfrac{140}{17} = 8.235$

If $n = 2$, $\mathrm{Nu}_1 = \dfrac{140}{8} = 17.5$ and $\mathrm{Nu}_2 = \dfrac{280}{43} = 6.51$

Case 3: Thermally and fully developed flow between parallel plates with a uniform velocity profile (slug flow).

Now we consider flow between two parallel plates with a uniform velocity profile (i.e., slug flow) with asymmetric heating. The upper plate is well-insulated, and the lower plate is kept at a uniform surface temperature. Using the momentum and energy differential equations, show the Nusselt number for the lower plate is:

$$\text{Nu} = \frac{\pi^2}{2}(2n-1)^2$$

8.4 FULLY DEVELOPED LAMINAR FLOW AND HEAT TRANSFER IN A RECTANGULAR CHANNEL

Rectangular channels have been used for many heat transfer designs. We will assume that flow is moving in the axial z-direction, both the velocity and temperature profiles are two dimensional as a function of (x, y). The 2-D velocity and temperature distributions in the rectangular channels can be solved by the method of separation of variables, as discussed in Chapter 3. The following has been adapted from reference [5].

8.4.1 FULLY DEVELOPED LAMINAR FLOW

For steady, fully developed, laminar flow of a fluid through a long rectangular channel ($0 \leq x \leq W$, $0 \leq y \leq H$), the momentum equation along the main flow direction (along z) may be written in dimensionless form as

$$\rho u \frac{\partial w}{\partial x} + \rho v \frac{\partial w}{\partial y} + \rho w \frac{\partial w}{\partial z} = \mu\left(\frac{\partial^2 w}{\partial x^2} + \frac{\partial^2 w}{\partial y^2} + \frac{\partial^2 w}{\partial z^2}\right) - \frac{\partial P}{\partial z}$$

$$\left(\frac{1}{D_h^2}\right)\left(-\frac{1}{\mu}\right)\left(\frac{dP}{dz}\right)(D_h^2)\left(\frac{\partial^2 \Omega}{\partial X^2} + \frac{\partial^2 \Omega}{\partial Y^2}\right) - \frac{1}{\mu}\frac{dP}{dz} = 0$$

$$\frac{\partial^2 \Omega}{\partial X^2} + \frac{\partial^2 \Omega}{\partial Y^2} + 1 = 0$$

where $\Omega = \Omega(X, Y)$, $\Omega \equiv \dfrac{w}{-\dfrac{1}{\mu}\left(\dfrac{dP}{dz}\right)D_h^2}$ is a dimensionless velocity in the z-direction,

and X and Y are coordinates normalized with the hydraulic diameter of the channel, D_h (that is, $X \equiv x/D_h$ and $Y \equiv y/D_h$).

The above equation may be solved using the method of separation of variables with appropriate boundary conditions (see Chapter 3, Section 3.4.2, Equation (3.19); replace temperature by velocity), and the average velocity, $\bar{\Omega}$, may be determined

from the velocity distribution. Thus, $f \cdot \text{Re}_{D_h} = \dfrac{2}{\bar{\Omega}}$ can be obtained as

$$f \cdot \mathrm{Re}_{D_h} = \left[\frac{\left(-\dfrac{dP}{dx} \right) D_h}{\dfrac{1}{2}\rho\bar{w}^2} \right] \left(\frac{\rho\bar{w}D_h}{\mu} \right) = \frac{2\left(-\dfrac{1}{\mu} \right)\left(\dfrac{dP}{dx} \right) D_h^2}{\bar{w}}$$

$$f \cdot \mathrm{Re}_{D_h} = \frac{2}{\bar{\Omega}}$$

where f and Re_{D_h} are the friction factor and the Reynolds number, respectively. Also:

$$\bar{\Omega} = \frac{\bar{w}}{\left(-\dfrac{1}{\mu} \right)\left(\dfrac{dP}{dz} \right) D_h^2}$$

The velocity profiles can also be determined using the matrix method shown in chapter 5.

8.4.2 THERMALLY AND FULLY DEVELOPED LAMINAR FLOW WITH UNIFORM WALL HEAT FLUX

After determining the velocity profile within the rectangular duct, we can now determine the Nusselt number within the duct. For steady, thermally fully developed, laminar flow of a fluid through a long rectangular channel ($0 \le x \le W$, $0 \le y \le H$) that is subjected to a uniform heat flux at the walls, the energy equation may be written in dimensionless form as

$$\rho c_p \left(u\frac{\partial \cancel{T}}{\cancel{\partial x}} + v\frac{\partial \cancel{T}}{\cancel{\partial y}} + w\frac{\partial T}{\partial z} \right) = k\left(\frac{\partial^2 T}{\partial x^2} + \frac{\partial^2 T}{\partial y^2} + \frac{\cancel{\partial^2 T}}{\cancel{\partial z^2}} \right) + \cancel{\dot{q}} + \cancel{\mu\Phi}$$

$$-\left(\frac{1}{D_h^2} \right)\left(\frac{\bar{w}D_h^2}{\alpha} \right)\left(\frac{dT_b}{dz} \right)\left(\frac{\partial^2 \theta}{\partial X^2} + \frac{\partial^2 \theta}{\partial Y^2} \right) - \frac{1}{\alpha}w\frac{dT_b}{dz} = 0$$

$$\frac{\partial^2 \theta}{\partial X^2} + \frac{\partial^2 \theta}{\partial Y^2} + \frac{w}{\bar{w}} = 0$$

where X and Y are coordinates normalized with the hydraulic diameter of the channel, D_h (that is, $X \equiv x/D_h$ and $Y \equiv y/D_h$), and $\theta = \theta(X, Y)$ is defined as

$$\theta = \frac{T_w - T}{\left(\dfrac{\bar{w}D_h^2}{\alpha} \right)\left(\dfrac{dT_b}{dz} \right)}$$

Also, w *and* $\bar{w}$ are the local velocity and the average velocity along the main flow direction (in the z-direction).

Following the same procedures as velocity, the temperature, and heat transfer coefficient can be solved as:

$$\text{Nu}_{D_h} = \frac{hD_h}{k} = \frac{q''_w}{T_w - T_b} \frac{D_h}{k}$$

$$\text{Nu}_{D_h} = \frac{\left[\dfrac{\left(\rho \bar{w} A_c\right) c_p}{P_w} \left(\dfrac{dT_b}{dz} \right) \right]}{\left(T_w - T_b\right)} \frac{D_h}{k} \left(\frac{D_h}{4A_c / P_w} \right) \left[\frac{k / \left(\rho c_p \right)}{\alpha} \right]$$

$$= \frac{1}{4} \left[\frac{\left(\bar{w} D_h^2 / \alpha \right)}{\left(T_w - T_b\right)} \left(\frac{dT_b}{dz} \right) \right]$$

Since

$$\theta_b \equiv \frac{T_w - T_b}{\left(\dfrac{\bar{w} D_h^2}{\alpha} \right) \left(\dfrac{dT_b}{dz} \right)}$$

$$\text{Nu}_{D_h} = \frac{1}{4\theta_b}$$

The temperature profiles can also be determined using the matrix method shown in chapter 5.

8.4.3   Thermally and Fully Developed Laminar Flow with Uniform Wall Temperature

For steady, thermally fully developed, laminar flow of a fluid through a long rectangular channel ($0 \leq x \leq W$, $0 \leq y \leq H$), the walls of which are maintained at a uniform wall temperature, the energy equation may be written in dimensionless form. To represent the energy equation in a dimensionless form, we must recall the definition of thermally developed flow:

$$\left(\frac{T_w - T}{T_w - T_b} \right) \neq \text{function } z$$

Now to develop the nondimensional energy equation:

$$\frac{T_w - T}{T_w - T_b} = \frac{\theta}{\theta_b}$$

$$T = T_w - \left(T_w - T_b\right)\left(\frac{\theta}{\theta_b} \right)$$

$$\frac{\partial T}{\partial z} = \left(\frac{\theta}{\theta_b}\right)\frac{dT_b}{dz}$$

$$\rho c_p \left(u\frac{\partial \cancel{T}}{\cancel{\partial x}} + v\frac{\partial \cancel{T}}{\cancel{\partial y}} + w\frac{\partial T}{\partial z}\right) = k\left(\frac{\partial^2 T}{\partial x^2} + \frac{\partial^2 T}{\partial y^2} + \frac{\partial^2 \cancel{T}}{\cancel{\partial z^2}}\right) + \cancel{\dot{q}} + \cancel{\mu \Phi}$$

$$-\left(\frac{1}{D_h^2}\right)\left(\frac{\bar{w}D_h^2}{\alpha}\right)\left(\frac{dT_b}{dz}\right)\left(\frac{\partial^2 \theta}{\partial X^2} + \frac{\partial^2 \theta}{\partial Y^2}\right) - \frac{1}{\alpha}w\left(\frac{\theta}{\theta_b}\right)\frac{dT_b}{dz} = 0$$

$$\frac{\partial^2 \theta}{\partial X^2} + \frac{\partial^2 \theta}{\partial Y^2} + \frac{w}{\bar{w}}\left(\frac{\theta}{\theta_b}\right) = 0$$

where X and Y are coordinates normalized with the hydraulic diameter of the channel, D_h (that is, $X \equiv x/D_h$ and $Y \equiv y/D_h$), and $\theta = \theta(X, Y)$ is defined as

$$\theta \equiv \frac{T_w - T}{\left(\dfrac{\bar{w}D_h^2}{\alpha}\right)\left(\dfrac{dT_b}{dz}\right)}$$

and

$$\theta_b \equiv \frac{T_w - T_b}{\left(\dfrac{\bar{w}D_h^2}{\alpha}\right)\left(\dfrac{dT_b}{dz}\right)}$$

Also, w and $\bar{w}$ are the local velocity and the average velocity along the main flow direction (in the z-direction).

Following the same procedures described for the velocity profile, the temperature and heat transfer coefficient can be solved as:

$$\mathrm{Nu}_{D_h} = \frac{hD_h}{k} = \frac{q_w''}{T_w - T_b}\frac{D_h}{k}$$

$$\mathrm{Nu}_{D_h} = \frac{\left[\dfrac{(\rho\bar{w}A_c)c_p}{P_w}\left(\dfrac{dT_b}{dz}\right)\right]}{(T_w - T_b)}\frac{D_h}{k}\left(\frac{D_h}{4A_c/P_w}\right)\left[\frac{k/(\rho c_p)}{\alpha}\right]$$

$$= \frac{1}{4}\left[\frac{(\bar{w}D_h^2/\alpha)}{(T_w - T_b)}\left(\frac{dT_b}{dz}\right)\right]$$

Since

$$\theta_b \equiv \frac{T_w - T_b}{\left(\dfrac{\overline{w}D_h^2}{\alpha}\right)\left(\dfrac{dT_b}{dz}\right)}$$

$$Nu_{D_h} = \frac{1}{4\theta_b}$$

Again, the temperature profiles can be determined using the matrix method shown in chapter 5.

8.5 THERMALLY DEVELOPING HEAT TRANSFER IN A CIRCULAR TUBE—SEPARATION OF VARIABLES TECHNIQUE

8.5.1 COMBINED HYDRODYNAMIC AND THERMAL ENTRY FLOW

In many applications, the length of the tube can be relatively short, so the flow does not reach the fully developed condition (hydrodynamically and/or thermally). Both friction and heat transfer coefficients are elevated in the entry region of the tube; therefore, to accurately predict the pressure drop and heat transfer, we need the velocity and temperature profiles in the entry region. When previously considering fully developed flow, the momentum and energy equations simplified to ordinary differential equations; the velocity and temperature profiles were only changing in the radial direction. In the entry region, these profiles are a function of both x and r, as shown in Figure 8.9. To solve for velocity and temperature profiles, we must solve the partial differential equations (PDEs). We will utilize the separation of variables technique to obtain solutions to the PDEs.

Let us first consider fluids with a Prandtl number of one:

$$\Pr \cong 1 = \frac{\nu}{\alpha} \cong \frac{\delta}{\delta_T}$$

The continuity equation is:

$$\frac{\partial u}{\partial x} + \frac{1}{r}\frac{\partial(vr)}{\partial r} = 0$$

In the flow direction, the x-momentum equation is simplified as:

$$u\frac{\partial u}{\partial x} + v\frac{\partial u}{\partial r} = -\frac{1}{\rho}\frac{\partial P}{\partial x} + v\frac{1}{r}\frac{\partial}{\partial r}\left(r\frac{\partial u}{\partial r}\right)$$

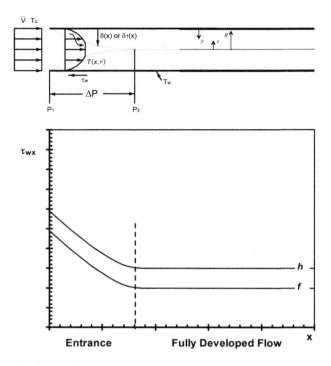

FIGURE 8.9 Developing flow in a round tube.

The momentum equation in the radial direction is:

$$u\frac{\partial v}{\partial x} + v\frac{\partial v}{\partial r} = -\frac{1}{\rho}\frac{\partial P}{\partial x} + v\frac{1}{r}\frac{\partial}{\partial r}\left(r\frac{\partial u}{\partial r} - \frac{v}{r}\right)$$

Finally, the energy equation:

$$u\frac{\partial T}{\partial x} + v\frac{\partial T}{\partial r} = \alpha\frac{1}{r}\frac{\partial}{\partial r}\left(r\frac{\partial T}{\partial r}\right)$$

For laminar flow, the hydrodynamic entry length is estimated as:

$$\frac{x}{D} = 0.05\mathrm{Re}_D$$

Similarly, the thermal entry length is:

$$\frac{x}{D} = 0.05\mathrm{Re}_D \cdot \mathrm{Pr}$$

Within the entrance region, a force balance yields:

$$\underbrace{(P_1 - P_2)\frac{\pi}{4}D^2}_{\text{Total Pressure Drop}} = \underbrace{\int \tau_{wx} dx}_{\text{Shear at Wall}} + \underbrace{\int \left(\rho u^2 dA - \rho \bar{V}^2 dA\right)}_{\text{Momentum Change}}$$

The friction factor can be written as:

$$f_{\text{total}} = \frac{D}{4}\frac{\Delta P}{\Delta x}\left(\frac{1}{\frac{1}{2}\rho \bar{V}^2}\right)$$

where

$$\frac{\Delta P}{\Delta x} = \text{total}$$

In the fully developed region, we have:

$$f_{\text{FDF}} = \frac{D}{4}\frac{\Delta P}{\Delta x}\left(\frac{1}{\frac{1}{2}\rho \bar{V}^2}\right)$$

where

$$\frac{\Delta P}{\Delta x} = \text{shear only}$$

We need to solve for the velocity and temperature profiles from both the momentum and energy equations using the finite-difference method (numerical computation) by Langhaar. Figure 8.10 shows how the Nusselt number varies along the length of the tube for various thermal boundary conditions.

As shown in the figure, the nondimensional x-coordinate is:

$$x^+ = \frac{x/R}{\text{Re}_D \text{Pr}} \equiv \frac{x/r_o}{\text{Re}_D \text{Pr}} = \frac{2(x/D)}{\text{Re}_D \text{Pr}}$$

$$= \frac{2}{Gz}$$

In this solution, additional unitless numbers are introduced. First, the Graetz number, Gz, is defined as:

$$Gz = \frac{\text{Re}_D \text{Pr}}{x/D} = \frac{\text{Pe}}{x/D}$$

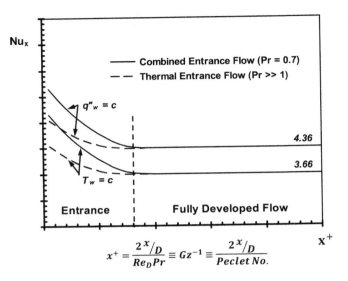

FIGURE 8.10   Nusselt number distribution along the length of a circular tube.

where Pe is the Peclet number:

$$Pe = Re_D Pr$$

The Peclet number is often used to indicate if axial conduction has a significant effect on the solution. Generally, if Pe > 100, axial conduction is considered negligible. In other words, conduction in the flow (axial) direction is significant for low Reynolds and/or Prandtl numbers.

8.5.2   THERMAL BOUNDARY CONDITION: GIVEN WALL TEMPERATURE

Case 1: Slug flow (or sharp entrance) with a constant wall temperature

Figure 8.11 contrasts the velocity and thermal boundary layers for this case. Assuming a uniform velocity profile across the tube implies a very thin hydrodynamic boundary layer, relative to the thickness of the thermal boundary layer. Therefore, this type of behavior would also be seen for fluids with low Prandtl numbers (Pr << 1).

With the uniform velocity profile, the continuity equation simplifies to:

$$v = 0$$

Similarly, the x-momentum equation becomes:

$$u \cong V_o = \text{constant}$$

The general form of the energy equation is:

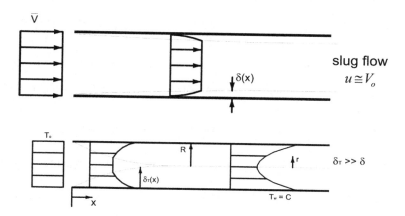

FIGURE 8.11   Thermal boundary layer development with a uniform (slug) velocity profile.

$$u\frac{\partial T}{\partial x} + v\frac{\partial T}{\partial r} = \alpha\left(\frac{\partial^2 T}{\partial x^2} + \frac{1}{r}\frac{\partial}{\partial r}r\frac{\partial T}{\partial r}\right)$$

where

$$u \cong V_o$$

$$v = 0$$

If the Peclet number is greater than 100, axial conduction is assumed negligible:

$$Re_D Pr > 100$$

or

$$Pe > 100$$

Neglecting axial conduction manifests itself in the energy equation as:

$$\frac{\partial^2 T}{\partial x^2} \to 0$$

The thermal boundary conditions are:

$$
\begin{aligned}
x = 0 \quad & T = T_o \\
r = 0 \quad & \frac{\partial T(x,0)}{\partial r} = 0 \\
r = R \quad & T(x,R) = T_w
\end{aligned}
$$

Now the separation of variables method can be applied. To being this process, let:

$$\theta(x,R) = T - T_w = X(x) \cdot R(r)$$

The assumed form of the solution, θ, can be substituted back into the energy equation:

$$\frac{V_o}{\alpha} \frac{X'}{X} = \frac{1}{r} \frac{1}{R} \frac{\partial}{\partial r}(rR') \equiv \text{constant} \equiv -n^2$$

The solution to the x-direction equation is an exponential decay function:

$$X = c_1 e^{-\frac{\alpha}{V_o} n^2 x}$$

In the radial direction, the ODE is in the form of the Bessel equation:

$$r^2 R'' + rR' + r^2 n^2 R = 0$$

With a Bessel function solution:

$$R = A \cdot J_0(nr) + B \cdot Y_0(nr)$$

Combining X(x) and R(r), we have the temperature profile within the entry region:

$$\theta(x,r) = c_1 e^{-\frac{\alpha}{V_o} n^2 x} \left[A \cdot J_0(nr) + B \cdot Y_0(nr) \right]$$

$$= e^{-\frac{\alpha}{V_o} n^2 x} \left[c_1 \cdot J_0(nr) + c_2 \cdot Y_0(nr) \right]$$

Applying the thermal boundary conditions, we can solve for the constants:
   At

$$r = 0, \quad \because Y_0(0) \to \infty \quad \therefore c_2 = 0$$

and

$$r = R, \quad \theta(x,r) = 0 \quad J_0(n_1 R) = 0$$

Refer to Figure 2.10 and Appendix A.3. J_o, the Bessel function of the first kind and zeroth order has multiple roots; therefore, the final solution for the temperature profile includes eigenvalues.

Entering the tube, $x = 0$, the temperature profile simplifies to:

$$\theta(0,r) = (T_o - T_w) = \sum_i c_i \cdot J_0(n_i r)$$

with

$$
c_i = \frac{(T_o - T_w)\displaystyle\int_0^R rJ_0(nr)\,dr}{\displaystyle\int_0^R rJ_0^2(nr)\,dr} = \ldots
$$

$$
= \frac{2}{n_i R J_1(n_i R)}(T_o - T_w)
$$

The general form of the temperature profile becomes:

$$
\frac{\theta}{\theta_0} = \frac{T - T_w}{T_o - T_w} = \sum_{i=1}^{\infty} \frac{2}{n_i R}\frac{J_0(n_i R)}{J_1(n_i R)} e^{-\frac{n_i^2 \alpha}{V_o}x}
$$

where

$$
n_1 R = 2.4048
$$
$$
n_2 R = 5.5207
$$
$$
\vdots
$$

Now, with an expression for the local temperature profile, the heat transfer coefficient distribution (as a function of x) can be determined:

$$
h = \frac{q_w'}{T_w - T_b}
$$

$$
q_w'' = -k\left.\frac{\partial T}{\partial y}\right|_0 = +k\left.\frac{\partial T}{\partial r}\right|_R = \ldots
$$

$$
T_b = \frac{\displaystyle\int_0^R T \cdot V_0 \cdot r\,dr}{\displaystyle\int_0^R V_0 \cdot r\,dr} = \ldots
$$

$$
\mathrm{Nu}_x = \frac{hD}{k} = \frac{h2R}{k} = \frac{\displaystyle\sum_i e^{-\frac{n_i^2 \alpha}{V_o}x}}{\displaystyle\sum_i \frac{1}{R^2 n_i^2} e^{-\frac{n_i^2 \alpha}{V_o}x}}
$$

$$
= \frac{e^{-\frac{(2.4048)^2 \alpha}{V_o R^2}x} + e^{-\frac{(5.5207)^2 \alpha}{V_o R^2}x} + \ldots}{\dfrac{e^{-\frac{(2.4048)^2 \alpha}{V_o R^2}x}}{(2.4048)^2} + \dfrac{e^{-\frac{(5.5207)^2 \alpha}{V_o R^2}x}}{(5.5207)^2} + \ldots}
$$

If x is large, $(5.5207)^2 \gg (2.4048)^2$, so:

$$\text{Nu} \approx (2.4048)^2 \cong 5.75 = \text{Nu}_D = \frac{hD}{k}$$

which is the fully developed result shown in Table 8.1.

Figure 8.12 shows the Nusselt number distribution through the tube. As the result converges to the fully developed value of Nu = 5.75, this magnitude is greater than the value that would exist for a flow that is both thermally and hydrodynamically developed (Nu = 3.66, $T_w = constant$). The thin boundary layer associated with the slug flow provides enhanced heat transfer compared to the fully developed profile.

Case 2: Parabolic flow (or a long, unheated tube upstream of the heated tube) with a constant wall temperature

Now we will consider the case shown in Figure 8.13. This is often referred to the as the Graetz problem and considers a fully developed (parabolic) velocity profile with a developing thermal boundary layer. In addition to a long, unheated entrance, this type of boundary layer development could be observed for high Prandtl number fluids (Pr $\gg$ 1).

For hydrodynamically, fully developed flow, the continuity equation yields:

$$v = 0$$

For fully developed flow through a circular tube, the velocity profile can be solved using the x-momentum equation. From Equations (8.8) and (8.10), the velocity profile is:

$$u = 2\bar{V}\left(1 - \frac{r^2}{R^2}\right)$$

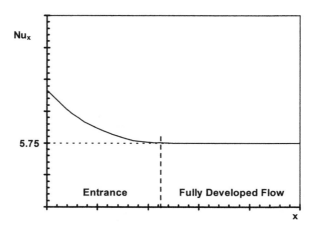

FIGURE 8.12 Nusselt number distribution for thermally developing flow with a uniform velocity profile.

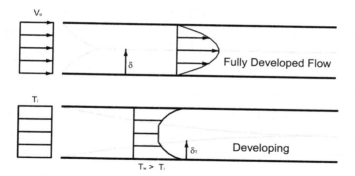

FIGURE 8.13   Thermal entry flow with a fully developed velocity profile.

The general form of the energy equation is:

$$u\frac{\partial T}{\partial x} + v\frac{\partial T}{\partial r} = \alpha\left(\frac{1}{r}\frac{\partial}{\partial r}r\frac{\partial T}{\partial r}\right)$$

with

$$v = 0$$

Again, we will utilize the separation of variables method. However, before solving the PDE, we will nondimensionalize all variables:

$$\theta(x,R) = \frac{T - T_w}{T_i - T_w}$$

$$r^+ = \frac{r}{R}$$

$$u^+ = \frac{u}{V_o}$$

$$x^+ = \frac{x/R}{Re \cdot Pr}$$

The nondimensional values are substituted into the energy equation, and the new equation is:

$$\left[1 - \left(r^+\right)^2\right]\frac{\partial\theta}{\partial x^+} = \frac{1}{r^+}\frac{\partial}{\partial r^+}\left(r^+\frac{\partial\theta}{\partial r^+}\right)$$

The thermal boundary conditions become:

$$\theta(0, r^+) = 1$$

$$\theta(x^+, 1) = 0$$

$$\frac{\partial \theta(x^+, 0)}{\partial r^+} = 0$$

With the separation of variables method, we assume the final form of the solution:

$$\theta(x^+, r^+) = X(x^+) \cdot R(r^+)$$

Substitute assumed solution back into the energy equation:

$$\frac{1}{X} \frac{\partial X}{\partial x^+} = \frac{1}{r^+ \left[1 - (r^+)^2\right]} \left(\frac{1}{R} \frac{\partial R}{\partial r^+} + \frac{r^+}{R} \frac{\partial^2 R}{\partial r^{+2}} \right) \equiv -\lambda^2$$

Again, the solution to the ODE in the x-direction is the exponential decay function:

$$\frac{dX}{dx^+} + \lambda^2 X = 0$$

In the radial direction, the ODE is the Sturm-Liouville type equation:

$$\frac{d^2 R}{dr^{+2}} + \frac{1}{r^+} \frac{dR}{dr^+} + \lambda^2 R \left[1 - (r^+)^2\right] = 0$$

The solution is obtainable as an infinite series. The λ values have been obtained by both numerical and approximate methods. The solution becomes:

$$\theta(x^+, r^+) = \sum c_n \cdot R_n(r^+) \exp(-\lambda^2 x^+)$$

where
c_n—constant
$R_n(r^+)$—eigen function
λ_n—eigen values

From the developing temperature profile, the local Nusselt number is:

$$\mathrm{Nu}_x = \frac{hD}{k} = \frac{h2R}{k} = \frac{\sum G_n \cdot \exp(-\lambda_n^2 x^+)}{2 \sum \dfrac{G_n}{\lambda_n^2} \cdot \exp(-\lambda_n^2 x^+)}$$

with

$$G_n = -\frac{C_n}{2} \cdot R_n'(1)$$

The eigen values are shown in Table 8.3, and these values are used to calculate the temperature and Nusselt number distributions provided in Table 8.4. These tables have been adapted from Kays and Crawford [3]. The Nusselt number distribution is plotted in Figure 8.14. As shown in the figure, for large values of x^+, the Nusselt number converges to the fully developed value corresponding to the constant surface temperature boundary condition.

For $n > 2$, $\lambda_n = 4n + \frac{8}{3}$, $G_n = 1.01276\lambda_n^{-1/3}$

TABLE 8.3
Infinite Series Solution Functions for Thermal Entry Flow in a Circular Tube with a Constant Surface Temperature (Fully Developed Velocity)

N	λ_n^2	G_n
0	7.313	0.749
1	44.61	0.544
2	113.9	0.463
3	215.2	0.415
4	348.6	0.383

TABLE 8.4
Nusselt Number and Mean Temperature Distribution in the Entry Region of a Circular Tube with a Constant Surface Temperature (Fully Developed Velocity)

x^+	Nu_x	Nu_m	θ_m
0	∞	∞	1.000
0.001	12.80	19.29	0.962
0.004	8.03	12.09	0.908
0.01	6.00	8.92	0.837
0.04	4.17	5.81	0.628
0.08	3.77	4.86	0.459
0.10	3.71	4.64	0.396
0.20	3.66	4.15	0.190
∞	3.66	3.66	0.0

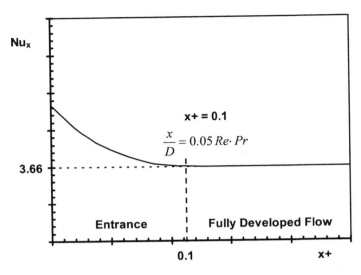

FIGURE 8.14   Nusselt number distribution for thermally developing flow with a parabolic velocity profile.

Case 3: Parabolic flow with a variable wall temperature

Building on Case 2, we now introduce a variable wall temperature boundary condition. Figure 8.15 shows the linear temperature distribution along the length of the tube, $T_w(x)$. As we evaluate this case, we will generally consider fluids with Prandtl numbers greater than unity (Pr > 1).

From Case 2:

$$\theta\left(x^+,r^+\right) = \sum c_n \cdot R_n\left(r^+\right)\exp\left(-\lambda^2 x^+\right)$$

To determine the heat flux and Nusselt number on the surface, we need to differentiate the temperature profile:

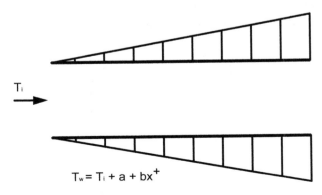

FIGURE 8.15   Variable wall temperature boundary condition.

$$\frac{\partial \theta}{\partial r^+} = \dots$$

$$q''_w = \frac{k}{R}\frac{\partial T}{\partial r^+}\bigg|_{\substack{r^+=1 \\ \text{at wall}}} = \frac{k}{R}\int_0^{x^+}\frac{\partial}{\partial r^+}\theta\left(x^+ - \xi^+, r^+\right)\frac{dT_w}{d\xi^+}d\xi^+$$

For example, for the linear temperature profile:

$$T_w = T_i + a + bx^+$$

$$\frac{dT_w}{d\xi^+} = b$$

Therefore:

$$q''_w = \frac{4\pi R^3 V\rho c_p}{k}\int_0^{x^+}q''_w dx^+ = \pi R^2 V\rho c_p\left(T_b - T_i\right)$$

Now we can solve for the bulk mean fluid temperature:

$$T_b = \dots$$

Along with the heat transfer coefficient and Nusselt number:

$$h_x = \frac{q''_w}{T_w - T_b}$$

$$\text{Nu}_x = \frac{hD}{k}$$

For the given wall temperature distribution, we have:

$$q''_w\left(x^+\right) = \frac{2k}{R}\left[\frac{b}{8} - b\sum\frac{G_n}{\lambda_n^2}\exp\left(-\lambda_n^2 x^+\right) + a\sum G_n\exp\left(-\lambda_n^2 x^+\right)\right]$$

$$T_b = T_i + bx^+ + a - 8b\sum\frac{G_n}{\lambda_n^4} + 8b\sum\frac{G_n}{\lambda_n^4}\exp\left(-\lambda_n^2 x^+\right) - 8a\sum\frac{G_n}{\lambda_n^2}\exp\left(-\lambda_n^2 x^+\right)$$

$$\mathrm{Nu}_x = \frac{b - 8b\sum \dfrac{G_n}{\lambda_n^2}\exp\left(-\lambda_n^2 x^+\right) + 8a\sum G_n \exp\left(-\lambda_n^2 x^+\right)}{16b\sum \dfrac{G_n}{\lambda_n^4} - 16b\sum \dfrac{G_n}{\lambda_n^4}\exp\left(-\lambda_n^2 x^+\right) + 16a\sum \dfrac{G_n}{\lambda_n^2}\exp\left(-\lambda_n^2 x^+\right)}$$

$$= \frac{1}{16\sum \dfrac{G_n}{\lambda_n^4}} = \frac{1}{16\cdot 0.01433} = 4.364 \qquad (\text{for } q_w'' = \text{constant})$$

For large values of x^+ $\left(x^+ \to \infty\right)$, all the summations approach zero $\left(\sum \to 0\right)$.

8.5.3 Thermal Boundary Condition: Given Wall Heat Flux

Case 1: Parabolic flow with a constant wall heat flux

As we consider thermally developing flow with a specified wall heat flux, we begin with a constant wall heat flux $\left(q_w'' = \text{constant}\right)$. Again, we will consider fluids with $Pr > 1$. With the fully developed (parabolic) velocity profile, the local Nusselt number in the entrance region is:

$$\mathrm{Nu}_x = \left[\frac{1}{\mathrm{Nu}_\infty} - \frac{1}{2}\cdot\sum_{m=1}^{m}\frac{\exp\left(-\lambda_m^2 x^+\right)}{A_m \lambda_m^4}\right]^{-1}$$

where

$$\mathrm{Nu}_\infty = 4.36$$

which is the Nusselt number corresponding to fully developed flow, both velocity and temperature profiles being fully developed. Therefore, the second term in the bracket represents the entrance effect on the local Nusselt numbers (elevated heat transfer in the entrance region). Table 8.5 provides the eigen values and functions needed to solve for the local Nusselt numbers, and Table 8.6 gives the local Nusselt

---

TABLE 8.5

Infinite Series Solution Functions for Thermal Entry Flow in a Circular Tube with a Constant Surface Heat Flux (Fully Developed Velocity)

m	λ_m^2	A_m
1	25.68	7.630×10^{-3}
2	83.86	2.053×10^{-3}
3	147.2	0.903×10^{-3}
4	196.5	0.491×10^{-3}
5	450.9	0.307×10^{-3}

TABLE 8.6

Nusselt Number Distribution in the Entry Region of a Circular Tube with a Constant Heat Flux (Fully Developed Velocity)

x^+	Nu_x
0	∞
0.002	12.00
0.004	9.93
0.010	7.49
0.020	6.14
0.040	5.19
0.100	4.51
∞	4.36

number as a function of location into the tube. Both Tables 8.5 and 8.6 have been adapted from Kays and Crawford [3]. Figure 8.16 shows the local Nusselt number distribution for developing flow with both constant surface temperature and constant heat flux boundary conditions. As with the fully developed values, the heat transfer within the entrance region of the tube is sensitive to the thermal boundary condition.

From the local Nusselt number, the wall temperature distribution can also be calculated:

$$T_w - T_b = \frac{q''_w}{h} = q''_w \frac{D}{Nu_x k}$$

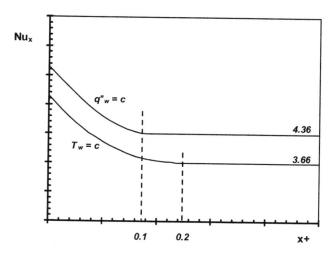

FIGURE 8.16 Nusselt number distributions for both constant surface temperature and constant wall heat flux boundary conditions (fully developed, parabolic velocity profile).

For larger m, $\lambda_m = 4m + \dfrac{4}{3}$, $A_m = 0.4165\lambda_m^{-7/3}$

Case 2: Parabolic flow with a variable wall heat flux

The final case we will present is commonly encountered with the fuel rods of nuclear reactors. For this Case 2, we will consider a variable wall heat flux, $q_w''(x)$. As with the previous cases, the solution shown below is not applicable to liquid metals and is generally acceptable for Pr > 1.

For the nuclear fuel rods, the local wall heat flux often displays a sinusoidal behavior:

$$q_w'' = q_{max}'' \sin\left(\frac{\pi x}{L}\right) \quad \text{or} \quad \cos\left(\frac{\pi x}{L}\right)$$

$$= a + b\sin\left(\frac{\pi x}{L}\right) \quad \text{or} \quad \cos\left(\frac{\pi x}{L}\right)$$

By integrating the wall heat flux, the local wall temperature can be determined as:

$$T_w - T_i = \frac{R}{k}\int_0^{x^+} g\left(x^+ - \xi\right)q_w''(\xi)d\xi$$

where

$$g(x^+) = 4 + \sum_m \frac{\exp\left(-\gamma_m^2 x^+\right)}{A_m\gamma_m^2}$$

and

$$T_b(x^+) - T_i = \frac{4R}{k}\int_0^{x^+} q_w''(\xi)d\xi$$

From the given heat flux and temperature distributions, the heat transfer coefficient is calculated:

$$h_x = \frac{q_w''}{T_w - T_b}$$

Given a wall heat flux similar to that shown in Figure 8.17, the local wall and bulk mean temperature distributions can be determined. These distributions are also shown in Figure 8.17.

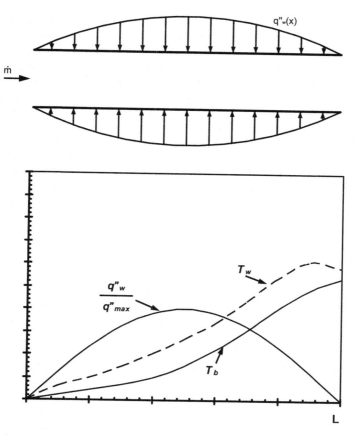

FIGURE 8.17 Heat flux, surface temperature, and bulk mean temperature distributions through a circular tube with a variable surface heat flux boundary condition (fully developed, parabolic velocity profile).

REMARKS

There are many engineering applications such as electronic equipment, small (milli-, micro-, or nano-) scale channels, and compact heat exchangers that required laminar flow heat transfer analysis and design. In the undergraduate-level heat transfer, students are expected to know many heat transfer relationships between Nusselt numbers and Reynolds and Prandtl numbers for developing and fully developed flows inside circular tubes with various thermal BCs. Students are expected to calculate heat transfer coefficients from these correlations at given Reynolds and Prandtl numbers.

In the intermediate-level heat transfer, this chapter focuses on how to solve fully developed heat transfer problems for flow between two parallel plates or inside circular tubes. Both uniform and asymmetric thermal boundary conditions have been considered. Students are expected to know how to analytically determine the velocity profile, the friction factor, the temperature profile, and the Nusselt number for these cases.

For advanced heat convection, students were shown how to analytically predict heat transfer in both developing flow and thermal entrance regions, with various thermal BCs such as variable surface heat flux as well as variable surface temperature BCs. Flow in rectangular channels with various aspect ratios with various thermal BCs was also presented. These scenarios require more complex mathematics; therefore, only a sample of these topics was provided.

PROBLEMS

8.1 Consider a steady, constant-property, laminar flow, between two parallel plates at $y = \pm\ell$. The plates are electrically heated to give a uniform wall heat flux. The differential equations for momentum and energy are listed here for reference:

$$u\frac{\partial u}{\partial x} + v\frac{\partial u}{\partial y} = -\frac{1}{\rho}\frac{\partial P}{\partial x} + v\left(\frac{\partial^2 u}{\partial x^2} + \frac{\partial^2 u}{\partial y^2}\right)$$

$$u\frac{\partial T}{\partial x} + v\frac{\partial T}{\partial y} = \alpha\left(\frac{\partial^2 T}{\partial x^2} + \frac{\partial^2 T}{\partial y^2}\right) + \frac{v}{c_p}\left(\frac{\partial u}{\partial y}\right)^2$$

   a. Assume a low-speed, slug flow velocity profile (i.e., a uniform velocity profile) between two parallel plates, and also, assume a thermally, fully developed condition. Write the simplified equations for momentum and energy and the associated BCs that can be used for this problem.

   b. Under the assumptions in (a), determine the Nusselt number on the plate.

   c. Consider a fully developed velocity profile (i.e., a parabolic velocity profile) between two parallel plates and a thermally, fully developed condition. Discuss whether the Nusselt number on the plate will be higher, the same, or lower than those of the slug flow velocity profile (uniform velocity profile). Explain why?

8.2 Consider a steady, constant-property, laminar flow, between two parallel plates at $y = \pm\ell$. The plates are electrically heated to give a uniform wall heat flux. The differential equations for momentum and energy are listed here for reference:

$$u\frac{\partial u}{\partial x} + v\frac{\partial u}{\partial y} = -\frac{1}{\rho}\frac{\partial P}{\partial x} + v\left(\frac{\partial^2 u}{\partial x^2} + \frac{\partial^2 u}{\partial y^2}\right)$$

$$u\frac{\partial T}{\partial x} + v\frac{\partial T}{\partial y} = \alpha\left(\frac{\partial^2 T}{\partial x^2} + \frac{\partial^2 T}{\partial y^2}\right) + \frac{v}{c_p}\left(\frac{\partial u}{\partial y}\right)^2$$

   a. Assume a low-speed, symmetrical linear velocity profile (i.e., $u = a + by$ with a maximum velocity at $y = 0$ and zero velocity at $y = \pm\ell$) between the two parallel plates. Also, assume a thermally, fully developed condition. Write the simplified equations for the momentum and energy equations and the associated BCs that can be used for this problem.

b. Under the assumption in (a), determine the Nusselt number on the plate.

c. Consider a fully developed velocity profile (i.e., a parabolic velocity profile) between two parallel plates and a thermally, fully developed condition, and comment on whether the Nusselt number on the plate will be higher, the same, or lower than those of the symmetrical linear velocity profile ($u = a + by$)? Explain why.

8.3 Internal flow, fully developed laminar forced convection: Consider a low-speed, constant-property, fully developed, laminar flow between two parallel plates, with one plate insulated and the other uniformly heated. Determine the Nusselt number on the heated surface.

8.4 Consider a concentric, circular-tube annulus with a radius ratio of $r_i/r_o = 0.6$. Let the inner tube wall be heated at a constant rate and the outer tube wall remains insulated. Let the fluid be a low-Prandtl number fluid and assume a slug flow inside of the annulus. Develop an expression for the Nusselt number at the inner surface by means of the following steps:

a. Discuss how the temperature and velocity profiles develop in the system by considering Pr and the type of flow present. Indicate the system conditions. (Is the flow and/or temperature developed?)

b. Let T_m be the mass averaged fluid temperature and $T_{s, ri}$ be the surface temperature at the inner surface with radius r_i. Form the thermal BCs, what can be said about $(T_m - T_{s, ri})$ and $\partial T/\partial x$?

c. Draw a diagram, indicating the BCs for both temperature and velocity and indicate the assumptions used to simplify the energy equation.

d. Solve for T using the simplified velocity profile, BCs, and the energy equation.

e. Find the heat transfer coefficient and the Nusselt number by calculating T_m, the mass averaged fluid temperature, and the heat flux at the inner surface.

Remember that the energy equation in cylindrical coordinates is given by

$$\rho c \left(\frac{\partial T}{\partial t} + v_r \frac{\partial T}{\partial r} + \frac{v_\theta}{r} \frac{\partial T}{\partial \theta} + v_x \frac{\partial T}{\partial x} \right) = \frac{1}{r} \frac{\partial}{\partial r} \left(rk \frac{\partial T}{\partial r} \right) + \frac{1}{r^2} \frac{\partial}{\partial \theta} \left(k \frac{\partial T}{\partial \theta} \right) + \frac{\partial}{\partial x} \left(k \frac{\partial T}{\partial y} \right) + \dot{q}$$

where x represents the axial direction of the cylinder.

8.5 Consider a low-speed, constant-property fluid, fully developed, laminar flow between two parallel plates located at $y = \pm b$. The plates are electrically heated to give a uniform heat flux.

a. Determine the velocity profile and define the bulk velocity (u_b) in terms of the pressure gradient driving the flow.

b. Using the hydraulic diameter $D_h = 4A_c/P$, show that the friction factor is given by $f = 24/Re_{Dh}$.

c. Obtain an expression for the connective heat transfer coefficient and the Nusselt number. (*Hint:* Consider the fully developed condition to approximate $\partial T/\partial x$ and define $T(y) = T - T_b$, where $T_b = 2 \int_0^b uTdy/2bu_b$)

8.6 Consider a steady constant-property, laminar flow, between two parallel plates at $y \pm \ell$. The plates are electrically heated to give a uniform wall heat flux. The differential equations for momentum and energy are listed here for reference:

$$u\frac{\partial u}{\partial x} + v\frac{\partial u}{\partial y} = \frac{1}{\rho}\frac{\partial P}{\partial x} + v\left(\frac{\partial^2 u}{\partial x^2} + \frac{\partial^2 u}{\partial y^2}\right)$$

$$u\frac{\partial T}{\partial x} + v\frac{\partial T}{\partial y} = \alpha\left(\frac{\partial^2 T}{\partial x^2} + \frac{\partial^2 T}{\partial y^2}\right) + \frac{v}{c_p}\left(\frac{\partial u}{\partial y}\right)^2$$

a. Assume a low-speed, linear velocity profile (i.e., $u = u_{max}(1 - y/\ell)$ with a maximum velocity at $y = 0$ and zero velocity at $y \pm \ell$) between the two parallel plates, and also assume a thermally, fully developed condition. Write the simplified equations for the momentum and energy equations and the associated BCs that can be used for this problem.

b. Under the assumption in (a), determine the Nusselt number on the plate.

c. Consider a fully developed velocity profile (i.e., a parabolic velocity profile) between two parallel plates and a thermally, fully developed condition, and comment on whether the Nusselt number on the plate will be higher, the same, or lower than those of symmetry linear velocity profile in (a)? Explain why.

8.7 Consider an incompressible, laminar, 2-D flow in a parallel plate channel. The top plate is pulled at a constant velocity U_T. The top and bottom plates are maintained at a constant heat flux, q_w''. Flow is both hydrodynamically and thermally fully developed. Assume that the pressure gradient is zero in the parallel plate channel.

a. Obtain the differential equations governing the velocity $U(Y)$ and temperature $T(Y)$ fields.

b. Use appropriate BCs to evaluate $U(Y)$. Obtain $T(Y)$. Do not attempt to evaluate the constants of integration for the temperature field.

8.8 Find the Nusselt number for the following problems.

a. Fully developed Couette flow (i.e., assume that velocity and temperature profiles do not change along the channel) with the lower plane wall at a uniform wall temperature, T_0, and the upper plane wall at T_1. If the velocity profile is a linear profile ($U = 0$ at the lower plane wall, $U = V$ at the upper plane wall), find the temperature profile from the energy equation.

b. Fully developed Poiseuille flow (i.e., assume that velocity and temperature profiles do not change along the channel) with the lower plane wall at uniform temperature, T_0, and the upper plane wall at T_1. If the velocity profile is a parabolic profile, find the temperature profile from the energy equation.

8.9 A 2-D channel flow is subjected to a uniform heat flux on one wall and insulated on the other wall. Assume that the viscous dissipation is negligible and the properties are constant. Determine the following.

a. The governing momentum and energy equations with the appropriate BCs for the developing region. Do not solve the equations; however, show the details of the simplified governing equations. Sketch the temperature and velocity profiles with respect to y at two x positions.

b. Repeat (a) for the fully developed region.

c. Sketch the temperature profile in the fully developed region if both walls are insulated. Consider two cases: (1) viscous dissipation is negligible and (2) viscous dissipation is not negligible.

8.10 Consider liquid metal flow in a parallel-plate channel at a uniform wall heat flux condition.

a. Using the momentum and energy differential equations, and making the appropriate assumptions, derive the surface Nusselt number if flow is laminar and the temperature profile is in a fully developed condition.

b. Using the momentum and energy differential equations and making the appropriate assumptions, outline the methods (no need to solve) to determine the surface Nusselt number if the flow is laminar and the temperature profile is in a developing condition. Describe if the surface Nusselt numbers for (b) will be higher or lower than those for (a). Explain why.

c. Plot Nu, T_w, and T_b versus x from the entrance to the fully developed region.

8.11 Consider a steady, constant-property, laminar flow, between two parallel plates at $y = \pm \ell$. The plates are electrically heated to give a uniform wall heat flux. The differential equations for momentum and energy are listed here for reference:

$$u \frac{\partial u}{\partial x} + v \frac{\partial u}{\partial y} = -\frac{1}{\rho} \frac{\partial P}{\partial x} + v \left(\frac{\partial^2 u}{\partial x^2} + \frac{\partial^2 u}{\partial y^2} \right)$$

$$u \frac{\partial T}{\partial x} + v \frac{\partial T}{\partial y} = \alpha \left(\frac{\partial^2 T}{\partial x^2} + \frac{\partial^2 T}{\partial y^2} \right) + \frac{v}{c_p} \left(\frac{\partial u}{\partial y} \right)^2$$

a. What is the physical meaning of the last term shown in the above differential energy equation? Explain under what conditions the last term should be included in order to solve the temperature distribution between the two parallel plates.

b. Assume a low-speed, slug-flow velocity profile (i.e., a uniform velocity profile) between two parallel plates and also assume a thermally, fully developed condition. Write the simplified equations for momentum and energy and the associated BCs that can be used for this problem.

c. Under the assumption in (b), determine the Nusselt number on the plate.

8.12 Water at 43°C enters a 5-cm-ID pipe at a rate of 6 kg/s. If the pipe is 9 m long and maintained at 71°C, calculate the exit water temperature and the total heat transfer. Assume that the Nusselt number of the flow in the pipe is 3.657. Also, the thermal properties of water are $\rho = 1000 \, \text{kg/m}^3$, $c_p = 4.18 \, \text{kJ/(kg K)}$, and $k = 0.6 \, \text{W/(mK)}$.

8.13 Consider a given fluid (ρ, cp, k, μ) suddenly flowing into a channel formed by two pieces of parallel plates at a uniform wall temperature, T_w. The inlet velocity and temperature are V_o and T_o, respectively. Due to a sudden contraction inlet condition, the velocity V_o can be assumed as uniform through the entire parallel-plate channel. This is a laminar internal flow heat transfer problem with an abrupt contraction entry flow condition.

a. Outline the solution procedures to determine the local heat transfer coefficient distribution along the channel. You need to provide a detailed, step-by-step procedure, but you do not need to obtain the final solutions.

b. If the temperature is increasing linearly with the plate length, and if its axially averaged value equals that in case (a), plot T_w, T_b, and h_x versus x, and discuss the differences of T_w, T_b, and h_x between cases (a) and (b) in the same plot.

8.14 Fully developed laminar flow heat transfer in a tube with a nonuniform, wall heat flux is commonly seen in nuclear fuel rod designs. Given a wall heat flux similar to that shown in Figure 8.17, determine the local wall and bulk mean temperature distributions. These distributions are shown in Figure 8.17. Assume the tube wall heat flux varies axially as

$$q_w'' = q_{max}'' \sin\left(\frac{\pi x}{L}\right)$$

8.15 We have fully developed laminar flow heat transfer in a tube with a nonuniform wall temperature. Assume the tube wall temperature varies axially as

$$T_w = T_i + \left(\frac{a}{b}\right)\left[\exp\left(bx^+\right) - 1\right]$$

where a and b are arbitrary constants. Derive the local Nusselt number as a function of x^+.

REFERENCES

1. W. Rohsenow and H. Choi, *Heat, Mass, and Momentum Transfer*, Prentice-Hall, Inc., Englewood Cliffs, NJ, 1961.
2. F. Incropera and D. Dewitt, *Fundamentals of Heat and Mass Transfer*, Fifth Edition, John Wiley & Sons, New York, 2002.
3. W.M. Kays and M.E. Crawford, *Convective Heat and Mass Transfer*, Second Edition, McGraw-Hill, New York, 1980.
4. A. Mills, *Heat Transfer*, Richard D. Irwin, Inc., Boston, MA, 1992.
5. S.C. Lau, *MEEN 619-Conduction and Radiation, Lecture Notes*, Texas A&M University, 2010.

9 Natural Convection

9.1 LAMINAR NATURAL CONVECTION ON A VERTICAL WALL: SIMILARITY SOLUTION

Natural convection can occur when the solid surface temperature is different from the surrounding fluid. For example, natural convection can take place between a heated (or cooled) vertical (or horizontal) plate or tube and the surrounding fluid. Figure 9.1 shows the velocity profile, temperature profile, and heat transfer from a hot vertical wall to a cold fluid due to natural convection. The hot vertical wall conducts heat to the fluid particle (fluid layer) next to the wall and the heated fluid particle (fluid layer) conducts heat to the next cooler fluid particle, and so on. Therefore, the fluid particle near the hot wall is lighter than that is away from the hot wall and natural circulation takes place (near the wall, the hot fluid moving up and away from the wall, cold fluid moving down) due to gravity. This buoyancy-driven natural convection flow is primarily due to density gradient (temperature gradient) from the hot vertical wall and cold surrounding fluid. The key parameter/driving force to determine natural convection is Grashof number, a ratio of buoyancy force to viscous force (buoyancy force tries to move the fluid up but viscous force tries to resist it from moving). Another parameter is Prandtl number, a fluid property showing the ratio of kinematic viscosity to thermal diffusivity. The product of Grashof number with Prandtl number is called Rayleigh number, another way of measuring the natural convection.

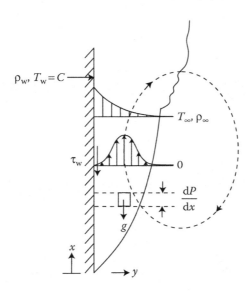

FIGURE 9.1 Natural convection boundary layer from a heated vertical wall.

DOI: 10.1201/9781003164487-9

The following shows the definition of Grashof number, Rayleigh number, and 2-D laminar natural convection boundary-layer equations from a heated vertical wall [1–4].

$$Gr_x = \frac{g\beta(T_w - T_\infty)x^3}{v^2} \tag{9.1}$$

$$Ra_x = Gr_x \cdot Pr \tag{9.2}$$

Continuity

$$\frac{\partial u}{\partial x} + \frac{\partial v}{\partial y} = 0 \tag{9.3}$$

Momentum

$$u\frac{\partial u}{\partial x} + v\frac{\partial v}{\partial y} = g\beta(T - T_\infty) + v\frac{\partial^2 u}{\partial y^2} \tag{9.4}$$

Energy

$$u\frac{\partial T}{\partial x} + v\frac{\partial T}{\partial y} = \alpha\frac{\partial^2 T}{\partial y^2} \tag{9.5}$$

The following shows how to derive the above natural convection momentum equation from the original x-momentum equation:

$$u\frac{\partial u}{\partial x} + v\frac{\partial u}{\partial y} = -\frac{1}{\rho}\frac{\partial P}{\partial x} - g + v\frac{\partial^2 u}{\partial y^2}$$

From outside of the boundary layer $0 = -(1/\rho_\infty)(\partial P/\partial x) - g$

$$\because \beta = -\frac{1}{\rho}\left(\frac{\partial \rho}{\partial T}\right)_P \cong -\frac{1}{\rho}\frac{\rho_\infty - \rho}{T_\infty - T}$$

$$\therefore \rho_\infty - \rho = \rho\beta(T - T_\infty)$$

$$\therefore \frac{\partial P}{\partial x} = -\rho_\infty g$$

$$-\frac{1}{\rho}\frac{\partial P}{\partial x} - g = -\frac{1}{\rho}(-\rho_\infty g) - g = \frac{g}{\rho}(\rho_\infty - \rho) = -g\beta(T_\infty - T)$$

From the above momentum equation, one can see that natural convection is due to temperature difference between the surface and fluid and the gravity force. This implies that there is no natural convection if there exists no temperature gradient or no gravity force. The larger delta T and gravity (means larger Grashof number) will cause larger natural circulation and results in thinner boundary-layer thickness and higher friction (shear) and higher heat transfer coefficient. The Grashof number in natural convection plays a similar role as Reynolds number does in forced convection; the larger Grashof number causes higher heat transfer in natural convection as the greater Reynolds number has higher heat transfer in forced convection. Prandtl number plays the same role in both natural and forced convection, basically the fluid property.

Just like in forced convection, both similarity and integral methods can be used to solve natural convection boundary-layer equations. The following only outlines the similarity method from Ostrach in 1953 [5].

Similarity variable:

$$\eta = y\left(\frac{g\beta(T_w - T_\infty)}{4v^2 x}\right)^{1/4} = \frac{y}{x}\left(\frac{\mathrm{Gr}_x}{4}\right)^{1/4} \tag{9.6}$$

where

$$\mathrm{Gr}_x = \frac{g\beta(T_w - T_\infty)x^3}{v^2}$$

Similarity functions for velocity and temperature:

$$f(\eta) = \frac{\Psi(x,\eta)}{4v(\mathrm{Gr}_x/4)^{1/4}} \tag{9.7}$$

$$\theta = \frac{T - T_\infty}{T_w - T_\infty} \tag{9.8}$$

Put them into the above momentum equation and energy equation, respectively:

$$u = \frac{\partial \Psi}{\partial y} = \cdots$$

$$v = \frac{\partial \Psi}{\partial x} = \cdots$$

$$\frac{\partial T}{\partial x} = \cdots$$

$$\frac{\partial T}{\partial y} = \cdots$$

The resultant similarity momentum and energy equations are

$$f''' + 3ff'' - 2f'^2 + \theta = 0 \tag{9.9}$$

$$\theta'' + 3\Pr f\theta' = 0 \tag{9.10}$$

The related BCs are

$$f(0) = f'(0) = 0, \quad f'(\infty) = 0 \tag{9.11}$$

$$\theta(0) = 1, \quad \theta(\infty) = 0 \tag{9.12}$$

These can be solved by the fourth-order Runge–Kutta method in order to obtain the velocity and temperature profiles across the natural convection boundary layer. Figure 9.2 shows typical dimensionless velocity and temperature profiles from the heated vertical wall, for different Prandtl fluids.

$$\begin{aligned} f' &= \frac{u}{2\sqrt{gx}}\sqrt{\frac{T_\infty}{T_w - T_\infty}} \\ &= \frac{ux}{2\nu}\,\mathrm{Gr}_x^{-1/2} \\ &= \frac{u}{\left(\mathrm{Gr}_x^{1/2}(2\nu/x)\right)} \end{aligned} \tag{9.13}$$

From the dimensionless velocity and temperature profiles, the associated heat flux and the heat transfer coefficient (or Nusselt number) can be determined.

$$q_w'' = -k\frac{\partial T}{\partial y}\Big|_0 = -\frac{k}{x}(T_w - T_\infty)\left(\frac{\mathrm{Gr}_x}{4}\right)^{1/4}\frac{d\theta}{d\eta}\Big|_{y=0} \tag{9.14}$$

where

$$\frac{d\theta}{d\eta}\Big|_{y=0} = \theta'(0) = f(\Pr)$$

$$\therefore h = \frac{q_w''}{T_w - T_\infty} = \frac{-k(\partial T/\partial y)_0}{T_w - T_\infty}$$

$$\mathrm{Nu} = \frac{hx}{k} = -\left(\frac{\mathrm{Gr}_x}{4}\right)^{1/4}\frac{d\theta}{d\eta}\Big|_{y=0} \tag{9.15}$$

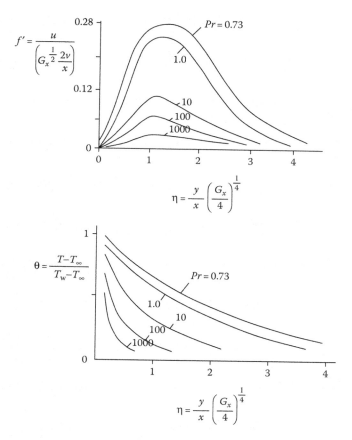

FIGURE 9.2 Dimensionless velocity and temperature profiles from heated vertical wall.

Numerical results: for laminar natural convection:

Pr	0.01	0.733	1	2	10	100	1000
$\dfrac{d\theta}{d\eta}\Big|_{y=0}$	0.081	0.508	0.567	0.716	1.169	2.191	3.966

For air, Pr = 0.733,

$$\mathrm{Nu}_x = \frac{hx}{k} = 0.359\mathrm{Gr}_x^{1/4} \tag{9.16}$$

$$\overline{\mathrm{Nu}_x} = \frac{\overline{h_x}L}{k} = \frac{4}{3}\mathrm{Nu}_L = 0.478\mathrm{Gr}_L^{1/4} \tag{9.17}$$

where

$$\bar{h}_x = \frac{1}{L}\int_0^L h_x \, dx = -\frac{4}{3}\frac{k}{L}\left(\frac{Gr_L}{4}\right)^{1/4}\frac{d\theta}{d\eta}\bigg|_{y=0} = \frac{4}{3}h_L$$

$$Gr_L = \frac{g\beta(T_w - T_\infty)L^3}{\nu^2} \tag{9.18}$$

$$Nu_x = \frac{3}{4}\left[\frac{2\,Pr}{5\left(1 + 2\,Pr^{1/2} + 2\,Pr\right)}\right]^{1/4}(Gr_x\,Pr)^{1/4} \tag{9.19}$$

$$Nu_x = 0.6\left(Gr_x\,Pr^2\right)^{1/4}, \quad \text{if } Pr \to 0 \tag{9.20}$$

$$Nu_x = 0.503(Gr_x\,Pr)^{1/4}, \quad \text{if } Pr \to \infty \tag{9.21}$$

In general,

$$Nu_x = a(Gr_x Pr)^b = a Ra_x^b$$

where Rayleigh number,

$$Gr_x Pr = Ra_x = \frac{g\beta(T_w - T_\infty)x^3}{\nu\alpha}$$

Compared to forced convection

$$Nu_x = a Re_x^m Pr^n$$

Note: The following is a simple guideline whether the problem can be solved by forced convection, natural convection, or mixed (combined forced and natural) convection.

If $Gr_x/Re_x^2 < 1$, the problem can be treated as forced convection.
If $Gr_x/Re_x^2 \cong 1$, the problem can be treated as mixed convection.
If $Gr_x/Re_x^2 > 1$, the problem can be treated as natural convection.

9.2 SIMILARITY SOLUTION FOR VARIABLE WALL TEMPERATURE

Let

$$T_w - T_\infty = Ax^n$$

where

$$n = \text{constant} = C$$

Follow the same procedures as outlined in 9.1, the resulting similarity momentum and energy equations are

$$f''' + (n+3)ff'' - (2n+2)f'^2 + \theta = 0$$

$$\theta'' + \Pr[(n+3)f\theta'] - 4nf'\theta = 0$$

If $n = 0$; $T_w = T_\infty + A = C$, a special case for constant wall temperature
  Boundary Conditions:

$$f(0) = f'(0) = 0, \quad f'(\infty) = 0$$
$$\theta(0) = 1, \quad \theta(\infty) = 0$$

These equations can be solved by the fourth-order Runge-Kutta method in order to obtain the velocity and temperature profiles across the natural convection boundary layer. From these dimensionless velocity and temperature profiles, the associated heat flux and the heat transfer coefficient or Nusselt number can be obtained.

$$q_w'' = -k\frac{\partial T}{\partial y}\bigg|_0 = \cdots = -\frac{k}{x}(T_w - T_\infty)\frac{\theta'(0)}{\sqrt{2}}\frac{1}{x}(Gr_x)^{1/4}$$

$$= k\left(\frac{g\beta}{v^2}\right)^{1/4}\left(\frac{-\theta'(0)}{\sqrt{2}}\right)(T_w - T_\infty)^{5/4}\frac{x^{3/4}}{x} = C(T_w - T_\infty)^{5/4} \cdot x^{-1/4}$$

$$\frac{h_x x}{k} = \frac{x}{k} \cdot \frac{q_w''}{T_w - T_\infty}$$

If $T_w - T_\infty = x^{1/5}$ then $q_w'' = \text{constant}$, a special case for constant wall heat flux

$$\mathrm{Nu}_x = \left(\frac{\Pr}{4 + 9\Pr^{1/2} + 10\Pr}\right)^{1/5} \cdot [Gr_x \cdot \mathrm{Nu}_x \cdot \Pr]^{1/5}$$

Note: $\mathrm{Nu}_x\left(\text{for } q_w'' = C\right) \cong 1.15\,\mathrm{Nu}_x\left(\text{for } T_w = C\right)$ for laminar natural convection
  But $\mathrm{Nu}_x\left(\text{for } q_w'' = C\right) \cong 1.36\,\mathrm{Nu}_x\left(\text{for } T_w = C\right)$ for laminar forced convection

9.3 LAMINAR NATURAL CONVECTION ON A VERTICAL WALL: INTEGRAL METHOD

Integral approximate solution by Pohlhausen 1921: We apply the momentum and energy balance to the control volume across the boundary layer as shown in Figure 9.3.

Net force = Momentum change

$$-\tau_w - \int \rho g\, dy - \int \frac{\partial P}{\partial x}dy = \frac{\partial}{\partial x}\int \rho uu\, dy$$

where $\partial P/\partial x = \partial P_\infty/\partial x = -\rho_\infty g$ (outside of the boundary layer).

Also $\rho - \rho_\infty = -\rho_\infty \beta(T - T_\infty)\beta$ from $\beta \equiv -1/\rho(\partial \rho/\partial T)_p$, assume $\rho \cong \rho_\infty$

$$\frac{\tau_w}{\rho_\infty} = \frac{\mu}{\rho_\infty}\frac{\partial u}{\partial y}|_0 = g\beta\int_0^\delta (T - T_\infty)dy - \frac{d}{dx}\int_0^\delta u^2 dy \qquad (9.22)$$

$$q'' = -k\frac{\partial T}{\partial y}|_0 = \frac{d}{dx}\int_0^{\delta_T}\rho c_p u(T - T_\infty)dy \qquad (9.23)$$

Boundary Conditions:

$$u(x,0) = 0 \qquad T(x,0) = T_w$$
$$u(x,\delta) = 0 \qquad T(x,\delta_T) = T_\infty$$
$$\frac{\partial u(x,\delta)}{\partial y} = 0 \qquad \frac{\partial T(x,\delta_T)}{\partial y} = 0$$

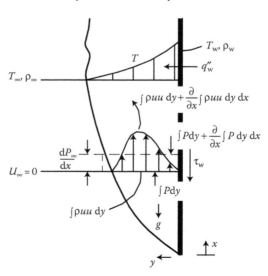

FIGURE 9.3 Integral method.

Assuming velocity and temperature profiles to satisfy boundary-layer conditions,

$$u = u(x,y) = u_1 \eta (1 - \eta)^2 \tag{9.24}$$

$$\frac{T - T_\infty}{T_w - T_\infty} = (1 - \eta)^2 = f(x,y) \tag{9.25}$$

where $\eta = y/\delta(x)$; $u_1 = c_1 x^m$; $\delta(x) = c_2 x^n$;

$$m = \frac{1}{2}; \quad n = \frac{1}{4}$$

Put this into momentum and energy integral equations:

$$u_1(x) = \left(\frac{80}{3}\right)^{1/2} \left(\frac{Gr_x}{(20/21) + \text{Pr}}\right)^{1/2} \frac{\nu}{x}$$

$$\frac{\delta(x)}{x} = \left[\frac{240(1 + (20/21\,\text{Pr}))}{\text{Pr} \cdot Gr_x}\right]^{1/4} \rightarrow \delta(x) \sim \left(\frac{x}{T_w - T_\infty}\right)^{1/4}$$

$$q_w = -k \frac{\partial T}{\partial y}\Big|_0 = \frac{2k(T_w - T_\infty)}{\delta(x)} = h(T_w - T_\infty) \tag{9.26}$$

$$\text{Nu}_x = \frac{h_x x}{k} = \frac{2x}{\delta(x)} = \left[\frac{\text{Pr}}{15((20/21) + \text{Pr})}\right]^{1/4} \cdot \text{Ra}_x^{1/4}$$

$$\text{Nu}_x \cong 0.413 \text{Ra}_x^{1/4} \quad \text{for} \quad \text{Pr} = 0.733 \tag{9.27}$$

Note: $\text{Nu}_x = 0.359 \text{Ra}_x^{1/4}$ for $\text{Pr} = 0.7333$ by using the exact similarity solution.

$$\text{Ra}_x = \frac{g\beta(T_w - T_\infty)x^3}{\nu\alpha} < 10^8 - 10^9 - \text{Laminar natural convection}$$

REMARKS

In the undergraduate-level heat transfer, we have heat transfer correlations of external natural convection for a vertical plate, an inclined plate, a horizontal plate, a vertical tube, and a horizontal tube as well as heat transfer correlations of internal natural convection for a horizontal tube, between two parallel plates, and inside a rectangular cavity with various aspect ratios. These correlations are important for many real-life engineering applications such as electronic components.

In the intermediate-level heat transfer, this chapter focuses on how to analytically solve the external natural convection from a vertical plate at a uniform surface temperature by using the similarity method as well as the integral method. In advanced heat transfer, these methods can be modified and extended to solve mixed convection (combined natural and forced convection) problems for vertical, horizontal, and inclined plates or tubes, respectively, for various Prandtl number fluids.

9.4   LAMINAR MIXED CONVECTION— NONSIMILARITY TRANSFORMATION

Many engineering applications involve laminar mixed convection. Mixed convection happens when Grashof number (natural convection) divided by Reynolds number square (forced convection) is around one. The mixed convection can happen along a vertical or horizontal flat plate. Both natural convection and forced convection make contributions to the surface heat transfer. The natural convection can assist or oppose the forced convection depending on the thermal and flow boundary conditions. Similarly, the forced convection can enhance or reduce the natural convection depending upon the flow and thermal boundary conditions. As mentioned before, either forced convection or natural convection boundary layer can be solved by the similarity method. But, the mixed convection boundary layer must be solved by non-similarity method. Velocity and temperature profiles can be solved by nonsimilarity transformation as outlined below. Refer to Chen et al. papers for the details [6–9].

9.4.1   MIXED CONVECTION ALONG A VERTICAL FLAT PLATE

Case 1—Forced Convection Dominated Regime to Study Buoyancy Effect on Forced Convection.

There are four possibilities, as shown in Figure 9.4.

1. Forced freestream flow moves along the vertical plate from bottom to top with wall temperature hotter than the freestream flow. In this case, velocity inside the boundary is assisted by the buoyancy force, both moving in the upward direction, thus, shear stress and heat transfer increase due to greater velocity and temperature gradients at surface.
2. Forced freestream flow moves along the vertical plate from bottom to top with wall temperature colder than the freestream flow. In this case, velocity inside the boundary is opposed by the buoyancy force, both moving in opposite direction, thus, shear stress and heat transfer decrease due to smaller velocity and temperature gradients at surface.
3. Forced freestream flow moves along the vertical plate from top to bottom with wall temperature hotter than the freestream flow. In this case, velocity inside the boundary is opposed by the buoyancy force, both moving in opposite direction, thus, shear stress and heat transfer decrease due to smaller velocity and temperature gradients at surface.

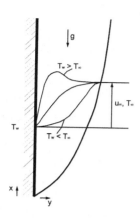
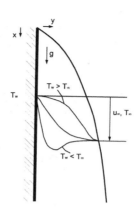

Upward Forced Flow

Downward Forced Flow

$T_w > T_\infty$ Buoyancy assists forced flow $T_w > T_\infty$ Buoyancy opposes forced flow

$T_w < T_\infty$ Buoyancy opposes forced flow $T_w < T_\infty$ Buoyancy assists forced flow

FIGURE 9.4 Forced convection along a vertical plate with buoyancy assists and buoyancy opposes forced flow.

4. Forced freestream flow moves along the vertical plate from top to bottom with wall temperature colder than the freestream flow. In this case, velocity inside the boundary is assisted by the buoyancy force, both moving downward direction, thus, shear stress and heat transfer increase due to greater velocity and temperature gradients at surface.

Conservation Equations   Boundary Conditions

$$\frac{\partial u}{\partial x} + \frac{\partial v}{\partial y} = 0$$

$$y = 0: u = v = 0, T = T_w$$

$$u\frac{\partial u}{\partial x} + v\frac{\partial u}{\partial y} = v\frac{\partial^2 u}{\partial y^2} \pm g\beta(T - T_\infty)$$

$$y \to \infty: u \to \mathcal{U}_\infty, T \to T_\infty$$

$$u\frac{\partial T}{\partial x} + v\frac{\partial T}{\partial y} = \alpha\frac{\partial^2 T}{\partial y^2}$$

$$+g\beta(T - T_\infty) \quad : \text{upward forced flow} (-\rho g)$$

$$-g\beta(T - T_\infty) \quad : \text{downward forced flow} (+\rho g)$$

Let $\eta = y\sqrt{\dfrac{\mathcal{U}_\infty}{vx}} = \dfrac{y}{x}\sqrt{\text{Re}_x}$

$\xi = \xi(x)$, nonsimilarity parameter to be determined

$$\psi(x, y) = \sqrt{v\mathcal{U}_\infty x} \cdot f(\xi, \eta) = v\sqrt{\text{Re}_x} \cdot f(\xi, \eta)$$

$$\theta(\xi,\eta) = \frac{T(x,y) - T_\infty}{T_w - T_\infty}$$

$$u = \frac{\partial \psi}{\partial y}, \quad v = -\frac{\partial \psi}{\partial x}$$

Follow the same procedures as outlined in 9.1; the transformation of momentum and energy equations leads to:

$$f''' + \frac{1}{2} ff'' \pm \frac{g\beta(T_w - T_\infty)x}{\mathcal{U}_\infty^2} \theta = x \frac{d\xi}{dx}\left[f' \frac{\partial f'}{\partial \xi} - f'' \frac{\partial f}{\partial \xi} \right]$$

$$\frac{\theta''}{Pr} + \frac{1}{2} f\theta' = x \frac{d\xi}{dx}\left[f' \frac{\partial \theta}{\partial \xi} - \theta' \frac{\partial f}{\partial \xi} \right]$$

$$f'(\xi,0) = 0, \quad f(\xi,0) + 2x \frac{d\xi}{dx}\frac{\partial f}{\partial \xi}(\xi,0) = 0, \quad f'(\xi,\infty) = 1$$

$$\theta(\xi,0) = 1, \quad \xi(\xi,\infty) = 0$$

where the primes stand for partial derivatives with respect to η

$$\text{Now } \xi(x) = \frac{g\beta(T_w - T_\infty)x}{\mathcal{U}_\infty^2} = \frac{g\beta(T_w - T_\infty)x^3}{v^2}\frac{v^2}{\mathcal{U}_\infty^2 x^2} = \frac{Gr_x}{Re_x^2}$$

$$\xi = ax, \quad x\frac{d\xi}{dx} = \xi \quad \xi(x) = \frac{Gr_x}{Re_x^2}; \text{ buoyancy force parameter}$$

$\xi = 0$; pure forced convection
$\xi = \infty$; pure free convection
   Case 2—Free Convection Dominated Regime to Study Forced Flow Effect on Free Convection.
   Consider the following two possibilities, as shown in Figure 9.5:

1. Free convection moves along the vertical plate from bottom to top with wall temperature hotter than the freestream flow which moves upward. In this case, velocity inside the boundary is assisted by the forced flow, both moving in the upward direction, thus, shear stress and heat transfer increase due to greater velocity and temperature gradients at surface.
2. Free convection moves along the vertical plate from top to bottom with wall temperature colder than the freestream flow which moves downward. In this case, velocity inside the boundary is assisted by the forced flow, both moving in downward direction, thus, shear stress and heat transfer increase due to greater velocity and temperature gradients at surface.

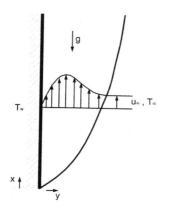 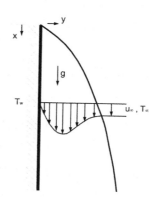

Upward Flow: $T_W > T_\infty$ Downward Flow: $T_w < T_\infty$

FIGURE 9.5   Free convection along a vertical plate with forced flow effect on free convection.

Conservation Equations	Boundary Conditions
$$\frac{\partial u}{\partial x} + \frac{\partial v}{\partial y} = 0$$	$y = 0 : u = v = 0, T = T_w$
$$u\frac{\partial u}{\partial x} + v\frac{\partial u}{\partial y} = v\frac{\partial^2 u}{\partial y^2} + g\beta(T - T_\infty)$$	$y \to \infty : u \to \mathcal{U}_\infty, T \to T_\infty$
$$u\frac{\partial T}{\partial x} + v\frac{\partial T}{\partial y} = \alpha\frac{\partial^2 T}{\partial y^2}$$	

Let $\eta = \dfrac{y}{x}\left(\dfrac{Gr_x}{4}\right)^{1/4} = cyx^{-1/4}, \quad c = \left[\dfrac{g\beta(T_w - T_\infty)}{4v^2}\right]^{1/4}$

$\xi = \xi(x)$, nonsimilarity parameter to be determined

$$\psi(x,y) = 4v\left(\frac{Gr_x}{4}\right)^{1/4} \cdot f(\xi,\eta) = 4vcx^{3/4} \cdot f(\xi,\eta)$$

$$\theta(\xi,\eta) = \frac{T(x,y) - T_\infty}{T_w - T_\infty}$$

$$u = \frac{\partial \psi}{\partial y}, \quad v = -\frac{\partial \psi}{\partial x}, \quad Gr_x = \frac{g\beta(T_w - T_\infty)x^3}{v^2}$$

Follow the same procedures as outlined in 9.1, the transformation of momentum and energy equations leads to:

$$f''' + 3ff' - 2(f')^2 + \theta = 4x\frac{d\xi}{dx}\left[f'\frac{\partial f'}{\partial \xi} - f''\frac{\partial f}{\partial \xi}\right]$$

$$\frac{\theta''}{\text{Pr}} + 3f\theta' = 4x\frac{d\xi}{dx}\left[f'\frac{\partial \theta}{\partial \xi} - \theta'\frac{\partial f}{\partial \xi}\right]$$

$$f'(\xi,0) = 0, \quad 3f(\xi,0) + 4x\frac{d\xi}{dx}\frac{\partial f}{\partial \xi}(\xi,0) = 0, \quad f'(\xi,\infty) = \frac{\mathcal{U}_\infty}{4vc^2x^{1/4}}$$

$$\theta(\xi,0) = 1, \quad \xi(\xi,\infty) = 0$$

Now $\xi(x) = \dfrac{\mathcal{U}_\infty}{4vc^2x^{1/2}} = \dfrac{\mathcal{U}_\infty}{4v\left[\dfrac{g\beta(T_w - T_\infty)}{4v^2}\right]^{1/2}x^{1/2}} = \dfrac{1}{4}\dfrac{\mathcal{U}_\infty x/v}{\left[\dfrac{g\beta(T_w - T_\infty)x^3}{4v^2}\right]^{1/2}} = \dfrac{1}{2}\dfrac{\text{Re}_x}{\text{Gr}_x^{1/2}}$

$$\xi = bx^{-1/2}, \quad x\frac{d\xi}{dx} = -\frac{1}{2}\xi$$

$\xi(x) = \dfrac{1}{2}\dfrac{\text{Re}_x}{\text{Gr}_x^{1/2}}$, forced flow parameter

$\xi = 0$; pure free convection

$\xi = \infty$; pure forced convection

9.4.2 MIXED CONVECTION OVER A HORIZONTAL FLAT PLATE

Free Convection Dominated Regime to Study Forced Flow Effect on Free Convection. Consider the following two possibilities, as shown in Figure 9.6:

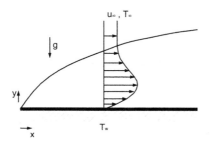
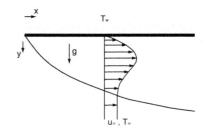

Flow above the plate: $T_w > T_\infty$,   Flow below the plate: $T_w < T_\infty$

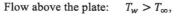

FIGURE 9.6   Free convection on a horizontal plate with forced flow effect on free convection.

1. Forced flow moves above the plate from left to right, the plate is hotter than the freestream. Both forced and natural flows move in the same direction to enhance heat transfer.
2. Forced flow moves below the plate from left to right, the plate is colder than the freestream. Both forced and natural flows move in the same direction to enhance heat transfer.

Conservation Equations   Boundary Conditions

$$\frac{\partial u}{\partial x} + \frac{\partial v}{\partial y} = 0$$ $y = 0 : u = v = 0, T = T_w$

$$u\frac{\partial u}{\partial x} + v\frac{\partial u}{\partial y} = v\frac{\partial^2 u}{\partial y^2} + g\beta \int_y^\infty (T - T_\infty)dy$$ $y \to \infty : u \to \mathcal{U}_\infty, T \to T_\infty$

$$u\frac{\partial T}{\partial x} + v\frac{\partial T}{\partial y} = \alpha\frac{\partial^2 T}{\partial y^2}$$

Let $\eta = \frac{y}{x}\left(\frac{Gr_x}{5}\right)^{\frac{1}{5}}, \quad Gr_x = \frac{g\beta(T_w - T_\infty)x^3}{v^2}$

$\xi = \xi(x)$, nonsimilarity parameter to be determined

$$\psi(x,y) = 5v\left(\frac{Gr_x}{5}\right)^{1/5} \cdot f(\xi,\eta)$$

$$\theta(\xi,\eta) = \frac{T(x,y) - T_\infty}{T_w - T_\infty}$$

$$u = \frac{\partial \psi}{\partial y}, \quad v = -\frac{\partial \psi}{\partial x},$$

Follow the same procedures as outlined in 9.1; the transformation of momentum and energy equations leads to:

$$f''' + 3ff'' - (f')^2 + \frac{2}{5}\left[\eta\theta - \int_\eta^\infty \theta d\eta + \frac{5}{2}x\frac{d\xi}{dx}\int_\eta^\infty \frac{\partial \theta}{\partial \xi}d\eta\right] = 5x\frac{d\xi}{dx}\left[f'\frac{\partial f'}{\partial \xi} - f''\frac{\partial f}{\partial \xi}\right]$$

$$\frac{\theta''}{Pr} + 3f\theta' = 5x\frac{d\xi}{dx}\left[f'\frac{\partial \theta}{\partial \xi} - \theta'\frac{\partial f}{\partial \xi}\right]$$

$$f'(\xi,0) = 0, \quad 3f(\xi,0) + 5x\frac{d\xi}{dx}\frac{\partial f}{\partial \xi}(\xi,0) = 0$$

$$f'(\xi,\infty) = \frac{\mathcal{U}_\infty}{5v\left[\dfrac{g\beta(T_w - T_\infty)}{5v}\right]^{2/5} x^{1/5}} = \frac{1}{5}\frac{\text{Re}_x}{\left(\dfrac{\text{Gr}_x}{5}\right)^{2/5}}$$

$$\theta(\xi,0) = 1, \quad \xi(\xi,\infty) = 0$$

Now $\xi(x) = \dfrac{1}{5}\dfrac{\text{Re}_x}{\left(\dfrac{\text{Gr}_x}{5}\right)^{2/5}}$, forced flow parameter

$$\xi = bx^{-1/5}, \quad x\frac{d\xi}{dx} = -\frac{1}{5}\xi$$

$\xi = 0$; pure free convection

$\xi = \infty$; pure forced convection

REMARKS

The above analysis shows an important nonsimilarity parameter $\xi = \xi(x)$ that needs to be determined for mixed convection cases. Heat transfer correlations for laminar, mixed convection flows on vertical, inclined, and horizontal flat plates are available, such as $\text{Nu}_x/\text{Re}_x^{1/2}$ vs. $\text{Gr}_x/\text{Re}_x^2$ for the entire mixed convection region, for both upward and downward cases, for a given Prandtl number fluid. These extensive, mixed convection, heat transfer correlations are completed with thermal-flow parameters. Refer to Chen et al. [9] for the details.

PROBLEMS

9.1   Consider the system of boundary-layer equations

$$\frac{\partial u}{\partial x} + \frac{\partial v}{\partial y} = 0$$

$$u\frac{\partial u}{\partial x} + v\frac{\partial v}{\partial y} = R(T - T_\infty) + v\frac{\partial^2 u}{\partial y^2}$$

$$u\frac{\partial(T - T_\infty)}{\partial x} + v\frac{\partial(T - T_\infty)}{\partial y} = \alpha\frac{\partial^2(T - T_\infty)}{\partial y^2}$$

subject to the BCs

$$y = 0 \quad u = 0 \quad v = 0 \quad q_w = \text{constant}$$
$$y = \infty \quad u = 0 \qquad\qquad T = T_\infty$$

where the quantities R, v, and α are constants. Determine the similarity variables that will transform the equations to two ODEs. Derive the resultant ODEs.

9.2 Consider a natural convection flow over a vertical heated plate at a uniform wall temperature T_o Let T_∞ be the free-stream temperature. The following correlation holds for the heat transfer coefficient h_x at height x:

$$\frac{h_x x}{k} = 0.443(Gr_x Pr)^{1/4}$$

The Grashof number is

$$Gr_x = g\beta x^3 (T(x) - T_\infty)/v^2$$

where g is the acceleration due to gravity, β is the coefficient of thermal expansion, and v is the kinematic viscosity.

a. Draw a diagram of the system.

b. Sketch a plot of h_x along the plate length.

c. Find the average heat transfer coefficient.

d. Show that the average heat transfer coefficient between heights 0 and L is given by $\overline{Nu} = 0.59(Gr_L Pr)^{1/4}$.

9.3 Derive similarity momentum and energy equations shown in Equations (9.9) and (9.10).

9.4 Derive Equations (9.16) and (9.17).

9.5 Derive Equations (9.20) and (9.21).

9.6 Derive Equations (9.22) and (9.23).

9.7 Derive Equations (9.26) and (9.27).

9.8 Consider natural convection along a vertical wall with uniform heat flux boundary condition and determine local heat transfer coefficient distribution on the wall.

9.9 Consider natural convection over the outer surface of a cooled inclined plate at T_w in a heating fluid at T_∞. The angle between the inclined plate and vertical direction is α.

a. Describe and explain whether the boundary layer thickness and the heat transfer coefficient will increase or decrease with an increase in $(T_\infty - T_w)$ for a given Prandtl number fluid. Consider upward and downward, forced flow in addition to the natural convection on the inclined plate, sketch velocity, and temperature profiles, respectively, and explain whether the heat transfer coefficient will increase or decrease compared with that the natural convection only.

b. Sketch $Nu_x/Re_x^{1/2}$ vs. Gr_x/Re_x^2 for the entire mixed convection region, for both upward and downward cases, for a given Prandtl number fluid [9].

9.10 Consider natural convection over the outer surface of a heated inclined plate at T_w in a cooling fluid at T_∞. The angle between the inclined plate and vertical direction is α.

a. Describe and explain whether the boundary layer thickness and the heat transfer coefficient will increase or decrease with an increase in $(T_w - T_\infty)$ for a given Prandtl number fluid. Consider upward and downward, forced flow in addition to the natural convection on the inclined plate, sketch the velocity and temperature profiles, respectively, and explain whether the heat transfer coefficient will increase or decrease compared with that the natural convection only.

b. Sketch $Nu_x/Re_x^{1/2}$ vs. Gr_x/Re_x^2 for the entire mixed convection region, for both upward and downward cases, for a given Prandtl number fluid [9].

REFERENCES

1. W. Rohsenow and H. Choi, *Heat, Mass, and Momentum Transfer*, Prentice-Hall, Inc., Englewood Cliffs, NJ, 1961.
2. F. Incropera and D. Dewitt, *Fundamentals of Heat and Mass Transfer*, Fifth Edition, John Wiley & Sons, New York, 2002.
3. W.M. Kays and M.E. Crawford, *Convective Heat and Mass Transfer*, Second Edition, McGraw-Hill, New York, 1980.
4. A. Mills, *Heat Transfer*, Richard D. Irwin, Inc., Boston, MA, 1992.
5. S. Ostrach, "An Analysis of Laminar Free-Convection Flow and Heat Transfer about a Flat Plate Parallel to the direction of the Generating Body Force," NACA Report No. 1111, 1953.
6. T.S. Chen, "Analysis of Mixed Convection in External Flows," Lecture Notes, University of Missouri – Rolla, 1990.
7. N. Ramachandran, B.F., Armaly, and T.S. Chen, "Measurements and predictions of laminar mixed convection flow adjacent to a vertical surface," *ASME Journal of Heat Transfer*, Vol. 107, pp. 636–641, 1985.
8. T.S. Chen, E.M. Sparrow, and A. Mucoglu, "Mixed convection in boundary layer flow on a horizontal plate," *ASME Journal of Heat Transfer*, Vol. 99, pp. 66–71, 1977.
9. T.S. Chen, B.F. Armaly, and N. Ramachandran, "Correlations for laminar mixed convection flows on vertical, inclined, and horizontal flat plates," *ASME Journal of Heat Transfer*, Vol. 108, pp. 835–840, 1986.

10 Turbulent Flow Heat Transfer

10.1 REYNOLDS-AVERAGED NAVIER–STOKES (RANS) EQUATIONS

When flow transitions into turbulence, both shear stress and heat transfer from the surface increase due to turbulent mixing. However, in a fully turbulent region, both shear stress and heat transfer slightly decrease again due to the turbulent boundary-layer thickness growing along the surface. In this section, we discuss external and internal flow and heat transfer problems in a fully turbulent region. Refer to Chapter 15 for transitional flow heat transfer. Figure 10.1 shows a sketch of a typical 2-D turbulent boundary layer for heated flow over a cooled flat surface and the fairly uniform velocity and temperature profiles across the boundary layer due to turbulent mixing. A laminar sublayer is developed at a very-near-wall region where turbulent mixing is damped due to viscous effect. This laminar sublayer thickness is the major resistance for velocity and temperature change from the free-stream value to the wall. In a fully turbulent region, velocity and temperature change with time at a given location inside the turbulent boundary layer, that is, $u(x, y, t)$, $v(x, y, t)$, $T(x, y, t)$. These time- and location-dependent behaviors make turbulent flow and heat transfer more difficult to analyze as compared to laminar boundary-layer flow and heat transfer problems.

The following is to show how to obtain the Reynolds-averaged Navier–Stokes (RANS) equations for a fully turbulent boundary-layer flow [1–6]. The idea is to treat fully turbulent flow as purely random motion superimposed on a steady (time-averaged) mean flow. In other words, the time-dependent value (such as instantaneous velocity and temperature, etc.) equals the time-averaged value (Reynolds-averaged value) plus the fluctuation value (due to random motion). For example, the steady (time-averaged) x-direction velocity over a period of time can be shown as

$$\bar{u} = \lim_{t \to \infty} \frac{1}{t} \int_0^t u(t)\,dt$$

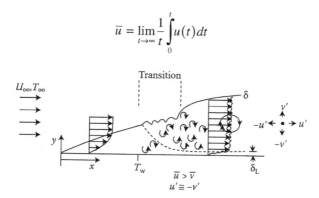

FIGURE 10.1   Typical turbulent boundary-layer flow velocity and temperature profile.

DOI: 10.1201/9781003164487-10

and

$$\overline{u'} = \lim_{t \to \infty} \frac{1}{t} \int_0^t u'(t) dt \cong 0$$

where t is a large enough time interval, so the time-averaged velocity, $\overline{u}$, is not affected by time, and the time-averaged fluctuation, u, is approximately zero. This can apply to the y-direction velocity as well as to temperature, pressure, and density.

Figure 10.2 shows the sketches for time-dependent (instantaneous) and time-averaged velocity and temperature over a period of time at a given location (x, y) inside a turbulent boundary layer. The positive and negative fluctuation values (up and down values) generally cancel each other; the average u velocity is much greater than the average v velocity; however, the fluctuating u value has approximately the same magnitude as the fluctuating v value, but is out of phase (when fluctuating u is positive, fluctuating v is negative); and the magnitude of the fluctuating T value is approximately the same as the fluctuating u value. This is under the condition or assumption of an isotropic turbulence case. In general, turbulent fluctuating quantities may not necessarily be the same value.

The following shows the step-by-step method to obtain the RANS equations.

Reynolds time-averaged method:

Time dependent (instantaneous) = time averaged (Reynolds averaged) + fluctuation.

$$u = \overline{u} + u' \equiv u + u'$$
$$v = \overline{v} + v' \equiv v + v'$$
$$T = \overline{T} + T' \equiv T + T'$$
$$P = \overline{P} + P' \equiv P + P'$$
$$\rho = \overline{\rho} + \rho' \equiv \rho + \rho'$$

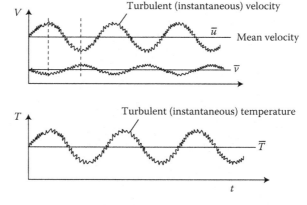

FIGURE 10.2 Instantaneous time-dependent velocity and temperature profiles inside a turbulent boundary layer.

Since

$$\int\limits_{t}^{t+\Delta t} u'dt = 0$$

$$\int\limits_{t}^{t+\Delta t} v'dt = 0$$

$$\int\limits_{t}^{t+\Delta t} P'dt = 0$$

$$\int\limits_{t}^{t+\Delta t} T'dt = 0$$

Therefore,

$$\overline{u'} = \overline{v'} = \overline{T'} = \overline{\rho'} = \overline{P'} \cong 0$$

Substitute the above instantaneous quantities into the original conservation of mass, momentum, and energy equations, respectively, and then perform the averaging over a period of time.

10.1.1 Continuity Equation

$$\lim_{t \to \infty} \frac{1}{t} \int\limits_{0}^{t} \frac{\partial}{\partial t}\rho + \frac{\partial}{\partial x}\left[(\overline{\rho} + \rho')(\overline{u} + u')\right] + \frac{\partial}{\partial y}\left[(\overline{\rho} + \rho')(\overline{v} + v')\right] = 0 \qquad (10.1)$$

$$\lim_{t \to \infty} \frac{1}{t} \int\limits_{0}^{t} \frac{\partial}{\partial t}\rho + \frac{\partial}{\partial x}\left[(\overline{\rho} + \rho')(\overline{u} + u')\right] + \frac{\partial}{\partial y}\left[(\overline{\rho} + \rho')(\overline{v} + v')\right] = 0$$

0 (steady)

$$\frac{\partial}{\partial x}\left(\overline{\rho}\overline{u} + \overline{\rho'u'}\right) + \frac{\partial}{\partial y}\left(\overline{\rho}\overline{v} + \overline{\rho'v'}\right) = 0$$

Therefore,

$$\frac{\partial}{\partial x}\left(\overline{\rho}\overline{u}\right) + \frac{\partial}{\partial y}\left(\overline{\rho}\overline{v}\right) = 0 \qquad (10.2)$$

and

$$\frac{\partial}{\partial x}\left(\overline{\rho' u'}\right) + \frac{\partial}{\partial y}\left(\overline{\rho' v'}\right) = 0 \tag{10.3}$$

10.1.2 MOMENTUM EQUATIONS: RANS

$$\lim_{t\to\infty}\frac{1}{t}\int_0^t \frac{\partial}{\partial t}\rho + \frac{\partial}{\partial x}\left[(\bar{\rho}+\rho')(\bar{u}+u')^2\right] + \frac{\partial}{\partial y}\left[(\bar{\rho}+\rho')(\bar{u}+u')(\bar{v}+v')\right]$$

$$= -\frac{\partial}{\partial x}\left(\bar{P}+P'\right) + (\bar{\rho}+\rho')\left(\bar{f}_x+f'_x\right) + \mu\left[\frac{\partial^2(\bar{u}+u')}{\partial x^2} + \frac{\partial^2(\bar{u}+u')}{\partial y^2}\right] \tag{10.4}$$

$$\lim_{t\to\infty}\frac{1}{t}\int_0^t \underset{\text{0 (steady)}}{\cancel{\frac{\partial}{\partial t}\rho u}} + \frac{\partial}{\partial x}[(\bar{\rho}+\rho')(\bar{u}+u')^2] + \frac{\partial}{\partial y}[(\bar{\rho}+\rho')(\bar{u}+u')(\bar{v}+v')] \tag{10.4}$$

$$= -\frac{\partial}{\partial x}\left(\bar{P}+P'\right) + (\bar{\rho}+\rho')\left(\bar{f}_x+f'_x\right) + \mu\left[\frac{\partial^2\left(\bar{u}+u'\right)}{\partial x^2} + \frac{\partial^2\left(\bar{u}+u'\right)}{\partial y^2}\right]$$

Therefore,

$$\frac{\partial}{\partial x}\bar{\rho}\bar{u}^2 + \frac{\partial}{\partial y}\bar{\rho}\overline{uv} + \frac{\partial}{\partial x}\bar{\rho}\overline{u'^2} + \frac{\partial}{\partial y}\bar{\rho}\overline{u'v'} + \frac{\partial}{\partial x}\overline{\rho'u'^2} + \frac{\partial}{\partial y}\overline{\rho'u'v'} + \frac{\partial}{\partial x}2\bar{u}\overline{\rho'u'}$$

$$+ \frac{\partial}{\partial y}\left(\bar{u}\overline{\rho'v'} + \bar{v}\overline{\rho'u'}\right) = -\frac{\partial\bar{P}}{\partial x} + \bar{\rho}\bar{f}_x + \overline{\rho'f'_x} + \mu\left[\frac{\partial^2\bar{u}}{\partial x^2} + \frac{\partial^2\bar{u}}{\partial y^2}\right]$$

Rearranging,

$$\left(\bar{\rho}\bar{u} + \overline{\rho'u'}\right)\frac{\partial\bar{u}}{\partial x} + \left(\bar{\rho}\bar{v} + \overline{\rho'v'}\right)\frac{\partial\bar{u}}{\partial y} = -\frac{\partial\bar{P}}{\partial x} + \bar{\rho}\bar{f}_x + \overline{\rho'f'_x} + \mu\left[\frac{\partial^2\bar{u}}{\partial x^2} + \frac{\partial^2\bar{u}}{\partial y^2}\right] \tag{10.5}$$

$$\left.\begin{array}{l} -\dfrac{\partial}{\partial x}\left(\bar{\rho}\overline{u'^2} + \overline{\rho'u'^2} + \bar{u}\overline{\rho'u'}\right) \\[2mm] -\dfrac{\partial}{\partial y}\left(\bar{\rho}\overline{u'v'} + \overline{\rho'u'v'} + \bar{v}\overline{\rho'u'}\right) \end{array}\right\} 6\,\text{Reynolds Stress}$$

For incompressible flow, $\rho' = 0$
  X-Momentum:

$$u\frac{\partial u}{\partial x} + v\frac{\partial u}{\partial y} = -\frac{1}{\rho}\frac{\partial P}{\partial x} + \frac{\mu}{\rho}\left(\frac{\partial^2 u}{\partial x^2} + \frac{\partial^2 u}{\partial y^2}\right) - \frac{\partial \overline{u'^2}}{\partial x} - \frac{\partial \overline{u'v'}}{\partial y} + f_x \qquad (10.6)$$

Y-Momentum:

$$u\frac{\partial v}{\partial x} + v\frac{\partial v}{\partial y} = -\frac{1}{\rho}\frac{\partial P}{\partial y} + \underbrace{\frac{\mu}{\rho}\left(\frac{\partial^2 v}{\partial x^2} + \frac{\partial^2 v}{\partial y^2}\right)}_{\text{viscous forces}} - \underbrace{\frac{\partial \overline{u'v'}}{\partial x} - \frac{\partial \overline{v'^2}}{\partial y}}_{\text{Reynolds stress, Turbulent stress}} + f_y \qquad (10.7)$$

From boundary-layer approximation, the Y-Momentum equation is not as important compared to the X-Momentum equation. In the x-momentum equation, $\dfrac{\partial^2 u}{\partial x^2} \ll \dfrac{\partial^2 u}{\partial y^2}$;
$\dfrac{\partial \overline{u'^2}}{\partial x} \ll \dfrac{\partial \overline{u'v'}}{\partial y}$.

10.1.3 ENTHALPY/ENERGY EQUATION

$$\lim_{t\to\infty}\frac{1}{t}\int (\overline{\rho} + \rho')c_p\left[\frac{\partial}{\partial t}T + (\overline{u} + u')\frac{\partial}{\partial x}(\overline{T} + T') + (\overline{v} + v')\frac{\partial}{\partial y}(\overline{T} + T')\right]$$

$$= \frac{\partial}{\partial x}k\frac{\partial}{\partial x}(\overline{T} + T') + \frac{\partial}{\partial y}k\frac{\partial}{\partial y}(\overline{T} + T') + \frac{\partial P}{\partial t} + (\overline{u} + u')\frac{\partial(\overline{P} + P')}{\partial x} \qquad (10.8)$$

$$+ (\overline{v} + v')\frac{\partial(\overline{P} + P')}{\partial y} + \Phi$$

Therefore, for steady flow,

$$c_p\left[\left(\overline{\rho u} + \overline{\rho' u'}\right)\frac{\partial \overline{T}}{\partial x} + \left(\overline{\rho v} + \overline{\rho' v'}\right)\frac{\partial \overline{T}}{\partial y}\right]$$

$$= \frac{\partial}{\partial x}k\frac{\partial T}{\partial x} + \frac{\partial}{\partial y}k\frac{\partial T}{\partial y} + \overline{u}\frac{\partial \overline{P}}{\partial x} + \overline{v}\frac{\partial \overline{P}}{\partial y} + \Phi + \rho\varepsilon$$

$$- c_p\frac{\partial}{\partial x}\left(\overline{\rho u'T'} + \overline{\rho' u'T'} + \overline{u}\,\overline{\rho'T'}\right) + \left(\frac{\partial \overline{u}}{\partial x}\right)^2 \dots \qquad (10.9)$$

$$- c_p\frac{\partial}{\partial y}\underbrace{\left(\overline{\rho v'T'} + \overline{\rho' v'T'} + \overline{v}\,\overline{\rho'T'}\right)}_{6\,\text{Reynolds Flux}} + \left(\frac{\partial u'}{\partial x}\right)^2 \dots$$

$$+ \frac{\partial}{\partial x}\overline{u'P'} + \frac{\partial}{\partial y}\overline{v'P'}$$

where Φ is the viscous dissipation and ε is the turbulent dissipation.

For steady, incompressible flow,

$$\rho' = 0$$
$$P' = 0$$
$$\Phi = 0$$
$$\varepsilon = 0$$

$$u\frac{\partial T}{\partial x} + v\frac{\partial T}{\partial y} = \underbrace{\alpha\left(\frac{\partial^2 T}{\partial x^2} + \frac{\partial^2 T}{\partial y^2}\right)}_{\text{molecular conduction}} + \underbrace{\frac{-\partial\overline{T'u'}}{\partial x} - \frac{\partial\overline{T'v'}}{\partial y}}_{\text{Reynolds flux, turbulent flux}} \tag{10.10}$$

From the boundary-layer approximation, $\dfrac{\partial^2 T}{\partial x^2} \ll \dfrac{\partial^2 T}{\partial y^2}; \dfrac{\partial\overline{T'u'}}{\partial x} \ll \dfrac{\partial\overline{T'v'}}{\partial y}.$

10.1.4 CONCEPT OF EDDY OR TURBULENT DIFFUSIVITY

The following is a summary from the above RANS equations. The continuity equation is not useful. The Y-momentum equation is small when compared to the X-momentum equation. Therefore, only one of six Reynolds stresses and one of six Reynolds fluxes remain in the RANS equations for a 2-D steady, incompressible, and constant property fully turbulent boundary-layer flow. In addition to a laminar-type shear stress due to the viscous effect, turbulent stress due to random velocity fluctuation is the major contributor to the total pressure loss over a turbulent boundary layer. Similarly, turbulent flux due to random velocity with temperature fluctuations dominates the total heat transfer over the turbulent boundary layer. The real challenge is how to quantify the time-averaged Reynolds stress and Reynolds flux because they are varying with the location (x, y) inside the turbulent boundary. It is assumed the Reynolds stress is proportional to the velocity gradient and the proportionality constant (actually it is not a constant value) is called eddy or turbulent diffusivity for momentum (turbulent viscosity divided by fluid density); similarly, the Reynolds flux is proportional to the temperature gradient and the proportionality constant (actually it is not a constant value) is called the eddy or turbulent diffusivity for heat. Therefore, the real turbulent flow problem is how to determine, or how to model, the turbulent diffusivity for momentum (turbulent viscosity) and turbulent diffusivity for heat because they are dependent upon the location (x, y). It is important to note that molecular Prandtl number is a ratio of fluid kinematic viscosity to thermal diffusivity and is a fluid property depending on what type of fluid is being considered (e.g., air or water has different molecular Prandtl numbers); however, the turbulent Prandtl number is a ratio of turbulent diffusivity for momentum to turbulent diffusivity for heat and is a flow structure behavior depending on how turbulent the flow is (e.g., air and water have the same turbulent Prandtl number at the same turbulent flow condition). For a simple turbulent flow problem, the turbulent Prandtl number is approximately one (say 0.9 for most of the models), which implies that one can solve for the turbulent diffusivity for heat if the turbulent diffusivity for momentum has been determined/modeled. Therefore, the first question is how to determine or model turbulent diffusivity for momentum (turbulent viscosity divided by fluid density). Once the turbulent viscosity is given, the turbulent flow and heat transfer problems can be solved.

$$u\frac{\partial u}{\partial x} + v\frac{\partial u}{\partial y} = \frac{1}{\rho}\frac{\partial}{\partial y}\left(\mu\frac{\partial u}{\partial y} - \rho\overline{u'v'}\right) = \frac{1}{\rho}\frac{\partial}{\partial y}(\tau_{\text{viscous}} + \tau_{\text{turb.}}) \tag{10.11}$$

$$u\frac{\partial T}{\partial x} + v\frac{\partial T}{\partial y} = \frac{1}{\rho C_p}\frac{\partial}{\partial y}\left(k\frac{\partial T}{\partial y} - \rho c_p\overline{v'T'}\right) = \frac{1}{\rho C_p}\frac{\partial}{\partial y}(q''_{\text{molecular}} + q''_{\text{turb.}}) \tag{10.12}$$

$$\tau_{\text{total}} = \tau_m + \tau_t = \mu\frac{\partial u}{\partial y} + \rho\varepsilon_m\frac{\partial u}{\partial y} \tag{10.13}$$

where $-\rho\overline{u'v'} = \rho\varepsilon_m(\partial u/\partial y)$, with the eddy diffusivity for momentum $\varepsilon_m = v_t = (\mu_t/\rho)$, and turbulent viscosity $\mu_t = \rho\varepsilon_m$

$$q''_{\text{total}} = q''_m + q''_t = k\frac{\partial T}{\partial y} + \rho c_p\varepsilon_h\frac{\partial T}{\partial y} \tag{10.14}$$

where $-\rho c_p\overline{v'T'} = \rho c_p\varepsilon_h(\partial T/\partial y)$, with the eddy diffusivity for heat $\varepsilon_h = \alpha_t$

$$\tau_{\text{total}} = \rho(v + \varepsilon_m)\frac{\partial u}{\partial y} \tag{10.15}$$

$$q''_{\text{total}} = \rho c_p(\alpha + \varepsilon_h)\frac{\partial T}{\partial y} = \rho c_p\left(\frac{v}{\text{Pr}} + \frac{\varepsilon_m}{\text{Pr}_t}\right)\frac{\partial T}{\partial y} \tag{10.16}$$

Combing the diffusivities for momentum and heat, we can define the turbulent Prandtl number

$$\text{Pr}_t = \frac{v_t}{\alpha_t} = \frac{\varepsilon_m}{\varepsilon_h} \sim 1 \tag{10.17}$$

So, the unknowns are

- Turbulent viscosity μ_t
- Turbulent diffusivity $v_t = (\mu_t/\rho) = \varepsilon_m$
- Turbulent diffusivity for heat $\varepsilon_h = \alpha_t$

The following shows how to determine the turbulent diffusivity for momentum using the method from Von Kármán.

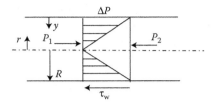

FIGURE 10.3 Force balance in a circular tube.

From the force balance in a circular tube as shown in Figure 10.3,

$$\frac{dP}{dx} = \frac{1}{r}\frac{\partial}{\partial r}(r\tau)$$

$$\int r\frac{dP}{dx} = \int \frac{d}{dr}(r\tau)$$

$$\int r\frac{dP}{dx}dr = \int \frac{d}{dr}(r\tau)dr$$

Therefore,

$$\tau = \frac{1}{2}r\frac{dP}{dx} \sim r$$

Using the following BCs,

$$r = 0, \qquad \tau = 0,$$
$$r = R, \qquad \tau = \tau_w$$

One obtains

$$\tau = \frac{r}{R}\tau_w \tag{10.18}$$

$$\tau = \frac{r}{R}\tau_w = \frac{R-y}{R}\tau_w = \left(1-\frac{y}{R}\right)\tau_w = \frac{\partial u}{\partial y}\rho(v + v_t)$$

$$v\frac{\partial u}{\partial y}\left(1+\frac{v_t}{v}\right) = \left(1-\frac{y}{R}\right)\frac{\tau_w}{\rho}$$

$$\left(1-\frac{y^+}{R^+}\right) = \left(1+\frac{v_t}{v}\right)\frac{\partial u}{\partial y}\frac{v}{\sqrt{(\tau_w/\rho)}\sqrt{\tau_w/\rho}} = \left(1+\frac{v_t}{v}\right)\frac{du^+}{dy^+}$$

where

$$u^+ = \frac{u}{u^*}, \qquad y^+ = \frac{yu^*}{v}, \qquad R^+ = \frac{Ru^*}{v}, \qquad u^* = \sqrt{\frac{\tau_w}{\rho}}$$

The above dimensionless parameters are derived from the Prandtl mixing length theory with more details presented in Section 10.2.

Therefore,

$$\frac{v_t}{v} = \frac{\varepsilon_m}{v} = \frac{1-\left(y^+/R^+\right)}{\left(du^+/dy^+\right)} - 1 \tag{10.19}$$

Similarly, the turbulent diffusivity for momentum for a boundary-layer flow can be obtained by replacing R^+ by δ^+ (where δ is turbulent boundary-layer thickness shown in Figure 10.4) as

$$\frac{v_t}{v} = \frac{\varepsilon_m}{v} = \frac{1-\left(y^+/\delta^+\right)}{\left(du^+/dy^+\right)} - 1 \tag{10.20}$$

where

$$\delta^+ = \frac{\delta u^*}{v}$$

From above, the turbulent viscosity depends on the dimensionless velocity profile across turbulent boundary layer. Thus, turbulent viscosity distributions can be predicted using the Martinelli Three-Region Velocity Profile (refer to Section 10.2 later), as shown in Figure 10.5. The procedures of calculating the turbulent viscosity for velocity and its gradient in regions 1, 2, and 3 are outlined below:

1. For the laminar sublayer region,

   $$y^+ < 5; \quad y^+ \ll R^+$$

   $$u^+ = y^+, \quad \frac{du^+}{dy^+} = 1$$

   $$\frac{v_T}{v} = \frac{1-0}{1} - 1 = 0 \text{ (viscous dominated)}$$

2. For the buffer zone,

   $$5 \le y^+ < 30; \quad y^+ \ll R^+$$

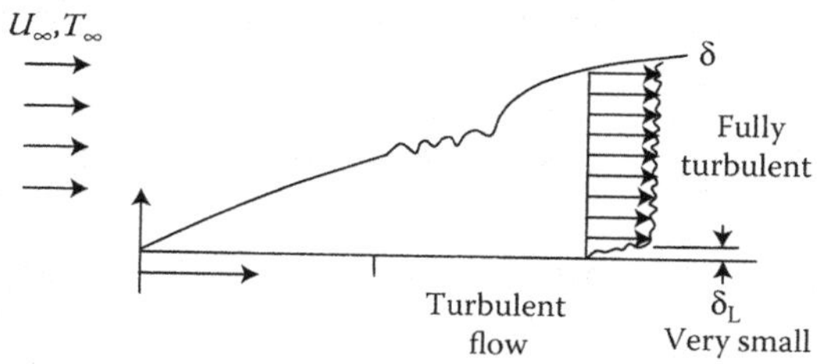

FIGURE 10.4 Concept of 2-D turbulent boundary layer flow.

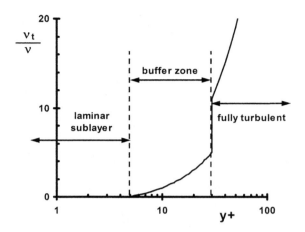

FIGURE 10.5 Turbulent viscosity distributions across turbulent boundary layer.

$$u^+ = 5\ln y^+ - 3.05, \quad \frac{du^+}{dy^+} = \frac{5}{y^+}$$

$$\frac{v_T}{v} = \frac{1-0}{\dfrac{5}{y^+}} - 1 = \frac{y^+}{5} - 1$$

3. For the turbulent region,

$$y^+ \geq 30$$

$$u^+ = 2.5\ln y^+ + 5.5, \quad \frac{du^+}{dy^+} = \frac{2.5}{y^+}$$

$$\frac{v_T}{v} = \frac{1 - \dfrac{y^+}{R^+}}{\dfrac{2.5}{y^+}} - 1$$

Note that Reichard proposed the below empirical correlation from near the wall to the centerline:

$$\frac{v_T}{v} = \frac{1}{6}\kappa R^+ \eta (2-\eta)\left(3 - 4\eta + 2\eta^2\right)$$

$$\sim \kappa y^+ \text{ near wall, } \eta = \frac{y}{R}$$

If $\dfrac{y}{R} < 0.2$ mixing length theory is okay (refer to Section 10.2)

$\dfrac{y}{R} > 0.2$ use experimental data

At $\dfrac{y}{R} = 0.2$, $\dfrac{v_T}{v} \cong 80$, but $\dfrac{du}{dy}$ is small, thus, $v_t \dfrac{du}{dy} =$ reasonable value

10.1.5 REYNOLDS ANALOGY FOR TURBULENT FLOW AND HEAT TRANSFER

The following outlines the simple Reynolds analogy between momentum and heat transfer for a turbulent boundary-layer flow.

$$\frac{\tau}{q''} = \frac{\rho(v + \varepsilon_m)(\partial u / \partial y)}{\rho c_p(\alpha + \varepsilon_h)(\partial T / \partial y)} = \frac{(\partial u / \partial y)}{C_p(\partial T / \partial y)} \tag{10.21}$$

If $v = \alpha \Rightarrow \mathrm{Pr} = (v / \alpha) \approx 1$, then $\varepsilon_m \approx \varepsilon_h \Rightarrow \mathrm{Pr}_t = (\varepsilon_m / \varepsilon_h) \approx 1$

The turbulent Prandtl number is dependent on the flow structure. It can be estimated as 0.9 for air flow.

Assuming a linear velocity and temperature profile, Equation 10.21 becomes

$$\frac{\tau_w}{q_w''} = \frac{1}{C_p} \frac{\Delta u}{\Delta T} \tag{10.22}$$

$$\frac{q_w''}{C_p(T_w - T_\infty)} = \frac{\tau_w}{\mathcal{U}_\infty}$$

where

$$\tau_w = \frac{1}{2} \rho \mathcal{U}_\infty^2 \cdot C_f \tag{10.23}$$

$$\mathrm{Nu} = \frac{1}{2} C_f \frac{\rho \mathcal{U}_\infty x}{\mu} \cdot \frac{\mu c_p}{k}$$

$$\frac{\mathrm{Nu}}{\mathrm{Re} \cdot \mathrm{Pr}} = \frac{1}{2} C_f \Rightarrow \frac{1}{2} C_f = \mathrm{St}$$

$$\frac{1}{2} C_f = \mathrm{St} \cdot \mathrm{Pr}^{2/3} \tag{10.24}$$

Based on experimental data, the Prandtl number effect is included in Equation 10.24. The equation is applicable for $0.7 \leq \mathrm{Pr} \leq 60$ (air, water, and oil).

where

$$\mathrm{St} = \frac{\mathrm{Nu}}{\mathrm{Re} \cdot \mathrm{Pr}} = \frac{(hx / k)}{(\rho \mathcal{U}_\infty x / \mu) \cdot (\mu C_p / k)} = \frac{h}{\rho C_p \mathcal{U}_\infty} = \frac{q_w''}{\rho C_p \mathcal{U}_\infty (T_w - T_\infty)} \tag{10.25}$$

There are two types of problems:

1. For given C_f determine St or h;
2. For given h or St, determine C_f

Apply Equation 10.24 for a turbulent pipe flow,

$$C_f = \frac{0.046}{\text{Re}_D^{0.2}} \tag{10.26}$$

$$\frac{1}{2} \frac{0.046}{\text{Re}_D^{0.2}} = \frac{Nu}{\text{Re} \cdot \text{Pr}} \text{Pr}^{2/3}$$

for cooling

$$\text{Nu}_D = \frac{hD}{k} = 0.023 \, \text{Re}_D^{0.8} \, \text{Pr}^{1/3} \tag{10.27}$$

$$\text{Nu}_D = \frac{hD}{k} = 0.023 \, \text{Re}_D^{0.8} \, \text{Pr}^{0.4} \quad \text{for heating} \tag{10.28}$$

For a turbulent boundary-layer flow,

$$C_f = \frac{0.0592}{\text{Re}_x^{0.2}} \tag{10.29}$$

$$\frac{1}{2} \frac{0.0592}{\text{Re}_x^{0.2}} = \frac{Nu_x}{\text{Re}_x \, \text{Pr}} \text{Pr}^{2/3}$$

$$\text{Nu}_x = \frac{hx}{k} = 0.0296 \, \text{Re}_x^{0.8} \, \text{Pr}^{1/3} \tag{10.30}$$

From the above simplified Reynolds analogy, if friction factors are given, heat transfer coefficients can be predicted, and vice versa.

Summary of Heat and Momentum Transfer Analogy: The following shows the proposed heat and momentum transfer analogy for various Prandtl number fluids.

1874 Reynolds Analogy: only good for air, Prandtl number ~ 1.0

1910 Prandtl–Taylor (1919) Analogy: reasonable match to the experimental data from Colburn

1939 Kármán–Martinelli (1947) Analogy: excellent match to the experimental data from Colburn

1933 Colburn Experimental Correlation for a wide range of Prandtl number fluids

Reynolds Analogy Method: Based on turbulent viscosity and shear stress of momentum transfer—Assumed analogy between heat and momentum transfer as shown in Figure 10.6.

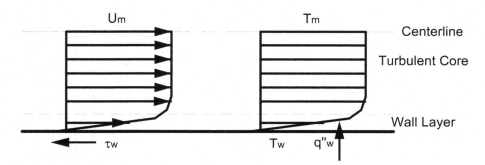

FIGURE 10.6   Concept of heat and momentum transfer analogy on turbulent boundary layer.

ε_m : eddy diffusivity of momentum

$$\tau_w = \rho(v + v_t)\frac{du}{dy},$$

v_t : turbulent viscosity

$$q_w'' = -\rho c_p(\alpha + \alpha_t)\frac{dT}{dy}$$

ε_h : eddy diffusivity of heat

α_t : turbulent diffusivity

$$\int \frac{c_p \tau_w}{q_w''} = -\frac{v + v_t}{\alpha + \alpha_t}\int \frac{du}{dT} = -\frac{v + v_t}{\alpha + \alpha_t}\cdot\frac{\mathcal{U}_m - 0}{T_m - T_w}$$

$$\frac{c_p f \frac{1}{2}\rho \mathcal{U}_m^2}{h(T_w - T_m)} = \frac{v + v_t}{\alpha + \alpha_t}\cdot\frac{\mathcal{U}_m - 0}{T_w - T_m} \rightarrow \frac{c_p f \frac{1}{2}\rho \mathcal{U}_m}{h} = \frac{v + v_t}{\alpha + \alpha_t}$$

$$\mathrm{Pr} = \frac{v}{\alpha}$$ molecular Prandtl number

$$\mathrm{Pr}_t = \frac{v_t}{\alpha_t}\left(=\frac{\varepsilon_m}{\varepsilon_h}\right)$$ turbulent Prandtl number

$$\mathrm{St} \equiv \frac{h}{\rho c_p \mathcal{U}_m} \equiv \frac{\mathrm{Nu}}{\mathrm{RePr}}$$ Stanton number

$$\mathrm{St} \equiv \frac{h}{\rho c_p \mathcal{U}_m} = \frac{1}{2}f\cdot\frac{\alpha + \alpha_t}{v + v_t}$$ if : $v \ll v_t, \alpha \ll \alpha_t, \mathrm{Pr}_t = \frac{v_t}{\alpha_t} \cong 1$

$$\mathrm{St} = \frac{h}{\rho c_p \mathcal{U}_m} = \frac{\mathrm{Nu}}{\mathrm{RePr}} = \frac{1}{2}f$$ by Reynolds Analogy (original) for Air (Pr $\approx$ 1)

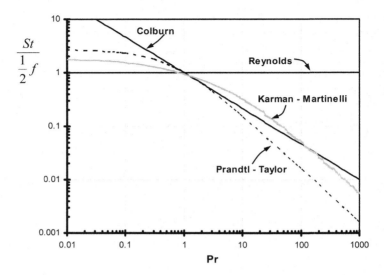

FIGURE 10.7   Reynolds analogy for a wide range of Prandtl number fluids.

Also, $St = \dfrac{1}{2} f \cdot Pr^{-\frac{2}{3}}$ by Colburn for $0.7 \leq Pr \leq 60$ (for air, water, organic, oil).

Experimental data for a wide range of Prandtl number fluids, although still referred to as the Reynolds analogy, is shown in Figure 10.7.

Reynolds Analogy Factor:

$$\frac{St}{\frac{1}{2}f} = 1 \text{ for turbulent air flow}$$

$$= Pr^{-2/3} \text{ for other fluid turbulent flow}$$

10.2   PRANDTL MIXING LENGTH THEORY AND LAW OF THE WALL FOR VELOCITY AND TEMPERATURE PROFILES

In a laminar boundary-layer flow, universal velocity and temperature profiles can be obtained by solving conservation equations for mass, momentum, and energy using the similarity method. In the turbulent flow boundary layer, we also hope to obtain universal velocity and temperature profiles. The following outlines step by step how the Prandtl mixing length theory can be applied to achieve the law of the wall for velocity and temperature profiles (a type of universal velocity and temperature profiles for turbulent boundary-layer flow). Unlike the laminar boundary, however, there is no complete analytical solution for the turbulent boundary layer due to turbulent random motion. The law of the wall for the velocity profile still requires a predetermined shear stress (or the friction factor from the experimental data) for a given turbulent flow problem. Therefore, we can only obtain a semi-theoretical (or semi-empirical) velocity profile for turbulent boundary-layer flow.

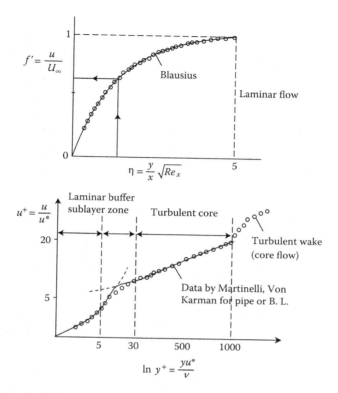

FIGURE 10.8 Analytical universal velocity profile for laminar boundary layer and semiempirical law of wall velocity profile for turbulent boundary layer.

Figure 10.8 shows the analytical universal velocity profile for a laminar boundary layer and the semiempirical law of the wall velocity profile for the turbulent boundary layer.

The following is the detail of the Prandtl mixing length theory and the law of the wall for velocity and temperature profiles [3–6]. First, we define the dimensionless x-direction velocity and y-direction wall coordinate.

$$u^+ = \frac{u}{u^*} \tag{10.31}$$

$$y^+ = \frac{yu^*}{\nu} \tag{10.32}$$

$$u^* = \sqrt{\frac{\tau_w}{\rho}} = \sqrt{\frac{(1/2)C_f \rho \mathcal{U}_\infty^2}{\rho}} = \mathcal{U}_\infty \sqrt{\frac{1}{2}C_f} \tag{10.33}$$

where u^* is the friction velocity and C_f is the experimentally, predetermined friction factor.

For example,

$$C_f = 0.046\,Re_D^{-0.2}, \qquad\qquad \text{for turbulent flow in a tube}$$
$$C_f = 0.0592\,Re_x^{-0.2}, \qquad \text{for turbulent flow over a flat plate}$$

and y^+ is the dimensionless wall coordinate or the roughness Reynolds number.

Consider the laminar sublayer region, very close to the wall region where viscosity dominates and turbulence is damped at the wall.

$$\tau = \rho v \frac{du}{dy} \tag{10.34}$$

$$\frac{\tau_w}{\rho} = v\frac{du}{dy}\Big|_w \approx v\frac{u}{y}$$

$$\sqrt{\frac{\tau_w}{\rho}}\sqrt{\frac{\tau_w}{\rho}} = v\frac{du}{dy}$$

$$u^* u^* = v\frac{du}{dy}$$

Therefore,

$$\frac{u}{u^*} = \frac{yu^*}{v}$$

and

$$u^+ = y^+ \tag{10.35}$$

Now, consider the fully turbulent region, as sketched in Figure 10.9, away from the wall where turbulence dominates. From the Prandtl mixing length theory, assuming velocity fluctuation is proportional to the velocity gradient and the mixing length is linearly increasing with distance from the wall,

$$u' \sim \frac{\partial u}{\partial y} = l\frac{\partial u}{\partial y} \tag{10.36}$$

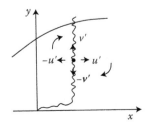

FIGURE 10.9 Concept of turbulence in 2-D turbulent boundary-layer flow.

$$l \approx y = \kappa y \tag{10.37}$$

where $\kappa = 0.4$, a universal constant from the experimental data by Von Kármán.
And

$$v' = -u' = -\kappa y \frac{\partial u}{\partial y} \tag{10.38}$$

Since wall shear is dominated by turbulence, it can be approximated as

$$\tau_w \approx -\rho u'v' = \rho \kappa^2 y^2 \left(\frac{\partial u}{\partial y} \right)^2 \tag{10.39}$$

Therefore, from the above Prandtl mixing length assumption,

$$\sqrt{\frac{\tau_w}{\rho}} = \kappa y \frac{\partial u}{\partial y}$$

$$u^* = \kappa y \frac{\partial u}{\partial y}$$

$$\frac{du}{u^*} = \int \frac{1}{\kappa y} dy$$

$$\int du^+ = \int \frac{1}{ky} dy = \int \frac{1}{ky^+} dy^+$$

Therefore, the law of wall velocity profile can be obtained as

$$u^+ = \frac{1}{\kappa} \ln y^+ + C \tag{10.40}$$

From experimental data curve fitting for $y^+ \geq 30$, $\kappa = 0.4$ or 0.41

$$u^+ = 2.5 \ln y^+ + 5.0 \tag{10.41}$$

or

$$u^+ = 2.44 \ln y^+ + 5.5 \tag{10.42}$$

Then, consider the buffer zone between the laminar sublayer and the turbulence region, $5 \leq y^+ \leq 30$, the velocity profile can be obtained as

$$u^+ = 5 \ln y^+ - 3.05 \tag{10.43}$$

It is important to mention that the above three-region velocity profile has been validated and can be applied for turbulent flow over a flat plate or in a tube with air, water, or oil as the working fluid. In addition, before showing the law of wall for temperature, it is interesting to point out, as with the laminar boundary-layer case, the law of wall temperature profile is identical to the law of wall velocity profile if the Prandtl number is unity ($Pr = 1$). Therefore, the above dimensionless velocity can be replaced with the appropriate dimensionless temperature with the same dimensionless y-direction, wall coordinate. The effect of Prandtl number on the law of wall for temperature will be discussed in Section 10.3.

The Law of the Wall Velocity Profile for tube or boundary layer flow was developed with the Prandtl Mixing Length Theory, discussed above. The Law of Wall Temperature Profile for tube or boundary layer flow can also be developed with the Heat and Momentum Transfer Analogy as follows:

1. Kármán–Martinelli Analogy (between Heat and Momentum)
   Assumed Three-Region Profile—1949
2. Kays Analogy (between Heat and Momentum)
   Assumed Two-Region Profile—1966
3. Kader–Yaglom Analogy (between Heat and Momentum)
   Assumed Three-Region Profile—1972
   Verified Pr effect by experimental data

Method:

From Law of the Wall Velocity Profile $u^+ \sim y^+$

Use Heat and Momentum Analogy to obtain Law of the Wall Temperature Profile $T^+ \sim y^+$

Use Heat and Momentum Analogy to obtain Heat Transfer Coefficients as discussed in the following sections.

10.3 TURBULENT FLOW HEAT TRANSFER

10.3.1 TURBULENT INTERNAL FLOW HEAT TRANSFER COEFFICIENT

Consider a fully turbulent flow in a circular tube with uniform wall heat flux $\left(q_w'' = C \right)$ as the thermal BC [4–6], as sketched in Figure 10.10, using the Kármán–Martinelli heat and momentum transfer analogy. From the energy equation,

$$u\frac{\partial T}{\partial x} + v\frac{\partial T}{\partial r} = \frac{1}{r}\frac{\partial}{\partial r}\left[r(\alpha + \varepsilon_h)\frac{\partial T}{\partial r} \right] + \alpha\frac{\partial^2 T}{\partial x^2} \qquad (10.44)$$

$$y = R - r$$
$$r = R - r$$
$$dr = -dy$$

For a fully developed flow $(v = 0)$ and assuming $\left(\partial^2 T / \partial x^2 \right) \ll (1/r)\left(\partial / \partial r \right)$ $\left[r(\alpha + \varepsilon_h)(\partial T / \partial r) \right]$, the energy equation becomes

$$u\frac{\partial T}{\partial x} = \frac{1}{R - y}\frac{\partial}{(-\partial y)}\left[(R - y)(\alpha + \varepsilon_h)\frac{\partial T}{(-\partial y)} \right]$$

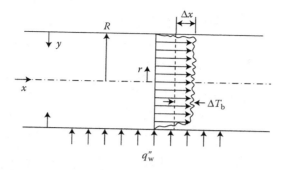

FIGURE 10.10 Turbulent flow heat transfer in a circular tube.

Assume

$$u = \bar{V}$$

$$\frac{\partial T}{\partial x} = \frac{\partial T_b}{\partial x}$$

From the energy balance,

$$q_w'' 2\pi R \, dx = \rho C_p \bar{V} A_c dT_b$$

$$\bar{V} \frac{dT_b}{dx} = \frac{q_w'' 2\pi R}{\rho C_p \pi R^2} = \frac{2q_w''}{\rho C_p R} \equiv C$$

And applying the BCs,

$$y = 0 \qquad T = T_w$$

$$y = R \qquad \frac{\partial T}{\partial y} = 0$$

The energy equation becomes

$$\bar{V} \frac{dT_b}{dx} (R - y) d(-y) = d\left[(R - y)(\alpha + \varepsilon_h) \frac{\partial T}{-\partial y} \right]$$

$$\bar{V} \frac{dT_b}{dx} \frac{1}{2} (R - y)^2 = (R - y)(\alpha + \varepsilon_h) \frac{dT}{-dy} + C$$

$$\frac{1}{2} (R - y) \bar{V} \frac{dT_b}{dx} = (\alpha + \varepsilon_h) \frac{dT}{-dy} + \frac{C}{R - y}$$

where $C = 0$ at $R - y = 0$, $(dT/dy) = 0$.

Therefore,

$$dT = \frac{1}{2}\bar{V}\frac{dT_b}{dx}\frac{y-R}{\alpha+\varepsilon_h}dy$$

$$\int_{T_w}^{T}dT = \frac{1}{2}\bar{V}\frac{dT}{dx}\int_0^y \frac{y-R}{\alpha+\varepsilon_h}dy$$

$$T - T_w = \frac{1}{2}\bar{V}\frac{\partial T_b}{\partial x}\int_0^y \frac{y-R}{\alpha+\varepsilon_h}dy = \frac{-q''_w}{\rho C_p\sqrt{\tau_w/\rho}}\int_0^{y^+}\frac{1-\left(y^+/R^+\right)}{(1/\Pr)+(\varepsilon_h/v)}dy^+$$

Therefore,

$$T^+ \equiv \frac{T_w - T}{q''_w/\rho C_p u^*} = \int_0^{y^+}\frac{1-\left(y^+/R^+\right)}{(1/\Pr)+(\varepsilon_h/v)}dy^+ \qquad (10.45)$$

The above equation can be integrated if one assumes

$$\frac{\varepsilon_h}{v} \approx \frac{\varepsilon_m}{v} \quad \text{or} \quad \Pr_t = \frac{\varepsilon_m}{\varepsilon_h} \approx 1$$

And from Equation (10.19), the turbulent viscosity can be determined as

$$\frac{\varepsilon_m}{v} = \frac{1-\left(y^+/R^+\right)}{\left(du^+/dy^+\right)} - 1$$

From Martinelli Three Region Velocity Profile:

For the laminar sublayer region,

$$0 \le y^+ \le 5, \quad y^+ \ll R^+, \quad u^+ = y^+, \quad \frac{du^+}{dy^+} = 1, \quad \frac{\varepsilon_m}{v} = 0$$

For the buffer zone,

$$5 \le y^+ \le 30, \quad u^+ = 5\ln y^+ - 3.05, \quad \frac{du^+}{dy^+} = \frac{5}{y^+}$$

For the turbulent region,

$$y^+ \ge 30, \quad u^+ = 2.5\ln y^+ + 5.0, \quad \frac{du^+}{dy^+} = \frac{2.5}{y^+}$$

The above energy integral Equation (10.45) can be obtained in the following three regions, and the Prandtl number effect is shown in Figure 10.11.

$$T^+ = \int_0^{y^+} \frac{1-\left(y^+/R^+\right)}{(1/\Pr)+\left(\varepsilon_h/v\right)} dy^+$$

$$\frac{\varepsilon_h}{v} = \frac{\varepsilon_m}{v}$$

For the region $0 \le y^+ \le 5$ with $y^+ \approx 0$ and $\left(\varepsilon_m/v\right) \approx 0$,

$$T^+ = \Pr \int_0^{y^+} dy^+ = \Pr y^+$$

$$u^+ = y^+, \quad T^+ = \Pr y^+ \tag{10.46}$$

For the region $5 \le y^+ \le 30$ with $u^+ = 5\ln y^+ - 3.05$ and $\left(y^+/R^+\right) \approx 0$,

$$\frac{\varepsilon_m}{v} = \frac{\varepsilon_h}{v} = \frac{y^+}{5} - 1$$

$$T^+ - T_5^+ = \int_5^{y^+} \frac{1}{(1/\Pr)+\left(y^+/5\right)-1} dy^+$$

$$= 5\int_5^{y^+} \frac{1}{(1/\Pr)+\left(y^+/5\right)-1} d\left(\frac{1}{\Pr}+\frac{y^+}{5}-1\right)$$

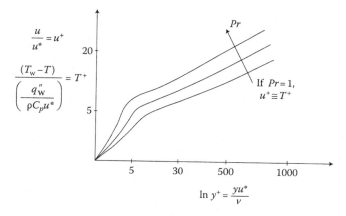

FIGURE 10.11   Law of wall temperature profile for turbulent boundary layer.

Therefore,

$$T^+ - T_5^+ = 5 \ln\left(\frac{1}{\text{Pr}} + \frac{y^+}{5} - 1\right) - 5 \ln\left(\frac{1}{\text{Pr}} + \frac{5}{5} - 1\right)$$

$$= 5 \ln\frac{(1/\text{Pr}) + (y^+/5) - 1}{(1/\text{Pr})}$$

$$= 5 \ln\left(1 + \frac{\text{Pr}\, y^+}{5} - \text{Pr}\right) \tag{10.47}$$

For the region $30 \le y^+$ with $u^+ = 2.5 \ln y^+ + 5.0$, $\left(du^+/dy^+\right) = \left(2.5/y^+\right)$ and

$$\frac{\varepsilon_h}{v} = \frac{\varepsilon_m}{v} = \frac{1 - \left(y^+ / R^+\right)}{\left(2.5 / y^+\right)} - 1$$

$$T^+ - T_{30}^+ = \int_{30}^{y^+} \frac{1 - \left(y^+/R^+\right)}{(1/\cancel{\text{Pr}}) + \left\{\left[1 - \left(y^+/R^+\right)\right]/\left(2.5/y^+\right)\right\} - 1}\, dy^+$$

$$= \int_{30}^{y^+} \frac{2.5}{y^+}\, dy^+ = 2.5 \ln y^+ \,\big|_{30}^{y^+}$$

Assume $(1/\text{Pr} - 1) << \left\{\left[1 - \left(y^+/R^+\right)\right]/\left(2.5/y^+\right)\right\}$ for the above integration. Therefore,

$$T^+ - T_{30}^+ = 2.5 \ln y^+ - 2.5 \ln 30 \tag{10.48}$$

The next question is how to determine the heat transfer coefficient from the law of wall temperature profile. This is shown below.

If one assumes the simple velocity and temperature profiles as 1/7 power law profiles,

$$\frac{u}{U_{\max}} \cong \left(\frac{y}{R}\right)^{1/7} = \left(1 - \frac{r}{R}\right)^{1/7}$$

$$\frac{T - T_w}{T_c - T_w} \cong \left(\frac{y}{R}\right)^{1/7} = \left(1 - \frac{r}{R}\right)^{1/7}$$

Therefore,

$$u_b \text{ or } \bar{V} = \frac{\int u 2\pi r \, dr}{\int 2\pi r \, dr} = 0.82 \mathcal{U}_{\max}$$

$$T_w - T_b = \frac{\int_0^R (T_w - T_c)\left(1 - (r/R)\right)^{1/7} \mathcal{U}_{\max}\left(1 - (r/R)\right)^{1/7} \cdot r \, dr}{\int_0^R \mathcal{U}_{\max}\left(1 - (r/R)\right)^{1/7} \cdot r \, dr}$$

$$\cong \frac{15}{18}(T_w - T_c)$$

$$= 0.833(T_w - T_c)$$

where

$$(T_w - T_c) = (T_w - T_5) + (T_5 - T_{30}) + (T_{30} - T_c)$$

$$= \frac{q_w''}{\rho V_p u^*}\left[5\Pr + 5\ln(5\Pr + 1) + 2.5\ln\left(\frac{R^+}{30}\right)\right]$$

Therefore,

$$h = \frac{q_w''}{T_w - T_b} = \frac{q_w''}{0.833\left(q''/\rho C_p u^*\right)\left[5\Pr + 5\ln(5\Pr + 1) + 2.5\ln\left(R u^*/\nu 30\right)\right]}$$

The final heat transfer coefficient and the Nusselt number are expressed as

$$\mathrm{Nu}_D \equiv \frac{hD}{k} = \frac{\mathrm{Re}\cdot\Pr\sqrt{C_f/2}}{0.833\left[5\Pr + 5\ln(5\Pr + 1) + 2.5\ln\left(\left(\mathrm{Re}_D\sqrt{C_f/2}\right)60\right)\right]} \quad (10.49)$$

where

$$C_f = \frac{0.046}{\mathrm{Re}_D^{0.2}} \quad \text{or} \quad C_f = \frac{0.079}{\mathrm{Re}_D^{0.25}}$$

For a given Re_D and $\Pr$, the above prediction is fairly close to the following experimental correlation:

$$\mathrm{Nu}_D \equiv \frac{hD}{k} \approx 0.023 \mathrm{Re}_D^{0.8}\Pr^{0.3} \quad \text{or} \quad \approx 0.023 \mathrm{Re}_D^{0.8}\Pr^{0.4}$$

Turbulent Flow Between Two Parallel Plates:
We will now consider fully developed turbulent flow between two parallel plates with a gap of b and a uniform wall heat flux $(q/A)_w$. Using the Kármán–Martinelli Analogy, determine the turbulent heat transfer coefficient. The result should be in a format such as $\mathrm{Nu} = \dfrac{h2b}{k} = \text{function of}\left(\mathrm{Re}, \mathrm{Pr}, f\right)$

Follow the same procedures as outlined above for the tube flow.

For flow between two parallel plates with a gap of b:

$$T^* = \frac{q_w''}{\rho c_p u^*}$$

$$\frac{u}{u_c} = \left(\frac{y}{\frac{1}{2}b}\right)^{1/7}, \quad \frac{T - T_w}{T_c - T_w} = \left(\frac{y}{\frac{1}{2}b}\right)^{1/7}$$

$$T_b = \frac{\displaystyle\int_0^{\frac{1}{2}b} uT dy}{\displaystyle\int_0^{\frac{1}{2}b} u dy}$$

$$\therefore T_w - T_b \cong 0.89\left(T_w - T_c\right)$$

$$= 0.89 T^*\left[5\mathrm{Pr} + 5\ln(5\mathrm{Pr} + 1) + 2.5\ln\left(\frac{\frac{1}{2}b^+}{30}\right)\right]$$

$$h = \frac{q_w''}{T_w - T_b}, \quad \mathrm{Nu} = \frac{hD_e}{k} = \frac{h2b}{k}$$

Thus

$$\mathrm{Nu} = \frac{h2b}{k} = \frac{\mathrm{Re}\,\mathrm{Pr}\sqrt{f/2}}{0.89\left[5\mathrm{Pr} + 5\ln(5\mathrm{Pr} + 1) + 2.5\ln\left(\dfrac{\frac{1}{2}b^+}{30}\right)\right]}$$

where

$$u^* = V\sqrt{\frac{1}{2}f}, \quad f = \frac{0.079}{Re^{1/4}}, \quad Re = \frac{V2b}{v}, \quad b^+ = \frac{bu^*}{v}$$

If $Re = \bar{V}2b/v = 2\times10^3, 2\times10^4, 2\times10^5$, Pr = 0.7, compare the result of Nu to those of empirical correlations, such that $Nu = \frac{h2b}{k} = 0.023\,Re^{0.8}\,Pr^{0.4}$.

For example, if the Reynolds number, Prandtl number, viscosity, and b are given, b^+ and f can be calculated. Next the Nu can be determined from above derived equation: $Re = 2\times10^3, 2\times10^2, 2\times10^5$, Pr = 0.7, $b = \sqrt{}$, $v = \sqrt{}$, $b^+ = \sqrt{}$, Nu = $\sqrt{}$.

The calculated results can be compared with experimental correlation, $Nu = 0.023\,Re^{0.8}\,Pr^{0.4} = \sqrt{}$

10.3.2 Turbulent External Flow Heat Transfer Coefficient

Consider a fully turbulent boundary-layer flow over a flat plate with uniform wall temperature as the thermal BC $(T_w = C)$ [4–6]. From the energy equation,

$$u\frac{\partial T}{\partial x} + v\frac{\partial T}{\partial y} = \frac{\partial}{\partial y}\left[(\alpha + \varepsilon_h)\frac{\partial T}{\partial y}\right] \tag{10.50}$$

For a fully turbulent boundary-layer flow $(v = 0)$, and assuming

$$\frac{\partial T}{\partial x} \sim \frac{\partial T_w}{\partial x} \approx 0$$

One can obtain the law of wall for the temperature profile as

$$T^+ = \int_0^{y^+} \frac{1}{(1/Pr)+(\varepsilon_h/v)}dy^+$$

Following a similar procedure as in the previous case, one can obtain the three-region temperature profile.

$$u\frac{\partial T}{\partial x} + v\frac{\partial T}{\partial y} = \frac{\partial}{\partial y}\underbrace{\left[(\alpha + \varepsilon_h)(\partial T/\partial y)\right]}_{const.=-q_w''/\rho C_p} = 0$$

$$(\alpha + \varepsilon_h)\frac{\partial T}{\partial y} = c$$

where $c = -\left(q_w''/\rho C_p\right)$

Therefore,

$$(\alpha + \varepsilon_h)\frac{\partial T}{\partial y} = \frac{-q_w''}{\rho C_p}$$

$$\int \frac{\partial T}{\partial y} = \frac{-q_w''/\rho C_p}{\alpha + \varepsilon_h}$$

$$T - T_w = \frac{-q_w''}{\rho C_p}\int_0^y \frac{1}{v\left((1/\Pr)+(\varepsilon_h/v)\right)}dy\frac{u^*}{u^*} \qquad (10.51)$$

$$T_w - T = \frac{q_w''}{\rho C_p u^*}\int_0^y \frac{1}{\left((1/\Pr)+(\varepsilon_h/v)\right)}dy^+$$

where $\varepsilon_h \simeq \varepsilon_m$.

And from Equation (10.20), the turbulent viscosity can be calculated,

$$\frac{\varepsilon_m}{v} = \frac{1-\left(y^+/\delta^+\right)}{\left(du^+/dy^+\right)} - 1$$

Following the same procedure as outlined previously for three regions:

$$0 < y^+ < 5, \quad u^+ = y^+$$
$$T^+ = \Pr y^+ \qquad (10.52)$$

$$5 < y^+ \le 30, \quad u^+ = 5\ln y^+ - 3.05$$
$$T^+ - T_5^+ = 5\ln\left(1+\frac{\Pr y^+}{5} - \Pr\right) \qquad (10.53)$$

$$30 \le y^+, \quad u^+ = 2.5\ln y^+ + 5.0$$
$$T^+ - T_{30}^+ = 2.5\ln y^+ - 2.5\ln 30 \qquad (10.54)$$

Note that $\Delta T^+ = \Delta u^+$ at $y^+ \ge 30$

$$T^+ - T_{30}^+ = u^+ - u_{30}^+$$
$$T_\infty^+ - T_{30}^+ = u_\infty^+ - u_{30}^+$$

where

$$u_{30}^+ = 14, \quad u_\infty^+ = \frac{u_\infty}{u^*} = \frac{1}{\sqrt{C_f/2}}$$

Therefore,

$$T_w - T_\infty = (T_w - T_5) + (T_5 - T_{30}) + (T_{30} - T_\infty)$$

$$= \frac{q_w''}{\rho C_p u^*}\left[5\,\mathrm{Pr} + 5\ln(5\,\mathrm{Pr}+1) + \left(T_\infty^+ - T_{30}^+\right)\right]$$

$$h = \frac{q_w''}{T_w - T_\infty} = \frac{q_w''}{\left(\dfrac{q_w''}{\rho C_p u^*}\right)\left[5\,\mathrm{Pr} + 5\ln(5\,\mathrm{Pr}+1) + \left(\left(1/\sqrt{C_f/2}\right) - 14\right)\right]}$$

The final heat transfer coefficient and the Stanton number can be obtained as

$$\mathrm{St} = \frac{\mathrm{Nu}_x}{\mathrm{Re}_x \mathrm{Pr}} = \frac{0.0296\mathrm{Re}_x^{-0.2}}{\underbrace{1 + 0.172\mathrm{Re}_x^{-0.1}\left[5\,\mathrm{Pr} + 5\ln(5\,\mathrm{Pr}+1) - 14\right]}_{=1}} \approx 0.0296\mathrm{Re}_x^{-0.2} \quad (10.55)$$

for $\mathrm{Pr} \cong 1$

where $\mathrm{Nu}_x = (hx/k)$, $\mathrm{Re}_x = (\rho \mathcal{U}_\infty x/\mu)$, $C_f = \left(0.0592/\mathrm{Re}_x^{1/5}\right)$.

For a given Re_x and Pr, the above prediction of the Nu_x value is very close to the following experimental correlation:

$$\mathrm{Nu}_x = 0.0296\mathrm{Re}_x^{0.8}\mathrm{Pr}^{1/3}$$

10.3.3 LAW OF THE WALL FOR VELOCITY AND TEMPERATURE PROFILES: TWO-REGION ANALYSIS

From Figure 10.8, a two-region (or two-layer) universal velocity profile can be assumed as

$$u^+ = y^+ \text{ for } 0 \; y^+ < 13.6$$

$$u^+ = 5.0 + 2.44 \ln y^+ \text{ for } 13.6 \le y^+$$

For fully developed turbulent pipe flow with a uniform q_w'', derive the corresponding two-layer university temperature profiles.

For $0 < y^+ < 13.6$: $T^* = \dfrac{q''}{\rho c_p u^*}, u^* = \sqrt{\dfrac{\tau_w}{\rho}}, u^+ = \dfrac{u}{u^*}, y^+ = \dfrac{yu^*}{v}$

$$T^+ = \int_0^{y^+} \frac{1 - \dfrac{y^+}{R^+}}{\dfrac{\varepsilon_h}{v} + \dfrac{1}{Pr}} dy^+ = Pr \cdot y^+ = \frac{T_w - T}{T^*}$$

For $13.6 < y^+$:

$$T^+ = \int_{13.6}^{y^+} \frac{\left(1 - \dfrac{y^+}{R^+}\right)}{\dfrac{\left(1 - \dfrac{y^+}{R^+}\right)}{2.44}} dy^+ = 2.44 \ln \frac{y^+}{13.6} = \frac{T_{13.6} - T}{T^*}$$

For example, predict the local u and T at $y = 0.05$ cm from the pipe wall under the following conditions:

friction velocity $u^* = 10$ m/s $\cong \sqrt{\dfrac{\tau_w}{\rho}}$

friction temperature $T^* = 3°C \cong \dfrac{q_w''}{\rho C_p u^*}$

pipe wall temperature $T_w = 100°C$

air flow Prandtl Pr = 0.7

at $y = 0.05$ cm, $u^* = 10$ m/s, $T^* = 3°C$, $T_w = 100°C$,

$$Pr = 0.7, v = 20.92 \times 10^{-6} \text{ m}^2/\text{s}$$

$$y^+ = \frac{yu^*}{v} = 239$$

$$u^+ = 18.36 \text{ m/s}, \quad u = 183.6 \text{ m/s}$$

$$T^+ = \frac{T_w - T_{13.6}}{T^*} = \frac{100 - T_{13.6}}{3} = 9.52, \quad T_{13.6} = 71.44°C$$

and $T^+ = \dfrac{T_{13.6} - T}{T^*} = \dfrac{71.44 - T}{3} = 6.994, \quad T = 50.46 \text{ °C}$

Following the same procedures as outlined above, one can obtain the internal heat transfer coefficient and Nusselt number by using these two-region temperature profiles as shown below.

$$\frac{T_b - T_w}{T_c - T_w} \cong 0.833$$

$$T_c - T_w = T_c - T_{13.6} + T_{13.6} - T_w = \left(2.5\ln\frac{R^+}{13.6} + 13.6\,\mathrm{Pr}\right)T^*$$

$$h = \frac{q_w''}{T_w - T_b}, \quad \mathrm{Nu}_D = \frac{hD}{k}, \quad T^* = \frac{q_w''}{\rho C_p u^*}$$

$$\therefore \mathrm{Nu}_D = \frac{\mathrm{Re}_D\,\mathrm{Pr}\,\sqrt{f/2}}{0.833\left[13.6\,\mathrm{Pr} + 2.5\ln\left(\frac{\mathrm{Re}_D}{27.2}\sqrt{f/2}\right)\right]}$$

where $f = \dfrac{0.079}{\mathrm{Re}_D^{1/4}}$

For example, $\mathrm{Re}_D = 10^4$, $\mathrm{Pr} = 1$, $\mathrm{Nu}_D = 35.5$

Compared with $\mathrm{Nu}_D = 0.023\,\mathrm{Re}_D^{0.8}\,\mathrm{Pr}^{0.4} = 36.5$, ~ 3% difference

Note: The calculation is dependent on how to interpolate the experimental data in the buffer zone, shown in Figure 10.8. This will not affect the heat transfer coefficient prediction using the two-region analysis. As the experimental data show in Figure 10.8, the dividing line of the two-regions can be $0 \le y^+ < 13.6$, $0 \le y^+ < 10.8$, $0 \le y^+ < 10$.

For external flow heat transfer, the above mentioned two-region velocity profiles can also be used to obtain the corresponding two-region temperature profiles using Equation (10.51). Finally, one can obtain the external heat transfer coefficient and Nusselt number by using these two-region temperature profiles.

REMARKS

In undergraduate heat transfer, students are expected to know how to calculate heat transfer coefficients (Nusselt numbers) for turbulent flows over a flat plate at uniform surface temperature and inside a circular tube at uniform surface heat flux, by using heat transfer correlations from experiments, that is, Nusselt numbers related to Reynolds numbers and Prandtl numbers. There are many engineering applications involving turbulent flow conditions. These turbulent flow heat transfer correlations are very useful for basic heat transfer calculations such as for heat exchangers design.

In intermediate-level heat transfer, this chapter focuses on how to derive the RANS equations; introduces the concept of turbulent viscosity and turbulent Prandtl number; Reynolds analogy; Prandtl mixing length theory; law of wall for velocity and temperature profiles; and turbulent flow heat transfer coefficients derived from the law of wall velocity and temperature profiles and their comparisons with the heat transfer correlations from experiments. This classic turbulent flow theory is important to provide students with a fundamental background in order to handle advanced turbulence models.

In advanced turbulent heat transfer, students will learn many more topics such as flow transition and transitional flow heat transfer and unsteady high turbulence flow and heat transfer (Chapter 15); surface roughness effect and heat transfer enhancement and rotating flow heat transfer (Chapter 16); high-speed flow and heat transfer (Chapter 7); and advanced turbulence models including the two-equation model and the Reynolds stress model (Section 10.5).

10.4 TURBULENT FLOW—SHEAR STRESS AND PRESSURE DROP

Theoretical Models for Calculation of Turbulent Flow

Three Regions in the Turbulent Flow:

1. Laminar sublayer—very thin, turbulence is suppressed: $u = \dfrac{\tau_o}{\mu} y$, $\tau = \tau_o = \mu \dfrac{du}{dy}$

2. Buffer zone—appears to be region where turbulence is generated; viscous and turbulence stresses are of same order of magnitude: $\tau = \mu \dfrac{du}{dy} + \left(-\rho \overline{u'v'}\right)$

3. Outer layer—turbulence dominates: $\tau = \tau_t = -\rho \overline{u'v'} = \rho \ell^2 \left(\dfrac{du}{dy}\right)^2$

Universal velocity profile for turbulent pipe or boundary layer flow.

$$v_t = \ell^2 \frac{du}{dy}$$

where $\ell = \kappa y$ from Prandtl Mixing-Length Theory

Further improved from Van Driest: $u' \to 0$, exponentially instead of linearly with y as

$$\ell = \kappa y \left[1 - \exp\left(-\frac{u^* y}{\nu A^+}\right)\right] = \kappa y \left[1 - \exp\left(-\frac{y^+}{A^+}\right)\right]$$

where $A^+ = 26$ experimentally determined by Van Driest where $y^+ > A^+$ is the fully turbulent region, similar to $\ell = \kappa y$, but at outer region $y/\delta > 0.2$, $\ell \cong$ constant $= \lambda \delta = 0.075 - 0.09\delta$.

10.4.1 TURBULENT INTERNAL FLOW FRICTION FACTOR

Kármán–Prandtl developed an equation to predict the friction factor for turbulent flow in a tube at a given Reynolds number. The procedure is outlined below.

Velocity Defect Law from Von Kármán:

From the Law of the Wall Velocity Profile:

$$\frac{u}{u^*} = 2.5\ln\frac{yu^*}{v} + 5.5$$

At the pipe centerline, $y = R$, or the boundary layer edge, $y = \delta$, the maximum velocity is

$$\frac{\mathcal{U}_{max}}{u^*} = 2.5\ln\frac{Ru^*}{v} + 5.5, \text{ subtract to above local velocity,}$$

$$\frac{\mathcal{U}_{max} - u}{u^*} = 2.5\ln\frac{R}{y} \quad \text{or} \quad \left(2.5\ln\frac{\delta}{y}\right)$$

where $y = R - r$ for pipe flow

The relationship between the maximum velocity and the mean velocity can be obtained:

$$\dot{m} = \rho\bar{V}\pi R^2 = \int_0^R \rho u 2\pi r dr$$

$$\bar{V} = \frac{2}{R^2}\int_0^R u r dr$$

From the above velocity defect law—for the majority of the flow domain, neglect the buffer and laminar sublayer regions

$$\frac{\mathcal{U}_{max} - u}{u^*} = 2.5\ln\left(\frac{R}{y}\right) = 2.5\ln\left(\frac{R}{R-r}\right)$$

$$\frac{u}{u^*} = \frac{\mathcal{U}_{max}}{u^*} - 2.5\ln\left(\frac{R}{R-r}\right), \text{ insert to above mean velocity,}$$

$$\bar{V} = \frac{2}{R^2}\int_0^R \left(\mathcal{U}_{max} - 2.5u^* \cdot \ln\left(\frac{R}{R-r}\right)\right) r dr$$

$$\bar{V} = \mathcal{U}_{max} - 3.75 \cdot u^*$$

Pipe Flow Friction Factor–Kármán–Prandtl Equation can be obtained as following
Since

$$u^* = \sqrt{\frac{\tau_o}{\rho}} = \sqrt{\frac{c_f \cdot \frac{1}{2}\rho\bar{V}^2}{\rho}} = \bar{V}\sqrt{\frac{c_f}{2}}, \text{ insert to the above maximum velocity,}$$

$$\frac{\mathcal{U}_{max}}{\bar{V}\sqrt{\frac{c_f}{2}}} = 2.5 \cdot \ln\left(\frac{R\bar{V}\sqrt{\frac{c_f}{2}}}{\nu}\right) + 5.5$$

$$\frac{\bar{V} + 3.75\bar{V}\sqrt{\frac{c_f}{2}}}{\bar{V}\sqrt{\frac{c_f}{2}}} = 2.5 \cdot \ln\left(\frac{R\bar{V}\sqrt{\frac{c_f}{2}}}{\nu}\right) + 5.5$$

$$\sqrt{\frac{2}{c_f}} + 3.75 = 2.5 \cdot \ln\left(\frac{D\bar{V}}{\nu} \cdot \frac{1}{2}\sqrt{\frac{c_f}{2}}\right) + 5.5$$

$$\sqrt{\frac{2}{c_f}} = 2.5 \cdot \ln\left(\mathrm{Re}_D \cdot \frac{1}{2}\sqrt{\frac{c_f}{2}}\right) + 1.75 \rightarrow \text{Karman–Prandtl Equation}$$

The Kármán–Prandtl equation had been approximated by Blasius as:

$$C_f = \frac{\tau_o}{\frac{1}{2}\rho\bar{V}^2} \cong \frac{0.079}{Re_D^{0.25}} \text{ if } \mathrm{Re}_D < 10^5$$

$$\cong \frac{0.046}{\mathrm{Re}_D^{0.2}} \text{ if } \mathrm{Re}_D > 10^5 \text{ for smooth pipes}$$

The friction factor versus Reynolds number can be seen from Figure 10.12, the Moody diagram for both smooth (from Blasius) and rough pipes (from Nikuradse sand grain roughness data, refer to Chapter 16). If we know the friction factor C_f, we can predict the pressure drop through the tube. From fully developed pipe flow, the pressure-drop and shear force can be balanced as:

$$(P_1 - P_2)\pi R^2 = \tau_o 2\pi R\Delta L$$

$$\Delta P = \tau_o \frac{4\Delta L}{D} = C_f \cdot \frac{1}{2}\rho\bar{V}^2 \cdot \frac{4\Delta L}{D}$$

Shear Stress Prediction from Power Law Velocity Profile—An Alternative Approach:
From the Power Law Velocity Profile, the wall shear stress can be in terms of $\mathcal{U}_{max}$:

If $\dfrac{u}{\mathcal{U}_{max}} = \left(\dfrac{y}{R}\right)^{\frac{1}{n}}$ for $\mathrm{Re} < 10^5, n \cong 7$

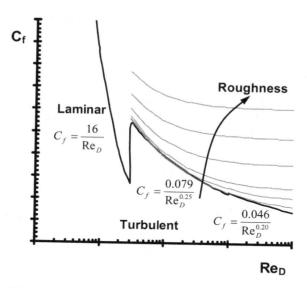

FIGURE 10.12   Moody diagram-friction factor versus Reynolds number.

Performing integration, $\bar{V} = 0.817\mathcal{U}_{max}$, therefore

$$c_f = \frac{0.079}{\text{Re}_D^{0.25}} = \frac{\tau_o}{\frac{1}{2}\rho\bar{V}^2} = \frac{0.079}{\left(\dfrac{\bar{V}D}{v}\right)^{0.25}}$$

$$\frac{\tau_o}{\frac{1}{2}\rho(0.817\mathcal{U}_{max})^2} = \frac{0.079}{\left(\dfrac{D}{v}\cdot 0.817\mathcal{U}_{max}\right)^{0.25}} \quad \text{where} \quad D = 2R, \text{ the wall shear stress}$$

becomes $\dfrac{\tau_o}{\rho\mathcal{U}_{max}^2} = \dfrac{0.0225}{\left(\dfrac{R\cdot\mathcal{U}_{max}}{v}\right)^{0.25}}$.

Effect of Roughness on Velocity Profile and Friction Factor:

The turbulent friction factor increases with surface roughness. It depends on the relative height of the roughness compared to the laminar sublayer thickness k/δ_ℓ. The laminar sublayer thickness decreases with increasing flow Reynolds number; therefore, k/δ_ℓ increases with increasing Reynolds number, i.e., the corresponding friction factor increases. Figure 10.13 shows the concept of k/δ_ℓ and the effect of roughness height-to-tube diameter ratio k/D on the friction factor. The following friction factor, with sand grain roughness data, was reported by Nikuradse, refer to Chapter 16 for details.

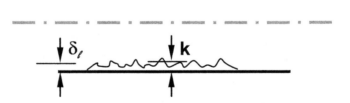

FIGURE 10.13 Concept of roughness height and effect on friction factor.

If $\quad \dfrac{k}{\delta_\ell} = \dfrac{\text{roughness height}}{\text{laminar sublayer}} < 1 \qquad \rightarrow \qquad$ no effect

$k < \delta_\ell < \dfrac{5v}{u^*}, \quad$ i.e. $\qquad \dfrac{ku^*}{v} < 5, \qquad \rightarrow \qquad$ no effect (sublayer)

$k \ge 14\delta_\ell \ge 14 \cdot \dfrac{5v}{u^*} \quad$ i.e. $\qquad \dfrac{ku^*}{v} > 70, \quad \rightarrow \quad$ has effect $k \ge 70 \cdot \dfrac{v}{u^*}$

If k is very large, $C_f \sim$ constant, the shear stress is dominated by Form Drag, $F_{\text{drag}} \sim \bar{V}^2$
The Law of Wall Velocity Profile for Rough Tube can be written as:

$$\frac{u}{u^*} = 2.5\ln\frac{yu^*}{v} + 5.5 - C\!\left(\frac{ku^*}{v}\right)$$

$$\frac{\mathcal{U}_{\max}}{u^*} = \ln\frac{Ru^*}{v} + 5.5 - C\!\left(\frac{ku^*}{v}\right)$$

$$\frac{u - \mathcal{U}_{\max}}{u^*} = 2.5\ln\frac{R}{y} \quad \left(\text{same as smooth pipe}\right)$$

The dimensionless velocity, u^+, decreases with increasing the roughness function C, as shown in Figure 10.14. This implies the friction factor increases with roughness as expected. Similarly, the dimensionless temperature T^+ decreases with C and the heat transfer coefficient increases with roughness, as well. Based on the friction similarity law and heat transfer similarity law, the roughness function and heat transfer function have been developed to correlate the friction factor and heat transfer coefficient for tube flows with various sand grain roughness sizes for given Reynolds and Prandtl numbers. Refer to Chapter 16 for details.

10.4.2 TURBULENT EXTERNAL FLOW FRICTION FACTOR

Integral Method—Momentum equation for pressure gradient flow can be written as

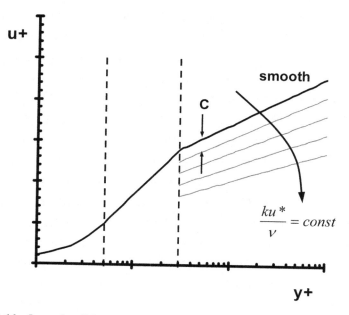

FIGURE 10.14   Law of wall for velocity profile for tube flows with roughness.

$$\frac{\tau_o}{\rho U^2} = \frac{d\theta}{dx} + \frac{1}{U}\frac{dU}{dx}\left(\delta^* + 2\theta\right)$$

$$= \frac{d\theta}{dx} + \frac{\theta}{U}\frac{dU}{dx}\left(2 + H\right)$$

where U = free stream velocity

In the laminar flow case (refer to Chapter 7 for the details):

Use a polynomial profile (4 terms) for the entire velocity profile by boundary conditions for the 4 constants.

Then

$$\tau_o = \mu\frac{\partial u}{\partial y}\bigg|_0$$

In the turbulent flow case:

Cannot represent all of the turbulent boundary layer with a polynomial profile. Options:

a. Divide the boundary layer into 3 regions

b. Use one expression for most of the flow plus an empirical relation for the wall shear stress

Using approach (b) for flat plate flow:

$$\frac{u}{U} = \left(\frac{y}{\delta}\right)^{1/7} \text{ for } \frac{U\delta}{v} < 10^5 \text{ okay}\left(\text{but not good at wall}\right)$$

The displacement thickness, momentum thickness, and shape factor can be determined by their definitions as:

$$\delta^* = \int_0^{\delta}\left(1 - \frac{u}{U}\right)dy$$

$$= \delta\int_0^{\delta}\left(1 - \frac{u}{U}\right)d\left(\frac{y}{\delta}\right) = \delta\int_0^{\delta}\left(1 - \left(\frac{y}{\delta}\right)^{\frac{1}{7}}\right)d\left(\frac{y}{\delta}\right)$$

$$\delta^* = \frac{1}{8}\delta$$

$$\theta = \int_0^{\delta}\frac{u}{U}\left(1 - \frac{u}{U}\right)d\left(\frac{y}{\delta}\right) = \delta\int_0^{\delta}\frac{u}{U}\left(1 - \frac{u}{U}\right)d\left(\frac{y}{\delta}\right)$$

$$= \delta\int_0^{\delta}\left(\frac{y}{\delta}\right)^{\frac{1}{7}}\left(1 - \left(\frac{y}{\delta}\right)^{\frac{1}{7}}\right)d\left(\frac{y}{\delta}\right)$$

$$\theta = \frac{7}{72}\delta$$

$$\frac{d\theta}{dx} = \frac{7}{72}\frac{d\delta}{dx}$$

$$H = \frac{\delta^*}{\theta} = \frac{1/8}{7/72} = \frac{9}{7} = 1.3 \rightarrow \text{much smaller than for the laminar boundary layer}$$

For flat plate flow:

$\dfrac{\tau_o}{\rho U^2} = \dfrac{d\theta}{dx}$, assume flat plate flow following the method for pipe flow with $u = U_{max}$, $\delta = R$.

From pipe flow as mentioned in the previous section, and apply to the external flow:

$$\frac{\tau_o}{\rho U_{max}^2} = \frac{0.0225}{\left(\dfrac{U_{max}R}{v}\right)^{0.25}}$$

$$\frac{\tau_o}{\rho U^2} = \frac{0.0225}{\left(\dfrac{U\delta}{\nu}\right)^{0.25}} = \frac{d\theta}{dx} = \frac{7}{72}\frac{d\delta}{dx}$$

$$0.0225\,dx = \left(\frac{U}{\nu}\right)^{0.25}\frac{7}{72}\delta^{1/4}\cdot d\delta$$

$$0.0225\,dx + \text{const} = \left(\frac{U}{\nu}\right)^{0.25}\frac{7}{72}\cdot\frac{4}{5}\cdot\delta^{5/4}$$

at $x = 0$, assume $\delta = 0$, constant $= 0$

$$\frac{\delta}{x} = \frac{0.37}{\left(\dfrac{Ux}{\nu}\right)^{1/5}}, \quad \delta(x) = \sqrt{\quad} \quad \delta^*(x) = \sqrt{\quad} \quad \theta(x) = \sqrt{\quad}$$

$$\tau_o \sim U^{9/5}$$

$$\tau_o = \rho U^2 \frac{d\theta}{dx} = \sqrt{\quad}$$

10.5 NUMERICAL MODELLING FOR TURBULENT FLOW HEAT TRANSFER

10.5.1 ZERO EQUATION MODEL–PRANDTL MIXING LENGTH THEORY (1925)

$$-\rho\overline{u'v'} = \rho\nu_t\frac{\partial u}{\partial y} \qquad \rightarrow \qquad \tau_w = \rho(\nu + \nu_t)\frac{\partial u}{\partial y}$$

$$-\rho c_p\overline{v'T'} = -\rho c_p\alpha_t\frac{\partial T}{\partial y} \qquad \rightarrow \qquad q_w'' = -\rho c_p(\alpha + \alpha_t)\frac{\partial T}{\partial y}$$

$$\nu_t = \ell^2\frac{\partial u}{\partial y} \quad \text{and} \quad \alpha_t \sim \nu_t, \qquad \ell \sim \kappa y$$

Good for: Pipe and boundary layer flat plate flow; thin shear layer flow.

Fails for: Pressure gradient flow; acceleration flow; curvature effect; recirculation flow; other complicated flows.

Prandtl Mixing Length Theory—assumes local equilibrium of turbulence; at each point the flow turbulent energy is dissipated at the same rate as it is produced. The theory cannot account for "Transport" and "History" effects of turbulence as sketched in Figure 10.15.

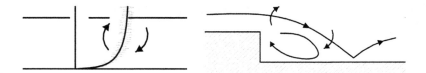

FIGURE 10.15   Concept of turbulence transport and history.

For example:

1. Flow in the pipe centerline, the predicted turbulent viscosity is zero which is wrong, $v_t = \ell^2 \dfrac{du}{dy} = \ell^2 \cdot 0 = 0$. Because the theory did not account the turbulence transport from the wall region to the centerline so that turbulent viscosity is not actually zero.
2. Flow over a backward facing step, the predicted turbulent viscosity is very small due to the small velocity gradient $\dfrac{du}{dy}$ in the recirculation zone, but turbulence is transported to this region, so the turbulent viscosity is actually very large.

10.5.2   Standard Two-Equation Models (κ-ε)

Considering both turbulent kinetic energy and dissipation rate can be transported and included in the momentum equation. Two-equation models have been developed.

Two equations: k equation for transport, k is turbulent kinetic energy

ε equation for transport, ε is turbulent dissipation rate, also

$$\varepsilon \sim \frac{k^{3/2}}{L}$$

where L is a length scale, can also transport and has a history in a similar way as the k equation.

Use the same turbulent viscosity v_T concept, turbulent viscosity is related to turbulent kinetic energy and length scale as:

$$v_t = c'_\mu \sqrt{k} \cdot L = c'_\mu \sqrt{k} \cdot C_D \frac{k^{3/2}}{\varepsilon} = c_\mu \frac{k^2}{\varepsilon}$$

Thus, the turbulent viscosity is related to the turbulent kinetic energy and dissipation rate. Consider turbulence production, P, is related to turbulent viscosity and velocity gradient as:

$$P = v_t \left(\frac{\partial u}{\partial y} \right)^2$$

and dissipation rate transport, one can obtain the following $k - \varepsilon$ transport equations. For High Reynolds Number Flow: Boundary Layer of $k - \varepsilon$ model:

1. $\dfrac{\partial u}{\partial x} + \dfrac{\partial v}{\partial y} = 0$

2. $u\dfrac{\partial u}{\partial x} + v\dfrac{\partial u}{\partial y} = -\dfrac{1}{\rho}\dfrac{\partial P}{\partial x} + \dfrac{\partial}{\partial y}(v + v_t)\dfrac{\partial u}{\partial y}$

3. $v_t = c_\mu \dfrac{k^2}{\varepsilon}$

4. $u\dfrac{\partial k}{\partial x} + v\dfrac{\partial k}{\partial y} = \dfrac{\partial}{\partial y}\left(\dfrac{v_t}{\sigma_k}\dfrac{\partial k}{\partial y}\right) + v_t\left(\dfrac{\partial u}{\partial y}\right)^2 - \varepsilon$

5. $u\dfrac{\partial \varepsilon}{\partial x} + v\dfrac{\partial \varepsilon}{\partial y} = \dfrac{\partial}{\partial y}\left(\dfrac{v_t}{\sigma_\varepsilon}\dfrac{\partial \varepsilon}{\partial y}\right) + c_{\varepsilon 1}\dfrac{\varepsilon}{k}v_t\left(\dfrac{\partial u}{\partial y}\right)^2 - c_{\varepsilon 2}\dfrac{\varepsilon^2}{k}$

Five Equations: 1—continuity; 2—momentum; 3—turbulent viscosity, relate to turbulent kinetic energy and dissipation rate; 4—kinetic energy, can be transported, including production and dissipation rate; 5—dissipation rate, can be transport.

Five Unknowns: u, v, v_t, k, ε

Five Experimental Constants: C_μ, σ_k, σ_ε, $c_{\varepsilon 1}$, $c_{\varepsilon 2}$

Experimental data fit for five constants;

$$C_\mu = 0.09 \qquad \sigma_k = 1.0 \qquad \sigma_\varepsilon = 1.3 \qquad c_{\varepsilon 1} = 1.45 \qquad c_{\varepsilon 2} = 2.0$$

Special treatment in the near wall region: In wall region (so called wall function), $k - \varepsilon$ model cannot be applied because molecular viscous effect dominates. At the edge of the laminar sublayer, y_δ, turbulence is at a local equilibrium, i.e., production = dissipation, $P = \varepsilon$.

Therefore, the law of the wall for velocity applied to the edge of the sublayer, y_δ, so $k - \varepsilon$ model applied to the edge of sublayer y_δ to bridge the laminar sublayer toward $y = 0$ at the wall. Using a wall function, the laminar sublayer thickness, y_δ^+, is around 50–100 depending upon the actual turbulent flow conditions.

$$u_\delta^+ = \frac{u_\delta}{u^*} = \frac{1}{\kappa}\ln\left(E \cdot y_\delta^+\right) \qquad k_\delta = \frac{u^{*2}}{\sqrt{c_\mu}} \qquad \varepsilon_\delta = \frac{u^{*3}}{k \cdot y_\delta}$$

$k - \varepsilon$ model for Low Reynolds Number Flow—Jones and Launder 1972

1. $\dfrac{\partial u}{\partial x} + \dfrac{\partial v}{\partial y} = 0$

2. $u\dfrac{\partial u}{\partial x} + v\dfrac{\partial u}{\partial y} = -\dfrac{1}{\rho}\dfrac{\partial P}{\partial x} + \dfrac{\partial}{\partial y}(v + v_t)\dfrac{\partial u}{\partial y}$

3. $v_t = c_\mu \dfrac{k^2}{\varepsilon}$

4. $u\dfrac{\partial k}{\partial x} + v\dfrac{\partial k}{\partial y} = \dfrac{\partial}{\partial y}\left[\left(v + \dfrac{v_t}{\sigma_k}\right)\dfrac{\partial k}{\partial y}\right] + v_t\left(\dfrac{\partial u}{\partial y}\right)^2 - \varepsilon - 2v\left(\dfrac{\partial k^{1/2}}{\partial y}\right)^2$

5. $u\dfrac{\partial \varepsilon}{\partial x} + v\dfrac{\partial \varepsilon}{\partial y} = \dfrac{\partial}{\partial y}\left[\left(v + \dfrac{v_t}{\sigma_\varepsilon}\right)\dfrac{\partial \varepsilon}{\partial y}\right] + c_{\varepsilon 1}\dfrac{\varepsilon}{k}v_t\left(\dfrac{\partial u}{\partial y}\right)^2 - c_{\varepsilon 2}\dfrac{\varepsilon^2}{k} + 2vv_t\left(\dfrac{\partial^2 u}{\partial y^2}\right)$

where

$c_\mu = 0.09\exp\left(\dfrac{-2.5}{1 + R_T/50}\right) \rightarrow= 0.09$ If R_T is very high

$\sigma_k = 1.0 \qquad \sigma_\varepsilon = 1.3 \qquad c_{\varepsilon 1} = 1.55$

$c_{\varepsilon 2} = 2.0\left[1 - 0.3\exp\left(-R_T^2\right)\right] \rightarrow= 2.0$ If R_T is very high

$R_T = \dfrac{k^2}{v\varepsilon} \sim \dfrac{v_T}{v}$ if very high $\rightarrow$ return to $k - \varepsilon$ model for high R_T

Low Reynolds number $k - \varepsilon$ model is an improved version of high Reynolds number $k - \varepsilon$ model. Equations 1–3 are the same. Equations 4 and 5 account for molecular viscous diffusion, nonisotropic dissipation, and constants depending on Reynolds number R_T as:

1. viscous diffusion of k, ε, must be included in transport Equations 4 and 5
2. constants become dependent on turbulence, R_T
3. additional term to account for nonisotropic dissipation

$R_T \sim \dfrac{v_t}{v}$ is the relative strength of the turbulent viscosity compared to the fluid viscosity. If R_T is very high, all constants come back to the standard $k - \varepsilon$ model for high Reynolds number flow.

Results of $k - \varepsilon$ prediction: Low Reynolds number $k - \varepsilon$ model (Jones and Launder, 1972) showed good prediction for boundary layer flow, even in the wake region, good prediction for pipe flow. Figure 10.16 shows good agreement between prediction and experimental data for velocity development in pipe flow with a sudden expansion. The low Reynolds number $k - \varepsilon$ model can catch the flow reversal immediately downstream of the step, but the Prandtl mixing length theory cannot.

Similarly, Figure 10.17 shows good agreement between the prediction and experimental data for velocity development in channel flow with an obstacle. The Low Reynolds number $k - \varepsilon$ model can catch the flow reversal downstream of the obstruction, but the Prandtl mixing length theory cannot.

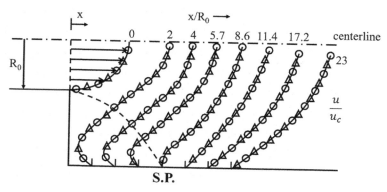

Velocity Development in a sudden pipe expansion
By Rodi (Germany) , O Pitot tube
△ Hot wire
— k-ε model

FIGURE 10.16 Velocity development in pipe flow with a sudden expansion.

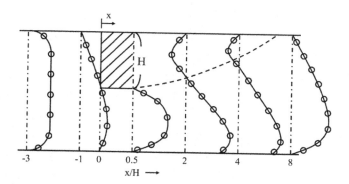

Channel Flow with An obstacle – Recirculation Flow

FIGURE 10.17 Velocity development in channel flow with an obstacle.

10.5.3 TWO EQUATION MODELS (K-ε) FOR THERMAL TRANSPORT

Use the same Equations 1, 2, 3, 4, and 5 for the $k - \varepsilon$ model.
Add 6 and 7 for thermal transport equations:

$$6.\ u\frac{\partial k_\theta}{\partial x} + v\frac{\partial k_\theta}{\partial y} = \frac{\partial}{\partial y}\left[\left(\alpha + \frac{\alpha_t}{\sigma_{k\theta}}\right)\frac{\partial k_\theta}{\partial y}\right] + \alpha_t\left(\frac{\partial T}{\partial y}\right)^2 - \varepsilon_\theta$$

7. ε_θ equation is as

$$\text{Assume:}\quad \frac{\varepsilon_\theta}{\varepsilon}\cdot\frac{k}{k_\theta} \sim 1 \rightarrow \varepsilon_\theta \cong \frac{k}{k_\theta}\varepsilon\ k \sim \frac{1}{\varepsilon}$$

$$\rightarrow \alpha_t = c_\mu \frac{k^2}{\varepsilon} \cdot \frac{1}{Pr_t}$$

$$Pr_t = \frac{v_t}{\alpha_t}$$

$$\text{and} \quad v_t = c_\mu \frac{k^2}{\varepsilon}$$

If using one-equation model:

$$\alpha_t = c_\mu' \frac{k^{1/2}L}{Pr_t} = \frac{\ell_1 \sqrt{k}}{Pr_t}$$

where ℓ_1 = constant.

Proposed $k - \varepsilon$ model for buoyancy-driven flow (natural convection):

$$4.\ u\frac{\partial k}{\partial x} + v\frac{\partial k}{\partial y} = \frac{\partial}{\partial y}\left[\left(v + \frac{v_t}{\sigma_k}\right)\frac{\partial k}{\partial y}\right] + v_t\left(\frac{\partial u}{\partial y}\right)^2 - \varepsilon - g\beta\overline{v'\theta'}$$

$$6.\ u\frac{\partial k_\theta}{\partial x} + v\frac{\partial k_\theta}{\partial y} = \frac{\partial}{\partial y}\left[\left(\alpha + \frac{\alpha_t}{\sigma_{k\theta}}\right)\frac{\partial k_\theta}{\partial y}\right] + \alpha_t\left(\frac{\partial T}{\partial y}\right)^2 - \varepsilon_\theta$$

where

$$g\beta\overline{v'\theta'} = g\beta\left(\frac{q_w''}{\rho c_p}\right) \cong g\beta\left(-\alpha_t \frac{\partial T}{\partial y}\right) \cong g\beta\left(\frac{v_t}{Pr_t}\frac{\partial T}{\partial y}\right)$$

v_t and α_t are the same as before. Equation 4 is for momentum and Equation 6 is for thermal transport.

REMARKS

Most common models are based on a two-equation turbulence model, namely, the $k - \varepsilon$ model, low Reynolds number $k - \varepsilon$ model, and low Reynolds number $k - \omega$ model. Advanced Reynolds stress models and the second-moment closure model are also employed [7,8]. The advanced second-order Reynolds stress (second-moment) turbulence models are capable of providing detailed three-dimensional velocity, pressure, temperature, Reynolds stresses, and turbulent heat fluxes that were not available in most of the experimental studies. Various turbulence models and CFD simulations have been applied for numerous engineering designs. For example, to improve gas turbine heat transfer and cooling systems, some of the above-mentioned turbulence models and CFD predictions are reviewed and documented in Chapter 7 of reference [9].

PROBLEMS

10.1 Consider a steady low-speed, constant-property, fully turbulent boundary-layer flow over a flat surface at constant wall temperature. Based on the Reynolds time-averaged concept, the following momentum and energy equations are listed for reference:

$$u\frac{\partial u}{\partial x} + v\frac{\partial u}{\partial y} = \frac{\partial}{\partial y}\left[(v + \varepsilon_m)\frac{\partial u}{\partial y}\right]$$

$$u\frac{\partial T}{\partial x} + v\frac{\partial T}{\partial y} = \frac{\partial}{\partial y}\left[\left(\frac{v}{\text{Pr}} + \frac{\varepsilon_m}{\text{Pr}_t}\right)\frac{\partial T}{\partial y}\right]$$

a. Define dimensionless parameters, u^+, T^+, and y^+, respectively, for universal velocity and temperature profiles for turbulent flow and heat transfer problems.
b. Based on the Prandtl's mixing length theory, derive and plot (u^+ versus y^+) the following universal velocity profile (make necessary assumptions):
$u^+ = y^+$ for a viscous sublayer region
$u^+ = 1/\kappa \ln y^+ + C$ for a turbulent layer (the law of the wall region)
c. Based on the heat and momentum transfer analogy, derive and plot (T^+ versus y^+) for various Pr numbers; the universal temperature profile (i.e., temperature law of the wall, T^+ (y^+, Pr)) for a turbulent boundary layer on a flat plate. Make necessary assumptions.

10.2 Consider a fully developed turbulent pipe flow with a uniform q_w''. If a two-layer universal velocity profiles can be assumed as

$$u^+ = y^+ \quad \text{for } 0 \le y^+ < 13.6$$

$$u^+ = 5.0 + 2.44 \ln y^+ \quad \text{for } 13.6 \le y^+$$

Derive the corresponding two-layer university temperature profiles.

Also, predict the local u and T at $y = 0.05$ cm from the pipe wall under the following conditions: friction velocity $u^* = 10$ m/s $\cong \sqrt{\tau_w/\rho}$, friction temperature $T^* = 3°C \cong q_w''/(\rho c_p u^*)$, pipe wall temperature $T_w = 100°C$, air flow Prandtl Pr $= 0.7$

10.3 Consider the Von Kármán–Martinelli heat–momentum analogy for a turbulent pipe flow:
a. If a two-layer universal velocity profile will be employed, that is,

$$\bar{u}^+ = y^+ \quad \text{for } 0 < y^+ < 10$$
$$\bar{u}^+ = 5.0 + 2.5\ell n y^+ \text{ for } 10 < y^+$$

For a constant wall heat flux, determine the universal temperature profiles at the corresponding two-layer region. Then determine the Nu_D where $\text{Nu}_D = $ function $(\text{Re}_D, \text{Pr}, f)$.

b. If $Re_D = 10^4$, $Pr = 1$, calculate Nu_D from (a) and then compare it with correlation $Nu_D = 0.023 Re_D^{0.8} \cdot Pr^{0.4}$. If velocity increases, the laminar sublayer thickness will be increased or decreased? Why? How about Nu_D?

10.4 Consider a fully developed turbulent flow between two parallel plates with a gap of b and a uniform wall heat flux $(q/A)_w$. Using the Kármán–Martinelli analogy, determine the turbulent heat transfer coefficient. The result should be in a format such as

$$Nu = \frac{h2b}{k} = \text{function of} \left(Re, Pr, f \right)$$

If $Re = \left(\bar{V} 2b/v \right) = 2 \times 10^3$, 2×10^4, 2×10^5, and $Pr = 0.7$, compare your result of Nu to those of semiempirical correlations, such that $Nu = (h2b/k) = 0.023 Re^{0.8} Pr^{0.4}$.

10.5   Consider the turbulent flow heat transfer.

a. Derive the following momentum and energy equations for a turbulent boundary-layer flow, a 2-D flat plate, incompressible, constant properties:

$$u \frac{\partial u}{\partial x} + v \frac{\partial u}{\partial y} = \frac{\partial}{\partial y} \left[(v + \varepsilon_m) \frac{\partial u}{\partial y} \right]$$

$$u \frac{\partial T}{\partial x} + v \frac{\partial T}{\partial y} = \frac{\partial}{\partial y} \left[\left(\frac{v}{Pr} + \frac{\varepsilon_m}{Pr_t} \right) \frac{\partial T}{\partial y} \right]$$

b. Derive the following momentum and energy equations for a fully developed turbulent flow in a circular tube, incompressible, constant properties:

$$\frac{1}{r} \frac{d}{dr} \left[r(v + \varepsilon_m) \frac{du}{dr} \right] = \frac{1}{\rho} \frac{dP}{dx}$$

$$u \frac{\partial T}{\partial x} = \frac{1}{r} \frac{\partial}{\partial r} \left[r(\alpha + \varepsilon_h) \frac{\partial T}{\partial r} \right]$$

10.6 Consider a fully developed turbulent pipe flow in a circular tube with a 5.0 cm I-D, constant properties.

a. Draw the velocity distribution from Martinelli Universal Velocity Profile (i.e., law of the wall) for air flow at 1 atm, 25°C, and $Re_D = 30,000$.

b. Calculate the laminar sublayer thickness.

10.7 Consider the turbulent boundary-layer flow heat transfer: air at 300 K, 1 atm, flows at 12 m/s along a flat plate maintained at 600 K. Plot the temperature profile $T(y)$ across the boundary layer for the following two cases:

a. At a location $x = 0.1$ m for a laminar boundary layer.

b. At a location $x = 1.0$ m if the transition Reynolds number is 10^5. Plot both profiles on the same graph to show significant differences.

c. Determine C_{fx}, τ_S, u^*, St_x, Nu_x, h_x, and q_s'' for case (b).

10.8 Consider a 2-D incompressible turbulent flow in a pipe.

a. Specialize (simplify) the given continuity and Navier–Stokes equations for a fully developed turbulent flow in a pipe. Write appropriate BCs to solve the flow equations for a fully developed turbulent flow. Do not attempt to solve the problem.

b. The velocity profile $(u(r))$ for a fully developed turbulent flow in a pipe is given by

$$\frac{u}{\mathcal{U}_{max}} = \left(\frac{R-r}{R}\right)^{1/7}$$

where R is the pipe radius, r is the radial distance measured from the pipe axis, and $\mathcal{U}_{max}$ is the maximum velocity. Calculate mean or the bulk velocity (U_m) for a fully developed flow in terms of $\mathcal{U}_{max}$.

c. Obtain an expression for the skin friction coefficient (Fanning friction factor) for a fully developed turbulent flow in terms of Re_D, where Re_D is the Reynolds number based on the pipe hydraulic diameter.

10.9 Consider a steady low-speed, constant-property, fully turbulent boundary-layer flow over a flat surface at constant wall temperature.

a. Based on the Reynolds time-averaged concept, derive the following momentum and energy equations (make necessary assumptions):

$$u\frac{\partial u}{\partial x} + v\frac{\partial u}{\partial y} = \frac{\partial}{\partial y}\left[(v + \varepsilon_m)\frac{\partial u}{\partial y}\right]$$

$$u\frac{\partial T}{\partial x} + v\frac{\partial T}{\partial y} = \frac{\partial}{\partial y}\left[\left(\frac{v}{Pr} + \frac{\varepsilon_m}{Pr_t}\right)\frac{\partial T}{\partial y}\right]$$

b. Define dimensionless parameters, u^+, T^+, and y^+, respectively, for universal velocity and temperature profiles for turbulent flow and heat transfer problems. Based on the Prandtl's mixing length theory, the following universal velocity profile has been derived:

$u^+ = y^+$ for the viscous sublayer region.

$u^+ = (1/\kappa)\ln y^+ + C$ for the turbulent layer (the law of the wall region).

Now, based on the heat and momentum transfer analogy, derive and plot (T^+ versus y^+ for various Pr) the universal temperature profile (i.e., temperature law of the wall, $T^+(y^+, Pr)$) for a turbulent boundary layer on a flat plate? Make the necessary assumptions.

10.10 Consider a steady low-speed, constant-property, fully turbulent boundary-layer flow over a flat surface at constant wall temperature. Based on the Reynolds time-averaged concept, the following momentum and energy equations can be derived:

$$u\frac{\partial u}{\partial x} + v\frac{\partial u}{\partial y} = \frac{\partial}{\partial y}\left[(v + \varepsilon_m)\frac{\partial u}{\partial y}\right]$$

$$u\frac{\partial T}{\partial x} + v\frac{\partial T}{\partial y} = \frac{\partial}{\partial y}\left[\left(\frac{v}{\mathrm{Pr}} + \frac{\varepsilon_m}{\mathrm{Pr}_t}\right)\frac{\partial T}{\partial y}\right]$$

a. Explain why the turbulent viscosity and turbulent Prandtl number should be included in the above equations. Explain the physical meaning and the importance of the turbulent viscosity and turbulent Prandtl number, respectively.

b. Based on the Prandtl's mixing length theory, the following universal velocity profile has been obtained:

$u^+ = y^+$ for the viscous sublayer region.

$u^+ = \left(\dfrac{1}{\kappa}\right)\ln y^+ + C$ for the turbulent layer (the law of the wall region).

Based on the heat and momentum transfer analogy, derive and plot (T^+ versus y^+ for various Pr) the universal temperature profile for a turbulent boundary layer on a flat plate. Make necessary assumptions.

10.11 Consider a turbulent flow over a flat plate at a constant wall temperature. From Prandtl mixing length theory, the two-region velocity profiles have been obtained as

$$\begin{array}{lll} u^+ = y^+ & \text{if} & 0 \le y^+ \le \delta_\ell^+ \\ u^+ = 1/\kappa \ln y^+ + C & \text{if} & \delta_\ell^+ \le y^+ \end{array}$$

where $\delta_\ell^+ = 11.6$, $\kappa = 0.41$, and $C = 5.0$

a. Based on heat and momentum transfer analogy, derive the two-region temperature profiles as

$$0 \le y^+ \le \delta_{t\ell}^+ \qquad\qquad T^+ = \mathrm{Pr}\, y^+$$

$$\delta_{t\ell}^+ \le y^+ \qquad T^+ = \frac{\mathrm{Pr}_t}{\kappa}\ln y^+ - \frac{\mathrm{Pr}_t}{\kappa}\ln\delta_{t\ell}^+ + T^+\left(\text{at } y^+ = \delta_{t\ell}^+\right)$$

where $\delta_{t\ell}^+ = 13.2$, $\kappa = 0.41$, and $\mathrm{Pr}_t = 0.9$.

b. Consider airflow (100 m/s, 25°C and 1 atmosphere) over a flat plate at a constant wall temperature of 100°C. Assume that the turbulent boundary layer starts from the leading edge of the plate. Using the above equations, predict the laminar "sublayer" thickness $\left(\delta_\ell^+\right)$ and the "thermal sublayer" thickness $\left(\delta_{t\ell}^+\right)$ at $x = 20$ cm from the leading edge of the plate.

c. At $x = 20$ cm and $y = 0.2$ cm, predict the local velocity u, and the local temperature T, respectively using the above log-law velocity and temperature profiles.

10.12 Derive the law of the wall u^+ vs. y^+ by using Van Driest mixing length equation.

10.13 Derive the turbulent pipe flow velocity profile and friction factor over a rough wall [see Chapter 16].

10.14 Derive the turbulent pipe flow temperature profile and heat transfer coefficient over a rough wall [see Chapter 16].

10.15 Derive the turbulent pipe flow heat transfer with a non-uniform wall heat flux boundary conditions (such as parabolic heat flux distribution along the pipe).

REFERENCES

1. W. Rohsenow and H. Choi, *Heat, Mass, and Momentum Transfer*, Prentice-Hall, Inc., Englewood Cliffs, NJ, 1961.

2. F. Incropera and D. Dewitt, *Fundamentals of Heat and Mass Transfer*, Fifth Edition, John Wiley & Sons, New York, 2002.

3. W.M. Kays and M.E. Crawford, *Convective Heat and Mass Transfer*, Second Edition, McGraw-Hill, New York, 1980.

4. A. Mills, *Heat Transfer*, Richard D. Irwin, Inc., Boston, MA, 1992.

5. H. Schlichting, *Boundary-Layer Theory*, Sixth Edition, McGraw-Hill, New York, 1968.

6. E. Levy, *Convection Heat Transfer, Class Notes*, Lehigh University, Bethlehem, PA, 1973.

7. W.P. Jones and B.E. Launder, "The prediction of laminarization with a two-equation model of turbulence," *International Journal of Heat and Mass Transfer*, Vol. 15(2), pp. 301–314, 1972.

8. B.E. Launder, G.J. Reece, and W. Rodi, "Progress in the development of a reynolds-stress turbulence closure," *Journal of Fluid Mechanics*, Vol. 68(3), pp. 537–566, 1975.

9. J.C. Han, S. Dutta, and S. Ekkad, *Gas Turbine Heat Transfer and Cooling Technology*, First Edition, CRC Press, Taylor & Francis Group, New York, 2000; second edition, 2013.

11 Fundamental Radiation

11.1 THERMAL RADIATION INTENSITY AND EMISSIVE POWER

Any surface can emit energy as long as its surface temperature is greater than absolute zero. Thermal radiation can refer to (1) surface radiation and (2) gas or volume radiation. Surface radiation is the radiation which comes from an opaque surface (such as a solid or liquid surface, penetrating approximately $1\,\mu m$ thick into the surface). Gas radiation is that radiation which comes from a volume of gas (such as CO_2, H_2O, CO, or NH_3). However, gas radiation is not associated with a volume of air, as air cannot emit or absorb radiation energy. Modern theory describes the nature of radiation in terms of electromagnetic waves that travel at the speed of light. The various forms of radiation differ only in terms of wavelength. In this chapter, the discussion will be confined to thermal radiation [1–4], as shown in Figure 11.1. Thermal radiation primarily depends on wavelength (spectral distribution, 0.1–$100\,\mu m$, from visible light to infrared (IR)), direction (directional distribution, θ, ϕ), material temperature (absolute temperature, °K or °R), and material radiation properties. Figure 11.2 shows the nature of spectral and directional distributions.

The following sections establish the relationship between surface radiation flux (i.e., radiation rate per unit surface area, also referred to as emissive power) and radiation intensity. Here we assume that radiation intensity from a surface is given. The later section will discuss how to obtain the radiation intensity from Planck. Figure 11.3 shows the conceptual view of hemispheric radiation from a surface (consider radiation from the upper surface only) [4]. Monochromatic directional radiation intensity (a function of wavelength and temperature in all directions) is defined as the differential radiation rate per unit surface area and unit solid angle,

$$I_\lambda\left(\theta,\varphi,\lambda,T\right) = \frac{dq}{dA_n d\omega} \tag{11.1}$$

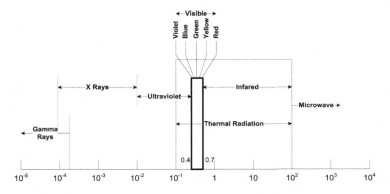

FIGURE 11.1 Spectrum of electromagnetic radiation.

DOI: 10.1201/9781003164487-11

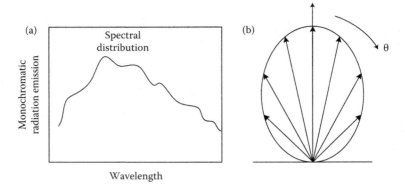

FIGURE 11.2 (a) Spectral radiation varies with wavelength. (b) Directional distribution.

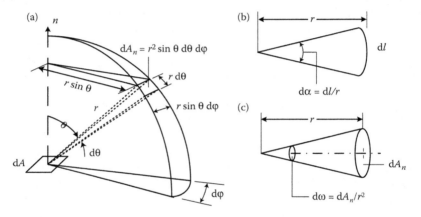

FIGURE 11.3 (a) Conceptual view of hemispheric radiation from a differential element area. (b) Definition of plane angle. (c) Definition of solid angle.

where $d\omega$ is the unit solid angle,

$$d\omega = \frac{dA_n}{r^2} = \frac{r d\theta \cdot r \sin\theta \cdot d\varphi}{r^2} = \sin\theta \, d\theta \, d\varphi$$

and

$$dA_n = dA \cos\theta$$

Therefore, the differential surface radiation flux (or differential emissive power) becomes

$$\frac{dq}{dA} \equiv dE_\lambda = I_\lambda\left(\theta,\varphi,\lambda,T\right)\sin\theta \cos\theta \, d\theta \, d\varphi$$

After integration in all directions, the monochromatic hemispherical emissive power is

$$E_\lambda = \int_0^{2\pi}\int_0^{\pi/2} I_\lambda\left(\theta,\varphi,\lambda,T\right)\sin\theta\cos\theta\,d\theta\,d\varphi \tag{11.2}$$

The total hemispherical emissive power (integration over wavelength) is

$$E(T) = \int_0^\infty E_\lambda\left(\lambda,T\right)d\lambda \tag{11.3}$$

For the isotropic surface (independent of circumferential direction angle), for most of the surfaces,

$$E_\lambda = 2\pi \int_0^{\pi/2} I_\lambda\left(\theta,\lambda,T\right)\sin\theta\cos\theta\,d\theta$$

For the isotropic and black surface (independent of vertical direction angle),

$$E_{b,\lambda} = 2\pi I_{b,\lambda}\left(\lambda,T\right)\int_0^{\pi/2}\sin\theta\cos\theta\,d\theta = \pi I_{b,\lambda} \tag{11.4}$$

11.2 SURFACE RADIATION PROPERTIES FOR BLACKBODY AND REAL-SURFACE RADIATION

Total emissive power for a black surface (ideal surface) is

$$E_b = \pi I_b = \sigma T^4 \tag{11.5}$$

Total emissive power for a real surface is

$$E = \pi I = \varepsilon\sigma T^4 \tag{11.6}$$

Monochromatic emissivity, a surface radiation property, is defined as monochromatic emissive power from a real surface to an ideal surface.

$$\varepsilon_{\lambda,\theta}\left(\theta,\varphi,\lambda,T\right) = \frac{E\left(\theta,\varphi,\lambda,T\right)}{E_b\left(\lambda,T\right)} \tag{11.7}$$

Therefore, the total hemispherical emissivity can be obtained as a ratio of total emissive power from a real surface to an ideal surface. The total emissivity varies from 0 to 1.

$$E(T) = \frac{E(T)}{E_b(T)} = 0 \sim 1 \tag{11.8}$$

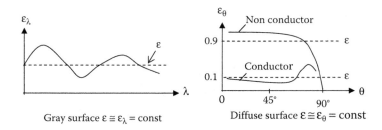

FIGURE 11.4 Definition of gray and diffuse surfaces.

A gray surface is defined as if the surface emissivity is independent of wavelength, as shown in Figure 11.4. A diffuse surface is defined as if the surface emissivity is independent of direction, as shown in Figure 11.4. In general, experimental data show that the emissivity of nonconductive materials ($\varepsilon \sim 0.9$ for nonmetals) is much higher than conductive materials ($\varepsilon \sim 0.1$ for metals). Emissivity is the most important radiation property and has a slight dependence on temperature. Emissivity is a property that can be experimentally measured and is available for various materials from any heat transfer textbook.

In addition to emission, a surface can reflect, absorb, or transmit any oncoming radiation energy (irradiation). Figure 11.5 sketches an energy balance between irradiation (radiation coming to the surface) and reflection, absorption, and transmission. Absorptivity, reflectivity, and transmissivity are defined as the fractions of irradiation that are absorbed, reflected, and transmitted, respectively. For many engineering gray and diffuse surfaces, we can assume that surface absorptivity is the same as surface emissivity (ε_λ, $\theta = \alpha_\lambda$, θ, $\varepsilon_\lambda = \alpha_\lambda$, then $\varepsilon = \alpha$, but in general $\varepsilon \neq \alpha$), and the transmissivity approaches zero (except window glass); therefore, the reflectivity can also be determined from the emissivity.

Absorptivity	$\alpha = G_a/G$
Reflectivity	$\rho = G_r/G$
Transmissivity	$\tau = G_t/G$
From radiation energy balance:	$\alpha + \rho + \tau = 1$
If $\tau = 0$, and assume $\alpha = \varepsilon$, then	$\rho = 1 - \alpha \approx 1 - \varepsilon$

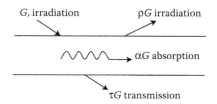

FIGURE 11.5 Radiation energy balance on a surface.

Blackbody radiation is the maximum radiation from an ideal surface. A blackbody is a diffuse surface and can emit the maximum radiation and can absorb the maximum radiation (i.e., $\alpha = 1$ and $\varepsilon = 1$). Therefore, blackbody radiation intensity is not a function of direction ($\neq (\theta, \varphi)$), but a function of wavelength and temperature ($= (\lambda, T)$). Planck obtained blackbody radiation intensity from quantum theory as

$$I_{\lambda,b}(\lambda,T) = \frac{2hC_0^2}{\lambda^5\left[\exp(hC_0/\lambda kT) - 1\right]} \tag{11.9}$$

where h is Planck's constant $= 6.626 \times 10^{-34}$ J s, C_0 is the speed of light in vacuum $= 2.998 \times 10^8$ m/s, k is the Boltzmann's constant $= 1.381 \times 10^{-23}$ J/K, and T is the absolute temperature, $°K$ or $°R$. It can also be shown as:

$$E_{\lambda,b} = \pi I_{\lambda,b} = \frac{C_1 \lambda^{-5}}{e^{(C_2/\lambda T)} - 1}$$

where

$$C_1 = 2\pi hC_0^2 = 3.742 \times 10^8 \ \text{W}\mu\text{m}^4/\text{m}^2$$

$$C_2 = hC_0/k = 1.4389 \times 10^4 \ \mu\text{m K}$$

Therefore, Planck's emissive power for the black surface can be shown as

$$E_{\lambda,b}(\lambda,T) = \pi I_{\lambda,b}(\lambda,T) = \pi\frac{2hC_0^2}{\lambda^5\left[\exp(hC_0/\lambda kT) - 1\right]} = \frac{C_1}{\lambda^5\left[\exp(C_2/\lambda t) - 1\right]}$$

Performing integration over the entire wavelength, one obtains the Stefan–Boltzmann law for blackbody radiation as

$$E_b(T) = \int_0^\infty E_{\lambda,b}(\lambda,T)d\lambda = \int_0^\infty \pi I_\lambda(\lambda,T)d\lambda = \sigma T^4 \ \text{W/m}^2 \tag{11.10}$$

with the Stefan–Boltzmann constant

$$\sigma = f(C_1,C_2) = 5.67 \times 10^{-8} \ \text{W/m}^2\text{K}^4$$

However, for a real surface, the emissive power is lower than Planck's blackbody radiation (because the emissivity for the real surface is less than unity). Therefore, the emissive power for the real surface is

$$E = \varepsilon\sigma T^4 \tag{11.11}$$

Figure 11.6 shows emissive power versus wavelength over a wide range of temperatures [1–4]. In general, emissive power increases with absolute temperature; emissive power for the black surface (solid lines) is greater than that for the real gray surface (lower than the solid lines, depending on emissivity) for a given temperature.

From Figure 11.6 we see that the blackbody emissive power distribution has a maximum and that the corresponding wavelength $\lambda_{\max}$ depends on temperature. Taking a derivative of Equation (11.9) with respect to λ and setting the result as equal to zero, we obtain Wien's displacement law as

$$\lambda_{\max} T = C_3 = 2898 \,\mu m\,K \qquad (11.12)$$

The focus of Wien's displacement law is also shown in Figure 11.6. According to this result, the maximum emissive power is displaced to shorter wavelengths with increasing temperature. For example, the maximum emission is in the middle of the visible spectrum ($\lambda_{\max} \approx 0.5\,\mu m$) for solar radiation at 5800 K; the peak emission occurs at $\lambda_{\max} = 1\,\mu m$ for a tungsten filament lamp operating at 2900 K emitting white light, although most of the emission remains in the IR region.

There are many engineering surfaces with diffuse, but not gray, behavior. In this case, surface emissivity is a function of wavelength and is not the same as absorptivity. Figure 11.7 shows the radiation problem between a hot coal bed and a cold brick wall with an emissivity as a function of wavelength [4]. To determine the emissive power from the cold brick wall, one needs to determine the average emissivity from the brick wall first. The following outlines a method to determine the average emissivity and absorptivity.

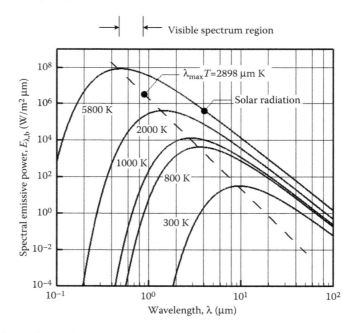

FIGURE 11.6   Spectral blackbody emissive power.

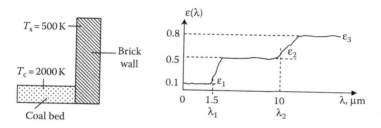

FIGURE 11.7 Radiation between a hot coal bed and a cold brick wall with nongray behavior.

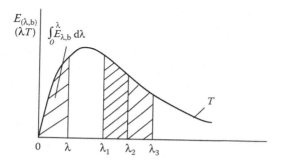

FIGURE 11.8 Concept of fraction method from a blackbody.

The following shows how to determine $\varepsilon(T_s)$, $E(T_s)$, and $\alpha(T_s)$. The average emissivity can be determined by combining the wavelengths of the three regions shown in Figure 11.7. Then each region is treated as a product of constant emissivity and the fraction of blackbody emissive power to the total blackbody emissive power, as shown in Figure 11.8. The fraction value is a function of wavelength and temperature and can be obtained from integration in each region (e.g., see Figure 11.9 or Table 11.1) [4].

$$E(T_s) = \frac{\int_0^\infty \varepsilon(\lambda) E_b \, d\lambda}{E_b} \tag{11.13}$$

$$= \varepsilon_1 \frac{\int_0^{\lambda_1} E_b \, d\lambda}{E_b} + \varepsilon_2 \frac{\int_{\lambda_1}^{\lambda_2} E_b \, d\lambda}{E_b} + \varepsilon_3 \frac{\int_{\lambda_2}^{\lambda_3} E_b \, d\lambda}{E_b} \tag{11.14}$$

$$= \varepsilon_1 F_{0-\lambda_1} + \varepsilon_2 \left[F_{0-\lambda_2} - F_{0-\lambda_1} \right] + \varepsilon_3 \left[1 - F_{0-\lambda_2} \right]$$
$$= 0.1 \times 0 + 0.5 \times 0.634 + 0.8 \times (1 - 0.634) \tag{11.15}$$

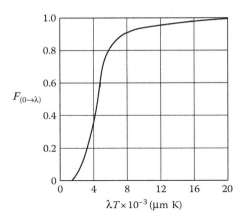

FIGURE 11.9 Fraction of the total blackbody emission in the spectral band from 0 to λ as a function of λT.

where

$$F_{0-\lambda} = \frac{\int_0^\lambda E_{\lambda,b}\,d\lambda}{\int_0^\infty E_{\lambda,b}\,d\lambda} = \frac{\int_0^\lambda E_{\lambda,b}\,d\lambda}{\sigma T^4} = \int_0^{\lambda T} \frac{E_{\lambda,b}\,d(\lambda T)}{\sigma T^5} = f(\lambda T) \qquad (11.16)$$

$$F_{\lambda_1-\lambda_2} = \frac{\int_0^{\lambda_2} E_{\lambda,b}\,d\lambda - \int_0^{\lambda_1} E_{\lambda,b}\,d\lambda}{\sigma T^4} = F_{0-\lambda_2} - F_{0-\lambda_1} \qquad (11.17)$$

From Table 11.1 or from Figure 11.9, $F_{0-\lambda}$ is a function of λT (μm K), with $T = T_s = 500$ K, emission from the brick wall.

The radiation constants used to generate these blackbody functions are

$$C_1 = 3.7420 \times 10^8 \ \mu\text{m}^4/\text{m}^2$$

$$C_2 = 1.4388 \times 10^4 \ \mu\text{m} \cdot \text{K}$$

$$\sigma = 5.670 \times 10^{-8} \ \text{W/m}^2 \cdot \text{K}^4$$

Therefore, the average emissivity can be calculated as

$$\varepsilon(T_s) = 0.61$$

and the total emissive power is

$$E(T_s) = \varepsilon(T_s)\sigma T_s^4 = 2161 \, \text{W/m}^2$$

The brick wall is not a gray surface, and hence $\alpha(T_s)$ is not equal to $\varepsilon(T_s)$. However, it is a diffuse surface, and hence $\alpha(\lambda) = \varepsilon(\lambda)$. In a window $0 \to \lambda_1$, $\lambda_1 \to \lambda_2$, $\lambda_2 \to \infty$, $\varepsilon_\lambda \approx$ constant (i.e., a gray surface), $\varepsilon_\lambda \approx \alpha_\lambda$. The irradiation from the black coal bed (at temperature $T_c = 2000$ K) to the brick wall is $G(\lambda) \propto E_b$. The following shows a similar way to determine the brick wall absorptivity.

$$\alpha(T_s) \equiv \frac{\int_0^\infty \alpha(\lambda)G(\lambda)d\lambda}{\int_0^\infty G(\lambda)d\lambda} = \frac{\int_0^\infty \varepsilon(\lambda)E_b\, d\lambda}{E_b} \tag{11.18}$$

$$= \varepsilon_1 F_{0-\lambda_1} + \varepsilon_2 \left[F_{0-\lambda_2} - F_{0-\lambda_1} \right] + \varepsilon_3 \left[1 - F_{0-\lambda_2} \right]$$
$$= 0.1 \times 0.275 + 0.5 \times (0.986 - 0.273) + 0.8 \times (1 - 0.986) \tag{11.19}$$

where the fractional values can be found in Table 11.1, or from Figure 11.9, with $T = T_c = 2000$ K, irradiation from the black coal bed.

Therefore, average absorptivity can be calculated as

$$\alpha(T_s) = 0.395 < 0.61 \text{ for } \varepsilon(T_s)$$

TABLE 11.1
Blackbody Radiation Functions

λT ($\mu m \cdot K$)	$F_{(0 \to \lambda)}$
200	0.000000
400	0.000000
600	0.000000
800	0.000016
1000	0.000321
1200	0.002134
1400	0.007790
1600	0.019718
1800	0.039341
2000	0.066728
2200	0.100888
2400	0.140256
2600	0.183120
2800	0.227897
2898	0.250108
3000	0.273232
3200	0.318102
3400	0.361735
3600	0.403607

(Continued)

TABLE 11.1 (*Continued*)
Blackbody Radiation Functions

λT (μm·K)	$F_{(0 \to \lambda)}$
3800	0.443382
4000	0.480877
4200	0.516014
4400	0.548796
4600	0.579280
4800	0.607559
5000	0.633747
5200	0.658970
5400	0.680360
5600	0.701046
5800	0.720158
6000	0.737818
6200	0.754140
6400	0.769234
6600	0.783199
6800	0.796129
7000	0.808109
7200	0.819217
7400	0.829527
7600	0.839102
7800	0.848005
8000	0.856288
8500	0.874608
9000	0.890029
9500	0.903085
10,000	0.914199
10,500	0.923710
11,000	0.931890
11,500	0.939959
12,000	0.945098
13,000	0.955139
14,000	0.962898
15,000	0.969981
16,000	0.973814
18,000	0.980860
20,000	0.985602
25,000	0.992215
30,000	0.995340
40,000	0.997967
50,000	0.998953
75,000	0.999713
100,000	0.999905

Source: Data from [4].

11.3 SOLAR AND ATMOSPHERIC RADIATION

Solar radiation is essential to all life on earth. Through the thermal and photovoltaic process, solar radiation is important for the design of solar collectors, air-conditioning systems for buildings and vehicles, temperature control systems for spacecraft, and photocells for electricity. The sun is approximated as a spherical radiation source with a diameter of 1.39×10^9 m and is located around 1.50×10^{11} m from the earth. The average solar flux (solar constant) incident on the outer edge of the Earth's atmosphere is approximately 1353 W/m². Assuming blackbody radiation, the sun's temperature can be estimated as 5800 K. Figure 11.10 shows the spectral distribution of solar radiation [2]. The radiation is concentrated in the low-wavelength region ($0.2 \leq \lambda \leq 3 \mu m$) with the peak value of $0.50 \mu m$.

The magnitudes of the spectral and directional distributions of solar flux change significantly as solar radiation passes through the Earth's atmosphere. The change is due to absorption and scattering of the radiation by atmospheric particles and gases. The effect of absorption by the atmospheric gases O_3 (ozone), H_2O vapor, O_2, and CO_2 is shown by the lower curve in Figure 11.10. Absorption by ozone is strong in the UV region, providing considerable attenuation below 0.3–0.4 μm. In the visible light region (0.4–0.7 μm), absorption is contributed by O_3 and O_2; in the IR region (0.7–3.0 μm), absorption is due to H_2O vapor and CO_2. The effect of scattering by particles and gases is that about half of the solar radiation goes back to atmosphere and approximately half reaches the earth surface. Therefore, the average solar flux incident on the Earth's surface is reduced to approximately 300–800 W/m², depending on the time of the day, the season, the latitude, and the weather conditions.

It is known that H_2O vapor and CO_2 gas in the atmosphere not only absorb solar radiation, but these gases also absorb radiation from the Earth's surface. Radiation is

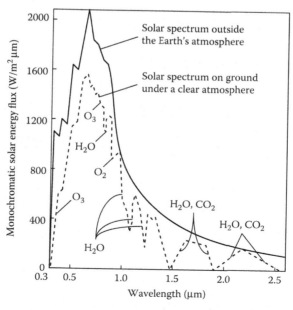

FIGURE 11.10 Solar spectra outside the Earth's atmosphere and on the ground.

emitted from the Earth's surface at approximately 300 K at wavelengths in the range of 10–20 μm with an emissivity approaching 1.0. In addition, H_2O vapor and CO_2 gas in the atmosphere (sky) can emit energy at wavelengths of 5–10 μm at the effective sky temperature around 250–270 K (with an assumed emissivity in the range of 0.8–1.0).

Figure 11.11 shows a typical setup for a solar collector. A special glass is used as a cover for the collector and a specialized coating is used on the collector plate and tubes where the solar energy is collected to maximize the performance of the collector.

Figure 11.12 shows a typical design for a house with a skylight. The thin glass of the skylight has a specific spectral emissivity or absorptivity distribution. For a given solar flux, atmospheric emission flux, interior surface emission flux, and convection inside and outside the house, the thin glass temperature or the inside house temperature can be predicted.

Examples

11.1 A simple solar collector plate without the cover glass has a selective absorber surface of high absorptivity α_1 (for $\lambda < 1$ μm) and low absorptivity α_2 (for $\lambda > 1$ μm). Assume the solar irradiation flux $= G_s$, the effective sky temperature $= T_{sky}$, the absorber surface temperature $= T_s$, and the ambient air temperature $= T_\infty$;

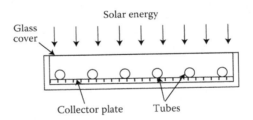

FIGURE 11.11 A typical setup for a solar collector.

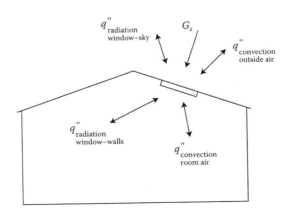

FIGURE 11.12 A typical design for a house with the skylight.

determine the useful heat removal flux $\left(q''_{useful}\right)$ from the collector under these conditions. What is the correspondent efficiency (η) of the collector?

Solution

Performing an energy balance on the absorber plate per unit surface area, we obtain

$$q''_{useful} = \alpha_s\,G_s + \alpha_{sky}G_{sky} - q''_{conv} - E$$

$$\eta = \frac{q''_{useful}}{G_s}$$

where

$$\alpha_s = \alpha_1$$

$$\alpha_{sky} = \alpha_2$$

$$G_{sky} = \sigma T_{sky}^4$$

$$E = \varepsilon\sigma T_s^4$$

$$\varepsilon \cong \alpha_2$$

$$q''_{conv} = h(T_s - T_\infty)$$

11.2 A thin glass is used on the roof of a greenhouse. The glass is totally transparent for $\lambda < 1\,\mu m$, and opaque with an absorptivity $\alpha = 1$ for $\lambda > 1\,\mu m$. Assume that solar flux = G_s, atmospheric emission flux = G_{atm}, thin glass temperature = T_g, and interior surface emission flux = G_i, where G_{atm} and G_i are concentrated in the far IR region ($\lambda > 10\,\mu m$), and determine the temperature of the greenhouse ambient air (i.e., inside room air temperature, $T_{\infty,\,i}$).

Solution

Performing an energy balance on the thin glass plate per unit surface area, and considering two convection processes (inside and outside greenhouse), two emissions (inside and outside the glass plate), and three absorbed irradiations (from solar, atmospheric, interior surface), we obtain $T_{\infty,\,i}$ from

$$\alpha_s G_s + \alpha_{atm} G_{atm} + h_o\left(T_{\infty,o} - T_g\right) + h_i\left(T_{\infty,i} - T_g\right) + \alpha_i G_i - 2\varepsilon\sigma T_g^4 = 0$$

where α_s = solar absorptivity for absorption of $G_{\lambda,\,s} \sim E_{\lambda,\,b}$ (λ, 5800 K)

$$\alpha_s = \alpha_1 F_{0-1\mu m} + \alpha_2\left[1 - F_{0-1\mu m}\right]$$

$$= 0 \times 0.72 + 1.0\left[1 - 0.72\right]$$

$$= 0.28$$

(Note: From Table 11.1, $\lambda T = 1\,mm \times 5800 K$, $F_{0-1\,\mu m} = 0.72$)

α_{atm} = absorptivity for $\lambda > 10\,\mu m = 1$
α_i = absorptivity for $\lambda > 10\,\mu m = 1$
$\varepsilon = \alpha\lambda$ for $\lambda \gg 1\,\mu m$, $= 1$
(emissivity of the glass for long wavelength emission)
h_o = convection heat transfer coefficient of the outside roof
h_i = convection heat transfer coefficient of the inside room

11.3 The glass of the skylight of a house, as shown in Figure 11.12, has a spectral emissivity, $\varepsilon_\lambda(\lambda)$ [or absorptivity, $\alpha_\lambda(\lambda)$] distribution as shown: $\varepsilon_\lambda = 0.9$ for $0 \le \lambda \le 0.3\,\mu m$, $\varepsilon_\lambda = 0$ for $0.3 \le \lambda \le 2.0\,\mu m$, $\varepsilon_\lambda = 0.9$ for $\lambda \ge 2.0\,\mu m$. During an afternoon when the solar flux is $900\,W/m^2$, the temperature of the glass is 27°C. The interior surfaces of the walls of the house and the air in the house are at 22°C, and the heat transfer coefficient between the glass of the skylight and the air in the house is $5\,W/(m^2 \times K)$.

    a. What is the overall emissivity, ε, of the skylight?
    b. What is the overall absorptivity, α, of the skylight for solar irradiation? You may assume the sun emits radiation as a blackbody at 6000 K.
    c. What is the convective heat flux on the outer surface of the skylight, in W/m^2? Is the temperature of the outside air higher or lower than the temperature of the skylight? Please assume that the sky is at 0°C.

Solution

a. $\varepsilon = 0.9F_{0-0.3} + 0(F_{0-2} - F_{0-0.3}) + 0.9(1 - F_{0-2}) \sim 0.9$

b. $\alpha_s = 0.9F_{0-0.3} + 0(F_{0-2} - F_{0-0.3}) + 0.9(1 - F_{0-2})$

$\qquad = 0.9 \times 0.0393 + 0.9(1 - 0.945) \sim 0.085$

c. Refer to Example 11–2:

$$\alpha_s = 0.085,\, G_s = 900\,\frac{W}{m^2},\, G_{atm} = \sigma T_{sur}^4,\, \alpha_{atm} \approx 0.9,$$

$$h_i = 5\,\frac{W}{m^2 K},\, T_{\infty,i} = 22°C,\, T_g = 27°C,\, \varepsilon = 0.9$$

$$\alpha_i \cong 0.9,\, G_i = \varepsilon_i \sigma T_i^4,\, \varepsilon_i \cong 0.9,\, T_i = 22°C,\, T_{sky} = 273K$$

$$\therefore h_0\left(T_{\infty,0} - T_g\right) = \sqrt{} = +,\, T_{\infty,0} > T_g-,\, T_{\infty,0} < T_g$$

11.4 Consider an opaque, horizontal plate with an electrical heater on its backside. The front side is exposed to ambient air that is at 20°C and provides a convection heat transfer coefficient of $10\,W/m^2 \cdot K$, solar irradiation (at 5800 K) of $600\,W/m^2$, and an effective sky temperature of −40°C. What is the electrical power (W/m^2) required to maintain the plate surface temperature at $T_s = 60°C$ (steady state) if the plate is diffuse and has designated spectral, hemispherical reflectivity (reflectivity = 0.2 for a wavelength less than $2\,\mu m$, reflectivity = 0.7 for wavelength greater than $2\,\mu m$)?

Solution

$$\alpha_{solar} = 0.8F_{0-2} + 0.3(1 - F_{0-2})$$

$$\approx 0.8 \quad \text{where } F_{0-2} \cong 0.94 \quad \text{at } \lambda T = 2\,\mu m \times 5800K$$

$$\alpha_{sky} \cong 0.3 \quad \text{where } F_{0-2} = 0 \quad \text{at } \lambda T = 2\,\mu m \times 233\,K$$

$$\varepsilon_{plate} \cong 0.3 \quad \text{where } F_{0-2} = 0 \quad \text{at } \lambda T = 2\,\mu m \times 333\,K$$

$$q''_{heater} = \varepsilon_p \sigma T_s^4 + h(T_s - T_\infty) - \alpha_{solar} \times 600 - \alpha_{sky}\sigma T_{sky}^4$$

$$\cong 97\,\frac{W}{m^2}$$

11.5 A diffuse surface has the following spectral characteristics ($\varepsilon_\lambda = 0.4$ for $0 \le \lambda \le 3\,\mu m$, $\varepsilon_\lambda = 0.8$ for $3\,\mu m \le \lambda$) is maintained at 500 K when situated in a large furnace enclosure whose walls are maintained at 1500 K.
a. Sketch the spectral distribution of the surface emissive power E_λ and the emissive power $E_{\lambda,b}$ that the surface would have if it were a blackbody.
b. Neglecting convection effects, what is the net heat flux to the surface for the prescribed conditions?
c. Plot the net heat flux as a function of the surface temperature for $500 \le T \le 1000$ K. On the same coordinates, plot the heat flux for a diffuse, gray surface with total emissivities of 0.4 and 0.8.
d. For the prescribed spectral distribution of ε_λ, how do the total emissivity and absorptivity of the surface vary with temperature in the range $500 \le T \le 1000$ K?

Solution

a. From Figure 11.6, for a blackbody

$$\lambda_{max} = \frac{2897.6}{500} = 5.8\mu m$$

$$E_\lambda = 0.4E_{\lambda,b} \quad \text{for } \lambda < 3\,\mu m = 0.8E_{\lambda,b} \quad \text{for } \lambda > 3\,\mu m$$

b. $\varepsilon = 0.4F_{0-3} + 0.8(1 - F_{0-3})$

$$= 0.795 \quad \text{where } F_{0-3} = 0.01376 \text{ at } \lambda T_s = 1500\,\mu m \cdot K$$

$$\alpha = 0.4F_{0-3} + 0.8(1 - F_{0-3})$$

$$= 0.574 \quad \text{where } F_{0-3} = 0.564 \text{ at } \lambda T_s = 4500\,\mu m \cdot K$$

$$q''_{net} = \alpha G - E = \alpha\sigma T_{sur}^4 - \varepsilon\sigma T_s^4$$

$$= 5.67 \times 10^{-8}\left[0.574(1500)^4 - 0.795(500)^4\right]$$

$$= 161.946 \times 10^3\,\frac{W}{m^2}$$

c. $q''_{net} \downarrow$ with $T_s \uparrow$

$q''_{net} |\varepsilon = 0.4 \gg q''_{net} |\varepsilon = 0.795 \geq q''_{net} |\varepsilon = 0.8$

d. $\varepsilon \downarrow$ with $T_s \uparrow$

$\alpha \approx$ constant for $T_s = 500 \sim 1000\,\text{K}$

REMARKS

This chapter covers the same topics as in undergraduate-level heat transfer. These include spectrum thermal radiation intensity and emissive power for a blackbody, as well as a real surface at elevated temperatures; surface radiation properties such as spectral emissivity and absorptivity for real-surface radiation; how to obtain the total emissivity or absorptivity from the fraction method; how to perform energy balance from a flat surface including radiation and convection; and solar and atmospheric radiation problems. This chapter provides fundamental thermal radiation and surface properties that are useful for many engineering applications such as surface radiators, space vehicles, and solar collectors.

PROBLEMS

11.1 A diffuse surface having the following spectral distributions ($\varepsilon_\lambda = 0.3$ for $0 \leq \lambda \leq 4\,\mu\text{m}$, $\varepsilon_\lambda = 0.7$ for $4\,\mu\text{m} \leq \lambda$) is maintained at 500 K when situated in a large furnace enclosure whose walls are maintained at 1500 K. Neglecting convection effects,
   a. Determine the surface's total hemispherical emissivity (ε) and absorptivity (α).
   b. What is the net heat flux to the surface for the prescribed conditions?
      Given: $\sigma = 5.67 \times 10^{-8}$ (W/m^{2}K^4)
11.2 An opaque, gray surface at 27°C is exposed to an irradiation of 1000 W/m^2, and 800 W/m^2 is reflected. Air at 17°C flows over the surface, and the heat transfer convection coefficient is 15 W/m^2 K. Determine the net heat flux from the surface.
11.3 A diffuse surface having the following spectral characteristics ($\varepsilon_\lambda = 0.4$ for $0 \leq \lambda \leq 3\,\mu\text{m}$, $\varepsilon_\lambda = 0.8$ for $3\,\mu\text{m} \leq \lambda$) is maintained at 500 K when situated in a large furnace enclosure whose walls are maintained at 1500 K:
   a. Sketch the spectral distribution of the surface emissive power E_λ and the emissive power $E_{\lambda,b}$ that the surface would have if it were a blackbody.
   b. Neglecting convection effects, what is the net heat flux to the surface for the prescribed conditions?
   c. Plot the net heat flux as a function of the surface temperature for $500 \leq T \leq 1000$ K. On the same coordinates, plot the heat flux for a diffuse, gray surface with total emissivities of 0.4 and 0.8.
   d. For the prescribed spectral distribution of ε_λ, how do the total emissivity and absorptivity of the surface vary with temperature in the range $500 \leq T \leq 1000$ K?

11.4 The spectral, hemispherical emissivity distributions for two diffuse panels to be used in a spacecraft are as shown.

For panel A: $\varepsilon_\lambda = 0.5$ for $0 \le \lambda \le 3\,\mu m$, $\varepsilon_\lambda = 0.2$ for $3\,\mu m < \lambda$.
For panel B: $\varepsilon_\lambda = 0.1$ for $0 \le \lambda \le 3\,\mu m$, $\varepsilon_\lambda = 0.01$ for $3\,\mu m \le \lambda$.

Assuming that the backsides of the panels are insulated and that the panels are oriented normal to the solar flux at $1300\,W/m^2$, determine which panel has the highest steady-state temperature.

11.5 From a heat transfer and engineering approach, explain how a glass green-house, which is used in the winter to grow vegetables, works. Include sketches of both the system showing energy flows and balances, and of radiation property data (radioactive properties versus wavelengths) for greenhouse components (glass and the contents inside the greenhouse). When applica-ble, show the appropriate equations and properties to explain the greenhouse phenomenon. When finished with the above for a glass greenhouse, extend your explanation to global warming, introducing new radioactive properties and characteristics if needed.

11.6 The glass of the skylight of a house as shown in Figure 11.12 has a spectral emissivity, $\varepsilon_\lambda(\lambda)$ [or absorptivity, $\alpha_\lambda(\lambda)$] distribution as shown: $\varepsilon_\lambda = 0.9$ for $0 \le \lambda \le 0.3\,\mu m$, $\varepsilon_\lambda = 0$ for $0.3 \le \lambda \le 2.0\,\mu m$, $\varepsilon_\lambda = 0.9$ for $\lambda \ge 2.0\,\mu m$. During an afternoon when the solar flux is $900\,W/m^2$, the temperature of the glass is $27°C$. The interior surfaces of the walls of the house and the air in the house are at $22°C$, and the heat transfer coefficient between the glass of the skylight and the air in the house is $5\ W/(m^2 K)$.
   a. What is the overall emissivity, ε, of the skylight?
   b. What is the overall absorptivity, α, of the skylight for solar irradiation? You may assume that the sun emits radiation as a blackbody at $6000\ K$.
   c. What is the convective heat flux on the outer surface of the skylight, $q''_{convection\ outside\ air}$, in W/m^2? Is the temperature of the outside air higher or lower than the temperature of the skylight? Please assume that the sky is at $0°C$.

11.7 Consider a typical setup for a solar collector as shown in Figure 11.11. A special glass is used as a cover for the collector and a specialized coating is used on the collector plate and tubes, where the solar energy is collected, to maximize the performance of the collector.
   a. If you had to specify the value of the glass transmissivity, τ_λ, as a func-tion of λ to maximize the performance of the collector, what would you choose and why? Explain. Use illustrations or sketches if needed to help explain your answer.
   b. If you had to specify the value of the collector plate and tube absorptivity, α_λ, as a function of λ to maximize the performance of the collector, what would you choose and why? Explain.
   c. A manufacturing process calls for heating a long aluminum rod that is coated with a thin film with an emissivity of ε. The rod is placed in a large convection oven whose surface is maintained at T_w (K). Air at T_∞ (K)

circulates in the oven at a velocity of u (m/s) across the surface of the rod and produces a convective heat transfer coefficient of h [W/(m² K)]. The rod has a small diameter of d (m) and has an initial temperature of T_i (K). Here, $T_i < T_\infty < T_w$. What is the rate of change of the rod temperature (K/s) when the rod is first placed in the oven?

11.8 A solar collector consists of an insulating back layer, a fluid conduit through which a water–glycol solution flows to remove heat, an absorber plate, and a glass cover plate. The external temperatures T_{air}, T_{sky}, and T_{ground} are known. Solar radiation of intensity q_s'' (W/m²) is incident on the collector and collected heat $q_c''\left(\text{W/m}^2\right)$ is removed by the fluid. The absorber plate is painted black with an average solar absorptivity of 0.8 and an average emissivity of 0.8. Assume that the collector plate is so large that you may treat the problems as 1-D heat flow with heat sources and/or sinks.

a. Identify and label all significant heat transfer resistances and flows and draw the steady-state thermal network diagram for the collector.

b. Write the heat balance equations needed to solve for q_c''. Do not solve.

c. If the absorber plate is replaced with a black chrome surface with an average solar absorptivity of 0.95 and an average emissivity of 0.1, what values will change in the thermal network diagram? How will q_c'' change and why?

11.9 A solar collector consists of an insulating back layer, a fluid conduit through which a water–glycol solution flows to remove heat, an absorber plate and a glass cover plate. The external temperatures T_{air}, T_{sky}, and T_{ground} are known. Solar radiation of intensity q_s'' (W/m²) is incident on the collector and collected heat q_c'' (W/m²) is removed by the fluid. The absorber plate is painted black with an average solar absorptivity of 0.95 and an average emissivity of 0.95. Assume that the collector is so large that you may treat the problems as 1-D heat flow with heat sources and/or sinks.

a. Identify and label all significant heat transfer resistances and flows and draw the steady-state thermal network diagram for this collector.

b. Write the heat balance equations needed to solve for q_c''. Do not solve.

c. If the absorber plate is replaced with a black chrome surface with an average solar absorptivity of 0.95 and an average emissivity of 0.1, what values will change in the thermal network diagram? How will q_c'' change and why?

11.10 An opaque, gray surface at 27°C is exposed to an irradiation of 1000 W/m², and 600 W/m² is reflected. Air at 20°C flows over the surface and the heat transfer convection coefficient is 20W/m²K. Determine the net heat flux from the surface.

11.11 Consider an opaque, horizontal plate with an electrical heater on its back-side. The front side is exposed to ambient air that is at 20°C and provides a convection heat transfer coefficient of 10 W/m² K, a solar irradiation (at 5800°K) of 600 W/m², and an effective sky temperature of −40°C. What is the electrical power (W/m²) required to maintain the plate surface temperature at $T_s = 60°C$ (steady state) if the plate is diffuse and has designated spectral, hemispherical reflectivity (reflectivity $= 0.2$ for wavelength less than $2\,\mu m$, reflectivity $= 0.7$ for wavelength greater than $2\,\mu m$)?

11.12 Consider a typical setup for a solar collector: A special glass is used as a cover for the collector and a specialized coating is used on the collector plate and tubes, where the solar energy is collected, to maximize the performance of the collector. The external temperatures T_{air}, T_{sky}, and T_{ground} are known. The solar radiation of intensity q_s'' (W/m^2) is incident on the collector and a collected heat flux, q_c'' (W/m^2), is removed by the fluid inside the tubes.

a. Write the energy balance equations that can be used to solve q_c'' for the solar collector, and write the energy balance equations that can be used to determine the glass cover temperature. You do not need to calculate the numerical answer. Make any necessary assumptions.

b. If you had to specify the value of the glass transmissivity, τ_λ, as a function of λ to maximize performance of the collector, what would you choose and why? Explain. Use illustrations or sketches if needed to help explain your answer. If you had to specify the value of the collector plate and tube absorptivity, α_λ, as a function of λ to maximize performance of the collector, what would you choose and why?

REFERENCES

1. W. Rohsenow and H. Choi, *Heat, Mass, and Momentum Transfer*, Prentice-Hall, Inc., Englewood Cliffs, NJ, 1961.

2. A. Mills, *Heat Transfer*, Richard D. Irwin, Inc., Boston, MA, 1992.

3. K.-F. Vincent Wong, *Intermediate Heat Transfer*, Marcel Dekker, Inc., New York, 2003.

4. F. Incropera and D. Dewitt, *Fundamentals of Heat and Mass Transfer*, Fifth Edition, John Wiley & Sons, New York, 2002.

12 View Factors

12.1 VIEW FACTORS

In addition to surface radiation properties such as emissivity, reflectivity, and absorptivity, the view factor is another important parameter for determining radiation heat transfer between two surfaces. The view factor is defined as the fraction of radiation energy (the so-called radiosity including emission and reflection) from a given surface that can be seen (viewed) by the other surface. It is purely a geometric parameter dependent upon the relative geometric configuration between two surfaces (how the surfaces can "see" each other). A view factor is also called an angle factor or shape factor. The following shows how to define the view factor between two surfaces [1–4].

Figure 12.1 shows radiation exchange between two diffuse isothermal surfaces (i.e., each surface has uniform emission and reflection). The differential radiation rate (including emission and reflection) from the unit surface i to j is proportional to its intensity and the unit solid angle as discussed previously.

$$dq_{i-j} = I_i dA_i \cos\theta_i \, dw_{j-i} = I_i \cos\theta_i \frac{\cos\theta_j}{R^2} dA_j \, dA_i$$

$$= \pi I_i \frac{\cos\theta_i \cos\theta_j}{\pi R^2} dA_i \, dA_j \tag{12.1}$$

Performing integration over surface area i and surface area j, one obtains the radiation rate from surface i to surface j as

$$q_{i-j} = J_i \int_{A_i} \int_{A_j} \frac{\cos\theta_i \cos\theta_j}{\pi R^2} dA_j \, dA_i \tag{12.2}$$

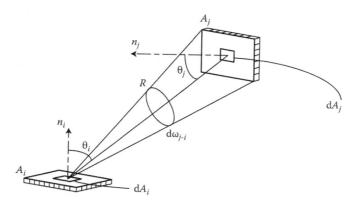

FIGURE 12.1 Radiation exchange between two diffuse isothermal surfaces.

DOI: 10.1201/9781003164487-12

where $\pi I_i \equiv J_i =$ radiosity (emission plus reflection).

If the radiosity J_i is uniform, that is, diffuse reflection and isothermal emission, then

$$F_{ij} = \frac{\text{Engery intercepted by } A_j}{\text{Radiosity leaving } A_i} = \frac{q_{ij}}{A_i J_i} = \frac{1}{A_i} \int_{A_i} \int_{A_j} \frac{\cos\theta_i \cos\theta_j}{\pi R^2} dA_j \, dA_i \quad (12.3)$$

Similarly,

$$F_{ij} = \frac{1}{A_j} \int_{A_j} \int_{A_i} \frac{\cos\theta_i \cos\theta_j}{\pi R^2} dA_i \, dA_j \quad (12.4)$$

Therefore, we obtain the Reciprocity Rule

$$A_i F_{ij} = A_j F_{ji} \quad (12.5)$$

For an enclosure with N surfaces, as shown in Figure 12.2,

$$\sum_{j=1}^{n} F_{ij} = 1 \quad (12.6)$$

That is,

$$F_{11} + F_{12} + F_{13} + \cdots = 1$$
$$F_{21} + F_{22} + F_{23} + \cdots = 1$$
$$\vdots$$
$$F_{N1} + F_{N2} + F_{N3} + \cdots = 1$$

Figure 12.3 shows the differential view factor between two differential areas i and j. Also, the view factor between area i and differential area j is shown. The differential view factor between differential area i and differential area j can be obtained as

$$dF_{dAi-dAj} = \frac{dq_{ij}}{J_i \, dA_i} = \frac{\cos\theta_i \cos\theta_j \, dA_j}{\pi R^2}$$

$$dF_{dAj-dAi} = \frac{dq_{ji}}{J_j \, dA_j} = \frac{\cos\theta_i \cos\theta_j \, dA_i}{\pi R^2}$$

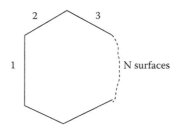

FIGURE 12.2 An N-surface enclosure.

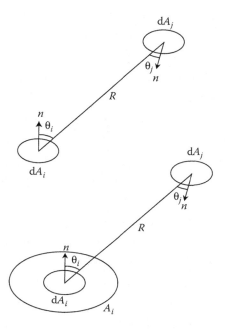

FIGURE 12.3 Concept of view factor between two differential areas.

Similarly, the differential view factor between area i and differential area j is

$$dF_{Ai-dAj} = \frac{\int J_i \left(\cos\theta_i \cos\theta_j / \pi R^2\right) dA_i \, dA_j}{J_i A_i} = \frac{dA_j}{A_i} \int_{A_i} \frac{\cos\theta_i \cos\theta_j}{\pi R^2} \, dA_i$$

The view factor between differential area j and area i is

$$F_{dAj-Ai} = \frac{\int J_j \left(\cos\theta_i \cos\theta_j / \pi R^2\right) dA_i \, dA_j}{J_j \, dA_j} = \int_{A_i} \frac{\cos\theta_i \cos\theta_j}{\pi R^2} \, dA_i$$

From the reciprocity rule for diffuse and isothermal surfaces:

$$A_i F_{A_i-A_j} = A_j F_{A_j-A_i}$$

$$dA_i \, dF_{dA_i-dA_j} = dA_j \, dF_{dA_j-dA_i}$$

$$A_i \, dF_{Ai-dAj} = dA_j \, dF_{dAj-Ai}$$

$$dA_i \, dF_{dA_i-A_j} = A_j \, dF_{A_j-dA_i}$$

Example 12.1

Determine the view factor between two parallel discs as shown in Figure 12.4. Assume that $A_i \ll A_j$, the distance between two surfaces is L, and the larger disc has a diameter D.

From Equation (12.3),

$$F_{ij} = \frac{1}{A_i} \int_{A_i} \int_{A_j} \frac{\cos\theta_i \cos\theta_j}{\pi R^2} dA_i \ dA_j$$

$$= \int_{A_j} \frac{\cos\theta_i \cos\theta_j}{\pi R^2} dA_j$$

where

$$A_i = \int_{A_i} dA_i$$

with

$$\theta_i = \theta_j$$

$$R^2 = r^2 + L^2$$

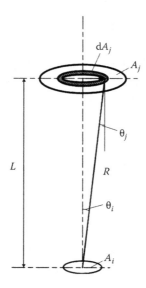

FIGURE 12.4   View factor between two parallel discs.

$$\cos \theta = L/R$$

$$dA_j = 2\pi r dr$$

$$F_{ij} = \int\limits_{A_j} \frac{\cos^2 \theta}{\pi R^2} dA_j$$

$$= 2L^2 \int\limits_{0}^{D/2} \frac{r\, dr}{\left(r^2 + L^2\right)^2} = \frac{D^2}{D^2 + 4L^2}$$

12.2 EVALUATION OF THE VIEW FACTOR

The view factor is only a function of geometry. The following shows several well-known methods to obtain the view factors for common geometries seen in radiation heat transfer applications [3].

Elongated surfaces – use Hottel's string method for 2-D geometries.
  Direct integration – need to perform double-area integration (higher level of difficulty).
  Contour integration – use Stoke's theorem to transform area-to-line integration.
  Algebraic method – determine the unknown view factor from the known value.

12.2.1 METHOD 1—HOTTEL'S CROSSED-STRING METHOD FOR 2-D GEOMETRIES

Hottel proposed the following process to determine the view factor between surface i and surface j for a 2-D geometry (with surfaces elongated in the direction normal to the paper, as shown in Figure 12.5). Although Figure 12.5 depicts flat surfaces, Hottel's crossed-strings method can also be used with curved or irregular surfaces.

The procedure begins by identifying the endpoints of each surface. Next the endpoints are connected using tightly stretched strings. After connecting the endpoints, strings will either cross other strings, or they will be uncrossed. Hottel showed the view factor between two surfaces can be determined in terms of the lengths of the crossed and uncrossed strings:

$$F_{i-j} = \frac{\sum \left(\text{crossed strings}\right) - \sum \left(\text{uncrossed strings}\right)}{2\left(\text{length of surface } i\right)}$$

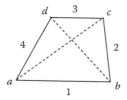

FIGURE 12.5   Concept of Hottel's cross-string method for a 2-D geometry.

To demonstrate the crossed-strings method, let us first consider how surface 1 separately views surfaces 2, 3, and 4. To find F_{1-2}, the end points are (a), (b), and (c). This is a unique application of Hottel's method, as surfaces 1 and 2 share the endpoint (b). With the shared endpoint, the method can still be used. To connect the endpoints, strings are stretched from a–c, a–b, and c–b. With the common edge, strings of b–a and b–c are duplicates of a–b and c–b, respectively, and are considered "crossed." Also, with the shared endpoint (b), the string connecting b–b has a "zero" length. Using these strings, F_{1-2} can be determined:

$$F_{1-2} = \frac{(L_{ab} + L_{cb}) - (L_{ac} + 0)}{2L_{ab}}$$

$$= \frac{L_1 + L_2 - L_{ac}}{2L_1} \tag{12.7}$$

From Figure 12.5, the view factor between surfaces 1 and 4, F_{1-4}, is similar:

$$F_{1-4} = \frac{(L_{ab} + L_{ad}) - (L_{bd} + 0)}{2L_{ab}}$$

$$= \frac{L_1 + L_4 - L_{bd}}{2L_1} \tag{12.8}$$

The final view factor, F_{1-3}, is also determined based on the endpoints (a), (b), (c), and (d):

$$F_{1-3} = \frac{(L_{ac} + L_{bd}) - (L_{ab} + L_{cd})}{2L_{ab}}$$

$$= \frac{L_{ac} + L_{bd} - (L_2 + L_4)}{2L_1} \tag{12.9}$$

The summation and reciprocity rules can be used to verify Hottel's cross-strings method. First, consider a surface represented by the string connecting (a) to (c). Now a three-surface enclosure has been created with surfaces (1), (2), and the new surface from (a–c). With surface (1) being flat ($F_{1-1} = 0$), the summation rule is:

$$F_{1-2} + F_{1-ac} = 1$$

Because the surfaces extend infinitely into the page, the area of each surface can be represented by its length (as every surface shares the common length into the page).

$$F_{1-2} = 1 - F_{1-ac}$$

$$= 1 - \frac{L_{ac}}{L_1} F_{ac-1}$$

$$= 1 - \frac{L_{ac}}{L_1}(1 - F_{ac-2})$$

$$= 1 - \frac{L_{ac}}{L_1} + \frac{L_{ac}}{L_1} \cdot \frac{L_2}{L_{ac}} \cdot F_{2-ac}$$

$$= 1 - \frac{L_{ac}}{L_1} + \frac{L_2}{L_1} \cdot (1 - F_{2-1})$$

$$= 1 - \frac{L_{ac}}{L_1} + \frac{L_2}{L_1} - \frac{L_2}{L_1} \cdot \frac{L_1}{L_2} \cdot F_{1-2}$$

$$= 1 - \frac{L_{ac}}{L_1} + \frac{L_2}{L_1} - F_{1-2}$$

$$\therefore F_{1-2} = \frac{L_1 + L_2 - L_{ac}}{2L_1}$$

Similarly:

$$F_{1-4} = \left((L_1 + L_4 - L_{bd})/2L_1\right)$$

Finally:

$$F_{1-2} + F_{1-3} + F_{1-4} = 1$$

$$F_{1-3} = 1 - F_{1-2} - F_{1-4}$$

$$= 1 - \frac{L_1 + L_2 - L_{ac}}{2L_1} - \frac{L_1 + L_4 - L_{bd}}{2L_1}$$

$$= \frac{2L_1 - L_1 - L_2 + L_{ac} - L_1 - L_4 + L_{bd}}{2L_1}$$

$$= \frac{L_{ac} + L_{bd} - L_2 - L_4}{2L_1}$$

The concept of the following examples (2) through (4) comes from [3].

Example 12.2

Determine the view factor between two parallel plates with partial blockages as shown in Figure 12.6.

$$\text{Length of each crossed string} = \sqrt{l^2 + c^2}$$

$$\text{Length of each uncrossed string} = 2\sqrt{b^2 + \left(\frac{c}{2}\right)^2}$$

From Hottel's crossed-string method, the view factor can be determined as

$$F_{1-2} = \frac{2\sqrt{l^2 + c^2} - 2 \cdot 2\sqrt{b^2 + (c/2)^2}}{2l} = \sqrt{1 + \left(\frac{c}{l}\right)^2} - \sqrt{\left(\frac{2b}{l}\right)^2 + \left(\frac{c}{l}\right)^2}$$

Example 12.3

Determine the view factor between two opposite circular tubes as shown in Figure 12.7. From Hottel's cross-string method, the view factor is

$$F_{1-2} = \frac{2L_1 - 2L_2}{2A_1} = \frac{L_1 - L_2}{\pi R}$$

where
    L_1 = crossed string abcde
    L_2 = uncrossed string ef
    $L_2 = D + 2R$

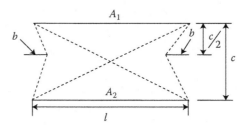

FIGURE 12.6   View factor between two parallel plates with partial blockages.

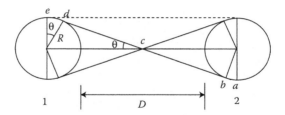

FIGURE 12.7   View factor between two opposite circular tubes.

Let $X = 1 + \dfrac{D}{2R}$

if $A_1 = \pi R$

$$F_{1-2} = \frac{2}{\pi}\left[\left(X^2 - 1\right)^{1/2} + \sin^{-1}\left(\frac{1}{X}\right) - X\right]$$

$$F_{1-2} = \frac{2}{\pi}\left[\left(X^2 - 1\right)^{1/2} + \frac{\pi}{2} - \cos^{-1}\left(\frac{1}{X}\right) - X\right]$$

if $A_1 = 2\pi R$

$$F_{1-2} = \frac{1}{\pi}\left[\left(X^2 - 1\right)^{1/2} + \frac{\pi}{2} - \cos^{-1}\left(\frac{1}{X}\right) - X\right]$$

Example 12.4

Determine the view factor between two circular tubes with partial blockage as shown in Figure 12.8.
From Hottel's cross-string method, the view factor can be determined as follows:
    The sum of the length of crossed strings: $2(L_{A-B-D-G-I} + L_{H-C-D-E-F})$
    The sum of the length of uncrossed strings: $2(L_{A-F} + L_{H-C-D-G-I})$
    Therefore:

$$F_{1-2} = 2\frac{\left(L_{A-B-D-G-I} + L_{H-C-D-E-F}\right) - \left(L_{A-F} + L_{H-C-D-G-I}\right)}{2L_{A-B-C-H}}$$

Table 12.1 shows many useful view factors for 2-D geometries that can be determined by using Hottel's cross-string method [2,4].

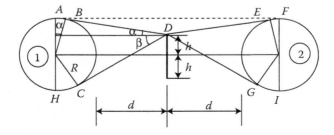

FIGURE 12.8 View factor between two circular tubes with partial blockage.

TABLE 12.1
View Factors for 2-D Geometries

Geometry	Relation

Parallel Plates with Midlines Connected by Perpendicular

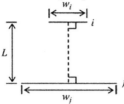

$$F_{i-j} = \frac{\left[\left(W_i + W_j\right)^2 + 4\right]^{1/2} - \left[\left(W_j + W_i\right)^2 + 4\right]^{1/2}}{2W_i}$$

$$W_i = w_i/L, \quad W_j = w_j/L$$

Inclined Parallel Plates of Equal Width and a Common Edge

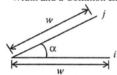

$$F_{i-j} = 1 - \sin\left(\frac{\alpha}{2}\right)$$

Perpendicular Plates with a Common Edge

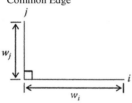

$$F_{i-j} = \frac{1 + \left(w_j/w_i\right) - \left[1 + \left(w_j/w_i\right)^2\right]^{1/2}}{2}$$

Three-sided Enclosure

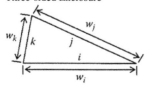

$$F_{i-j} = \frac{w_i + w_j - w_k}{2w_i}$$

Parallel Cylinders of Different Radii

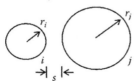

If $A_i = 2\pi r_i$

$$F_{i-j} = \frac{1}{2\pi}\left\{\pi + \left[C^2 - (R+1)^2\right]^{1/2} - \left[C^2 - (R-1)^2\right]^{1/2}\right.$$

$$+ (R-1)\cos^{-1}\left[\left(\frac{R}{C}\right) - \left(\frac{1}{C}\right)\right]$$

$$\left. - (R+1)\cos^{-1}\left[\left(\frac{R}{C}\right) - \left(\frac{1}{C}\right)\right]\right\}$$

$$R = r_j/r_i, \quad S = s/r_i, \quad C = 1 + R + S$$

(Continued)

TABLE 12.1 (*Continued*)
View Factors for 2-D Geometries

Geometry	Relation

If $A_i = \pi r_i$

$$F_{i-j} = \frac{1}{\pi}\left\{\pi + \left[C^2 - (R+1)^2\right]^{1/2} - \left[C^2 - (R-1)^2\right]^{1/2}\right.$$

$$+ (R-1)\cos^{-1}\left[\left(\frac{R}{C}\right) - \left(\frac{1}{C}\right)\right]$$

$$\left. - (R+1)\cos^{-1}\left[\left(\frac{R}{C}\right) - \left(\frac{1}{C}\right)\right]\right\}$$

Cylinder and Parallel Rectangle

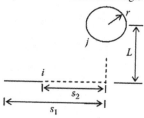

$$F_{i-j} = \frac{r}{s_1 - s_2}\left[\tan^{-1}\left(\frac{s_1}{L}\right) - \tan^{-1}\left(\frac{s_2}{L}\right)\right]$$

Infinite Plane and Row of Cylinders

$$F_{i-j} = 1 - \left[1 - \left(\frac{D}{s}\right)^2\right]^{1/2} + \left(\frac{D}{s}\right)\tan^{-1}\left(\frac{s^2 - D^2}{D^2}\right)^{1/2}$$

Concentric Cylinders

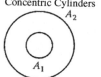

$$F_{1-2} = 1$$

$$F_{2-1} = \frac{A_1}{A_2}$$

$$F_{2-2} = 1 - F_{2-1} = 1 - \frac{A_1}{A_2}$$

Long Duct with Equilateral
Triangular Cross-Section

$$F_{1-2} = F_{1-3} = \frac{1}{2}$$

Long Parallel Plates of Equal Width

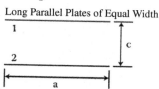

$$F_{1-2} = F_{2-1} = \left[1 + \left(\frac{c}{a}\right)^2\right]^{1/2} - \left(\frac{c}{a}\right)$$

(*Continued*)

TABLE 12.1 (*Continued*)
View Factors for 2-D Geometries

Geometry	Relation
Long Cylinder Parallel to a Large Plane Area 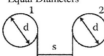	$F_{1-2} = \dfrac{1}{2}$
Long Adjacent Parallel Cylinders of Equal Diameters	If $A_i = \pi d$ $$F_{1-2} = F_{2-1} = \dfrac{1}{\pi}\left[\left(X^2 - 1\right)^{1/2} + \sin^{-1}\left(\dfrac{1}{X}\right) - X\right]$$ If $A_i = \pi\left(\dfrac{d}{2}\right)$ $$F_{1-2} = F_{2-1} = \dfrac{2}{\pi}\left[\left(X^2 - 1\right)^{1/2} + \sin^{-1}\left(\dfrac{1}{X}\right) - X\right]$$ $$X = 1 + \dfrac{s}{d}$$
Concentric Spheres 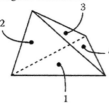	$F_{1-2} = 1$ $$F_{2-1} = \dfrac{A_1}{A_2}$$ $$F_{2-2} = 1 - F_{2-1} = 1 - \dfrac{A_1}{A_2}$$
Regular Tetrahedron 	$F_{1-2} = F_{1-3} = F_{1-4} = \dfrac{1}{3}$
Sphere Near a Large Plane Area 	$F_{1-2} = \dfrac{1}{2}$
Small Area Perpendicular to the Axis of a Surface of Revolution	$F_{1-2} = \sin^2\theta$

(*Continued*)

TABLE 12.1 (Continued)
View Factors for 2-D Geometries

Geometry   **Relation**

Area on the Inside of a Sphere

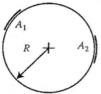

$$F_{1-2} = \frac{A_2}{4\pi R^2}$$

Source: Data from A. Mills, *Heat Transfer*, Richard D. Irwin, Inc., Boston, MA, 1992;
F. Incropera and D. Dewitt, *Fundamentals of Heat and Mass Transfer*, John Wiley &
Sons, Fifth Edition, 2002.

12.2.2 METHOD 2—DOUBLE-AREA INTEGRATION

Use direct integration to determine the view factor between two adjacent surface
areas i and j, as shown in Figure 12.9.

$$F_{Ai-Aj} = \frac{1}{A_i} \int\limits_{A_i} \int\limits_{A_j} \frac{\cos\theta_i \cos\theta_j}{\pi R^2} dA_i \, dA_j$$

where

$$R^2 = \left(x_i - x_j\right)^2 + \left(y_i - y_j\right)^2 + \left(z_i - z_j\right)^2$$

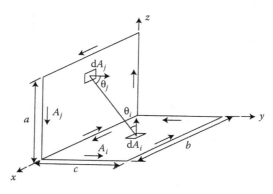

FIGURE 12.9 View factor between two adjacent surfaces.

$$\cos\theta_i = \frac{1}{R}\Big[l_i\big(x_j - x_i\big) + m_i\big(y_j - y_i\big) + n_i\big(z_j - z_i\big)\Big]$$

$$\cos\theta_j = \frac{1}{R}\Big[l_j\big(x_i - x_j\big) + m_j\big(y_i - y_j\big) + n_j\big(z_i - z_j\big)\Big]$$

Therefore, the view factor can be determined by performing the following integration:

$$F_{Ai-Aj} = \frac{1}{bc}\int_0^b dx_j \int_0^b dx_i \int_0^a z_j\, dz_j \int_0^c \frac{y_i\, dy_i}{\pi\Big[\big(x_i - x_j\big)^2 + y_i^2 + z_j^2\Big]^2} \tag{12.10}$$

12.2.3 Method 3—Contour Integration

Use Stokes' theorem to transform area integration into line integration.

Again, determine the view factor between two adjacent surfaces as shown in Figure 12.9.

$$F_{Ai-Aj} = \frac{1}{2\pi A_i}\left[\oint_{C_i}\oint_{C_j}\big(\ln R\, dx_i dx_j + \ln R\, dy_i\, dy_j + \ln R\, dz_i\, dz_j\big)\right] \tag{12.11}$$

Apply Stokes' theorem to reduce quadric to double integrations as

$$F_{Ai-Aj} = \frac{1}{2\pi bc}\left[\int_0^b dx_j \left\{\int_0^b \ln\Big[\big(x_i - x_j\big)^2 + a^2\Big]^{1/2} dx_i + \int_b^0 \ln\Big[\big(x_i - x_j\big)^2 + c^2 + a^2\Big]^{1/2} dx_i\right\} \right.$$
$$\left. + \int_b^0 dx_j \left\{\int_0^b \ln\Big[\big(x_i - x_j\big)^2 + 0^2\Big]^{1/2} dx_i + \int_b^0 \ln\Big[\big(x_i - x_j\big)^2 + c^2 + 0^2\Big]^{1/2} dx_i\right\}\right]$$

$$\tag{12.12}$$

The results can be obtained from integration tables or numerical integration.

The following shows how to use Stokes' theorem to determine the view factor between two opposite surfaces [3] as shown in Figure 12.10.

$$F_{Ai-Aj} = \frac{1}{2\pi A_i}\left[\oint_{c_i}\oint_{c_j}\big(\ln R\, dx_i\, dx_j + \ln R\, dy_i\, dy_j + \ln R\, dz_i\, dz_j\big)\right]$$

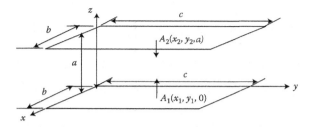

FIGURE 12.10 View factor between two opposite surfaces.

$$F_{Ai-Aj} = \frac{1}{2\pi bc} \oint_{c_i} \left\{ \int_0^c \ln\left[x_i^2 + (y_j - y_i)^2 + a^2 \right]^{1/2} dy_j \right\} dy_i \qquad (1)$$

$$+ \frac{1}{2\pi bc} \oint_{c} \left\{ \int_0^b \ln\left[(x_j - x_i)^2 + (c - y_i)^2 + a^2 \right]^{1/2} dx_j \right\} dx_i \qquad (2)$$

$$+ \frac{1}{2\pi bc} \oint_{c_i} \left\{ \int_c^0 \ln\left[(b - x_i)^2 + (y_j - y_i)^2 + a^2 \right]^{1/2} dy_j \right\} dy_i \qquad (3)$$

$$+ \frac{1}{2\pi bc} \oint_{c_i} \left\{ \int_b^0 \ln\left[(x_j - x_i)^2 x_i^2 + y_i^2 + a^2 \right]^{1/2} dx_j \right\} dx_i \qquad (4)$$

$$F_{Ai-Aj} = (1) + (2) + (3) + (4)$$

$$F_{Ai-Aj} = \frac{1}{2\pi bc} \int_c^0 \int_0^c \left\{ \ln\left[(y_j - y_i)^2 + a^2 \right]^{1/2} \right.$$

$$\left. + \ln\left[b^2 + (y_j - y_i)^2 + a^2 \right]^{1/2} \right\} dy_j \, dy_i + (2) + (4)$$

$$F_{Ai-Aj} = \frac{2a^2}{\pi bc} \left\{ \ln\left[\frac{(1+(b/a)^2)(1+(c/a)^2)}{1+(b/a)^2+(c/a)^2} \right]^{1/2} + \frac{b}{a}\left[1+(c/a)^2 \right]^{1/2} \tan^{-1}\left[\frac{b/a}{\left[1+(c/a)^2\right]^{1/2}} \right] \right.$$

$$\left. + \frac{c}{a}\left(1+\left(\frac{b}{a}\right)^2 \right)^{1/2} \tan^{-1}\left[\frac{c/a}{\left[1+(b/a)^2\right]^{1/2}} \right] - \frac{b}{a}\tan^{-1}\left(\frac{b}{a}\right) - \frac{c}{a}\tan^{-1}\left(\frac{c}{a}\right) \right\}$$

$$(12.13)$$

Again, the results can be obtained from integration tables or numerical integration.

Table 12.2 shows several useful view factors for 3-D geometries [4] that can be determined by using Stoke's theorem to transform area-to-line integration.

TABLE 12.2
View Factors for 3-D Geometries

Geometry	Relation

Aligned Parallel Rectangles

$$F_{i-j} = \frac{1}{\pi \bar{X} \bar{Y}} \left\{ \ln \left[\frac{\left(1+\bar{X}^2\right)\left(1+\bar{Y}^2\right)}{1+\bar{X}^2+\bar{Y}^2} \right]^{\frac{1}{2}} \right.$$

$$+ \bar{X}\left(1+\bar{Y}^2\right)^{1/2} \tan^{-1}\left(\frac{\bar{X}}{\left(1+\bar{Y}^2\right)^{1/2}} \right)$$

$$\left. + \bar{Y}\left(1+\bar{X}^2\right)^{1/2} \tan^{-1}\left(\frac{\bar{Y}}{\left(1+\bar{X}^2\right)^{1/2}} \right) - \bar{X}\tan^{-1}\left(\bar{X}\right) - \bar{Y}\tan^{-1}\left(\bar{Y}\right) \right\}$$

$$\bar{X} = X/L, \quad \bar{Y} = Y/L$$

Coaxial Parallel Discs

$$F_{i-j} = \frac{1}{2} \left\{ S - \left[S^2 - 4\left(\frac{r_j}{r_j}\right)^2 \right]^{1/2} \right\}$$

$$R_i = r_i / L, R_j = r_j / L$$

$$S = 1 + \frac{1+R_j^2}{R_i^2}$$

Perpendicular Rectangles with a Common Edge

$$F_{i-j} = \frac{1}{\pi W} \left\{ W \tan^{-1}\left(\frac{1}{W}\right) + H \tan^{-1}\left(\frac{1}{H}\right) \right.$$

$$- \left(H^2+W^2\right)^{\frac{1}{2}} \tan^{-1}\left(\frac{1}{\left(H^2+W^2\right)^{\frac{1}{2}}} \right)$$

$$+ \frac{1}{4} \ln \left[\left(\frac{\left(1+W^2\right)\left(1+H^2\right)}{1+W^2+H^2} \right) \left(\frac{W^2\left(1+W^2+H^2\right)}{\left(1+W^2\right)\left(W^2+H^2\right)} \right)^{W^4} \right.$$

$$\left. \left. \times \left(\frac{H^2\left(1+H^2+W^2\right)}{\left(1+H^2\right)\left(H^2+W^2\right)} \right)^{H^2} \right] \right\}$$

$$H = Z/X, \quad W = Y/X$$

Source: F. Incropera and D. Dewitt, *Fundamentals of Heat and Mass Transfer*, John Wiley & Sons, Fifth Edition, New York, NY, 2002.

Example 12.5

Determine the view factors F_{1-2} and F_{2-1} for the geometries shown in Figure 12.11. For geometries (a) and (b),

$$F_{1-1} + F_{1-2} + F_{1-3} = 1$$

$$F_{1-1} = 0$$

$$\therefore F_{1-2} = 1 - F_{1-3}$$

$$A_1 F_{1-2} = A_2 F_{2-1}$$

$$\therefore F_{2-1} = \frac{A_1}{A_2} F_{1-2}$$

F_{1-3} can be obtained for each geometry from Table 12.2.

12.2.4   METHOD 4—ALGEBRAIC METHOD

Determine the unknown view factor between surface areas A_1 and A_2, as shown in Figure 12.12, from the known values of view factors.

$$A_1 F_{1-2} = A_1 F_{1-j} - A_1 F_{1-4}$$
$$= A_j \left(F_{j-i} - F_{j-3} \right) - A_4 \left(F_{4-i} - F_{4-3} \right) \tag{12.14}$$

where F_{j-i}, F_{j-3}, F_{4-i}, and F_{4-3} are available from formulas or charts.

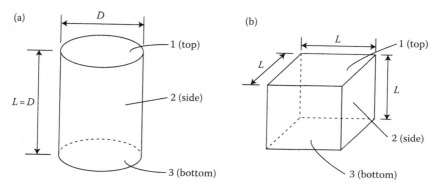

(a)

D

1 (top)

$L = D$

2 (side)

3 (bottom)

(b)

L

L

1 (top)

L

L

2 (side)

3 (bottom)

FIGURE 12.11   (a) A cylindrical furnace. (b) A cubic furnace.

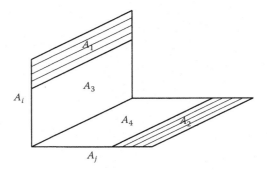

FIGURE 12.12 Algebraic method.

Example 12.6

Determine view factors F_{1-4} and F_{4-1} from Figures 12.13a and b.

$$A_1 F_{1-4} = A_2 F_{2-3}$$

$$
\begin{aligned}
A_i F_{i-j} &= A_1 F_{1-j} + A_2 F_{2-j} \\
&= A_1 F_{1-3} + A_1 F_{1-4} + A_2 F_{2-3} + A_2 F_{2-4} \\
&= A_1 F_{1-3} + 2 A_1 F_{1-4} + A_2 F_{2-4}
\end{aligned}
$$

Hence,

$$A_1 F_{1-4} = \frac{1}{2}\left[A_i F_{i-j} - A_1 F_{1-3} - A_2 F_{2-4} \right] \tag{12.15}$$

where F_{i-j}, F_{1-3}, and F_{2-4} are available from Table 12.2 formulas or charts. Also, $A_1 F_{1-4} = A_4 F_{4-1}$.

Example 12.7

Using area integration, determine the view factor between the two surfaces shown in Figure 12.14. This is a special case, similar to Figure 12.9.

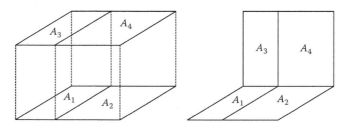

FIGURE 12.13 Applications of shape factor algebra to opposing and adjacent rectangles.

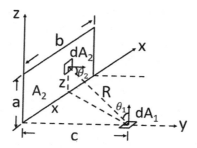

FIGURE 12.14 View factor between a small area and a larger area

$$F_{dA_1-A_2} = \int_{A_2} \frac{\cos\theta_1 \cdot \cos\theta_2}{\pi R^2} dA_2$$

where

$$R^2 = z^2 + x^2 + c^2$$

$$\cos\theta_1 = \frac{z}{R}$$

$$\cos\theta_2 = \frac{c}{R}$$

$$F_{dA_1-A_2} = \frac{1}{\pi} \int_{x=0}^{b} \int_{z=0}^{a} \frac{zc}{\left(z^2 + x^2 + c^2\right)^2} \cdot dzdx$$

$$= -\frac{c}{2\pi} \int_{x=0}^{b} \left[\frac{1}{z^2 + x^2 + c^2} \right]_{z=0}^{a} \cdot dx$$

$$= \frac{c}{2\pi} \left[\int_{0}^{b} \frac{dx}{x^2 + c^2} - \int_{0}^{b} \frac{dx}{\left(c^2 + a^2\right) + x^2} \right]$$

$$= \frac{1}{2\pi} \left[\tan^{-1}\left(\frac{b}{c}\right) - \frac{c}{\sqrt{a^2 + c^2}} \cdot \tan^{-1}\left(\frac{b}{\sqrt{a^2 + c^2}}\right) \right]$$

Example 12.8

Consider two plate fins on a tube with an angle α as shown in Figure 12.15 (a 2-D configuration elongated into the page). Determine the view factor between fin elements i and j using the crossed strings method.

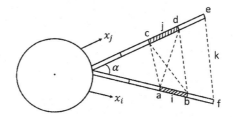

FIGURE 12.15 View factor between two plate fins.

From the cross-string method:

$$F_{i-j} = \frac{L_{ad} + L_{bc} - \left(L_{ac} + L_{bd}\right)}{2L_{ab}}$$

From the triangular geometry:

$$L_{ad}^2 = L_a^2 + L_d^2 - 2L_a \cdot L_d \cdot \cos\alpha$$

$$L_{bc}^2 = L_b^2 + L_c^2 - 2L_b \cdot L_c \cdot \cos\alpha$$

$$L_{ac}^2 = L_a^2 + L_c^2 - 2L_a \cdot L_c \cdot \cos\alpha$$

$$L_{bd}^2 = L_b^2 + L_d^2 - 2L_b \cdot L_d \cdot \cos\alpha$$

The view factor between fin element i and surrounding element k

$$F_{i-k} = \frac{L_{af} + L_{be} - \left(L_{ae} + L_{bf}\right)}{2L_{ab}}$$

From the triangular geometry:

$$L_{be}^2 = L_b^2 + L_e^2 - 2L_b \cdot L_e \cdot \cos\alpha$$

$$L_{ae}^2 = L_a^2 + L_e^2 - 2L_a \cdot L_e \cdot \cos\alpha$$

$$L_{af} = L_f - L_a$$

$$L_{bf} = L_f - L_b$$

Example 12.9

Determine the view factor between coaxial parallel discs, as shown in Figure 12.16. From Table 12.2, assume

$$r_i = r_j = R, D = 2R = \text{diameter}$$

$$L = \text{distance between two discs.}$$

$$F_{i-j} = 1 + 2\frac{L}{D}\left[\frac{L}{d} - \sqrt{1 + \left(\frac{L}{D}\right)^2}\right]$$

If an enclosure surface, s, is placed between the two discs, the view factor between disk i and the enclosure surface s can be determined as

$$F_{i-s} = 1 - F_{i-j} = \frac{2L}{D}\left[\sqrt{1 + \left(\frac{L}{D}\right)^2} - \frac{L}{D}\right]$$

How to determine the view factor F_{s-s}?

$$F_{s-s} = 1 - F_{s-i} - F_{s-j}$$

$$= 1 - 2F_{s-i}$$

$$\pi DL F_{s-i} = \frac{\pi}{4}D^2 \cdot F_{i-s}$$

$$F_{s-i} = \frac{1}{4}\frac{D}{L} \cdot \frac{2L}{D}\left[\sqrt{1 + \left(\frac{L}{D}\right)^2} - \frac{L}{D}\right]$$

$$= \frac{1}{2}\left[\sqrt{1 + \left(\frac{L}{D}\right)^2} - \frac{L}{D}\right]$$

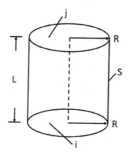

FIGURE 12.16 View factor between coaxial parallel discs.

Thus

$$F_{s-s} = 1 - \left[\sqrt{1 + \left(\frac{L}{D} \right)^2} - \frac{L}{D} \right]$$

Example 12.10

Now, consider the view factor between two elements within the wall of a tube, i, j, as shown in Figure 12.17.
    Tube diameter $= D$; tube length $= L$; two elements $= s_i$, s_j
    $x =$ distance; $s =$ surface

Element i: has axial distance $= x_{iR} - x_{iL}$

has inlet cross-sectional area $= s_{iL} = \dfrac{\pi}{4} D^2$

has outlet cross-sectional area $= s_{iR} = \dfrac{\pi}{4} D^2$

has surface area $= s_i = \pi D (x_{iR} - x_{iL})$

Element j has axial distance $= x_{jR} - x_{jL}$

has inlet cross-sectional area $= s_{jL} = \dfrac{\pi}{4} D^2$

has outlet cross-sectional area $= s_{jR} = \dfrac{\pi}{4} D^2$

has surface area $= s_j = \pi D (x_{jR} - x_{jL})$

1. How to determine the view factor between the tube inlet and surface element, s_i, $F_{\text{inlet}-s_i}$?
    From previous Example 12.9, the view factor between disc i and enclosure surface s, F_{i-s}, can be obtained as:
    Since tube surface $s_i =$ tube surface $s_{x_{iR}} -$ tube surface $s_{x_{iL}}$

$$F_{\text{inlet}-s_i} = F_{\text{inlet}-s_{x_{iR}}} - F_{\text{inlet}-s_{x_{iL}}}$$

$$= 2\frac{x_{iR}}{D} \cdot \left[\sqrt{1 + \left(\frac{x_{iR}}{D} \right)^2} - \frac{x_{iR}}{D} \right] - 2\frac{x_{iL}}{D} \cdot \left[\sqrt{1 + \left(\frac{x_{iL}}{D} \right)^2} - \frac{x_{iL}}{D} \right]$$

2. How to determine the view factor between tube element surface s_i and element surface s_j, $F_{s_i-s_j}$?

$$A_{s_i} \cdot F_{s_i-s_j} = A_{s_i} \cdot \left[F_{s_i-s_{jL}} - F_{s_i-s_{jR}} \right], \text{ by reciprocity rule,}$$

$$= A_{s_{jL}} \cdot F_{s_{jL}-s_i} - A_{s_{jR}} \cdot F_{s_{jR}-s_i}$$

From the previous discussion, as shown in Figure 12.16:

$$F_{s_{jL}-s_i} = F_{s_{jL}-s_{xiL}} - F_{s_{jL}-s_{xiR}}$$

$$F_{s_{jR}-s_i} = F_{s_{jR}-s_{xiL}} - F_{s_{jR}-s_{xiR}}$$

where

$$F_{s_{jL}-s_{xiL}} = \frac{2(x_{jL} - x_{iL})}{D} \cdot \left[\sqrt{1 + \left(\frac{x_{jL} - x_{iL}}{D}\right)^2} - \frac{x_{jL} - x_{iL}}{D}\right]$$

$$F_{s_{jL}-s_{xiR}} = \frac{2(x_{jL} - x_{iR})}{D} \cdot \left[\sqrt{1 + \left(\frac{x_{jL} - x_{iR}}{D}\right)^2} - \frac{x_{jL} - x_{iR}}{D}\right]$$

$$F_{s_{jR}-s_{xiL}} = \frac{2(x_{jR} - x_{iL})}{D} \cdot \left[\sqrt{1 + \left(\frac{x_{jR} - x_{iL}}{D}\right)^2} - \frac{x_{jR} - x_{iL}}{D}\right]$$

$$F_{s_{jR}-s_{xiR}} = \frac{2(x_{jR} - x_{iR})}{D} \cdot \left[\sqrt{1 + \left(\frac{x_{jR} - x_{iR}}{D}\right)^2} - \frac{x_{jR} - x_{iR}}{D}\right]$$

Therefore, $F_{s_i-s_j}$ can be obtained by substituting these view factors.
3. How to determine the view factor for an element surface of a circular tube? $F_{s_i-s_i} = ?$

$$F_{s_i-s_i} + F_{s_i-s_{iL}} + F_{s_i-s_{iR}} = 1$$

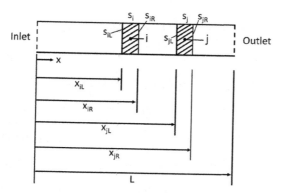

FIGURE 12.17 View factor between two element surfaces of a tube.

where

$$\pi D\left(x_{iR} - x_{iL}\right)F_{si-siL} = \frac{\pi}{4}D^2 \cdot F_{siL-si}$$

$$= \frac{\pi}{4}D^2 \cdot \left\{ 2\frac{x_{iR}-x_{iL}}{D}\left[\sqrt{1+\left(\frac{x_{iR}-x_{iL}}{D}\right)^2} - \frac{x_{iR}-x_{iL}}{D}\right]\right\}$$

$$F_{si-siL} = \frac{1}{2}\left[\sqrt{1+\left(\frac{x_{iR}-x_{iL}}{D}\right)^2} - \frac{x_{iR}-x_{iL}}{D}\right]$$

and

$$F_{si-siR} = F_{si-siL}$$

Thus,

$$F_{si-si} = 1 - 2F_{si-siL}$$

$$= 1 - \left[\sqrt{1+\left(\frac{x_{iR}-x_{iL}}{D}\right)^2} - \frac{x_{iR}-x_{iL}}{D}\right]$$

Example 12.11

Determine the view factor between two elements on the inside of a sphere, as shown in Figure 12.18

$$F_{dA_1-A_2} = \int_{A_2} \frac{\cos\theta_1 \cos\theta_2}{\pi S^2} dA_2$$

where

$$\theta_1 = \theta_2 = \theta$$

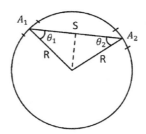

FIGURE 12.18 View factor between two elements on the inside of a sphere.

$$F_{dA_1 - A_2} = \int\limits_{A_2} \frac{\cos\theta \cdot \cos\theta}{\pi (R\cos\theta + R\cos\theta)^2} \, dA_2$$

$$F_{dA_1 - A_2} = \frac{1}{4\pi R^2} \int\limits_{A_2} dA_2 = \frac{A_2}{4\pi R^2} = \frac{A_2}{A_{\text{sphere}}}$$

Thus,

$$F_{A_1 - A_2} = \frac{1}{A_1} \int\limits_{A_1} F_{dA_1 - A_2} \, dA_1 = \frac{A_1}{A_1} \cdot \frac{A_2}{A_{\text{sphere}}} = \frac{A_2}{A_{\text{sphere}}}$$

REMARKS

This chapter covers the same information as in undergraduate heat transfer. In the undergraduate-level heat transfer, students are expected to know how to use those view factors available from tables or charts in order to calculate radiation heat transfer between two surfaces for many engineering applications. However, at the intermediate-level heat transfer, students are expected to focus on how to derive view factors instead of simply using them. In particular, students are expected to know how to determine the view factors using Hottel's string method for many 2-D geometries. View factors for 3-D geometries require double-area integration and are explored in more detail in advanced radiation texts.

PROBLEMS

12.1 Determine the view factors for Examples 1, 2, 3, and 4.
12.2 Determine the view factors shown in Table 12.1.
12.3 Determine the view factors shown in Table 12.2.

REFERENCES

1. W. Rohsenow and H. Choi, *Heat, Mass, and Momentum Transfer*, Prentice-Hall, Inc., Englewood Cliffs, NJ, 1961.
2. A. Mills, *Heat Transfer*, Richard D. Irwin, Inc., Boston, MA, 1992.
3. K.-F.V. Wong, *Intermediate Heat Transfer*, Marcel Dekker, Inc., New York, 2003.
4. F. Incropera and D. Dewitt, *Fundamentals of Heat and Mass Transfer*, John Wiley & Sons, Fifth Edition, New York, 2002.

13 Radiation Exchange in a Nonparticipating Medium

13.1 RADIATION EXCHANGE BETWEEN GRAY DIFFUSE ISOTHERMAL SURFACES IN AN ENCLOSURE

Since we know how to determine surface radiation properties such as emissivity, reflectivity, and absorptivity and how to calculate the view factor between two surfaces, the following shows how to determine radiation heat transfer between surfaces in an enclosure [1–4]. Assume that there are N gray, diffuse, and isothermal surfaces. This implies that each surface at T_i has a uniform radiosity, J_i (emission plus reflection). Figure 13.1 shows an energy balance on each surface i and an energy balance between surface i and the other of enclosure surfaces, j.

Based on the assumptions, radiation properties of each surface can be given as follows.

$\alpha_i = \varepsilon_i$, for gray and diffuse surfaces
$\tau_i = 0$, for the opaque body
$\rho_i = 1 - \alpha_i = 1 - \varepsilon_i$

There are three types of radiation problems for electric furnace applications.

1. Given each surface temperature, T_i, determine each surface heat flux, $(q/A)_i$
2. Given each surface heat flux, $(q/A)_i$, determine each surface temperature, T_i
3. A combination of 1 and 2: some surfaces of the enclosure have a specified temperature, but the heat flux is unknown, and other surfaces have a given heat flux while the temperature is unknown.

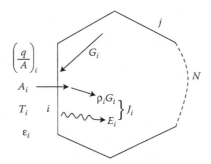

FIGURE 13.1 Radiation heat transfer between N surfaces in an enclosure.

DOI: 10.1201/9781003164487-13

Begin the analysis by performing an energy balance on the i surface:
net energy = energy out (radiosity) – energy in (irradiation)

$$q_i = A_i (J_i - G_i) \tag{13.1}$$

also

$$J_i = \varepsilon_i E_{bi} + (1 - \varepsilon_i) G_i \tag{13.2}$$

Therefore,

$$q_i = A_i \left(J_i - \frac{J_i - \varepsilon_i E_{bi}}{1 - \varepsilon_i} \right) = A_i \left(\frac{J_i - \varepsilon_i J_i - J_i + \varepsilon_i E_{bi}}{1 - \varepsilon_i} \right)$$

$$= A_i \left(\frac{\varepsilon_i (E_{bi} - J_i)}{1 - \varepsilon_i} \right)$$

Now the energy from surface i

$$q_i = \frac{E_{bi} - J_i}{(1 - \varepsilon_i)/(\varepsilon_i A_i)} \tag{13.3}$$

Next perform an energy exchange between surface i and all the other surfaces, j: net energy = energy out (radiosity) – energy in (irradiation)

$$q_i = A_i (J_i - G_i)$$

where

$$A_i G_i = \sum_{j=1}^{N} F_{ji} A_j J_j = \sum_{j=1}^{N} F_{ij} A_i J_j \tag{13.4}$$

Therefore,

$$q_i = A_i \left(J_i - \sum_{j=1}^{N} F_{ij} J_j \right)$$

By using $\sum_{j=1}^{N} F_{ij} = 1$, and multiplying to J_i,

$$q_i = A_i \left(\sum_{j=1}^{N} F_{ij} J_i - \sum_{j=1}^{N} F_{ij} J_j \right)$$

The energy transfer between surface i and the other enclosure surfaces, j, becomes

$$q_i = \sum_{j=1}^{N} A_i F_{ij} \left(J_i - J_j \right) = \sum_{j=1}^{N} q_{ij} \tag{13.5}$$

Combining Equations (13.3) and (13.5), we have

$$q_i = \frac{E_{bi} - J_i}{\left(1 - \varepsilon_i\right)/\varepsilon_i A_i} = \sum_{j=1}^{N} A_i F_{ij} \left(J_i - J_j \right) = \sum_{j=1}^{N} q_{ij}$$

$$q_i = \underbrace{\frac{E_{bi} - J_i}{\left(1 - \varepsilon_i\right)/\left(\varepsilon_i A_i\right)}}_{\substack{\text{surface resistance} \\ \text{due to emissivity}}} = \sum_{j=1}^{N} \underbrace{\frac{\left(J_i - J_j \right)}{1/\left(A_i F_{ij} \right)}}_{\substack{\text{geometrical resistance} \\ \text{due to view factor}}} = \sum_{j=1}^{N} q_{ij} \tag{13.6}$$

In addition, combining Equations (13.2) and (13.4), we get

$$J_i = \varepsilon_i E_{bi} + \left(1 - \varepsilon_i\right) \sum_{j=1}^{N} F_{ij} J_j \tag{13.7}$$

$$= \text{emission from surface } i + \text{reflection from surface } i$$

13.1.1 METHOD 1: ELECTRIC NETWORK ANALOGY

The electric network analogy [3] can be used to solve the aforementioned radiation heat transfer problem, as shown in Figure 13.2. The following shows a few special cases for radiation heat transfer applications.

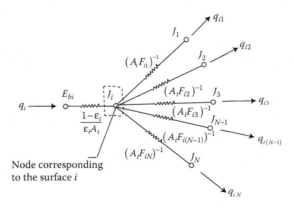

FIGURE 13.2 Network representation of the radiative exchange between one surface (i) and the remaining surfaces of an enclosure.

Special case 1: Radiation within a two-surface enclosure, as shown in Figure 13.3: (a) hemi-cylinder, (b) parallel plates, (c) rectangular channel, (d) long concentric cylinders, (e) concentric spheres, or (f) small convex object within a large enclosure:

$$q_1 = q_{12} = -q_2 = \frac{\sigma\left(T_1^4 - T_2^4\right)}{(1-\varepsilon_1)/(A_1\varepsilon_1) + 1/(A_1 F_{12}) + (1-\varepsilon_2)/(A_2\varepsilon_2)} \tag{13.8}$$

If for a blackbody, $\varepsilon_1 = \varepsilon_2 = 1$, then

$$q_1 = A_1 F_{12}\sigma\left(T_1^4 - T_2^4\right) \tag{13.9}$$

Special case 2: Radiation between two parallel surfaces with middle shields as shown in Figure 13.4:

$$q_1 = q_{12} = \frac{A_1\sigma\left(T_1^4 - T_2^4\right)}{\dfrac{(1-\varepsilon_1)}{\varepsilon_1} + \dfrac{1}{F_{13}} + \dfrac{(1-\varepsilon_{31})}{\varepsilon_{31}} + \dfrac{(1-\varepsilon_{32})}{\varepsilon_{32}} + \dfrac{1}{F_{32}} + \dfrac{(1-\varepsilon_2)}{\varepsilon_2}}$$

$$= \frac{A_1\sigma\left(T_1^4 - T_2^4\right)}{\dfrac{1}{\varepsilon_1} + \dfrac{(1-\varepsilon_{31})}{\varepsilon_{31}} + \dfrac{(1-\varepsilon_{32})}{\varepsilon_{32}} + \dfrac{1}{\varepsilon_2}} = -q_2 \tag{13.10}$$

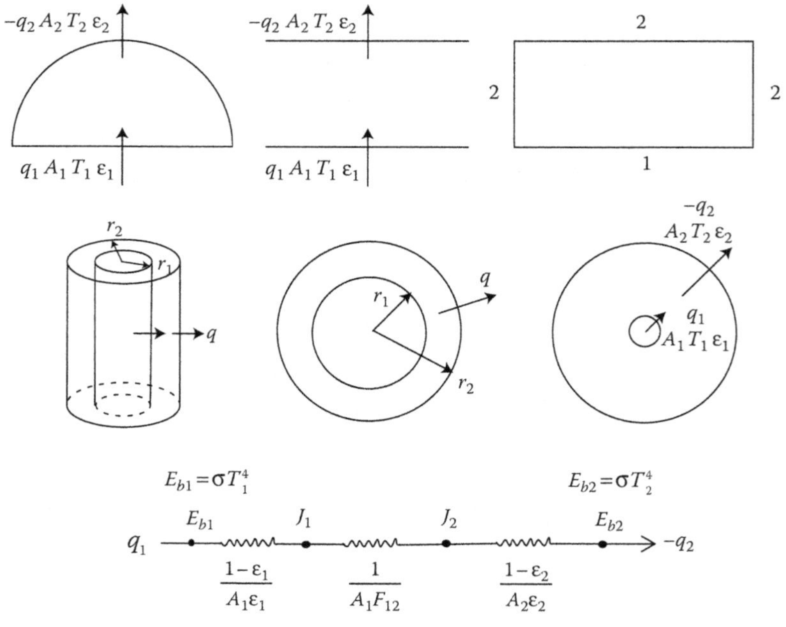

FIGURE 13.3 Radiation within a two-surface enclosure.

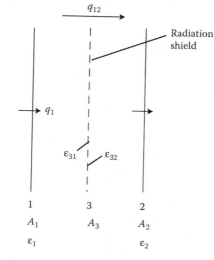

FIGURE 13.4 Radiation between two parallel surfaces with a middle shield.

where $F_{13} = F_{32} = 1$.

To cut down radiation heat loss, ε_{31} and ε_{32} should be small, that is, ρ_{31} is large.

Special case 3: Reradiating surfaces (insulated surface, $q_R = 0$): The following electric furnaces, as shown in Figure 13.5, can be modeled as radiation heat transfer between two opposite surfaces (hot and cold) with a third re-radiation side surface (perfect reflection and perfect insulation).

With $q_R = 0$, $q_1 = -q_2$

$$q_1 = -q_2 = \frac{\sigma T_1^4 - \sigma T_2^4}{\dfrac{(1-\varepsilon_1)}{A_1\varepsilon_1} + \dfrac{1}{A_1F_{12} + 1 \Big/ \left[\dfrac{1}{A_1F_{1R}} + \dfrac{1}{A_2F_{2R}}\right]} + \dfrac{(1-\varepsilon_2)}{A_2\varepsilon_2}} \qquad (13.11)$$

where T_1 and T_2 are given

$$A_R F_{R2} = A_2 F_{2R}$$

The surface emissivity, area, and view factors are also given or predetermined.

If $q_1 = -q_2$ is determined as shown above, and if $q_R = 0$, how do we determine the re-radiation surface temperature $T_R =$?

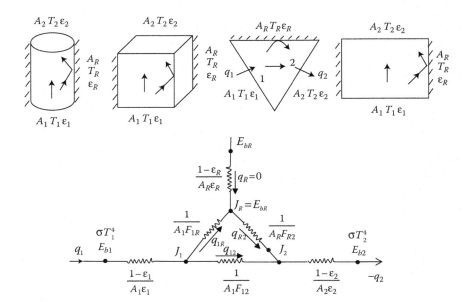

FIGURE 13.5   Electric furnaces with a reradiating surface.

From energy balance on the re-radiation surface,

$$q_{1R} = \frac{J_1 - J_R}{1/A_1 F_{1R}} = q_{R2} = \frac{J_R - J_2}{1/A_R F_{R2}}$$

and

$$q_1 = \frac{E_{b1} - J_1}{(1-\varepsilon_1)/A_1\varepsilon_1} \Rightarrow J_1 = \sigma T_1^4 - q_1 \frac{1-\varepsilon_1}{A_1\varepsilon_1}$$

$$q_2 = \frac{J_1 - E_{b2}}{(1-\varepsilon_2)/A_2\varepsilon_2} \Rightarrow J_2 = q_2 \frac{1-\varepsilon_2}{A_2\varepsilon_2} + \sigma T_2^4$$

from

$$\frac{J_1 - J_R}{1/A_1 F_{1R}} = \frac{J_R - J_2}{1/A_R F_{R2}} \Rightarrow J_R = \frac{A_1 F_{1R} J_1 + A_R F_{R2} J_2}{A_1 F_{1R} + A_R F_{R2}}$$

Therefore,

$$J_R = E_{b,R} = \sigma T_R^4 \Rightarrow T_R = \left[\frac{J_R}{\sigma}\right]^{1/4} \tag{13.12}$$

Special case 4: A radiant heater panel problem: A long radiant heater panel consists of a row of cylindrical electrical heating elements, as shown in Figure 13.6.

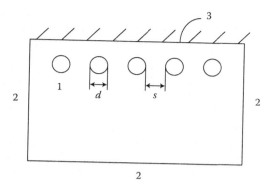

FIGURE 13.6 A radiant heater panel model.

The above Equations (13.11) and (13.12) can be used to determine the heat transfer rate $q_1 = -q_2$, and T_R. However, we need to calculate the view factors F_{11}, F_{12}, and F_{1R} (or F_{13}). In Table 12.1:

$$F_{11} = \frac{1}{\pi}\left[\left(X^2 - 1\right)^{1/2} + \sin^{-1}\frac{1}{X} - X\right]$$

with $X = 1 + (s/d)$. Assume $F_{12} \cong F_{13}$ for symmetry and $F_{11} + F_{12} + F_{13} = 1$. Therefore, $F_{12} = 1/[2(1 - F_{11})]$.

13.1.2 Method 2: Matrix Linear Equations

Applying the energy balance to each surface, one can obtain N radiosity linear equations for N surfaces in an enclosure. The matrix and its inverse matrix can be used to solve these N radiosity linear algebraic equations. The following shows how to solve this type of problem for either a given surface temperature or surface heat flux in an enclosure with N surfaces [3,4].

Case A: Given each surface temperature to determine the corresponding heat flux (T_i given $\Rightarrow q_i =?$).

Use the energy balance on surface i and the energy exchange between surface i and the other surfaces j, from Equation (13.6):

$$\frac{E_{bi} - J_i}{(1 - \varepsilon_i)/(\varepsilon_i A_i)} = \sum_{j=1}^{N}\frac{J_i - J_j}{1/A_i F_{ij}} \tag{13.13}$$

Applying the above equation to each surface (1, 2, 3, …, to N), one obtains the following N radiosity linear equations (after rearranging them).

$$a_{11}J_1 + a_{12}J_2 + \cdots + a_{1N}J_N = c_1$$
$$a_{21}J_1 + a_{22}J_2 + \cdots + a_{2N}J_N = c_2$$
$$\vdots$$
$$a_{N1}J_1 + a_{N2}J_2 + \cdots + a_{NN}J_N = c_N$$

The coefficient matrix $[A]$, column matrix $[J]$, and column matrix $[C]$ can be formed to satisfy the N linear equations. Therefore, the unknown radiosity matrix $[J]$ can be determined by solving the given inverse matrices $[A]$ and $[C]$.

$$[A][J] = [C]$$

$$[J] = [A]^{-1}[C] = \ldots$$

$$[A] = \begin{bmatrix} a_{11} & a_{12} & \cdots & a_{1N} \\ a_{21} & a_{22} & \cdots & a_{2N} \\ \cdots & \cdots & \cdots & \cdots \\ a_{N1} & a_{N2} & \cdots & a_{NN} \end{bmatrix} \quad [J] = \begin{bmatrix} J_1 \\ J_2 \\ \cdots \\ J_N \end{bmatrix} \quad [C] = \begin{bmatrix} c_1 \\ c_2 \\ \cdots \\ c_N \end{bmatrix}$$

Once the unknown radiosity, J, from each surface, i, has been determined from the aforementioned matrix relation, the radiation heat transfer from each surface can be shown from Equation (13.6) as

$$q_i = \frac{E_{bi} - J_i}{(1 - \varepsilon_i)/\varepsilon_i A_i} = \frac{\sigma T_i^4 - J_i}{(1 - \varepsilon_i)/\varepsilon_i A_i} \tag{13.14}$$

Special example of a three-surface enclosure problem: If the surface temperatures shown in Figure 13.7 are given (T_1, T_2, T_3), how do we determine the surface heat transfer rates (q_1, q_2, q_3)?

From Equation (13.13),

$$\frac{E_{bi} - J_i}{(1 - \varepsilon_i)/A_i \varepsilon_i} = \sum_{j=1}^{N} \frac{J_i - J_j}{1/A_i F_{ij}}$$

Apply for surface 1:

$$\frac{E_{b1} - J_1}{(1 - \varepsilon_1)/A_1 \varepsilon_1} = \frac{J_1 - J_1}{1/A_1 F_{11}} + \frac{J_1 - J_2}{1/A_1 F_{12}} + \frac{J_1 - J_3}{1/A_1 F_{13}}$$

$$\frac{A_1 \varepsilon_1}{1 - \varepsilon_1} E_{b1} - \left(\frac{A_1 \varepsilon_1}{1 - \varepsilon_1} \right) J_1 = (A_1 F_{12}) J_1 + (-A_1 F_{12}) J_2 + (A_1 F_{13}) J_1 + (-A_1 F_{13}) J_3$$

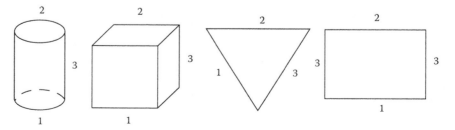

FIGURE 13.7   Radiation heat transfer among a three-surface enclosure.

$$\therefore \left(\frac{A_1 \varepsilon_1}{1 - \varepsilon_1} + A_1 F_{12} + A_1 F_{13} \right) J_1 + \left(-A_1 F_{12} \right) J_2 + \left(-A_1 F_{13} \right) J_3 = \frac{A_1 \varepsilon_1}{1 - \varepsilon_1} E_{b1} = \frac{A_1 \varepsilon_1}{1 - \varepsilon_1} \sigma T_1^4$$

$$\Rightarrow a_{11} J_1 + a_{12} J_2 + a_{13} J_3 = c_1$$

where $a_{11} = (A_{1\varepsilon1}/(1 - \varepsilon_1) + A_1 F_{12} + A_1 F_{13})$, $a_{12} = -A_1 F_{12}$, $a_{13} = -A_1 F_{13}$, $c_1 = (A_1 \varepsilon_1 / (1 - \varepsilon_1)) \sigma T_1^4$.

Take a similar approach for surface 2:

$$\frac{E_{b2} - J_2}{(1 - \varepsilon_2)/A_2 \varepsilon_2} = \frac{J_2 - J_1}{1/A_2 F_{21}} + \frac{J_2 - J_2}{1/A_2 F_{22}} + \frac{J_2 - J_3}{1/A_2 F_{23}}$$

$$\frac{A_2 \varepsilon_2}{1 - \varepsilon_2} E_{b2} - \left(\frac{A_2 \varepsilon_2}{1 - \varepsilon_2} \right) J_2 = \left(A_2 F_{21} \right) J_2 + \left(-A_2 F_{21} \right) J_1 + \left(A_2 F_{21} \right) J_2 + \left(-A_2 F_{23} \right) J_3$$

$$\therefore \left(-A_2 F_{21} \right) J_1 + \left(\frac{A_2 \varepsilon_2}{1 - \varepsilon_2} + A_2 F_{21} + A_2 F_{23} \right) J_2 + \left(-A_2 F_{23} \right) J_3 = \frac{A_2 \varepsilon_2}{1 - \varepsilon_2} E_{b2} = \frac{A_2 \varepsilon_2}{1 - \varepsilon_2} \sigma T_2^4$$

$$\Rightarrow a_{21} J_1 + a_{22} J_2 + a_{23} J_3 = c_2$$

where $a_{21} = -A_2 F_{21}$, $a_{22} = ((A_2 \varepsilon_2 / 1 - \varepsilon_2) + A_2 F_{21} + A_2 F_{23})$, $a_{23} = -A_2 F_{23}$, $c_2 = (A_2 \varepsilon_2 / (1 - \varepsilon_2)) \sigma T_2^4$.

Now for surface 3:

$$\frac{E_{b3} - J_3}{(1 - \varepsilon_3)/A_3 \varepsilon_3} = \frac{J_3 - J_1}{1/A_3 F_{31}} + \frac{J_3 - J_2}{1/A_3 F_{32}} + \frac{J_3 - J_3}{1/A_3 F_{33}}$$

$$\frac{A_3 \varepsilon_3}{1 - \varepsilon_3} E_{b3} - \left(\frac{A_3 \varepsilon_3}{1 - \varepsilon_3} \right) J_3 = \left(A_3 F_{31} \right) J_3 + \left(-A_3 F_{31} \right) J_1 + \left(A_3 F_{32} \right) J_3 + \left(-A_3 F_{32} \right) J_2$$

$$\therefore \left(-A_3 F_{31} \right) J_1 + \left(-A_3 F_{32} \right) J_2 + \left(\frac{A_3 \varepsilon_3}{1 - \varepsilon_3} + A_3 F_{31} + A_3 F_{32} \right) J_3 = \frac{A_3 \varepsilon_3}{1 - \varepsilon_3} E_{b3} = \frac{A_3 \varepsilon_3}{1 - \varepsilon_3} \sigma T_3^4$$

$$\Rightarrow a_{31} J_1 + a_{32} J_2 + a_{33} J_3 = c_3$$

where

$$a_{31} = -A_3 F_{31}$$

$$a_{32} = -A_3 F_{32}$$

$$a_{33} = \left(A_3 \varepsilon_3 / (1 - \varepsilon_3) + A_3 F_{31} + A_3 F_{32} \right)$$

$$c_3 = \left(A_3 \varepsilon_3 / (1 - \varepsilon_3) \right) \sigma T_3^4$$

From the above three linear equations, the following matrix can be formed:

$$A = \begin{bmatrix} a_{11} & a_{12} & a_{13} \\ a_{21} & a_{22} & a_{23} \\ a_{31} & a_{32} & a_{33} \end{bmatrix} \quad J = \begin{bmatrix} J_1 \\ J_2 \\ J_3 \end{bmatrix} \quad C = \begin{bmatrix} C_1 \\ C_2 \\ C_3 \end{bmatrix}$$

$$[A][J] = [C]$$

$$[J] = [A]^{-1}[C]$$

Alternatively, we can apply Equation (13.7) to each surface and get

$$J_1 = \varepsilon_1 E_{b1} + (1 - \varepsilon_1)\left[F_{11}J_1 + F_{12}J_2 + F_{13}J_3\right]$$

$$J_2 = \varepsilon_2 E_{b2} + (1 - \varepsilon_2)\left[F_{21}J_1 + F_{22}J_2 + F_{23}J_3\right]$$

$$J_3 = \varepsilon_3 E_{b3} + (1 - \varepsilon_3)\left[F_{31}J_1 + F_{32}J_2 + F_{33}J_3\right]$$

Similarly, the above three linear equations can be rearranged in order to obtain the matrix relation as $[A][J] = [C]$ and then solve for $[J] = [A]^{-1}[C]$. The matrix can be solved numerically using tools such as MATLAB®.

Once matrix $[J]$ is determined, that is, J_1, J_2, and J_3 have been determined, then use Equation (13.14) to find the surface heat transfer rates as

$$q_1 = \frac{E_{b1} - J_1}{(1 - \varepsilon_1)/(A_1\varepsilon_1)} = \frac{\sigma T_1^4 - J_1}{(1 - \varepsilon_1)/(A_1\varepsilon_1)}$$

$$q_2 = \frac{E_{b2} - J_2}{(1 - \varepsilon_2)/(A_2\varepsilon_2)} = \frac{\sigma T_2^4 - J_2}{(1 - \varepsilon_2)/(A_2\varepsilon_2)}$$

$$q_3 = \frac{E_{b3} - J_3}{(1 - \varepsilon_3)/(A_3\varepsilon_3)} = \frac{\sigma T_3^4 - J_3}{(1 - \varepsilon_3)/(A_3\varepsilon_3)}$$

Case B: Given each surface heat flux to determine the corresponding temperature (q_i given, $\Rightarrow T_i = ?$).

Use the energy exchange between surface i and the other surfaces j, from Equation (13.6),

$$q_i = \sum_{j=1}^{N} \frac{J_i - J_j}{1/A_i F_{ij}} \tag{13.15}$$

Apply the above equation to each surface (1, 2, 3, ..., and N) and obtain N radiosity linear equations as

$$a_{11}J_1 + a_{12}J_2 + \cdots + a_{1N}J_N = c_1$$
$$a_{21}J_1 + a_{22}J_2 + \cdots + a_{2N}J_N = c_2$$
$$\vdots$$
$$a_{N1}J_1 + a_{N2}J_2 + \cdots + a_{NN}J_N = c_N$$

The following matrix can be used to solve for $[J]$:

$$[A][J] = [C]$$

$$[J] = [A]^{-1}[C]$$

Once matrix $[J]$ has been solved, use Equation (13.14) on each surface, i, to determine the temperature on each surface

$$q_i = \frac{E_{bi} - J_i}{(1 - \varepsilon_i)/\varepsilon_i A_i} = \frac{\sigma T_i^4 - J_i}{(1 - \varepsilon_i)/\varepsilon_i A_i}$$

or

$$E_{bi} = \sigma T_i^4 = q_i \frac{1 - \varepsilon_i}{A_i \varepsilon_i} + J_i$$

Therefore,

$$T_i = \left(\frac{q_i(1 - \varepsilon_i)/A_i \varepsilon_i + J_i}{\sigma} \right)^{1/4} \tag{13.16}$$

Case C: Combine Case A and Case B

- Some surfaces are given temperatures, but heat fluxes are unknown.
- Some surfaces are given heat fluxes, but temperatures are unknown.
- Use the same procedure shown for Case A and Case B in order to form the matrix $[A]$, column matrix $[J]$, and column matrix $[C]$.
- After determining matrix $[J]$, either the heat fluxes or temperatures can be determined.

Special case for the blackbody radiation problem: Use the aforementioned results for any blackbody surface with unity emissivity ($\varepsilon_i = 1$).

13.2   RADIATION EXCHANGE BETWEEN GRAY DIFFUSE NONISOTHERMAL SURFACES

The following section shows how to solve radiation heat transfer between nonisothermal surfaces [4–5]. In this case, one shall consider radiation exchange between two differential surfaces (which can be assumed as a uniform temperature over each differential element), as shown in Figure 13.8. Then the aforementioned analysis method can be applied.

Consider radiosity from a differential element i:

$$J_i\left(\overline{r_i}\right) = \varepsilon_i \sigma T_i^4\left(\overline{r_i}\right) + \left(1 - \varepsilon_i\right) G_i\left(\overline{r_i}\right)$$

$$= \varepsilon_i \sigma T_i^4\left(\overline{r_i}\right) + \left(1 - \varepsilon_i\right) \sum_{j=1}^{N} \int_{A_j} J_j\left(\overline{r_j}\right) dF_{dA_i - dA_j} \tag{13.17}$$

Define the known quantity:

$$K\left(\overline{r_i}, \overline{r_j}\right) \equiv \frac{dF_{dA_i - dA_j}}{dA_j}$$

Therefore,

$$J_i\left(\overline{r_i}\right) = \varepsilon_i \sigma T_i^4\left(\overline{r_i}\right) + \left(1 - \varepsilon_i\right) \sum_{j=1}^{N} \int_{A_j} J_j\left(\overline{r_j}\right) K\left(\overline{r_i}, \overline{r_j}\right) dA_j \tag{13.18}$$

For the Case A problem, obtain N equations, where T_i is known, solve the integral, and $J_i(r_i)$ can be obtained.

When $J_i\left(\overline{r_i}\right)$ is determined using the matrix method for a given $T_i\left(\overline{r_i}\right)$, then

$$\left(\frac{q}{A}\right)_i \left(\overline{r_i}\right) = \frac{\varepsilon_i}{1 - \varepsilon_i}\left[\sigma T_i^4\left(\overline{r_i}\right) - J_j\left(\overline{r_j}\right)\right] \tag{13.19}$$

can be solved.

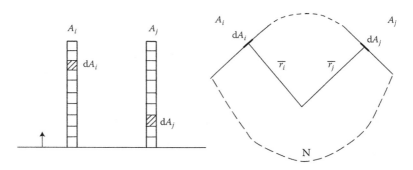

FIGURE 13.8   Radiation exchange between nonisothermal surfaces.

Example

Apply the numerical method—Simpson's rule (Trapezoidal rule)—for the noniso-thermal surfaces shown in Figure 13.9.

Given: $\varepsilon_a = 0.9$, $T_a = 1000-1500°C$

$E_b = 0.2$, $T_b = 300°C$

a. $\bar{q}_a =$?

b. Compare $q_a = q_{a1} + q_{a2} =$?

Solution

For the case B problem, if $(q/A)_i$, is given, then

$$\left(\frac{q}{A}\right)_i = \sum_{j=1}^{N} \frac{J_i\left(\bar{r}_i\right) - J_j\left(\bar{r}_j\right)}{1/dF_{dA_i - dA_j}}$$

$$\left(\frac{q}{A}\right)_i = \sum_{j=1}^{N} \left(J_i\left(\bar{r}_i\right) - J_j\left(\bar{r}_j\right)\right) K\left(\bar{r}_i, \bar{r}_j\right) dA_j$$

when $J_i\left(\bar{r}_i\right)$ is determined by the matrix method, and for given $\left(q/A\right)_i \left(\bar{r}_i\right)$, then

$$\sigma T_i^4\left(\bar{r}_i\right) = \frac{1-\varepsilon_i}{\varepsilon_i}\left(\frac{q}{A}\right)_i\left(\bar{r}_i\right) + J_i\left(\bar{r}_i\right)$$

Note: The view factor between any two elements in y_a and y_b can also be deter-mined using the cross-string method discussed in Chapter 12.

13.2.1 RADIATION EXCHANGE BETWEEN NONGRAY DIFFUSE ISOTHERMAL SURFACES

The following (a) integral model or (b) band model can be used to determine radiation exchange among isothermal diffuse nongray surfaces [4,5] as sketched in Figure 13.10.

a. The Integral Model

$$\left(\frac{q}{A}\right)_i = \int_0^\infty \left(\frac{q}{A}\right)_i d\lambda \qquad (13.20)$$

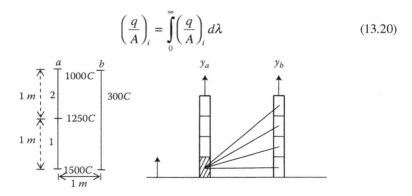

FIGURE 13.9 Radiation between nonisothermal surfaces.

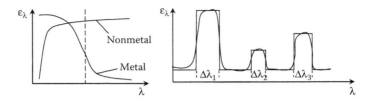

FIGURE 13.10 Radiation exchange among diffuse, isothermal, nongray surfaces.

b. The Band Model

$$\left(\frac{q}{A}\right)_i = \sum_{k=1}^{N} \left(\frac{q}{A}\right)_i \Delta\lambda_k \tag{13.21}$$

13.2.2 RADIATION INTERCHANGE AMONG DIFFUSE AND NONDIFFUSE (SPECULAR) SURFACES [4,5]

The exchange factor is obtained using the image method, as shown in Figure 13.11.

$$E_{A1-A4} = \underbrace{F_{A1-A4}}_{\text{diffuse}} + \underbrace{\rho_3^s F_{A1(3)-A4}}_{\text{specular reflection}} \tag{13.22}$$

$$E_{A2-A4} = F_{A2-A4} + \rho_3^s F_{A2(3)-A4} \tag{13.23}$$

$$E_{A1-A1} = F_{A1-A1} + \rho_3^s F_{A1(3)-A1} \tag{13.24}$$

Reciprocity Rules: $A_i E_{Ai-Aj} = A_j E_{Aj-Ai}$

13.2.3 ENERGY BALANCE IN AN ENCLOSURE WITH A DIFFUSE AND SPECULAR SURFACE

N_d is the diffuse reflection surface,

$N - N_d$ is the specular reflection surface.

Assume all the surfaces shown in Figure 13.12 are diffuse emitting, gray, and isothermal; then

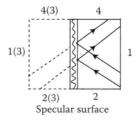

Specular surface

FIGURE 13.11 Radiation exchange among diffuse and specular surfaces.

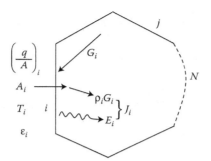

FIGURE 13.12 Radiation heat transfer between N surfaces (including diffuse and specular) in an enclosure.

$$J_i = \varepsilon_i \sigma T_i^4 + (1 - \varepsilon_i) G_i$$

where

$$G_i = G_i^d + G_i^s.$$

From the diffuse surface,

$$G_i^d = \sum_{j=1}^{Nd} J_j E_{Ai-Aj}$$ (13.25)

where

$$E_{Ai-Aj} = F_{Ai-Aj} + \sum_k \rho_k^s F_{Ai(k)-Aj}$$

From the specular surface,

$$G_i^s = \sum_{j=Nd+1}^{N} \varepsilon_j \sigma T_j^4 E_{Ai-Aj}$$ (13.26)

for a given T_i, from the above equations, J_i can be solved.
If T_i is given, G_i can be solved $\left(G_i = G_i^d + G_i^s\right)$.
For the N_d diffuse surfaces,

$$\left(\frac{q}{A}\right)_i = \frac{1}{(1-\varepsilon_i)/\varepsilon_i}\left(\sigma T_i^4 - J_i\right)$$ (13.27)

or for the $N - N_d$ specular surfaces,

$$\left(\frac{q}{A}\right)_i = \varepsilon_i\left(\sigma T_i^4 - G_i\right)$$ (13.28)

Examples

13.1 A rectangular oven is 1 m wide, 0.5 m tall, and very deep into the paper. It is used to bake a carbon-fiber cloth with an electric heater at the top. All vertical walls are reradiating (reflectory and insulated). Take $\varepsilon_1 = 0.7$, $\varepsilon_2 = 0.9$, and $\varepsilon_3 = 0.8$. When 20 kW of power are supplied, the heater temperature is 650°C. Neglecting convection, what is the cloth temperature?

Solution

From Hottel's crossed-string method,

$$F_{12}\left(\text{top to bottom wall}\right) = 0.618 = F_{21},$$

$$F_{1R}\left(\text{top to side wall}\right) = 0.382 = F_{2R}$$

From Equation (13.11), $T_1 = 923\,\text{K}$, $q = 20\,\text{kW}$, $T_2 \cong 560K$

13.2 We have a cubic furnace (1 m × 1 m × 1 m). During steady-state operation, the top surface ($\varepsilon_2 = 0.9$) is cooled at 250°C and the bottom floor ($\varepsilon_1 = 0.7$) is heated at 1000°C. The side walls are insulated refractory surfaces. The view factor between the top and bottom surfaces is 0.2.

a. Determine the net radiation transfer between the top and bottom surfaces.
b. Determine the temperature of the insulated refractory surfaces.
c. Comment on what effect changing the values of emissivities of top, bottom, and refractory surfaces would have on the results of parts (a) and (b).

Solution

a. Refer to Figure 13.5 and Equation (13.11):

$$q_{12}\left(\text{bottom to top}\right) \cong 25.5 \text{ kW}$$

b. $q_{12} = q_1 = \dfrac{E_{b1} - J_1}{\dfrac{1 - \varepsilon_1}{A_1 \varepsilon_1}}$ to get $J_1 = \sqrt{}$

$q_{12} = q_2 = \dfrac{J_2 - E_{b2}}{\dfrac{1 - \varepsilon_2}{A_2 \varepsilon_2}}$ to get $J_2 = \sqrt{}$

Then use $q_{1R} = \dfrac{J_1 - J_R}{\dfrac{1}{A_1 F_{1R}}} = q_{R2} = \dfrac{J_R - J_2}{\dfrac{1}{A_R F_{R2}}}$ to get $J_R = \sqrt{}$

$\therefore J_R = E_{b,R} = \sigma T_R^4$ to get $T_R \cong 1225K$

c. $\varepsilon_1 (\text{bottom}) \downarrow, \quad q_{12} \downarrow, \quad T_R \downarrow$

$\varepsilon_2 (\text{top}) \uparrow, \quad q_{12} \uparrow, \quad T_R \uparrow$

$\varepsilon_R \uparrow \text{ or } \downarrow, \quad$ No effect on q_{12} or T_R

13.3 Consider two, aligned, parallel, square planes $(0.5\,\text{m} \times 0.5\,\text{m})$ spaced 0.5 m apart and maintained at $T_1 = 500$ K and $T_2 = 1000$ K. Calculate the net radiative heat transfer from surface 1 for the following special conditions:

a. Both planes are black, and the surroundings are at 0 K.
b. Both planes are black with connecting, reradiating walls.
c. Both planes are diffuse and gray with $\varepsilon_1 = 0.6$, $\varepsilon_2 = 0.8$, and the surroundings at 0 K.
d. Both planes are diffuse and gray ($\varepsilon_1 = 0.6$ and $\varepsilon_2 = 0.8$) with connecting, reradiating walls.

Solution

a. $q_1 (\text{bottom}) = q_{12} (\text{bottom to top}) + q_{1s} (\text{bottom to surroundings})$

$$= A_1 F_{12} (E_{b1} - E_{b2}) + A_1 F_{1s} (E_{b1} - E_{bs})$$

$$= \sqrt{}, \text{ where } F_{12} = \sqrt{}, \ F_{1s} = \sqrt{}$$

b. $q_1 = \dfrac{A_1 (E_{b1} - E_{b2})}{F_{12} + \left[\left(\dfrac{1}{F_{1R}} + \dfrac{1}{F_{2R}} \right)^{-1} \right]^{-1}} = \sqrt{}$

c. Solve J_1 and J_2 from

$$\frac{E_{b1} - J_1}{\dfrac{1 - \varepsilon_1}{\varepsilon_1}} = F_{12} (J_1 - J_2) + F_{13} (J_1 - J_3)$$

$$\frac{E_{b2} - J_2}{\dfrac{1 - \varepsilon_2}{\varepsilon_2}} = F_{21} (J_2 - J_1) + F_{23} (J_2 - J_3)$$

Then

$$q_1 = A_1 F_{12} (J_1 - J_2) + A_1 F_{13} (J_1 - J_3)$$

where

$$J_3 = 0, \ F_{12} = \sqrt{}, \ F_{13} = \sqrt{}$$

d. From Equation (13.11), $q_1 = \sqrt{}$, where $F_{12} = \sqrt{}$, $F_{1R} = \sqrt{}$

13.3 COMBINED MODES OF HEAT TRANSFER

13.3.1 RADIATION WITH CONDUCTION

Case 1: A typical fin problem involving conduction, convection, and radiation (refer to Chapter 2). Consider radial plate fins on a circular tube shown in Figure 13.13.

Begin by assuming the outer wall temperature of the tube at T_b is hotter than the surrounding fluid temperature at T_∞, ignoring the convection effect, only considering conduction through the fin with radiation.

Determine the plate fin temperature distribution, $T(x)$, using finite-difference energy balance method (Chapter 5).

Given (while specific values are shown below, other numbers can be substituted as needed):

plate fin length $= L = 8\,\mathrm{cm}$
plate fin thickness $= t = 1\,\mathrm{cm}$
angle between two plate fins $= \alpha = 45°$
finite difference $= \Delta x = 2\,\mathrm{cm}$
number of elements $= N = 5$
plate fin conductivity $= k = 20$ W/m-K or 200 W/m-K
plate fin emissivity $= \varepsilon = 1.0$ or 0.5
tube wall temperature $= T_b = 500$ K
surrounding temperature $= T_\infty = 300$ K
convection coefficient $= h = 0$ or 20 W/m²-K

Begin by determining the view factor between any two elements on the two radial fins:

Referring to Hottel's crossed-string method for a 2-D geometry (Chapter 12, Example 12.8). For example, as shown in Figure 13.13, the view factor between element #4 of the two fins, i, j, can be obtained as

$$F_{4-4} = \frac{l_{4-5} + l_{4-5} - (l_{4-4} + l_{5-5})}{2\Delta x}$$

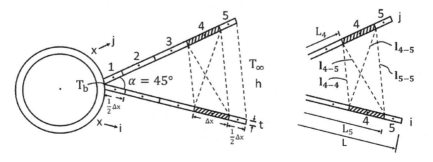

FIGURE 13.13 Radial plate fins on a circular tube and view factor determination.

From geometrical relationships:

$$l_{4-5} = \sqrt{L_4^2 + L_5^2 - 2L_4 \cdot L_5 \cdot \cos\alpha}$$

$$l_{4-4} = \sqrt{L_4^2 + L_4^2 - 2L_4 \cdot L_4 \cdot \cos\alpha}$$

$$l_{5-5} = \sqrt{L_5^2 + L_5^2 - 2L_5 \cdot L_5 \cdot \cos\alpha}$$

Similarly, the view factors between any two elements of fin i and j can be obtained, such as F_{4-5}, F_{4-3}, F_{4-2}, F_{4-1}, F_{3-3}, F_{3-4}, F_{3-5}, etc.

For view factor between an element and the surrounding such as from element #4 to the surrounding:

$$F_{4-\infty} = \frac{l_{4-\infty} + l_{4-\infty} - (l_{4-4} + l_{\infty-\infty})}{2\Delta x}$$

where

$$l_{4-\infty} = \sqrt{L^2 + L_5^2 - 2L \cdot L_5 \cdot \cos\alpha}$$

$$l_{4-\infty} = L - L_4$$

$$l_{4-4} = \sqrt{L^2 + L_4^2 - 2L \cdot L_4 \cdot \cos\alpha}$$

$$l_{\infty-\infty} = L - L_5$$

Similarly, $F_{3-\infty}$, $F_{2-\infty}$, $F_{1-\infty}$, etc., can be determined.

Now, perform the energy balance on any element of a radial fin:

For example, element # 4 of an interior node:
  Net conduction = Radiation

$$\frac{k \cdot t}{\Delta x}\left[(T_3 - T_4) + (T_5 - T_4)\right] = 2 \cdot q_4'' \, \Delta x$$

Similarly, the energy balance can be obtained for any interior nodes such as T_1, T_2, T_3, T_4, T_5. The tip node depends on the boundary conditions such as insulated, convection, radiation, etc., as indicated in Chapter 2. If $N = 1, 2, 3, 4, 5$, five energy balance equations can be derived, so that T_1, T_2, T_3, T_4, T_5 can be obtained. Here $T_1 = T_b =$ given. However, we need to determine q_4'', radiation term, in the above energy balance equation, and q_1'', q_2'', q_3'', q_5''.

From the radiation energy exchange between element # 4 of fin i and elements #1, #2, #3, #4, #5 of fin j, and the surrounding element:

$$q_4'' = (J_4 - J_1)F_{4-1} + (J_4 - J_2)F_{4-2} + (J_4 - J_3)F_{4-3} + (J_4 - J_4)F_{4-4}$$
$$+ (J_4 - J_5)F_{4-5} + (J_4 - J_\infty)F_{4-\infty}$$

Also,

$$q_4'' = \frac{\sigma T_4^4 - J_4}{\dfrac{1-\varepsilon}{\varepsilon}}$$

Similarly, q_1'', q_2'', q_3'', q_5'' of fin i can be obtained.

From these $q_1'', \ldots, q_5''$ equations, $J_1, J_2, \ldots, J_5$ can be determined. Substitute these $q_1'', \ldots, q_5''$, i.e., $J_1, \ldots, J_5$ into the above-mentioned five energy balance equations; thus, T_1, T_2, T_3, T_4, T_5 can be solved.

Note that $J_\infty =$ surrounding radiosity $= \sigma T_\infty^4 =$ surrounding element and $T_1 = T_b =$ given.

Case 2: A typical fin problem involving conduction, convection, and radiation (refer to Chapter 2). With Case 1, we ignored convection. Now we will consider convection loss, and the above-mentioned energy balance equation becomes:

$$\frac{k \cdot t}{\Delta x}\left[(T_3 - T_4) + (T_5 - T_4)\right] = 2 \cdot q_4'' \, \Delta x + h \, \Delta x \cdot 2 \cdot (T_4 - T_\infty)$$

Following the same procedure above, T_1, T_2, T_3, T_4, T_5 can be determined.

Note that the predicted fin temperature will be lower for the fin with $\varepsilon = 1$ than $\varepsilon = 0.5$; also, the fin temperature will decrease with the inclusion of both radiation and convective heat losses.

13.3.2 RADIATION WITH CONVECTION

Case 1: Consider flow through a circular tube at constant heat flux q_s'', as shown in Figure 13.14 [6].

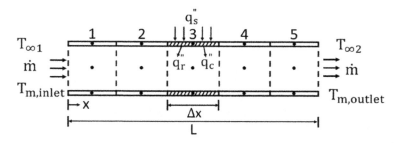

FIGURE 13.14 Flow through a tube with radiation and convection.

Predict the tube wall temperature distribution, $T(x)$, using the finite difference energy balance method. $q_s'' = $ given $=$ constant. The tube wall temperatures T_1, T_2, T_3, T_4, $T_5 = $ unknown. The fluid bulk mean temperature T_{m1}, T_{m2}, T_{m3}, T_{m4}, $T_{m5} = $ unknown.

Consider an energy balance on the tube wall element T_3:

$$q_s'' = q_r'' + q_c'' = \frac{\varepsilon}{1-\varepsilon}\left(\sigma T_3^4 - J_3\right) + h\left(T_3 - T_{m3}\right) \tag{1}$$

where

$$J_3 = \varepsilon T_3^4 + (1-\varepsilon)\cdot\left[\sigma T_{\infty 1}^4 \cdot F_{3-\infty 1} + \sigma T_{\infty 2}^4 \cdot F_{3-\infty 2} + \sum_{i=1}^{5} J_{i-3}F_{3-i}\right]$$

Similarly, the energy balance on the tube wall elements T_1, T_2, T_4, T_5 can be derived. The view factors, F_{3-i}, can be obtained from Chapter 12, Example 12.10, such as $F_{3-\infty 1}$, F_{3-1}, F_{3-2}, F_{3-3}, F_{3-4}, F_{3-5}, $F_{3-\infty 2}$. Similarly, the energy balance, view factors, and radiosity at elements 1, 2, 3, 4, and 5 can be derived.

Since the problem involves convection, assume the heat transfer coefficient inside the tube is $h = $ constant $=$ given. We need to determine T_{m3} in the above energy balance Equation 1.

Consider an energy balance in the fluid element T_{m3}:

$$q_c = h\pi D\Delta x\left(T_3 - T_{m3}\right)$$

$$= \dot{m}C_P\left(T_{m3} - T_{m2}\right)$$

Thus

$$T_{m3} = T_{m2} + \frac{h\pi D\Delta x}{\dot{m}C_P}\left(T_3 - T_{m3}\right) \tag{2}$$

Similarly, the fluid bulk mean temperatures, T_{m1}, T_{m2}, T_{m4}, T_{m5} can be derived.

From (1) and (2), the unknown wall temperatures, T_1, T_2, T_3, T_4, T_5, and unknown fluid temperatures, T_{m1}, T_{m2}, T_{m3}, T_{m4}, and T_{m5}, can be obtained through iteration. Assume a set of tube wall temperature distributions, $T_1, T_2, \ldots, T_5$, to get a set of fluid temperature distributions, $T_{m1}, T_{m2}, \ldots, T_{m5}$. Then try a new set of $T_1, T_2, \ldots, T_5$ to get a new set of $T_{m1}, T_{m2}, \ldots, T_{m5}$ until convergence.

Given values for the following quantities, the temperature distributions can be calculated:

Number of tube elements, $N = 1, 2, 3, 4, 5, \ldots$, etc.; tube diameter, D; tube length, L; mass flow rate, $\dot{m}$; fluid specific heat, C_P; fluid inlet temperature, $T_{m, inlet}$; surface emissivity, ε; surrounding fluid temperatures, $T_{\infty 1}$, $T_{\infty 2}$; surface heat flux (constant), q_s''; element spacing, Δx; the outlet temperature, $T_{m, outlet}$, is unknown and must be determined.

Case 2: Now we will extend case 1 to also include axial conduction in the tube wall.

The above energy balance equation on the tube wall element T_3 becomes:

$$q_s + q_{cond2-3} + q_{cond4-3} = q_r + q_c$$

$$q_s = \pi D\Delta x \frac{\varepsilon}{1-\varepsilon}\left(\sigma T_3^4 - J_3\right) + \pi D\Delta x h\left(T_3 - T_{m3}\right) - \frac{A_c k}{\Delta x}\left[\left(T_2 - T_3\right) + \left(T_4 - T_3\right)\right] \qquad (3)$$

where k = tube wall conductivity = given; $q_s = q_s'' \cdot \pi D\Delta x$ = given; A_c = tube wall cross-sectional area $= \frac{\pi}{4}\left(D_o^2 - D_i^2\right)$; D_i = tube inner diameter = given; D_o = tube outer diameter = given.

Following the above outlined procedures, $T_1, \ldots, T_5$, and $T_{m1}, \ldots, T_{m5}$ can be determined by solving (2) and (3).

REMARKS

In an undergraduate-level heat transfer course, students are expected to know how to calculate radiation heat transfer between two surfaces or between two surfaces with a third reradiating surface. Students are taught to use the electric network analogy for many engineering applications such as electric heaters, radiation shields, and electric furnaces with insulating side walls, and so on. At the intermediate-level, this chapter focuses on how to analyze and solve radiation heat transfer problems in an N-surface enclosure by using the matrix linear equations method for more complicated electric or combustion furnaces applications. Students are expected to know how to set up a matrix from linear equations by applying the energy balance on each of the N-surfaces with given surface temperature or surface heat flux BCs. Here we assume that each N-surface has gray and diffuse properties and is kept at an isothermal condition. For more details of N-surface enclosures with nongray, nondiffuse (specular), or nonisothermal surfaces, readers are referred to more advanced radiation texts.

PROBLEMS

13.1 A rectangular oven 1 m wide, 0.5 m tall, and 2 m deep into the paper and is used to bake a carbon-fiber cloth with an electric heater at the top. All vertical walls are re-radiating (reflectory and insulated). Take $\varepsilon_1 = 0.7$, $\varepsilon_2 = 0.9$, and $\varepsilon_3 = 0.8$. The heater temperature is 650°C when 20 kW of power is supplied. Convection is negligible. Given: $\sigma = 5.67 \times 10^{-8} \frac{W}{m^2 K^4}$

   a. Based on the analogy of electric resistance network, draw radiation heat transfer network from surface 1 to surface 2.

   b. Determine the cloth (surface 2) temperature.

13.2 A rectangular oven 1 m wide, 0.5 m tall, and very deep into the paper is used to bake a carbon-fiber cloth with an electric heater at the top. All vertical walls are reradiating (reflectory and insulated). Take $\varepsilon_1 = 0.7$, $\varepsilon_2 = 0.9$, and $\varepsilon_3 = 0.8$. When 20 kW of power is supplied, the heater temperature is 650°C. Neglecting convection, what is the cloth temperature?

13.3 A cubic furnace (1 m×1 m×1 m) is used at steady-state operation. The top surface ($\varepsilon_2 = 0.9$) is cooled at 250°C and the bottom floor ($\varepsilon_1 = 0.7$) is heated at 1000°C. The side walls are insulated, refractory surfaces. The view factor between the top and bottom surfaces is 0.2.

   a. Determine the net radiation transfer between the top and bottom surfaces.

   b. Determine the temperature of the insulated refractory surfaces.

   c. Comment on what effect changing the values of emissivities of top, bottom, and refractory surfaces would have on the results of (a) and (b).

13.4 Consider two aligned, parallel, square planes (0.5 m×0.5 m) spaced 0.5 m apart and maintained at $T_1 = 500$ K and $T_2 = 1000$ K. Calculate the net radiative heat transfer from surface 1 for the following special conditions:

   a. Both planes are black, and the surroundings are at 0 K.

   b. Both planes are black with connecting, reradiating walls.

   c. Both planes are diffuse and gray with $\varepsilon_1 = 0.6$, $\varepsilon_2 = 0.8$, and the surroundings at 0 K.

   d. Both planes are diffuse and gray ($\varepsilon_1 = 0.6$ and $\varepsilon_2 = 0.8$) with connecting, reradiating walls.

13.5 A room is 3 m square and 3 m high. The walls can be taken as adiabatic and isothermal. The ceiling is at 35°C and has an emittance of 0.8, while the floor is at 20°C and has an emittance of 0.9. Denote the ceiling as surface 1, the floor 2, and the walls 3.

   a. Set up the radiosity equations. Determine and evaluate all the view factors and tabulate as a 3×3 array. Solve these equations to determine the heat flow into the floor, q_2.

   b. Draw the radiation network. Use the network to obtain an expression for q_2, and solve for q_2 again.

13.6 A thin plate (surface area A_1, emissivity ε_1, absorptivity α_1) is mounted horizontally facing above a larger horizontal surface (area A_2, emissivity ε_2, absorptivity α_2).

   a. Give the corresponding thermal radiation network associated to the problem.

   b. Develop an expression without the radiosities for the radiation heat transfer rate from 1 to 2.

   c. What is the limit of this expression when the second surface is infinite?

   d. Let surface 2 be the sky which is a blackbody at a temperature 15°C cooler than ambient air, which is at 2°C. The plate is well insulated at the bottom. Let h be the average convective heat transfer coefficient between the surface and ambient air. Given $h = 10\,\text{W/m}^2\,\text{K}$, $\varepsilon_1 = 0.54 = \alpha_1$, $\sigma = 5.67 \times 10^{-8}\,\text{W/m}^2\,\text{K}^4$. Determine the equilibrium plate temperature.

13.7 Consider two very large parallel plates with diffuse, gray surfaces. Determine the irradiation and radiosity for the upper plate (at $T_1 = 1000°\text{K}$, $\varepsilon_1 = 1$). What is the radiosity for the lower plate (at $T_2 = 500°\text{K}$, $\varepsilon_2 = 0.8$)? What is the net radiation exchange between the plates per unit area of the plates?

13.8 A 0.25-m-diameter sphere (surface 1) is located inside of a 0.5-m-diameter sphere (surface 2). Surface 1 is 200°C and surface 2 is 100°C. Determine all the view factors and calculate the net heat transfer rate (W) between the two spherical surfaces. Show all work and list all assumptions. (*Note:* $\sigma = 5.67 \times 10^{-8}\,\text{W/m}^2\,\text{K}^4$)

13.9 For a three-surface enclosure problem, as shown in Figure 13.7, determine the following quantities using the matrix method.
   a. Given T_1, T_2, T_3, determine q_1, q_2, q_3.
   b. Given q_1, q_2, q_3, determine T_1, T_2, T_3.
   c. Given T_1, T_2, T_3, determine q_1, q_2, T_3.
   d. Given T_1, q_2, q_3, determine q_1, T_2, T_3.
   e. Given q_1, q_2, T_3, determine T_1, T_2, q_3.

13.10 A rectangular oven is 1 m wide, 0.5 m tall, and very deep into the paper, and is used to bake a carbon-fiber cloth at the bottom with an electric heater at the top. All vertical walls are at same temperature during operation. Take the heater $\varepsilon_1 = 0.7$, carbon-fiber cloth $\varepsilon_2 = 0.9$, all vertical walls $\varepsilon_3 = 0.8$. The heater temperature is 650°C when 20 kW per length of power are supplied. Convection is negligible.

Given: $\sigma = 5.67 \times 10^{-8}\,\dfrac{\text{W}}{\text{m}^2\text{K}^4}$

   a. If all vertical walls are reradiating surfaces (perfect reflection and insulation), determine the cloth (surface 2) temperature.
   b. If all vertical walls are gray and diffuse surfaces, to keep the same cloth temperature, comment on whether the required heater power will be the same, lower, or higher than that in part (a)? Why? Write down equations that can be used to solve this problem, but you do not need to calculate the numerical answer.

REFERENCES

1. W. Rohsenow and H. Choi, *Heat, Mass, and Momentum Transfer*, Prentice-Hall, Inc., Englewood Cliffs, NJ, 1961.
2. A. Mills, *Heat Transfer*, Richard D. Irwin, Inc., Boston, MA, 1992.
3. F. Incropera and D. Dewitt, *Fundamentals of Heat and Mass Transfer*, Fifth Edition, John Wiley & Sons, New York, 2002.
4. R. Siegel and J. Howell, *Thermal Radiation Heat Transfer*, McGraw-Hill, New York, 1972.
5. J. Chen, *Conduction and Radiation Heat Transfer, Class Notes*, Lehigh University, Bethlehem, PA, 1973.
6. E.M. Sparrow and R.D. Cess, *Radiation Heat Transfer*, Hemisphere Publishing, New York, 1978.

14 Radiation Transfer through Gases

14.1 GAS RADIATION PROPERTIES

A volume of gases, such as CO_2, H_2O (water vapor), CO, NO, NH_3, SO_2, HCl, the hydrocarbons, and the alcohols, can emit and absorb energy at a given temperature and pressure. It is important to determine gas radiation properties such as emissivity and absorptivity for combustion furnace designs [1–5]. The combustion products (CO_2, water vapor, CO, NO, and NH_3) have radiation properties but air (oxygen and nitrogen gases), helium, and hydrogen have no radiation properties (transparent to radiation). We will assume that there is no scattering effect, in order to simplify the analysis. In general, gas emissivity and absorptivity increase with pressure and volume, but decrease with temperature; gas emissivity and absorptivity vary with wavelength [1] as shown in Figure 14.1. Gases absorb and emit radiation in rather narrow wavelength bands rather than in the continuous spectrum exhibited by solid surfaces. For real gases, the gas absorptivity is not the same as emissivity. Under the gray gas assumption, the absorptivity can be equivalent to emissivity.

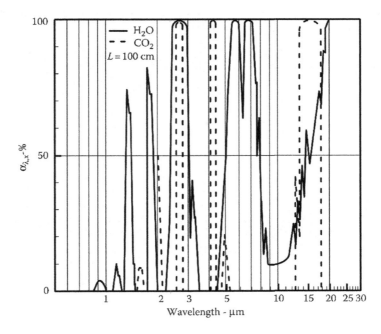

FIGURE 14.1 Band emission of carbon dioxide and water vapor.

DOI: 10.1201/9781003164487-14

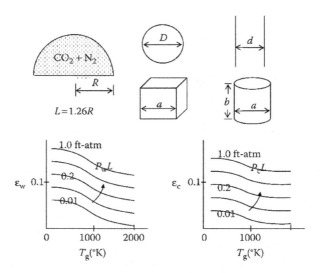

FIGURE 14.2 Gas radiation properties for water vapor and carbon dioxide.

Figure 14.2 shows that gas radiation properties (carbon dioxide, water vapor) increase with their partial pressure and geometric mean bean length (four times volume divided by surface area) and decrease with temperature [1]. The results were obtained by applying hemispherical gas radiation to an element area at the center of the base, as shown in Figure 14.3. Several other geometries that contain gases are also sketched in the figure. In general, gas emissivity and absorptivity are relatively low, approximately equivalent to an order of magnitude 0.1.

$$\varepsilon_c = \varepsilon_c\left(P_c L, T_g\right) \cong 0.1 \tag{14.1}$$

$$\varepsilon_w = \varepsilon_w\left(P_w L, T_g\right) \cong 0.1 \tag{14.2}$$

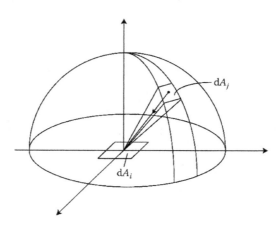

FIGURE 14.3 Hemispherical gas radiation to an element area at the center of base.

At 1 atm total pressure, CO_2 has partial pressure P_c and water vapor has partial pressure P_w. The geometric mean beam length, L, is defined as

$$L = \frac{4V}{A_s} \cong 0.9 \frac{4V}{A_s} \tag{14.3}$$

where V is the volume and A_s is the surface area.

The concept of geometric mean beam length for other gas geometries will be discussed in a later section.

The total gas emissivity for combined CO_2 and H_2O can be obtained as

$$\varepsilon_g = \varepsilon_c + \varepsilon_w - \Delta\varepsilon \tag{14.4}$$

where $\Delta\varepsilon \cong 0.01$ is a correction factor of emissivity for overlap of λ for CO_2 and H_2O.

For gray gas,

$$\varepsilon_g = \alpha_g \tag{14.5}$$

14.1.1 VOLUMETRIC ABSORPTION

Consider radiation heat transfer between two parallel plates at T_1 and T_2, filled with an absorption gas at a uniform temperature T_g. Spectral radiation absorption in a gas is proportional to the absorption coefficient $k_\lambda (1/m)$ and the thickness L of the gas. The radiation intensity decreases with increasing distance due to absorption [3], as shown in Figure 14.4.

$$dI_\lambda(x) = -k_\lambda I_\lambda(x)\, dx \tag{14.6}$$

If k_λ is a constant value for a given gas, we obtain

$$\frac{dI_\lambda(x)}{I_\lambda(x)} = -k_\lambda\, dx$$

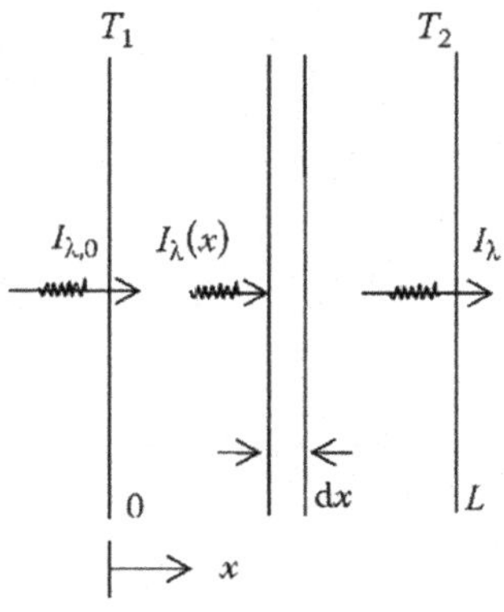

FIGURE 14.4 Absorption in a gas.

Performing integration, we obtain

$$\ln I_\lambda(x) = -k_\lambda x + C_1$$

$$I_\lambda(x) = e^{(-k_\lambda x + C_1)} = Ce^{-k_\lambda x}$$

at $x = 0$, $C = I_{\lambda,0}$

$$I_\lambda(x) = I_{\lambda,0}e^{-k_\lambda x}$$

at $x = L$

$$I_{\lambda,L} = I_{\lambda,0}e^{-k_\lambda L} \tag{14.7}$$

This exponential decay is called Beer's law. One can define the transmissivity as

$$\tau_\lambda = \frac{I_{\lambda,L}}{I_{\lambda,0}} = e^{-k_\lambda L} \tag{14.8}$$

The absorptivity is

$$\alpha_\lambda = 1 - \tau_\lambda = 1 - e^{-k_\lambda L}$$

For gases, $\alpha_\lambda = \varepsilon_\lambda =$ emissivity.

If we consider both gas emission and the absorption effect, the intensity of the beam is attenuated due to absorption and is augmented due to gas emission along the distance [2]. Assume a local thermodynamic equilibrium, absorption coefficient will equal the emission coefficient, and Equation (14.6) becomes

$$dI_\lambda(x) = \left[-k_\lambda I_\lambda(x) + k_\lambda I_{b\lambda}\right]dx \tag{14.9}$$

where $k_\lambda I_{b\lambda}$ is the intensity gained due to gas emission and $-kI_\lambda(x)$ is the intensity attenuated due to gas absorption.

Performing integration, we obtain:

$$dI_\lambda(x) = -k_\lambda\left[I_\lambda(x) - I_{b\lambda}\right]dx$$

$$\frac{d\left[I_\lambda(x) - I_{b\lambda}\right]}{\left[I_\lambda(x) - I_{b\lambda}\right]} = -k_\lambda dx$$

$$\ln\left[I_\lambda(x) - I_{b\lambda}\right] = -k_\lambda x + C_1$$

$$I_\lambda(x) - I_{b\lambda} = e^{(-k_\lambda x + C_1)} = Ce^{-k_\lambda x}$$

at $x = 0$, $C = I_{\lambda,0} - I_{b\lambda}$

$$I_\lambda(x) = I_{b\lambda} + \left(I_{\lambda,0} - I_{b\lambda}\right) - e^{-k_\lambda x}$$
$$I_\lambda(x) = I_{\lambda,0} e^{-k_\lambda x} + I_{b\lambda}\left(1 - e^{-k_\lambda x}\right)$$

at $x = L$, and we obtain

$$I_\lambda(L) = I_{\lambda,0} e^{-k_\lambda L} + I_{b\lambda}\left(1 - e^{-k_\lambda L}\right) = I_{\lambda,0}\tau_\lambda + I_{b\lambda}\varepsilon_\lambda \qquad (14.10)$$

In general, the absorption coefficient k_λ is strongly dependent on wavelength. If we use the averaged overall wavelength of total properties, we obtain: $k_\lambda = k$, $\alpha = 1 - \tau$, $\alpha = \varepsilon$, $I_\lambda = I$,

$$I_{b\lambda} = I_b = \sigma T_g^4$$

14.1.2 Geometry of Gas Radiation: Geometric Mean Beam Length

The typical gas emissivity data were obtained by applying radiation to the hemispherical (of radius R) collection of gases radiating to an element of area at the center of the base, as shown in Figure 14.3. For other furnace shapes, there exists an equivalent mean beam length (L), defined as the radius of a gas hemisphere which radiates to unit area at the center of its base the same as the average radiation over the area from the actual gas volume shape [2]. Consider two surface elements dA_i and dA_j of an enclosure containing an isothermal gray gas at temperature T_g, use total properties, as shown in Figure 14.5. The irradiation dG_i coming to surface dA_i from surface dA_j is

$$dG_i = I_i^- \cos\theta_i \, dw_j \qquad (14.11)$$

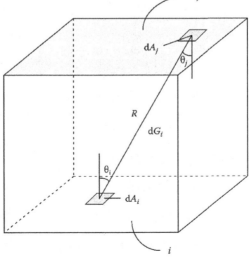

FIGURE 14.5 Elemental surface for radiation in an enclosure containing an isothermal gray gas.

where I_i^- is the intensity approaching surface dA_i, using the concept developed in Equation (14.10), replacing L by $R = I_j^+ e^{-kR} + I_{bg}\left(1 - e^{-kR}\right)$, I_j^+ the intensity leaving surface $dA_j = J_j/\pi$, J_j the radiosity leaving surface dA_j = emission and reflection from surface dA_j, I_{bg} the intensity from blackbody gas emission = $E_{bg}/\pi = \sigma T_g^4/\pi$, E_{bg} the blackbody gas emission = σT_g^4, $dw_j = dA_j \cos \theta_j/R^2$, R the distance (beam length) between surface dA_i and dA_j, and k the gas absorption coefficient.

Therefore,

$$dG_i = \int_{A_j} \left[J_i e^{-kR} + E_{bg}\left(1 - e^{-kR}\right) \right] \frac{\cos \theta_i \cos \theta_j}{\pi R^2} dA_j dA_i$$

$$G_i = \frac{1}{A_i} \int dG_i dA_i \qquad (14.12)$$

$$= \frac{1}{A_i} \int\int_{A_i A_j} \left[J_j e^{-kR} + E_{bg}\left(1 - e^{-kR}\right) \right] \frac{\cos \theta_i \cos \theta_j}{\pi R^2} dA_j dA_i$$

The distance (beam length) R varies over the surface. For convenience, we define a mean surface (mean beam length) L_{ij}, such that

$$G_i = \left[J_j e^{-kL_{ij}} + E_{bg}\left(1 - e^{-kL_{ij}}\right) \right] \frac{1}{A_i} \int\int_{A_i A_j} \frac{\cos \theta_i \cos \theta_j}{\pi R^2} dA_j dA_i$$

$$= \left[J_j e^{-kL_{ij}} + E_{bg}\left(1 - e^{-kL_{ij}}\right) \right] F_{ij} \qquad (14.13)$$

Comparing Equations (14.12) and (14.13), we obtain

$$e^{-kL_{ij}} F_{ij} = \frac{1}{A_i} \int\int_{A_i A_j} \frac{\cos \theta_i \cos \theta_j}{\pi R^2} e^{-kR} dA_j dA_i \qquad (14.14)$$

If kL_{ij} is small, for optically thin gases at low pressure, $e^{-kL_{ij}} \approx 1 - kL_{ij}$, Equation (14.14) becomes

$$L_{ij} = \frac{1}{A_i F_{ij}} \int\int_{A_i A_j} \frac{\cos \theta_i \cos \theta_j}{\pi R} dA_j dA_i \qquad (14.15)$$

$$L_{ij} A_i F_{ij} = L_{ji} A_j F_{ji} \qquad (14.16)$$

If a furnace or combustion chamber can be modeled as a single-surface enclosure, that is, with a uniform wall temperature and emission (uniform radiosity J_s), $A_i = A_s$, $F_{ij} = 1$, $L_{ij} = L_{ji} = L$ Equations (14.13) and (14.14) become

$$G_s = J_s e^{-kL} + E_{bg}\left(1 - e^{-kL}\right) \tag{14.17}$$

$$e^{-kL} = \frac{1}{A_s} \iint\limits_{A_s A_s} e^{-kR} \frac{\cos\theta\cos\theta}{\pi R^2} dA_s dA_s \tag{14.18}$$

If kL is small, for optically thin gases at low pressure, $e^{-kL} \approx 1 - kL$, Equation (14.18) becomes

$$L = \frac{1}{A_s} \iint\limits_{A_s A_s} \frac{\cos\theta\cos\theta}{\pi R} dA_s dA_s$$

$$= \frac{4V}{A_s} \tag{14.19}$$

where V is the volume of the gas in the enclosure and A_s is the enclosure surface area.

In general, the geometric mean beam length (L_{ij}) between surfaces i and j of an enclosure should be determined from Equation (14.4), and can be determined by Equation (14.15) for the optically thin gas (i.e., small absorption coefficient k, low pressure, and small enclosure L_{ij}, kL_{ij} is small). In addition, it can be determined from Equations (14.18) and (14.19), respectively, for a single-surface enclosure with a uniform temperature and emissivity. However, in some problems, for the optically thick gases (i.e., kL is not small, high pressure), the geometric mean beam length (L) is less than the above-mentioned values. From experience, the geometric mean beam length has proven to be a good approximation for the actual mean beam length. For practice, Equation (14.3), $L \cong 0.9(4V/A_s)$, can be used.

14.2   RADIATION EXCHANGE BETWEEN AN ISOTHERMAL GRAY GAS AND GRAY DIFFUSE ISOTHERMAL SURFACES IN AN ENCLOSURE

Since we know how to obtain gas radiation properties such as emissivity and absorptivity and how to determine the view factor between two surfaces, the following shows how to determine radiation heat transfer between surfaces in an enclosure with radiation gases. Assume that there are N gray diffuse and isothermal surfaces [2,4,5]. This implies that each surface at T_i has uniform radiosity J_i (emission plus reflection). Also assume that radiation gases are gray gases at uniform pressure and temperature (emissivity = absorptivity) and have no scattering effect. Figure 14.6 shows an energy balance on surface i and an energy balance between surface i and the rest of enclosure surfaces j through gases.

If given surface i temperature (T_i) and gas temperature (T_g), the following shows how to determine the heat transfer rate from surface i (q_i) and from the radiation gases (q_g). Gas transmissivity is inversely proportional to the gas absorption coefficient as

$$\tau_g = e^{-kL} \tag{14.20}$$

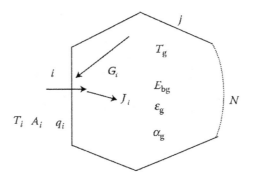

FIGURE 14.6   Radiation heat transfer through gases in an enclosure.

where k is the total absorption coefficient (predetermined), for example, $k = 0.3\,\text{m}^{-1}$, L is the mean beam length (predetermined from Equation 14.3).

Since gas absorptivity + gas transmissivity = 1, $\alpha_g + \tau_g = 1$.

Therefore,

$$1 - \alpha_g = \tau_g = e^{-kL} \tag{14.21}$$

For given T_i, T_g, how to obtain q_i, q_g?

Perform an energy balance on surface i, net heat transfer rate = radiosity (energy out) – irradiation (energy in)

$$q_i = A_i\left(J_i - G_i\right) \tag{14.22}$$

and

$$J_i = \varepsilon_i E_{bi} + \left(1 - \varepsilon_i\right)G_i \tag{14.23}$$

Therefore,

$$q_i = \frac{E_{bi} - J_i}{\left(1 - \varepsilon_i\right)/\left(A_i \varepsilon_i\right)} \tag{14.24}$$

Performing an energy balance between surface i and the rest of surfaces j through radiation gases,

$$
\begin{aligned}
q_i &= A_i\left(J_i - G_i\right) \\
&= A_i\left[J_i - \sum F_{ij}J_i\left(1 - \alpha_{ij,g}\right) - E_{bg}\varepsilon_{i,g}\right] + A_i\varepsilon_{i,g}J_i - A_i\varepsilon_{i,g}J_i \tag{14.25} \\
&= A_i\varepsilon_{i,g}\left(J_i - E_{bg}\right) + \sum A_i F_{ij}\left(1 - \alpha_{ij,g}\right)\left(J_i - J_j\right)
\end{aligned}
$$

where

$$A_i G_i = \sum A_j F_{ji} J_j \left(1 - \alpha_g\right) + A_i \varepsilon_{i,g} E_{bg}$$

$$= \sum A_i F_{ij} J_i \left(1 - \alpha_g\right) + A_i \varepsilon_{i,g} E_{bg}$$

(14.26)

And from the following relationships:

$$A_i \left(J_i - \varepsilon_{i,g} J_i\right) = A_i J_i \left(1 - \varepsilon_{i,g}\right)$$

$$= A_i J_i \left(1 - \alpha_{i,g}\right)$$

$$= \sum A_i F_{ij} J_i \left(1 - \alpha_{i,g}\right)$$

Therefore,

$$q_i = \underbrace{\frac{J_i - E_{bg}}{\left(\dfrac{1}{A_i \varepsilon_{i,g}}\right)}}_{\substack{\text{resistance due to} \\ \text{gas emissivity}}} + \sum_{j=1}^{N} \underbrace{\frac{J_i - J_j}{\left(\dfrac{1}{A_i F_{ij} \left(1 - \alpha_{ij,g}\right)}\right)}}_{\substack{\text{resistance due to view factor} \\ \text{and gas absorptivity}}}$$

(14.27)

If the gas has no radiation properties, that is, $\varepsilon_g = \alpha_g = 0$, $\tau_g = 1$, then the above equation returns to the one we have seen before as

$$q_i = \sum_{j=1}^{N} \frac{J_i - J_j}{1/A_i F_{ij}}$$

14.2.1 MATRIX LINEAR EQUATIONS

Method 2—The matrix method for an N surface enclosure with participating gases. For case A, we are given temperatures to determine heat fluxes. Let the right side of Equation (14.24) equals the right side of Equation (14.27) to form the matrix as before:

$$a_{11} J_1 + a_{12} J_2 + \cdots + a_{1N} J_N = c_1$$
$$a_{21} J_1 + a_{22} J_2 + \cdots + a_{2N} J_N = c_2$$
$$a_{31} J_1 + a_{32} J_2 + \cdots + a_{3N} J_N = c_3$$
$$a_{N1} J_1 + a_{N2} J_2 + \cdots + a_{NN} J_N = c_N$$

Therefore,

$$[A][J] = [C]$$

$$[J] = [A]^{-1}[C]$$

In addition, combining Equations (14.23) and (14.26), we obtain J_i = emission from surface i + reflection from surface j.

$$J_i = \varepsilon_i E_{bi} + (1 - \varepsilon_i) \left[\sum_{j=1}^{N} F_{ij} J_j (1 - \alpha_g) + \varepsilon_{i,g} E_{bg} \right] \qquad (14.28)$$

Similarly, Equation (14.28) can be used to form the matrix $[A][J] = [C]$ as follows:

$$J_1 = \varepsilon_1 E_{b1} + (1 - \varepsilon_1) \left[F_{11} J_1 (1 - \alpha_g) + \varepsilon_{1,g} E_{bg} + F_{12} J_2 (1 - \alpha_g) + \cdots \right]$$

$$J_2 = \varepsilon_2 E_{b2} + (1 - \varepsilon_2) \left[F_{21} J_1 (1 - \alpha_g) + \varepsilon_{2,g} E_{bg} + F_{22} J_2 (1 - \alpha_g) + \cdots \right]$$

$$\vdots$$

$$J_N = \varepsilon_N E_{bN} + (1 - \varepsilon_N) \left[F_{N1} J_1 (1 - \alpha_g) + \varepsilon_{N,g} E_{bg} + F_{N2} J_2 (1 - \alpha_g) + \cdots \right]$$

Once matrix $[J]$ has been solved, then surface heat transfer rate can be determined from Equation (14.24) as

$$q_i = \frac{E_{bi} - J_i}{(1 - \varepsilon_i)/(\varepsilon_i A_i)}$$

If we consider an energy balance between gases and enclosure surfaces i, the heat transfer rate (energy releases) from the gases to the enclosure is

$$q_g = \sum_{i=1}^{N} A_i \varepsilon_{i,g} (E_{bg} - J_i) = \sum_{i=1}^{N} \frac{E_{bg} - J_i}{1/A_i \varepsilon_{i,g}} \qquad (14.29)$$

However, we still need Equations (14.24) and (14.27) or Equation (14.28) to solve J_i using the matrix method. The aforementioned gas radiation problems can also be solved by Method 1—The electric network analogy method.

14.2.2 ELECTRIC NETWORK ANALOGY

Special case 1: Figure 14.7a shows several combustion furnaces that can be modeled as radiation between two surface enclosures containing hot radiation gases, if $T_g > T_1 > T_2$: By using Equations (14.24), (14.27), and (14.29), Figure 14.7b shows the associated electric network from the hot gas to surfaces 1 and 2. Hot gases release energy to surfaces 1 and 2 through their resistances (with gas emissivity less than unity); each surface has its own resistance (with surface emissivity less than unity). There is a reduced view factor between two surfaces because the gas is not completely transparent between the two surfaces (due to the gas absorption effect). If gas absorptivity is zero, the view factor is the same as the one with nonparticipating gases (such as air). An energy balance on surfaces 1 and 2 must be performed in order to solve for radiosities J_1 and J_2, respectively. Then, energy releases from hot gases, and heat transfer to surfaces 1 and 2 can be determined.

Special case 2: Figure 14.8 shows that several combustion furnaces can be modeled as heat transfer between hot gases and a single gray surface enclosure (assume an enclosure at a uniform temperature). The simple electric network can be used to solve this type of problem.

From Equations (14.24) and (14.28), solve for q_1 as

$$J_1 = \varepsilon_1 E_{b1} + (1 - \varepsilon_1)\left[J_1(1 - \alpha_g) + \varepsilon_g E_{bg}\right] \tag{14.30}$$

$$q_1 = (E_{b1} - J_1)A_1\varepsilon_1 / (1 - \varepsilon_1) \tag{14.31}$$

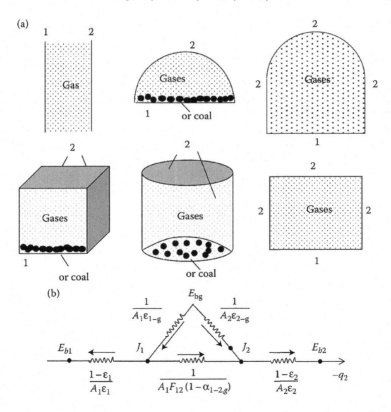

FIGURE 14.7 (a) Radiation between hot gases and two-surface enclosures; (b) electric network for radiation from hot gas to two-surface enclosure.

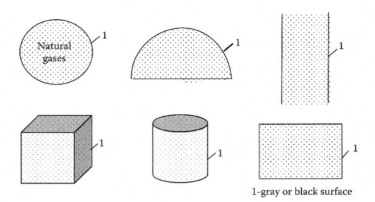

FIGURE 14.8 Radiation between hot gases and single-surface enclosure.

Substituting J_1 into q_1, we obtain

$$q_1 = \frac{A_1\varepsilon_1\alpha_g\sigma T_s^4}{1-(1-\varepsilon_1)(1-\alpha_g)} - \frac{A_1\varepsilon_1\varepsilon_g\sigma T_g^4}{1-(1-\varepsilon_1)(1-\alpha_g)}$$

$$= \frac{A_1\varepsilon_1}{1-(1-\varepsilon_1)(1-\alpha_g)}\left(\alpha_g\sigma T_s^4 - \varepsilon_g\sigma T_g^4\right)$$

(14.32)

Special case 3: Now we have net radiation heat transfer between nongray gases and a single black enclosure. To further simplify the problem, assume that the whole furnace surface is a blackbody at a uniform temperature as shown in Figure 14.9. The net heat transfer rate equals the hot gas emission and absorbed by the black surface minus the black surface emission and absorbed by gases [1].

$$q_{net} = \underbrace{\left(\varepsilon_g A_s\sigma T_g^4\right)\cdot\alpha_s}_{\substack{\text{emission}\\\text{from the gas}}} - \underbrace{\left(\varepsilon_s A_s\sigma T_s^4\right)\cdot\alpha_g}_{\substack{\text{emission}\\\text{from the surface}}} = A_s\sigma\left(\varepsilon_g T_g^4 - \alpha_g T_s^4\right)$$

(14.33)

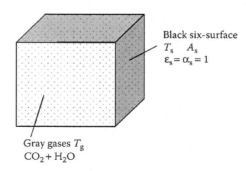

Black six-surface
T_s A_s
$\varepsilon_s = \alpha_s = 1$

Gray gases T_g
$CO_2 + H_2O$

FIGURE 14.9 Radiation between hot gases and single black enclosure.

For gray gas,

$$\varepsilon_g = \alpha_g \quad \left(\text{Otherwise}, \varepsilon_g \neq \alpha_g\right)$$

and for nongray gas

$$\varepsilon_g = \varepsilon_c + \varepsilon_w - \Delta\varepsilon$$

$$\alpha_g = \alpha_c + \alpha_w - \Delta\alpha$$

$$\Delta\alpha = \Delta\varepsilon$$

For water vapor,

$$\alpha_w = C_w \left(\frac{T_g}{T_s}\right)^{0.45} \cdot \varepsilon_w \left(T_s, P_w L \frac{T_s}{T_g}\right) \tag{14.34}$$

For carbon dioxide,

$$\alpha_c = C_c \left(\frac{T_g}{T_s}\right)^{0.65} \cdot \varepsilon_c \left(T_s, P_c L \frac{T_s}{T_g}\right) \tag{14.35}$$

Note that the problem will be more complicated if we consider radiation exchange between nongray gases and a single gray enclosure, or radiation exchange between nongray gases and an N-surfaces enclosure, where $N = 1, 2, 3,\ldots, N$.

Special case 4: Now consider a gray enclosure filled with a gray gas. Figure 14.10 shows a furnace consisting of a hot or a cold gray surface (1), a refractory surface R, and a gray gas, g, where each element is assumed to be at a uniform temperature: T_1, T_r, and T_g; determine the radiation heat transfer between the surface (1) and gas as

$$q_{1g} = \frac{\sigma\left(T_1^4 - T_g^4\right)}{(1 - \varepsilon_1)/A_1\varepsilon_1 + 1/\left\{A_1\varepsilon_1 + 1/\left[1/\left(A_R\varepsilon_{gR}\right) + 1/\left(A_1 F_{1R}\tau_{1gR}\right)\right]\right\}} \tag{14.36}$$

Special case 5: Increasing the number of surfaces, we can consider two gray surfaces with a gray gas. Figure 14.11 shows a furnace consisting of a hot gray surface (1),

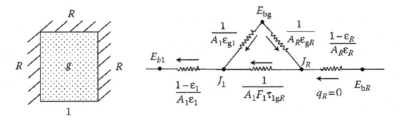

FIGURE 14.10 A gray enclosure and a refractory surface filled with a gray gas.

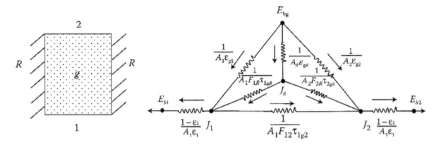

FIGURE 14.11 An enclosure of a gray hot surface, a gray cold surface, and a refractory surface.

cold gray surface (2), a refactory surface R, and a gray gas g; determine the radiation heat transfer [1].

Special case 6: Finally, we can include three gray surfaces with a gray gas. Figure 14.12 shows a furnace consisting of three gray surfaces, 1, 2, and 3, and a gray gas, g, determines the radiation heat transfer.

Real furnace applications—The zone method: Figure 14.13 shows the concept of the zone method for real-furnace applications proposed by Hottel (MIT) [5]. In a real furnace, combustion gases, as well as furnace surface temperatures, are non-uniform. The problem can be solved by dividing gases and surfaces into a number of gas zones and surface zones, respectively. An energy balance can be performed on each subsurface (each zone) and between each subsurface and the rest of subsurfaces (zones) through gas zones. View factors also need to be calculated between subsurfaces. The solution procedures are quite complicated.

Zone for gases $T_{g,i}$ = number of gas zones, a uniform temperature in each zone

Zone for surfaces T_i = number of surface zones, a uniform temperature in each zone

The problem will be even more complicated if convection effects are considered, which is up to 20% of heat transfer rate for circulation in a well-mixed furnace. In reality, soot formation and radiation can further make the problem harder to model.

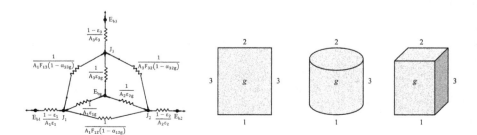

FIGURE 14.12 An enclosure of three gray surfaces.

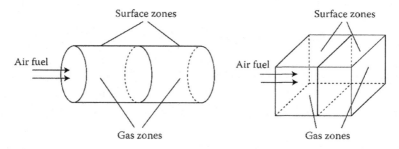

FIGURE 14.13 Concept of zone method for real furnace heat transfer problem.

14.3 RADIATION TRANSFER THROUGH GASES WITH NONUNIFORM TEMPERATURE

14.3.1 CRYOGENIC THERMAL INSULATION

In some applications, such as cryogenic thermal insulation, radiation heat is transferred from surface 1 to surface 2 through participating gases with varying temperature as shown in Figure 14.14. For a simple 1-D problem, the gray gas temperature changes from gray diffuse surface 1 to surface 2 [4–7].

14.3.2 RADIATION TRANSPORT EQUATION IN THE PARTICIPATING MEDIUM

Figure 14.14 contains the isotropic medium between two parallel flat plates at temperature T_1, and T_2. How to determine net radiation heat flux q between two plates? First, determine the radiation intensity in the positive y-direction (I^+) and in the negative y-direction (I^-). The intensity in either the positive or negative direction will be attenuated due to gas absorption and scattering, and will be augmented due to gas emission and scattering into the travelling path. Consider a radiation intensity path as ds, gas particles (such as CO_2) can absorb the photons emitting from the flat plates into ds, can scatter the photons away from ds, meanwhile, can emit photons from ds, and can absorb photons scattering from the surrounding particles into ds. Absorption, emission, and scattering into ds or away from ds, depend on the gas path location, and the gas path location is a function of (y, θ, λ, T_g). That means, $I^+ = I^+(y, \theta, \lambda, T_g)$ and $I^- = I^-(y, \theta, \lambda, T_g)$. Once I^+ and I^- can be obtained, the net radiation flux q from T_1 to T_2 can be determined. The radiation transport equation in the participating medium between two plates will be briefly outlined below, please refer to references [4,6,7] for the details.

From Figure 14.14, consider radiation intensity travelling in the positive s-direction as I^+. Consider radiation intensity change through ds, due to photon absorption and scattering away by the particles, and due to particles' emission and photons scattering into the ds, as mentioned above. From energy balance along the ds:

$$dI_\lambda^+(s,\theta) = -\alpha_\lambda I_\lambda^+(s,\theta) \cdot ds - r_\lambda I_\lambda^+(s,\theta) \cdot ds + \varepsilon_\lambda \frac{e_{b\lambda}}{\pi} \cdot ds + r_\lambda \frac{G_\lambda}{4\pi} \cdot ds \quad (14.37)$$

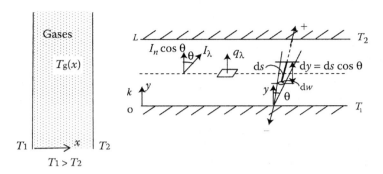

FIGURE 14.14 Radiation heat transfer through gases with varying temperature.

$dI_\lambda^+(s,\theta)$ = intensity change along s-direction

$-\alpha_\lambda I_\lambda^+(s,\theta)\cdot ds$ = intensity absorbed by the particles within ds

$-r_\lambda I_\lambda^+(s,\theta)\cdot ds$ = intensity scattered away by the particles within ds

$\varepsilon_\lambda \dfrac{e_{b\lambda}}{\pi}\cdot ds$ = intensity gained by the particles emission within ds

$r_\lambda \dfrac{G_\lambda}{4\pi}\cdot ds$ = intensity gained by the surrounding particles scattering into the ds

where

α_λ = absorption coefficient of the particles
r_λ = scattering coefficient of the particles
$e_{b\lambda} = \sigma T_g^4$ = emission energy of the black particles
G_λ = scattering energy from the surrounding particles into ds
Let $k_\lambda = \alpha_\lambda + r_\lambda$ = extinction coefficient = $k_\lambda(s)$
$\alpha_\lambda = \varepsilon_\lambda$ for the particles

To simplify the problem, assume

$$\varepsilon_\lambda \frac{e_{b\lambda}}{\pi} + r_\lambda \frac{G_\lambda}{\pi} = k_\lambda \cdot I_{b\lambda}^+(s,\theta)$$

The above energy equation along ds becomes

$$dI_\lambda^+(s,\theta) = -k_\lambda(s)I_\lambda^+(s,\theta)\cdot ds + k_\lambda(s)I_{b\lambda}^+(s,\theta)\cdot ds \qquad (14.38)$$

intensity change = intensity attenuation + intensity augmentation

For convenience, define an optical depth k_λ along s-direction as $k_\lambda(s)$

$$k_\lambda(s) = \int_0^s k_\lambda(s)\,ds$$

where $k_\lambda(s) = 0$ at $s = 0$, no extinction effect

$k_\lambda(s) = k_\lambda \cdot S$ at $s = S$, maximum extinction effect

Thus $dk_\lambda(s) = k_\lambda(s) \cdot ds$. Insert into the above Equation (14.38):

$$dI_\lambda^+(s,\theta) = -I_\lambda^+(s,\theta) \cdot dk_\lambda(s) + I_{b\lambda}^+(s,\theta) \cdot dk_\lambda(s) \tag{14.39}$$

To solve the problem, multiply both sides by a factor $e^{k_\lambda(s)}$, and move the right-side first term to left-side, and then divide both sides by $dk_\lambda(s)$, it becomes:

$$\frac{dI_\lambda^+(s,\theta)}{dk_\lambda(s)} e^{k_\lambda(s)} + I_\lambda^+(s,\theta) e^{k_\lambda(s)} = I_{b\lambda}^+(s,\theta) e^{k_\lambda(s)}$$

$$\frac{d}{dk_\lambda(s)} \left[I_\lambda^+(s,\theta) \cdot e^{k_\lambda(s)} \right] = I_{b\lambda}^+(s,\theta) e^{k_\lambda(s)} \tag{14.40}$$

Integrate from $k_\lambda(s) = 0$ to $k_\lambda(s)$:

$$I_\lambda^+(k_\lambda,\theta) \cdot e^{k_\lambda(s)} - I_\lambda^+(0) = \int_0^{k_\lambda(s)} I_{b\lambda}(k_\lambda',\theta) \cdot e^{k_\lambda'(s)} dk_\lambda'(s)$$

$$I_\lambda^+(k_\lambda,\theta) = I_\lambda^+(0) \cdot e^{-k_\lambda(s)} + \int_0^{k_\lambda(s)} I_{b\lambda}(k_\lambda',\theta) \cdot e^{k_\lambda'(s)-k_\lambda(s)} \cdot dk_\lambda'(s) \tag{14.41}$$

where $k_\lambda'(s)$ is a dummy variable of integration.

Now, relate the optical depth in s-direction to optical depth in y-direction:
From Figure 14.14,

$dy = ds \cdot \cos\theta$ in the positive direction

$dy = -ds \cdot \cos(\pi - \theta) = ds \cdot \cos\theta$ in the negative direction

The optical depth along y-direction:

$$k_\lambda(y) = \int_0^y k_\lambda(y) \cdot dy, \quad dk_\lambda(y) = k_\lambda(y) \cdot dy$$

$$k_\lambda(s) = \int_0^s k_\lambda(s) \cdot ds, \quad dk_\lambda(s) = k_\lambda(s) \cdot ds$$

Then $k_\lambda(y) = k_\lambda(s) \cdot \cos\theta$

Thus, the above energy balance equation along y-direction is

$$I_\lambda^+(k_\lambda,\theta) = I_\lambda^+(0)\cdot e^{-k_\lambda/\cos\theta} + \int_0^{k_\lambda} I_{b\lambda}(k_\lambda',\theta)\cdot e^{(k_\lambda'-k_\lambda)/\cos\theta}\cdot d\frac{k_\lambda'}{\cos\theta} \qquad (14.42)$$

for $0 \leq \theta \leq \dfrac{\pi}{2}$

Similarly, intensity change along the negative y-direction can be written as

$$I_\lambda^-(k_\lambda,\theta) = I_\lambda^-(k_\lambda)\cdot e^{-k_\lambda/\cos\theta} + \int_{k_\lambda}^0 I_{b\lambda}(k_\lambda',\theta)\cdot e^{(k_\lambda'-k_\lambda)/\cos\theta}\cdot d\frac{k_\lambda'}{\cos\theta} \qquad (14.43)$$

for $\dfrac{\pi}{2} \leq \theta \leq \pi$

The above $I^+(k_\lambda,\theta)$ and $I^-(k_\lambda,\theta)$ can be solved because the boundary conditions $I_\lambda^+(0)$ and $I_\lambda^-(k_\lambda)$ are known quantities. Therefore, the net radiation heat flux in y-direction can be determined.

$$q(y) = \int_{\lambda=0}^\infty q_\lambda(y)d\lambda = \int_{\lambda=0}^\infty q_\lambda(k_\lambda)d\lambda \qquad (14.44)$$

where

$$q_\lambda(y) = q_\lambda^+(y) - q_\lambda^-(y)$$

with

$$q_\lambda^+(y)d\lambda = d\lambda \int_{\theta=0}^{\theta=\frac{\pi}{2}} I_\lambda^+(k_\lambda,\theta)\cdot\cos\theta\cdot d\omega$$
$$= 2\pi\, d\lambda \int_0^{\pi/2} I_\lambda^+(k_\lambda,\theta)\cdot\cos\theta\cdot\sin\theta\cdot d\theta \qquad (14.45)$$

and

$$q_\lambda^-(y)d\lambda = 2\pi\, d\lambda \int_{\pi-\theta=0}^{\pi-\theta=\frac{\pi}{2}} I_\lambda^-(k_\lambda,\theta)\cdot\cos(\pi-\theta)\cdot\sin(\pi-\theta)\cdot d(\pi-\theta)$$
$$= -2\pi\, d\lambda \int_{\pi/2}^{\pi} I_\lambda^-(k_\lambda,\theta)\cdot\cos\theta\cdot\sin\theta\cdot d\theta \qquad (14.46)$$

where $d\omega = 2\pi \sin\theta d\theta$ for an isotropic gas

Since $I_\lambda^+(k_\lambda,\theta)$ and $I_\lambda^-(k_\lambda,\theta)$ can be solved from the above mentioned intensity balance Equations (14.42) and (14.43). Thus, $q_\lambda^+(y)d\lambda$ and $q_\lambda^-(y)d\lambda$ can be integrated from (Equation 14.45) and (Equation 14.46); therefore, $q(y)$ can be obtained from (Equation 14.44).

14.3.3 RADIATION TRANSFER THROUGH GRAY GAS BETWEEN TWO GRAY AND DIFFUSE PARALLEL PLATES

To further simplify the problem, assume two flat plates are gray, diffuse and isotropic surfaces at temperatures T_1 and T_2, and the participating gases are gray and isotropic at $T_g(y)$. In this case, all radiation properties such as k_λ, I_λ^+, I_λ^- are independent of λ. The above intensity equations and heat flux equation can be re-written as

$$I^+(k,\theta) = I^+(0) \cdot e^{-k/\cos\theta} + \int_0^k I_b^+(k') \cdot e^{-(k'-k)/\cos\theta} \cdot d\frac{k'}{\cos\theta} \qquad (14.47)$$

$$I^-(k,\theta) = I^-(k_\lambda) \cdot e^{-k/\cos\theta} + \int_k^0 I_b^-(k') \cdot e^{-(k'-k)/\cos\theta} \cdot d\frac{k'}{\cos\theta} \qquad (14.48)$$

where the boundary conditions:

$$I^+(0) = \frac{J_1}{\pi} = \frac{\varepsilon_1\sigma T_1^4 + (1-\varepsilon_1)\cdot I^-(0)}{\pi}$$

$$= \frac{\sigma T_1^4}{\pi} \quad (\text{if plate 1 is a black surface}) \qquad (14.49)$$

$$I^-(L) = \frac{J_2}{\pi} = \frac{\varepsilon_2\sigma T_2^4 + (1-\varepsilon_2)\cdot I^+(L)}{\pi}$$

$$= \frac{\sigma T_2^4}{\pi} \quad (\text{if plate 2 is a black surface}) \qquad (14.50)$$

However, in order to solve the above $I^+(k,\theta)$ and $I^-(k,\theta)$ from (Equation 14.47) and (Equation 14.48), one needs to solve I_b^+ and I_b^- first, where $I_b^+ = \sigma T_g^4(k)$ and $I_b^- = \sigma T_g^4(k)$. Under the condition of radiation equilibrium, the following equality at any optical depth k can be applied:

$$\sigma T_g^4(k) = \pi\bar{I}(k)$$

where $\bar{I}(k)$ is the mean radiation intensity

$$\bar{I}(k) = \frac{1}{4\pi}\int_0^{4\pi} I(\omega,k)d\omega$$

where $d\omega = 2\pi \sin\theta d\theta$

Thus,

$$\sigma T_g^4(k) = \frac{1}{2}\int I^+(k,\theta)\sin\theta d\theta + \frac{1}{2}\int I^-(k,\theta)\sin\theta d\theta \qquad (14.51)$$

Insert the intensity equations $I^+(k,\theta)$ and $I^-(k,\theta)$ from Equation (14.47) and Equation (14.48) into the gas temperature Equation (14.51), and $T_g(k)$ can be solved; thus insert $T_g(k)$ back to Equation (14.47) and Equation (14.48), $I^+(k,\theta)$ and $I^-(k,\theta)$ can be obtained. Once $I^+(k,\theta)$, $I^-(k,\theta)$, and $T_g(k)$ have been obtained, then the net radiation heat flux from the plate 1 to plate 2 can finally be determined.

Since $\dfrac{dq(y)}{dy} = 0$, that is, $q(y) = $ constant, if only consider radiation, no conduction, no convection in the participating medium:

Let $q(y) = $ constant $= q(0)$ at $y = 0$, at the plate 1. From Equations (14.44), (14.45), and (14.46):

$$q(0) = q^+(0) - q^-(L) \qquad (14.52)$$

$$q^+(0) = 2\pi \int_0^{\pi/2} I^+(0,\theta)\sin\theta\cos\theta d\theta \qquad (14.53)$$

$$q^-(L) = -2\pi \int_{\pi/2}^{\pi} I^-(L,\theta)\sin\theta\cos\theta d\theta \qquad (14.54)$$

where $I^+(0,\theta)$ and $I^-(L,\theta)$ can be obtained from Equations (14.47) and (14.48) at $k = 0$ and $k = L$ as

$$I^+(0,\theta) = I^+(0) \qquad (14.55)$$

$$I^-(L,\theta) = I^-(L) \cdot e^{-L/\cos\theta} + \int_L^0 I_b^-(k') \cdot e^{-(k'-L)/\cos\theta} \cdot d\frac{k'}{\cos\theta} \qquad (14.56)$$

Now, $q^+(0)$ and $q^-(L)$ can be solved with the intensity Equations (14.49), (14.50) and (14.55), (14.56) by the exponential integral functions or numerical method as detailed in reference [4,6]. Finally, the gas temperature T_g can be solved from Equation (14.51), and the heat flux $q(0)$ solved with Equation (14.52).

Note: In open literature, the volumetric extinction coefficient is the same as the optical depth; for example, $k = 100\,\mathrm{m^2/m^3} = 100$ per m for an optically thick gas, $k = 10^{-6}\mathrm{m^2/m^3} = 10^{-6}$ per m for an optically thin gas. L is the distance between two plates. $k_L = k \cdot L$, for example, $k_L = 100 \times 0.1 = 10$ if $k = 100$ per m, $L = 0.1$ m; $k_L = 10^{-6} \times 0.1 = 10^{-7}$ if $k = 10^{-6}$ per m, $L = 0.1$ m.

Results of simplified but important applications are outlined below.

For 1-D, gray gases, gray and diffuse surface, radiation only, the heat flux

$$q = \frac{\sigma\left(T_1^4 - T_2^4\right)Q}{1 + Q\left(\left(1/\varepsilon_1\right) + \left(1/\varepsilon_2\right) - 2\right)} \tag{14.57}$$

$$Q \equiv \frac{q}{J_1 - J_2} \equiv \frac{1}{1 + \left(3/4\right)k_L} \tag{14.58}$$

$$\phi(k) \equiv \frac{\sigma T^4(k) - J_2}{J_1 - J_2} \tag{14.59}$$

where Q is the nondimensional heat flux, $\phi(k)$ is the nondimensional temperature profile, as sketched in Figure 14.15.

Temperature profile:

$$\phi(k) = 1 - \frac{1}{2}Q - \frac{3}{4}Qk \tag{14.60}$$

Physical significances:

Special case (a)—*Optical thin medium:* $k_L = k \cdot L \ll 1 \approx 0$, then, $Q \rightarrow 1$ and the physical heat flux is

$$q = \frac{\sigma\left(T_1^4 - T_2^4\right)}{\left(1/\varepsilon_1\right) + \left(1/\varepsilon_2\right) - 1} \tag{14.61}$$

This equals a surface radiation problem.

Special case (b)—*Optical thick medium:* k is large or $k_L \rightarrow \infty$, then $Q = (4/3)(1/k_L)$ is small,

$$q = \sigma\left(T_1^4 - T_2^4\right)\frac{4}{3}\frac{1}{k_L} \tag{14.62}$$

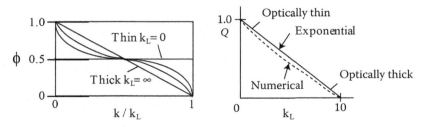

FIGURE 14.15 Nondimensional temperature and heat flux profiles.

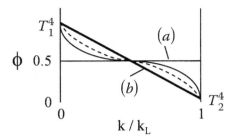

FIGURE 14.16   Temperature profile between two black surfaces through participating gas.

If considering black surfaces, $\varepsilon_1 = \varepsilon_2 = 1$, from Equations (14.58)–(14.60), the temperature profile between two surfaces with participating medium can be obtained and sketched in Figure 14.16,

a. $\phi = 1/2 = \left(T^4 - T_2^4\right)/\left(T_1^4 - T_2^4\right)$ for the optical thin medium.

b. $\phi = 1 - (k/k_L) = \left(T^4 - T_2^4\right)/\left(T_1^4 - T_2^4\right)$ for the optical thick medium.

If considering conduction and radiation,

$$q = q_c + q_r = \left(k_{cond}\frac{T_1 - T_2}{L}\right) + \left(\frac{\sigma\left(T_1^4 - T_2^4\right)Q}{1 + Q\left((1/\varepsilon_1) + (1/\varepsilon_2) - 2\right)}\right) \qquad (14.63)$$

If considering convection and radiation,

$$q = q_{conv} + q_r = \bar{h}\Delta\bar{T} + \frac{\sigma\left(T_1^4 - T_2^4\right)Q}{1 + Q\left(\left(\dfrac{1}{\varepsilon_1}\right) + \left(\dfrac{1}{\varepsilon_2}\right) - 2\right)} \qquad (14.64)$$

Examples

14.1 A hemispherical furnace is shown in Figure 14.7.

    a. If the furnace contains N_2 gas at 5 atm pressure, determine the net radiation heat transfer from surface 1 to surface 2 (assume $T_1 > T_2$).

    b. If the furnace contains $CO_2 + N_2$ gases at 5 atm pressure and temperature T_g, determine total radiation heat transfer from gases to surfaces 1 and 2 (assume $T_g > T_1 > T_2$).

    c. Reconsider (b), if surface 1 now is a reradiating surface, determine the total radiation heat transfer to the surface 2 of the furnace. In this new condition, comment on whether the radiation transfer to surface 2 will be higher, the same, or lower than that of (b).

Solution

a. $q_{12} = \dfrac{\sigma T_1^4 - \sigma T_2^4}{\dfrac{1-\varepsilon_1}{A_1\varepsilon_1} + \dfrac{1}{A_1 F_{12}} + \dfrac{1-\varepsilon_2}{A_2\varepsilon_2}}$ where $F_{12} = 1$

b. From Equation (14.29)

$$q_g = q_{g1} + q_{g2} = \frac{E_{bg} - J_1}{\dfrac{1}{A_1\varepsilon_g}} + \frac{E_{bg} - J_2}{\dfrac{1}{A_2\varepsilon_g}}$$

where J_1 and J_1 can be obtained from Equation (14.27),

$$q_1 = \frac{E_{b1} - J_1}{\dfrac{1-\varepsilon_1}{A_1\varepsilon_1}} = \frac{J_1 - E_{bg}}{\dfrac{1}{A_1\varepsilon_g}} + \frac{J_1 - J_2}{\dfrac{1}{A_1 F_{12}\left(1-\alpha_g\right)}}$$

$$q_2 = \frac{E_{b2} - J_2}{\dfrac{1-\varepsilon_2}{A_2\varepsilon_2}} = \frac{J_2 - E_{bg}}{\dfrac{1}{A_2\varepsilon_g}} + \frac{J_2 - J_1}{\dfrac{1}{A_2 F_{21}\left(1-\alpha_g\right)}}$$

where $A_1 F_{12} = A_2 F_{21}$ and assuming a gray gas, $\varepsilon_g = \alpha_g$

c. From Figure 14.10 and Equation (14.36),

$$q_{g2} = \frac{E_{bg} - \sigma T_2^4}{\left[A_2\varepsilon_g + \dfrac{1}{\dfrac{1}{A_1\varepsilon_g} + \dfrac{1}{A_1 F_{12}\left(1-\alpha_g\right)}} \right]^{-1} + \dfrac{1-\varepsilon_2}{A_2\varepsilon_2}}$$

$$q_{g2}\big|_{(c)} > q_{g2}\big|_{(b)}$$

14.2 A hemispherical furnace, with a re-radiating floor and a water-cooled ceiling, contains CO_2 and N_2 gases at 1 atm pressure and 1000°C. Take $\varepsilon_1 = 0.8$, $\varepsilon_2 = 0.7$, $D = 1$ m, and $T_2 = 500$°C. Determine the radiant heat transfer to the ceiling of the furnace. Assume gray gases. Given: $\sigma = 5.67\times10^{-8}$ (W/m²K⁴). Volume of a sphere $= (4/3)\pi((1/2)D)^3$, surface area of a sphere $= 4\pi((1/2)D)^2$.

Solution

From Figure 14.10 and Equation (14.36),

$$q_{g2} = \frac{\sigma T_g^4 - \sigma T_2^4}{\dfrac{1-\varepsilon_2}{A_2\varepsilon_2} + \left\{ A_2\varepsilon_g + \left[\dfrac{1}{A_1\varepsilon_g} + \dfrac{1}{A_1 F_{12}\left(1-\alpha_g\right)} \right]^{-1} \right\}^{-1}}$$

$$L = 0.9 \cdot \frac{4V}{A_s} = 0.6$$

$P_c = 0.5$ atm, from Figure 14.2: $\varepsilon_g = \alpha_g \approx 0.18$
$F_{12} = 1, A_1, A_2, T_g,$ and T_2 are given

$\therefore q_{g2} \approx 51.4\,\text{kW}$

14.3 For a cylindrical furnace (top wall 1, bottom wall 2, side wall 3) with hot gray gases, solve the following problems by using the matrix method.

a. Given T_1, T_2, T_3, T_g, determine q_1, q_2, q_3, q_g.
b. Given q_1, q_2, q_3, q_g, determine T_1, T_2, T_3, T_g.
c. Given T_1, T_2, q_3, T_g, determine q_1, q_2, T_3, q_g.

Solution

a. Follow the procedure outlined in Section 14.2:
  Set (Equation 14.24) = (Equation 14.27)
  With $i = 1$ for surface 1, and $j = 1, 2, 3$ to get

$$a_{11}J_1 + a_{12}J_2 + a_{13}J_3 = c_1$$

With $i = 2$ for surface 2, and $j = 1, 2, 3$ to get

$$a_{21}J_1 + a_{22}J_2 + a_{23}J_3 = c_2$$

With $i = 3$ for surface 2, and $j = 1, 2, 3$ to get

$$a_{31}J_1 + a_{32}J_2 + a_{33}J_3 = c_3$$

From $[A][J] = [C]$, solve for $[J]$, i.e., J_1, J_2, J_3.
  From Equation (14.24), solve for q_i, where $i = 1, 2, 3$.
  Use Equation (14.29) to solve for q_g.
  Note: You can provide numerical values for T_1, T_2, T_3, T_g to calculate q_1, q_2, q_3, q_g.
b. Similar to part (a). Apply Equation (14.27) from surface 1 ($i = 1$) and $j = 1$, 2, 3, where E_{bg} can be obtained from Equation (14.29):

$$q_g = \frac{E_{bg} - J_1}{\dfrac{1}{A_1 \varepsilon_g}} + \frac{E_{bg} - J_2}{\dfrac{1}{A_2 \varepsilon_g}} + \frac{E_{bg} - J_3}{\dfrac{1}{A_3 \varepsilon_g}}$$

Solve for E_{bg} and insert E_{bg} into Equation (14.27), then get

$$a_{11}J_1 + a_{12}J_2 + a_{13}J_3 = c_1$$

Repeat for surface 2 ($i = 2$) and $j = 1, 2, 3$, and insert E_{bg} into Equation (14.27) to obtain

$$a_{21}J_1 + a_{22}J_2 + a_{23}J_3 = c_2$$

Again, repeat for surface 3 ($i = 3$) and $j = 1, 2, 3$, and insert E_{bg} into Equation (14.27) to obtain

$$a_{31}J_1 + a_{32}J_2 + a_{33}J_3 = c_3$$

From $[A][J] = [C]$, solve for $[J]$, i.e., J_1, J_2, J_3.

From Equation (14.24), solve for E_{bi}, where $i = 1, 2, 3$: $E_{bi} = \sigma T_i^4$, and solve for T_i.

From Equation (14.29), solve for E_{bg}, where $i = 1, 2, 3$: $E_{bg} = \sigma T_g^4$, and solve for T_g.

Note: You can provide numerical values for q_1, q_2, q_3, q_g to calculate T_1, T_2, T_3, T_g.

c. Combine parts (a) and (b).

Note: You can provide numerical values for T_1, T_2, q_3, T_g to calculate q_1, q_2, T_3, q_g.

14.4 A gas turbine combustion chamber may be approximated as a long tube of 0.4 m diameter. The combustion gas is at a pressure and temperature of 1 atm and 1000°C, respectively, while the chamber surface temperature is 500°C. If the combustion gas contains CO_2 and water vapor, each with a mole fraction of 0.15, what is the net radiative heat flux between the gas and the chamber surface, which may be approximated as a blackbody?

Solution

From Equation (14.33)

$$q_{net} = A_s \sigma \left(\varepsilon_g T_g^4 - \alpha_g T_s^4 \right)$$

where

$$\varepsilon_g = \varepsilon_c + \varepsilon_w - \Delta\varepsilon \quad \text{with} \quad \Delta\varepsilon \approx 0.01$$

$$\alpha_g = \alpha_c + \alpha_w - \Delta\alpha \quad \text{with} \quad \Delta\alpha \approx 0.01$$

Use Figure 14.2 to determine ε_c and ε_w at $L = 0.9D = 0.36\text{m} = 1.15\text{ft}$

$$P_c L = P_w L = 1.15 \cdot 0.15 = 0.1725 \text{ atm-ft}$$

$$T_g = 1000 + 273 = 1273 \text{ K}$$

$$\therefore \varepsilon_c \approx 0.069, \varepsilon_w \approx 0.085, \varepsilon_g \approx 0.069 + 0.085 - 0.01 = 0.144$$

Use Equations (14.34) and (14.35) to determine α_w and α_c

$$P_c L \frac{T_s}{T_g} = P_w L \frac{T_s}{T_g} \approx 0.6 \cdot 0.1725 = 0.105 \text{atm-ft} \left(\text{at } T_s = 500 + 273 = 773 \text{K}\right)$$

$$\therefore \varepsilon_c \approx 0.08, \varepsilon_w \approx 0.083$$

$$\alpha_c = 1\left(\frac{1273}{773}\right)^{0.65} \times 0.08 = 0.111$$

$$\alpha_w = 1\left(\frac{1273}{773}\right)^{0.45} \times 0.083 = 0.104$$

$$\therefore \alpha_g \approx 0.111 + 0.104 - 0.01 = 0.205$$

$$q' = \pi(0.4)(5.67 \times 10^{-8})\left[0.144(1273)^4 - 0.205(773)^4\right] = 21.9\frac{W}{m}$$

14.5 A cryogenic storage chamber has double walls for the purpose of insulation against heat loss. The gap between the walls is filled with a gas whose properties are
Thermal conductivity: $kcond(T) = 2 \times 10{-}7 \times T°K$ (KW/(m – °C))
Volumetric radiation extinction coefficient: $k = 10^{-6}$ (m^2/m^3)

a. Determine the rate of heat loss if one wall is at 500K and the other wall is at 100K. Take $\varepsilon_1 = \varepsilon_2 = 0.1$, $L = 0.2$ m.
b. If $k = 100$ (m^2/m^3), what would be the result in (a)?

Solution

$$Lq_c = \int_0^L q_c dx = \int_{T_1}^{T_2} -k_{cond}(T)dT = 2 \times 10^{-7} \times \left(T_1^2 - T_2^2\right)$$

$$q_c = \frac{10^{-7}}{0.2}\left(5^2 - 1^2\right) \times 10^4 = 120\frac{W}{m^2}$$

For an optically thin gas, from Equation (14.61),

$$q_R = \frac{\sigma\left(T_1^4 - T_2^4\right)}{\dfrac{1}{\varepsilon_1} + \dfrac{1}{\varepsilon_2} - 1} = \frac{5.67 \times 10^{-8}\left(5^4 - 1^4\right) \times 10^8}{\dfrac{1}{0.1} + \dfrac{1}{.01} - 1} = 186.2\frac{W}{m^2}$$

$$\therefore q = q_c + q_R = 306.2\frac{W}{m^2}$$

For an optically thick gas, from Equation (14.62)

$$q_R = \sigma\left(T_1^4 - T_2^4\right)\frac{4}{3}\frac{1}{Lk}$$

$$= 5.67 \times 10^{-8}\left(5^4 - 1^4\right) \times 10^8 \cdot \frac{4}{3} \cdot \frac{1}{0.2 \times 100} = 78.6\frac{W}{m^2}$$

$$\therefore q = q_c + q_R = 198\frac{W}{m^2}$$

$$q\big|_{(b)} < q\big|_{(a)}$$

REMARKS

In the undergraduate-level heat transfer, from charts, students are expected to know how to read the emissivity and absorptivity values of water vapor and carbon dioxide in a furnace at given size, temperature, and pressure, and then apply these properties to calculate radiation transfer between these gases and the furnace wall, assuming the blackbody furnace wall at a uniform temperature.

In the intermediate-level heat transfer, this chapter focuses on how to derive volumetric absorption, geometric mean beam length, radiation transfer between gray gases at a uniform temperature and an N-surfaces furnace with each surface at different gray diffuse uniform temperature conditions. Students are expected to know how to analytically solve gas radiation problems by using the matrix linear equations method for an N-surfaces furnace with various surface thermal boundary conditions. By using the electric network analogy, this chapter has also provided several relevant engineering applications such as radiation transfer between gas at a high uniform temperature and one-surface furnace assuming the gray diffuse surface at a low uniform temperature; or between gas at a high uniform temperature and two-surfaces furnace assuming gray diffuse surfaces with each surface at different low uniform temperatures.

This chapter does not go into detail on real-furnace applications with varying gas temperature and surface temperature using the zoning method. We only dealt with the 1-D varying gas temperature, steady-state, and gray diffuse surface problem. However, in real-life applications, there are many gas radiation problems involving flow convection; 2-D or 3-D varying gas temperature; cylindrical or spherical furnace geometry; nongray gases; nongray nondiffuse surfaces; gases with scattering particulates and soot formation; and sphere fillet or fiber porous medium. These topics belong to advanced radiation transfer.

PROBLEMS

14.1 A hemispherical furnace is shown in Figure 14.7. If the furnace contains $CO_2 + N_2$ gases at 1 atm pressure and temperature T_g, determine the total radiation heat transfer from gases to surfaces 1 and 2 (assume $T_g > T_1 > T_2$, and make other necessary assumptions).

a. Based on the analogy of electric resistance network, draw a radiation heat transfer network from gases to surfaces 1 and 2.

b. Determine the total radiation from gases to surfaces 1 and 2. The final solutions should be the function of given temperatures, surface area, and radiation properties.

14.2 A hemispherical furnace is shown in Figure 14.7.

a. If the furnace contains N_2 gas at 5 atm pressure, determine the net radiation heat transfer from surface 1 to surface 2 (assume $T_1 > T_2$, and make other necessary assumptions).

b. If the furnace contains $CO_2 + N_2$ gases at 5 atm pressure and temperature T_g, determine total radiation heat transfer from gases to surfaces 1 and 2 (assume $T_g > T_1 > T_2$, and make other necessary assumptions).

    c. Reconsider (b), if surface 1 now is a reradiating surface, determine the total radiation heat transfer to the surface 2 of the furnace. In this new condition, comment on whether the radiation transfer to surface 2 will be higher, the same, or lower than that of (b) (make necessary assumptions).

14.3 A long hemicylindrical furnace is shown.

    a. Determine the net radiation heat transfer from surface 1 to surface 2, q_{12}.

    b. If $T_2 = T_2(\theta)$, describe how to determine q_{12}.

    c. Consider combustion gray gas with a uniform temperature T_g and emissivity, for example, inside the furnace, and determine the total radiation heat transfer from gas to surfaces 1 and 2, q_g. Assume T_1, T_2 constant.

14.4 A hemispherical furnace, with a reradiating floor and a water-cooled ceiling, contains CO_2 and N_2 gases at 1 atm pressure and 1000°C. Take $\varepsilon_1 = 0.8$, $\varepsilon_2 = 0.7$, $D = 1$m, and $T_2 = 500$°C. Determine the radiant heat transfer to the ceiling of the furnace. Assume gray gases.

Given: $\sigma = 5.67 \times 10^{-8} (W/m^2K^4)$.
Volume of a sphere $= (4/3)\pi((1/2)D)^3$
Surface of a sphere $= 4\pi((1/2)D)^2$

14.5 A hemispherical furnace, with a reradiating floor and a water-cooled ceiling, contains $2CO_2$ and $8N_2$ gases at 1 atm pressure and 1000°C. Take $\varepsilon_1 = 0.8$, $\varepsilon_2 = 0.7$, $D = 1$m, and $T_2 = 500$°C. Determine the radiant heat transfer to the ceiling of the furnace.

14.6 A cryogenic storage chamber has double walls for the purpose of insulation against heat loss. The gap between the walls is filled with a gas whose properties are thermal conductivity: $k_{cond}(T) = 2 \times 10^{-7} \cdot T°K$ ($KW/(m - °C)$); volumetric radiation extinction coefficient: $k = 10^{-6}$ (m^2/m^3).

    a. Determine the rate of heat loss if one wall is at 500°K and the other wall is at 100°K. Take $\varepsilon_1 = \varepsilon_2 = 0.1$, $L = 0.2$ m.

    b. If $k = 100$ (m^2/m^3), what would be the result in (a)?

14.7 A cryogenic storage chamber has double walls for the purpose of insulation against heat loss. The gap between the walls is filled with a gas whose properties are

Thermal conductivity: $k_{cond}(T) = 1 \times 10^{-7} \times T°K (KW/(m - °C))$
Volumetric radiation extinction coefficient: $k = 10^{-6}$ (m^2/m^3)

The walls are made of a polished metal, with an emissivity of 0.2. The gap between the walls is 0.5 m.

    a. Determine the rate of heat loss if one wall is at 300K and the other wall is at 100K.

    b. If $k = 100$ (m^2/m^3), what would be the result in (a)?

14.8 A cryogenic storage chamber has double walls for the purpose of insulation against heat loss. The gap between the walls is filled with a gas whose properties are

Thermal conductivity: $k_{cond}(T) = 3 \cdot 10^{-7} \times T°K(KW/(m - °C))$
Volumetric radiation extinction coefficient: $k = 10^{-6} \, (m^2/m^3)$

The walls are made of a polished metal, with an emissivity of 0.1.
The gap between the walls is 0.3 m.

a. Determine the rate of heat loss if one wall is at 400°K and the other wall is at 100°K.
b. If $k = 100 \, (m^2/m^3)$, what would be the result in (a)?

14.9 A gas turbine combustion chamber may be approximated as a long tube of 0.4 m diameter. The combustion gas is at a pressure and temperature of 1 atm and 1000°C, respectively, while the chamber surface temperature is 500°C. If the combustion gas contains CO_2 and water vapor, each with a mole fraction of 0.15, what is the net radiative heat flux between the gas and the chamber surface, which may be approximated as a blackbody?

14.10 Consider a hemispherical furnace, with a reradiating floor and a water-cooled ceiling, contains $2CO_2 + 8N_2$ gases at 1 atm pressure and 1200°C. Take $\varepsilon_1 = 0.9$, $\varepsilon_2 = 0.6$, $D = 1.5$ m, and $T_2 = 350°C$. Determine the radiant heat transfer to the ceiling of the furnace.

14.11 Consider a hemispherical furnace radiation heat transfer problem. The furnace floor (surface 1) has area A_1 and emissivity ε_1 at temperature T_1, whereas the furnace enclosure (surface 2) has area A_2 and emissivity ε_2 at temperature T_2. If the furnace contains $CO_2 + N_2$ gases at 1 atm pressure and temperature T_g, determine the total radiation heat transfer from gases to surfaces 1 and 2 (assume $T_g > T_1 > T_2$, and make other necessary assumptions).
a. Based on the analogy of electric resistance network, draw a radiation heat transfer network from gases to surfaces 1 and 2.
b. Determine the total radiation from gases to surfaces 1 and 2. The final solutions should be the function of given temperatures, surface area, and radiation properties.

14.12 Consider a hemispherical furnace. The hemispherical furnace wall has surface area A_1, emissivity ε_1, at temperature T_1, whereas the furnace floor has surface area A_2, emissivity ε_2, at temperature T_2. If the furnace contains $CO_2 + N_2$ gases at 10 atm pressure and temperature T_g, determine the total radiation heat transfer from gases to surfaces 1 and 2 (assume $T_g > T_2 > T_1$, and make other necessary assumptions). Assume that the gas emissivity is ε_g.
a. Based on the analogy of electric resistance network, draw a radiation heat transfer network from gases to surfaces 1 and 2.
b. Determine the total radiation from gases to surfaces 1 and 2. The final solutions should be the function of given temperatures, surface area, and radiation properties.

14.13 For a cylindrical furnace (top wall 1, bottom wall 2, side wall 3) with hot gray gases, solve the following problems by using the matrix method.
a. Given T_1, T_2, T_3, T_g, determine q_1, q_2, q_3, q_g.
b. Given q_1, q_2, q_3, q_g, determine T_1, T_2, T_3, T_g.

    c. Given T_1, T_2, T_3, T_g, determine q_1, q_2, T_3, q_g.

    d. Given q_1, q_2, T_3, T_g, determine T_1, T_2, q_3, q_g.

    e. Given T_1, T_2, T_3, q_g, determine q_1, q_2, q_3, T_g.

14.14 For a cubic furnace (top wall 1, bottom wall 2, four side wall 3) with hot gray gases, solve the following problems by using the matrix method if the side wall is a reradiating surface.

    a. Given T_1, T_2, T_R, T_g, determine q_1, q_2, q_R, q_g.

    b. Given q_1, q_2, q_R, q_g, determine T_1, T_2, T_R, T_g.

    c. Given T_1, T_2, T_R, T_g, determine q_1, q_2, T_R, q_g.

    d. Given q_1, q_2, T_R, T_g, determine T_1, T_2, q_R, q_g.

    e. Given T_1, T_2, T_R, q_g, determine q_1, q_2, q_R, T_g.

14.15 Consider a cubic furnace: bottom wall (surface 1) at temperature T_1, top wall (surface 2) at temperature T_2 and four side walls (surface 3) at temperature T_3. If the furnace contains $2CO_2 + 8N_2$ gases at 1 atm pressure and temperature T_g, determine the total radiation heat transfer from the gases to surfaces 1, 2, and 3 (assume $T_g > T_1 > T_2 > T_3$, and make other necessary assumptions).

    a. Based on the analogy of electric resistance network, draw a radiation heat transfer network from the gases to surfaces 1, 2, and 3.

    b. Write down the equations that can be used to solve the radiation heat transfer from the gases to surfaces 1, 2, and 3, respectively, but you do not need to calculate the numerical answer. Your final solutions should be the function of given temperatures, surface area, and radiation properties.

    c. Reconsider item (b) above, if surface 2 and surface 3 (top wall and four side walls) now are reradiating surfaces, determine the total radiation heat transfer from gases to surface 1. In this new condition, can you comment if the radiation transfer from the gases to surface 1 will be higher, the same, or lower than that of item (b) (make necessary assumptions)? Why?

REFERENCES

1. W. Rohsenow and H. Choi, *Heat, Mass, and Momentum Transfer*, Prentice-Hall, Inc., Englewood Cliffs, NJ, 1961.
2. A. Mills, *Heat Transfer*, Richard D. Irwin, Inc., Boston, MA, 1992.
3. F. Incropera and D. Dewitt, *Fundamentals of Heat and Mass Transfer*, Fifth Edition, John Wiley & Sons, New York, 2002.
4. R. Siegel and J. Howell, *Thermal Radiation Heat Transfer*, McGraw-Hill, New York, 1972.
5. H. Hottel and A. Sarofim, *Radiative Transfer*, McGraw-Hill, New York, 1967.
6. J. Chen, *Conduction and Radiation Heat Transfer, Class Notes*, Lehigh University, Bethlehem, PA, 1973.
7. E.M. Sparrow and R.D. Cess, *Radiation Heat Transfer*, Hemisphere Publishing, New York, 1978.

15 Laminar–Turbulent Transitional Heat Transfer

15.1 TRANSITION PHENOMENA

Turbulent flow heat transfer is very much different from laminar flow heat transfer. There exists a transitional region between fully laminar flow and fully turbulent flow as sketched in Figure 15.1. Based on the small perturbation theory, a Tollmien-Schlichting wave instability was developed to predict the transition phenomena [1]. It shows that laminar boundary layer becomes unstable at critical Reynolds number, and it implies laminar flow transitioning to turbulent flow. Transition caused by classic Tollmien–Schlichting wave instability theory is called Natural or Normal Transition with shorter transition length from fully laminar to fully turbulent flow. The transition location or critical Reynolds number predicted by Tollmien–Schlichting wave instability theory has been closely validated by many experimental results.

In addition to critical Reynolds number from instability theory, free-stream turbulence, unsteadiness, pressure gradient, surface curvature, surface roughness, film cooling, etc., affect the flow transition behaviors. In general, for a given Reynolds number flow, results showed that elevated free-stream turbulence, unsteadiness, concave surface, surface roughness, film cooling, and negative pressure gradient (flow deceleration) would promote earlier boundary layer transition, however, positive pressure gradient (flow acceleration) and convex surface would delay boundary layer transition. These additional free stream flow and surface parameters make the prediction of transition phenomena beyond the scope of classic instability theory. Therefore, prediction of transition location still relies mainly on the experimental data for many real engineering applications.

Transition caused by high free stream turbulence (for turbulence intensity Tu greater than 5%) and unsteadiness is called Bypass Transition with longer transition length from

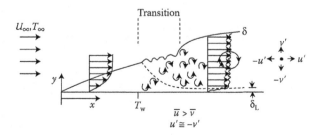

FIGURE 15.1 Concept of laminar boundary layer transition to turbulent boundary layer flow.

DOI: 10.1201/9781003164487-15

fully laminar to fully turbulent flow. Bypass means transition happened and bypassed from Normal Tollmien-Schlichting wave instability, due to high turbulent spots (turbulent vortices) production inside the boundary layer during the transition. For transition at high free-stream turbulence levels, the natural transition by Tollmien-Schlichting instability wave is completely bypassed such that turbulent spots are directly produced within the boundary layer by the influence of the free-stream disturbances.

In this bypass transition condition, Emmons [2] provided a theory to address the processes involved in the production, growth, and convection of turbulent spots which can be correlated to the free stream turbulence level. Emmons presented a statistical theory for transition and provided an expression for the fraction of time the flow is turbulent at any location within the transition region. They introduced the concept of intermittency, γ, the friction of time the flow is turbulent. The intermittency γ is 0 for fully laminar region, unity for fully turbulent region, between 0 and 1 for transition from fully laminar to fully turbulent flow. Therefore, transitional flow behaviors can be quantified by determining the intermittency values which can be directly correlated to turbulent spot production rate or free stream turbulence intensity level.

In general, it is still difficult to accurately predict the starting and ending points of boundary layer transition under various free-stream flow and surface conditions. However, based on fully laminar and fully turbulent results, several transitional flow heat transfer correlations have been reported in open literature if the critical Reynolds number (starting point of transition) and free stream flow and surface conditions are given or predetermined.

This chapter begins with the classic Tollmien–Schlichting wave instability theory in order to understand basic Normal Transition phenomena for low free stream turbulence conditions. Then follow the bypass transition phenomena of turbulent spot production and transport theory for high free stream turbulence or unsteadiness conditions. And provide laminar–turbulent transitional heat transfer correlations for various engineering applications.

15.2 NATURAL TRANSITION–TOLLMIEN SCHLICHTING WAVE INSTABILITY THEORY

15.2.1 SMALL DISTURBANCE STABILITY THEORY

The following is a brief outline of small disturbance stability theory. Refer to *Boundary Layer Theory* [1] for the complete presentation in more detail. Consider 2-D boundary layer flow over a flat surface as shown in Figure 15.1.

Let

$$u = \overline{u}(x,y) + u'(x,y,t)$$

where

$$u' \ll \overline{u}$$

Similarly

$$\upsilon = \overline{u}(x,y) + \upsilon'(x,y,t)$$

$$P = \bar{P}(x,y) + P'(x,y,t)$$

with

$$v' \ll \bar{v}$$

and

$$P' \ll \bar{P}$$

Assume

$$\bar{u} = \bar{u}(y)$$

$$\bar{v} = 0$$

From the Navier–Stokes x- and y-momentum equations:

$$\frac{\partial u'}{\partial t} + (\bar{u} + u')\frac{\partial u'}{\partial x} + v'\frac{\partial}{\partial y}(\bar{u} + u') = -\frac{1}{\rho}\frac{\partial}{\partial x}(\bar{P} + P') + v\left[\frac{\partial^2}{\partial x^2}u' + \frac{\partial^2}{\partial y^2}(\bar{u} + u')\right]$$

$$\frac{\partial v'}{\partial t} + (\bar{u} + u')\frac{\partial v'}{\partial x} + v'\frac{\partial}{\partial y}(v') = -\frac{1}{\rho}\frac{\partial}{\partial y}(\bar{P} + P') + v\left[\frac{\partial^2}{\partial x^2}v' + \frac{\partial^2}{\partial y^2}v'\right]$$

From boundary layer analysis

$$0 = -\frac{1}{\rho}\frac{\partial \bar{P}}{\partial x} + v\frac{\partial^2 \bar{u}}{\partial y^2}$$

also,

$$u' \ll \bar{u}, \quad 0 = \frac{\partial \bar{P}}{\partial y}, \quad v' \cdot \frac{\partial v'}{\partial y} \sim \text{small}$$

The above momentum equations become:

$$\frac{1}{\rho}\frac{\partial P'}{\partial x} = v\left(\frac{\partial^2 u'}{\partial x^2} + \frac{\partial^2 u'}{\partial y^2}\right) - \frac{\partial u'}{\partial t} - \bar{u}\frac{\partial u'}{\partial x} - v'\frac{\partial \bar{u}}{\partial y} \tag{15.1}$$

$$\frac{1}{\rho}\frac{\partial P'}{\partial y} = v\left(\frac{\partial^2 v'}{\partial x^2} + \frac{\partial^2 v'}{\partial y^2}\right) - \frac{\partial v'}{\partial t} - \bar{u}\frac{\partial v'}{\partial x} \tag{15.2}$$

$$\frac{\partial u'}{\partial x} + \frac{\partial v'}{\partial y} = 0 \tag{15.3}$$

where

15.1 is the small disturbance form of the x-momentum equation
15.2 is the small disturbance form of the y-momentum equation
15.3 is the small disturbance form of the continuity equation

Define the stream function, ψ, for u' and v' to satisfy Equation (15.3):

$$u' = \frac{\partial \psi}{\partial y} \quad \text{and} \quad v' = -\frac{\partial \psi}{\partial x}$$

Eliminate "pressure' by differentiating Equation (15.1) with respect to y, Equation (15.2) with respect to x, and subtract the equations. Replace u' and v' in terms of ψ, and equation in ψ as function of x, y, t, also involving $\bar{u}$. Try solution of form for stream function ψ :

$$\psi = \varphi(y) e^{i(\alpha x - \beta t)}$$

where

$$\beta = \beta_r + i\beta_i$$

Therefore,

$$\psi = \varphi(y) e^{\beta_i t} \left[\cos(\alpha x - \beta_r t) + i \sin(\alpha x - \beta_r t) \right] \tag{15.4}$$

The real part of the solution is a cosine wave traveling with velocity β_r / α and wavelength $\lambda = 2\pi / \alpha$, either amplifying or decaying depending on whether β_i is positive (+) or negative (−) in the exponential term of Equation (15.4) as shown in Figure 15.2.
  Try solution on stream function equation:

$$\left(\bar{u_1} - c \right)\left(\varphi'' - \alpha^2 \varphi \right) - \bar{u_1''}\varphi = \frac{-i}{\alpha \text{Re} \left(\varphi''' \right)^{24}} \tag{15.5}$$

Differential equation for φ, where

$$\bar{u_1} = \frac{\bar{u}}{u_m} = \text{function of} \left(\frac{y}{\delta} \right)$$

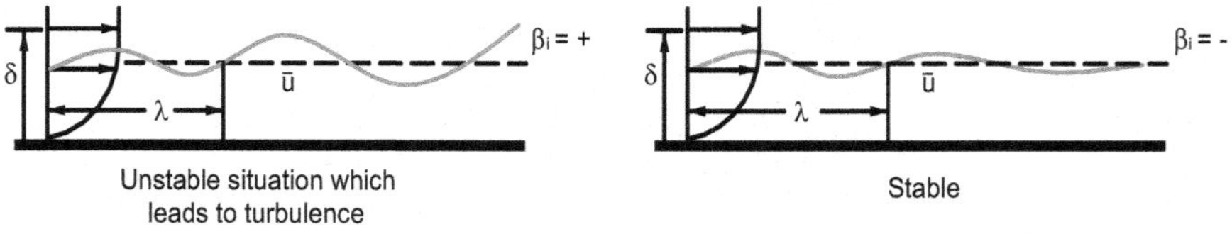

Unstable situation which   Stable
leads to turbulence

FIGURE 15.2   Concept of wave instability.

$$\varphi' = \frac{d\varphi}{d\left(\dfrac{y}{\delta}\right)}$$

$$\varphi'''' = \frac{d^4\varphi}{d\left(\dfrac{y}{\delta}\right)^4}$$

$$\mathrm{Re} = \frac{\overline{u_m}\delta}{\nu}$$

Fourth-order equation with $\overline{u_1}$ and $\overline{u_1''}$ variable coefficients: Orr–Sommerfeld Equation Solutions give $\varphi(y)$ when particular values of α and Re are chosen and a specific shape for $\dfrac{\overline{u}}{u_m}$ (function of y/δ) is chosen, also define a specific value of c necessary for a solution.

$$c = \frac{\beta}{\alpha} = \frac{\beta_r + i\beta_i}{\alpha} = c_r + ic_i \tag{15.6}$$

Select a value of Re, try different values of α, find corresponding values of β_r and β_i, as sketched in Figures 15.3 and 15.4. It implies that, for a given flow Reynolds number, laminar boundary layer can maintain stable condition for short or long wave length region but will become unstable for mediate wave length region. Therefore, laminar boundary layer can potentially become unstable (i.e., unstable wave, wave amplification, transition) at certain value of critical Reynolds number. It is important to find out this critical Reynolds number.

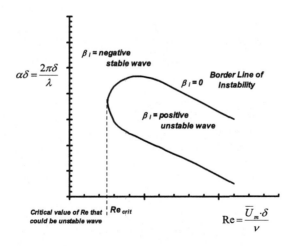

FIGURE 15.3 Unstable wave related to critical Reynolds number.

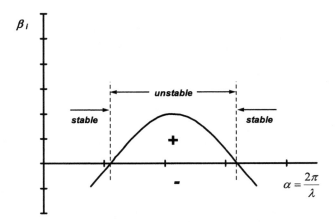

FIGURE 15.4   Concept of stable and unstable related to wave length.

For Blasius flow, it is at:

$$\text{Re}_{\text{crit}} = \left(\frac{\overline{u_m}\delta}{\nu} \right)_{\text{crit}} = 1260 \qquad (15.7)$$

$$\because \delta = 3\delta^* \left(\text{begin to amplify} \right)$$

$$\text{Re}_{\text{crit}} = \left(\frac{\overline{u_m}\delta^*}{\nu} \right)_{\text{crit}} = 420$$

where δ is boundary layer thickness and δ^* is displacement thickness. Note that the above value is based on the instability theoretical prediction. However, the observed transition is at

$$\text{Re} = \left(\frac{\overline{u_m}\delta}{\nu} \right)_{\text{crit}} = 3000 \qquad (15.8)$$

instead of 1260. The maximum value of α for unstable waves which correspond to wave with shortest wave length gives $\lambda = 6\delta$.

Effect of Pressure Gradient on Transition

The critical Reynolds number decreases (i.e., earlier transition) for deceleration flow (i.e., diffuser, positive or favorable pressure gradient) but increases (i.e., delay transition) for acceleration flow (i.e., nozzle, negative or adverse pressure gradient) as sketched in Figure 15.5.

15.2.2   CRITICAL REYNOLDS NUMBER FOR NATURAL TRANSITION

For pipe flow by Reynolds Experiment (1883), critical Reynolds number is around

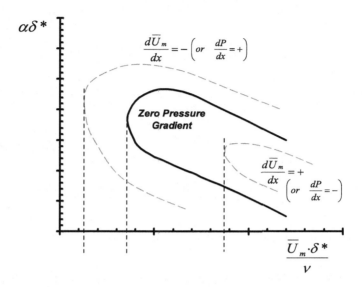

FIGURE 15.5 Critical Reynolds related to pressure gradient.

$$\text{Re}_D = \frac{VD}{\nu} = 2300$$

For flow over a flat plate, critical Reynolds number is around (depends on turbulence intensity):

$$\text{Re}_x = \frac{U_\infty x}{\nu} = 3.5 \times 10^5 - 10^6 \sim 5 \times 10^5 \qquad (15.9)$$

From Blasius

$$\frac{\delta}{x} = \frac{5}{\sqrt{\dfrac{U_\infty x}{\nu}}}$$

$$\text{Re}_\delta = \frac{U_\infty \delta}{\nu} = \frac{U_\infty x}{\nu}\frac{\delta}{x} = 3000 \qquad (15.10)$$

Transition occurs in a short distance from fully laminar flow to fully turbulent flow, so momentum thickness changes only a very little during transition, $\theta_1 = \theta_2$, but displacement thickness decreases greatly, $\because \ \delta_2^* < \delta_1^*$, refer to Figure 15.1. Therefore, shape factor, $H \equiv \dfrac{\delta^*}{\theta}$, changes largely. Shape factor of laminar boundary layer is around 2.6; however, it drops to about 1.4 in turbulent boundary layer as shown in Figure 15.6. Note that critical Reynolds number can be based on distance x, boundary layer thickness δ, displacement thickness δ^*, or momentum thickness θ. From Equations (15.9) and (15.10), and $H = 2.6$, therefore, $\text{Re}_\theta = \dfrac{U_\infty \theta}{\nu} = 400$.

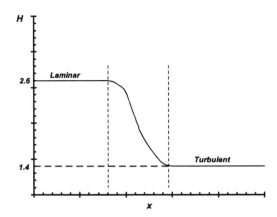

FIGURE 15.6 Shape factor variation from laminar to turbulent flow.

15.2.3 EFFECT OF FREE STREAM TURBULENCE LEVEL ON TRANSITION

Define Tu = turbulence intensity of free stream

$$\text{Tu} = \frac{\sqrt{\left(\overline{u'^2} + \overline{v'^2} + \overline{w'^2}\right) \cdot \frac{1}{3}}}{\overline{u_m}} \quad (\text{RMS\%})$$

As a frame of reference, Tu ~ 0.01 or 1% for typical wind tunnel. Transition is caused by the Tollmien–Schlichting wave instability at certain critical Reynolds number if free stream turbulence intensity is lower than 0.001 (0.1%).

The critical Reynolds number decreases (i.e., x decreases, earlier transition for a given velocity) if free stream turbulence intensity Tu is greater than 0.1% (Tu = 0.001) as shown in Figure 15.7. Note that free stream turbulence intensity Tu is greater than 1% (Tu = 0.01) for most of real engineering applications.

Example for Free Stream Turbulence-Induced Transition

In 1983 Blair [3] presented the heat-transfer distributions along the flat test wall at five increasing free-stream turbulence levels from Tu = 0.25% up to Tu = 6%. As indicated earlier, increase in free-stream turbulence induces earlier boundary layer transition. Figure 15.8 presents the heat-transfer distributions for five different turbulence levels. As seen clearly, transition location moves progressively upstream with increasing turbulence level. For higher levels of turbulence, the boundary layer is already fully turbulent. It showed that heat transfer in the fully turbulent region increased up to 36%, for an increase of free-stream turbulence from 0.25% to 6%.

Blair [4] also presented the influence of free-stream turbulence on boundary layer transition in favorable pressure gradients. He reported that transition location delayed due to flow acceleration in positive pressure gradient conditions.

The above natural transition theory can be found in Schlichting [1]. At a critical value of boundary layer thickness (or momentum thickness) Reynolds number, the

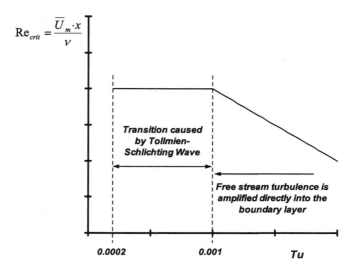

FIGURE 15.7 Critical Reynolds number decreases with turbulence intensity.

laminar boundary layer becomes susceptible to small disturbances and develops an instability in the form of a two-dimensional Tollmien–Schlichting wave. Then the instability wave amplifies within the layer to a point where three-dimensional instabilities grow and develop into random vortices or turbulent eddies (or turbulent spots) with large fluctuations. Finally, the highly fluctuating turbulent spots merge into a fully turbulent boundary layer with 3-D random fluctuations. This natural transition model predicts fairly well for the boundary layer flow with lower free stream turbulence intensity conditions as shown in Figure 15.8, for example, $Re_{xt} = 1,000,000$ for natural transition with very low free-stream turbulence Tu ~ 1/4%; $Re_{xt} = 350,000$ for natural transition with low free-stream turbulence Tu ~ 1%. This confirms that the Tollmien–Schlichting instability wave theory predicts the critical Reynolds number

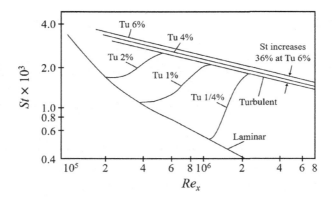

FIGURE 15.8 Heat transfer distributions along the flat surface for five freestream turbulence levels.

ranging from 1,000,000 to 350,000 for free stream turbulence intensity varying from 0.25% to 1%. For Tu~1%, $Re_{xt} = 350,000$, the corresponding momentum thickness Reynolds number $Re_\theta = \dfrac{U_\infty \theta}{\nu} = 400$.

15.3 BYPASS TRANSITION

For transition at high free-stream turbulence levels, the natural transition by Tollmien–Schlichting instability wave is completely bypassed such that turbulent spots (turbulent eddies, turbulent vortices) are directly produced within the boundary layer by the influence of the freestream disturbances. In this bypass transition condition, Emmons [2] provided a theory to address the processes involved in the production, growth, and convection of turbulent spots. The following is a brief outline; refer to the role of laminar-turbulent transition in gas turbine engines (Mayle [5]) for the details.

15.3.1 TURBULENT SPOT PRODUCTION, GROWTH, AND CONVECTION

From flow visualization in a simple water channel, Emmons [2] discovered that transition occurs through a random production (in time and position) of "turbulent spots" within the laminar boundary layer which subsequently grow as they propagate downstream until the flow becomes completely turbulent. Figure 15.9 shows the concept of turbulent spot production in transitional boundary layer. Based on this observation, Emmons presented a statistical theory for transition and provided an expression for the fraction of time the flow is turbulent at any location within the transition region. The concept of intermittency, γ, is that the friction of time the flow is turbulent. The intermittency γ is 0 for fully laminar region, unity for fully turbulent region, between 0 and 1 for transition from fully laminar to fully turbulent flow. Transitional flow behaviors can be quantified by determining the intermittency value. This implied that the time-averaged condition of the boundary layer at any streamwise location may be obtained from superposition as

$$f = (1 - \gamma)\, f_L + \gamma\, f_T \qquad (15.11)$$

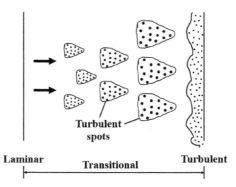

FIGURE 15.9 Turbulent spot geometry and emergence of a turbulent boundary layer through the growth and propagation of turbulent spots (plan view).

where f is a boundary layer flow-related quantity such as surface friction factor (f, Cf) and heat transfer coefficient (h, Nu, St), f_L is its laminar value, f_T is its turbulent value, and γ, the intermittency, is the fraction of time the flow is turbulent. The intermittency value γ is zero at the beginning of transition location, i.e., fully laminar region, and unity at the end of transition, i.e., fully turbulent region. It implies that transition region heat transfer coefficients and friction factors can also be predicted from the above-mentioned intermittency theory as

$$C_f = (1-\gamma)C_{fL} + \gamma C_{fT} \tag{15.12}$$

$$\text{Nu} = (1-\gamma)\text{Nu}_L + \gamma \text{Nu}_T \tag{15.13}$$

$$\text{St} = (1-\gamma)\text{St}_L + \gamma \text{St}_T \tag{15.14}$$

where the intermittency γ is directly correlated to free stream turbulence intensity levels Tu% or unsteady wake passing period or frequency.

From many previous experimental studies, intermittency in the normal transitional region is obtained as

$$\gamma_n = 1 - \exp\left[-\hat{n}\sigma(\text{Re}_x - \text{Re}_{xt})^2\right] \tag{15.15}$$

where $\hat{n} = nv^2/U^3$ is the dimensionless turbulent spot production parameter, Re_x is the local-flow Reynolds number, $\text{Re}_x = Ux/v$, and Re_{xt} is the Reynolds number at the location for onset of normal transition, U is the free stream velocity and σ is Emmons' dimensionless spot propagation parameter which depends on the shape and velocity of the turbulent spot. Measurements of the spot and its propagation velocity indicate that σ is constant and has a value of about 0.27. The intermittency increases with increasing turbulent spot production parameter. It is seen that an increase in the spot production rate decreases the transition length, $(\text{Re}_x - \text{Re}_{xt})$. It is seen that the real problem of transition is how to predict or determine the onset of transition, Re_{xt}, and the spot production parameter. That is, what are the effects of various flow and thermal parameters on both the onset of transition and spot production rate?

Therefore, the most important thing is to know where the laminar boundary layer begins to transition Re_{xt} for a given flow Reynolds number Re_x. Several important parameters affect the transition location and the transition distance such as free stream turbulence intensity, length scale, pressure gradient, surface curvature, surface roughness, film cooling, etc. From literature, high free stream turbulence, smaller length scale, unsteadiness, flow deceleration, concave surface, roughness, and film cooling promote earlier laminar boundary layer transition with longer transition distance.

15.3.2 TURBULENT SPOT INDUCED TRANSITION MODEL

Turbulent spot production rate is not easy to obtain, instead, turbulence intensity is relatively simple to measure by using the hot wire anemometry. From many previous

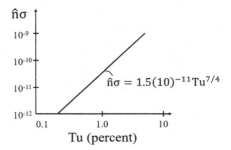

FIGURE 15.10 Turbulent spot production rate as a function of the free stream turbulence level for zero pressure gradient flows.

experimental studies, turbulent sport production rate has been correlated to the free stream turbulence intensity as shown in Figure 15.10

$$\hat{n}\sigma = 1.5 \times 10^{-11} \cdot \mathrm{Tu}^{7/4} \tag{15.16}$$

where Tu is in percent. It implies the turbulent spot production rate increases with increasing free stream turbulence intensity, subsequently, increasing the normal transition intermittency. From the above turbulent spot production theory, for the flat-plate boundary layer flow with zero pressure gradient, the normal transition intermittency value γ can be predicted at a given free stream turbulence intensity Tu, if we know the beginning of transition location Re_{xt}.

For example, assume transition at $\mathrm{Re}_{xt} = 200{,}000$, Tu = 2%, and transitional boundary layer location at $\mathrm{Re}_x = 400{,}000$, the intermittency value γ can be predicted from above equations (15.15) and (15.16) as 0.67 at that location. It implies 67% time the flow is turbulent (and 33% time laminar) at the transitional boundary layer location at $\mathrm{Re}_x = 400{,}000$. Heat transfer coefficients (or friction factors) can be calculated from the above equation (15.14) at this $\gamma = 0.67$ if fully laminar and fully turbulent heat transfer (or friction factor) values are given or predetermined, as shown in Figure 15.8. Note that the intermittency $\gamma = 0$ is at the beginning of transition (fully laminar flow) and $\gamma = 1$ is at the end of transition (fully turbulent flow).

The effect of free stream turbulence is to reduce the critical Reynolds number Re_{xt} at which transition begins as shown in Figure 15.8. The critical momentum thickness Reynolds number has often been used for predicting flow transition. The effect of free-stream turbulence level on the critical momentum thickness Reynolds number at transition $\mathrm{Re}_{\theta t}$, as shown in Figure 15.11, has also been obtained as

$$\mathrm{Re}_{\theta t} = 400 \cdot \mathrm{Tu}^{-5/8} \tag{15.17}$$

It shows the critical momentum thickness Reynolds number decreases with increasing free stream turbulence intensity, implies high free stream turbulence promotes earlier laminar boundary layer transition. Note that $\mathrm{Re}_{\theta t} = 400$, 146, and 95 if Tu = 1%, 5%, and 10%, respectively. Equation (15.17) can be derived from Equations (15.9) and (15.10) with the estimation of momentum thickness is around $1/2.6$ of

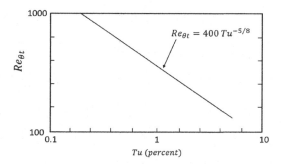

FIGURE 15.11 Momentum thickness Reynolds number at the onset of transition as a function of free-stream turbulence level for zero pressure gradient flows.

displacement thickness which is around 1/3 of boundary layer thickness. Therefore, either Equations (15.9), (15.10), or (15.17) can be used to predict the onset of critical Reynolds number for transition.

Intermittency and Turbulent Fluctuation Model:

The transitional flow at any time may be divided into two distinct domains within the boundary layer. Figure 15.12 shows the domain contains fluid where the flow is turbulent and the other outside this is nonturbulent, which changes with time and location (Mayle [5]).

To characterize these domains, an intermittency function, $I(x, y, z, t)$, may be defined which has the value of unity when the flow is turbulent, and zero when it is not. For a given location in transition boundary layer, the ensemble average of the intermittency function yields the intermittency factor,

$$\gamma\left(x,y,z\right) = \frac{1}{N}\sum_{1}^{N} I\left(x,y,z,t\right)$$

For two-dimensional flows in the x-y plane, y is independent of z, the streamwise velocity component at any time may be expressed as

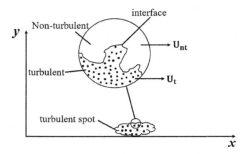

FIGURE 15.12 Supposed interface between the turbulent fluid within and nonturbulent fluid outside a turbulent spot.

$$U(x,y,t) = (1-I)\,U_{nt}(x,y,t) + I\,U_t(x,y,t) \qquad (15.18)$$

$$I(x,y,t) = 1 \qquad \text{turbulent}$$
$$= 0 \qquad \text{non-turbulent}$$

where U_t and U_{nt} are the velocities in the turbulent and nonturbulent domains, respectively. Decomposing the velocity in each region into an ensemble-averaged mean velocity, denoted by an overbar, and a fluctuating component denoted by a prime, the instantaneous velocities in the two domains can be written as

$$U_{nt} = \bar{U}_{nt} + u'_{nt}$$

$$U_t = \bar{U}_t + u'_t$$

Contrary to the usual averaging process, the ensemble-averaged mean velocities are obtained by averaging only during the time spent in the particular regime. Thus, if N_t is the number of occurrences of turbulent flow in N data samples, the mean velocities in each portion are

$$\bar{U}_{nt}(x,y) = \frac{1}{N-N_t}\sum_1^N (1-I)\,U(x,y,t) \qquad (15.19)$$

and

$$\bar{U}_t(x,y) = \frac{1}{N_t}\sum_1^N I\,U(x,y,t) \qquad (15.20)$$

With these definitions, the mean velocity at any position in the flow is given by

$$\bar{U}(x,y) = \frac{1}{N}\sum_1^N U(x,y,t)$$

$$\bar{U}(x,y) = (1-\gamma)\,\bar{U}_{nt}(x,y) + \gamma\,\bar{U}_t(x,y) \qquad (15.21)$$

Similar expressions may be written for the lateral velocities, $V(x, y)$.

The ensemble-averaged mean velocities can be predicted from intermittency which correspond to turbulence and nonturbulence portion velocity, respectively. Therefore, the transitional shear stress and heat flux that appear in the ensemble-averaged equations are now given by

$$\tau_{xy} = \mu\frac{\partial\bar{U}}{\partial y} - (1-\gamma)\left(\overline{u'v'}\right)_{nt} - \gamma\left(\overline{u'v'}\right)_t - \gamma(1-\gamma)\left(\bar{U}_t - \bar{U}_{nt}\right)\left(\bar{V}_t - \bar{V}_{nt}\right) \qquad (15.22)$$

$$q_{xy} = -k\frac{\partial \overline{T}}{\partial y} - (1-\gamma)\left(\overline{v'T'}\right)_{nt} - \gamma\left(\overline{v'T'}\right)_{t} - \gamma(1-\gamma)\left(\overline{V}_{t} - \overline{V}_{nt}\right)\left(\overline{T}_{t} - \overline{T}_{nt}\right) \qquad (15.23)$$

where V and T are the ensemble-averaged values of the lateral velocity and temperature, and v' and t' are their corresponding fluctuations. The first term is the molecular stress and heat flux component, while the second and third are the components caused by fluctuations in the nonturbulent and turbulent portions of the flow, respectively. The second term, $\left(\overline{u'v'}\right)_{nt}$, or $\left(\overline{v'T'}\right)_{nt}$, is very small and negligible. The third terms $\left(\overline{u'v'}\right)_{t}$ stress and $\left(\overline{v'T'}\right)_{t}$ flux components may be considered as the real turbulent shear stress and heat flux. These components are produced by the motion of eddies having various scales, but primarily at scale small compared to boundary layer thickness δ (the small-scale eddy components). The fourth term which is zero for either a completely nonturbulent $\gamma = 0$ or completely turbulent portion of the flow $\gamma = 1$ arises from the mean momentum and thermal exchange between the two regions. These components, $\left(\overline{U}_{t} - \overline{U}_{nt}\right)\left(\overline{V}_{t} - \overline{V}_{nt}\right), \left(\overline{V}_{t} - \overline{V}_{nt}\right)\left(\overline{T}_{t} - \overline{T}_{nt}\right)$, are produced by motion of various eddy scales, but primary at scale roughly same order of boundary layer thickness δ. These components may be called the large-scale eddy components and have their greatest value when $\gamma = 0.5$. It accounts for about 30% of the total stress in the wake flow and 15% for the boundary layer flow. The above intermittency analysis can be useful for developing computational models of transitional boundary layer flow and heat transfer predictions [5].

15.3.3 UNSTEADY WAKE-INDUCED TRANSITION MODEL

One of the primary effects on vane heat transfer is the free-stream turbulence generated at the combustor exit. Combustor-generated turbulence contributes to significant heat-transfer enhancement. Typically, combustor-generated, free-stream turbulence levels are around 15%–20% at the first-stage vane leading edge, and due to acceleration of flow in the vane passage, the turbulence intensity at the first-stage rotor blade leading edge is typically around 5%–10%. The reduced free-stream turbulence effects on the rotor blade heat transfer are not as significant as that of the high free-stream turbulence effects on the vane heat transfer. However, the rotor blade heat transfer is affected by another important parameter, the effect of unsteadiness in the flow.

The unsteadiness of flow arises from the relative motion of the rotor blade rows with reference to alternate stationary vane rows. Figure 15.13 shows a conceptual view of the unsteady wake propagation through a rotor blade row [6]. The shaded regions indicate where unsteadiness is caused by the upstream airfoils. For a first-stage blade, the principal component of unsteadiness is due to wake passing. These wakes impose the free stream with a periodic unsteady velocity, temperature, and turbulence intensity. The wakes cause an early unsteady laminar-to-turbulent boundary layer transition to occur on the blade surface. The heat transfer associated with the unsteady flow effects clearly indicates the early boundary layer transition.

There are three modes of transition: natural transition, bypass transition, and separated-flow transition: Natural transition begins with a weak instability in the

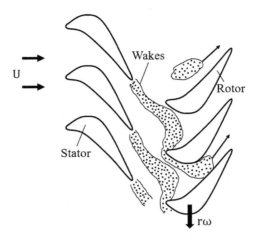

FIGURE 15.13 Unsteady wake propagation through a rotor blade row.

laminar boundary layer and proceeds through various stages of amplified instabil
ity to fully turbulent flow. This was first described by Tollmien and Schlichting [1
Bypass transition is caused by large disturbances in the external flow and completel
bypasses the Tollmien–Schlichting mode of instability. This is typical of gas tur
bines. Separated-flow transition occurs in a separated laminar boundary layer an
may or may not involve the Tollmien–Schlichting mode. This occurs mostly in com
pressors and low-pressure turbines.

The above-mentioned turbulent spot production theory was extended to includ
two transition modes simultaneously on the same surface as shown in Figure 15.1
[6]. It shows that normal transition was caused by turbulent spot production, but th
production of turbulent spots due to wake passing impingement was so intense tha
the spots immediately coalesced into turbulent strips that then propagated and grev
along the surface within the laminar boundary layer. In this case, transition is mainl
caused by unsteady passing wake impingement, named bypass transition.

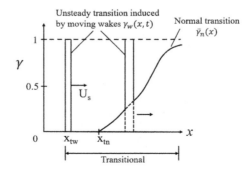

FIGURE 15.14 Multiple modes of transition on a surface as the results of an unsteady peri
odic passing of a wake.

Assuming that the turbulent spots produced normally or wake-induced are independent of each other, the time-averaged intermittency can be defined as

$$\tilde{\gamma}(x) = 1 - \left[1 - \gamma_n(x)\right]\left[1 - \tilde{\gamma}_w(x)\right] \tag{15.24}$$

where γ_n and $\tilde{\gamma}_w$ are the normal and wake-induced intermittencies, respectively. The tilde over the intermittency value indicates a time-averaged quantity over a wake-passing period. The normal mode intermittency has been discussed previously as

$$\gamma_n = 1 - \exp\left[-\hat{n}\sigma(\mathrm{Re}_x - \mathrm{Re}_{xt})^2\right] \tag{15.15}$$

$$\hat{n}\sigma = 1.5 \times 10^{-11} \cdot Tu^{7/4} \tag{15.16}$$

where $\hat{n} = nv^2/U^3$ is the dimensionless spot production parameter, σ is the turbulent spot propagation parameter, Re_x is the local-flow Reynolds number, Re_{xt} is the Reynolds number at the location for onset of normal transition, and Tu is free stream turbulence intensity.

Assumed a square wave distribution for the turbulent strip production function and evaluated a production rate from various experiments. A simple expression for the time-averaged, wake-induced intermittency was obtained as shown in Figure 15.15.

$$\tilde{\gamma}_w(x) = 1 - \exp\left[-1.9\left(\frac{x - x_{tw}}{U\tau}\right)\right] \tag{15.25}$$

where U is the airfoil's incident velocity, and τ is the wake-passing period. It shows that the time-averaged wake-induced intermittency increases with decreasing the wake-passing period (i.e., increasing the wake-passing frequency). It also shows the wake-induced transition length decreases with decreasing the wake-passing period for a given airfoil incident velocity. The effect of unsteady wake on surface heat transfer coefficients of a gas turbine blade was experimentally simulated using a spoked wheel type wake generator [7]. The above calculated wake-induced intermittency

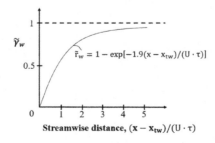

FIGURE 15.15 Time-averaged intermittency distribution as a function of the reduced streamwise distance.

values were employed to predict transitional region heat transfer coefficients and compared well with experimental measured results [5].

Example of Unsteady Wake-Induced Transition:

Here is an example to validate the wake-induced intermittency theory. The effect of unsteady wake on surface heat transfer coefficients of a gas turbine blade was experimentally simulated using a spoked wheel type wake generator by Han et al. [8]. The experiments were performed with a five airfoil, linear cascade in a low-speed wind tunnel facility. The cascade inlet Reynolds number based on the blade chord varied from 100,000 to 300,000. The wake passing Strouhal number, $S = 2\pi Ndn /(60V_1)$, varied between 0 and 1.6 by changing the rotating wake passing frequency (rod speed, N, and rod number, n), rod diameter d, and cascade inlet velocity V_1. Figure 15.16 shows the conceptual view of the effect of the unsteady wake on the blade mode. A hot wire anemometer system was located at the cascade inlet to detect the instantaneous velocity induced by the passing wake, and a blade instrumented with thin foil heaters and thermocouples was used to measure the surface heat transfer coefficients on the blade surface.

Figure 15.17 shows typical, instantaneous velocity profiles for S = 0.1, 0.2, and 0.4, respectively. The Strouhal number (S) was varied by increasing the rotating rod speed *(N)* for a given rod diameter ($d = 0.63$ cm), rod number ($n = 16$), and cascade inlet velocity ($V_i = 21$ m/s or Re $= 3 \times 10^5$). The instantaneous velocity profile shows the periodic unsteady fluctuations caused by the upstream passing wake, and the periodic fluctuations increase with the Strouhal number. The phase-averaged profile shows the time-dependent, mean velocity defect caused by the upstream passing wake, whereas the wake width increases with the Strouhal number. The phase-averaged turbulence intensity reaches 20% inside the wake. The time-averaged turbulence intensity Tu is about 8%, 10%, and 15%, for S = 0.1, 0.2, and 0.4, respectively. The background turbulence intensity is only about 0.75% for the case of no rotating rods in the wind tunnel.

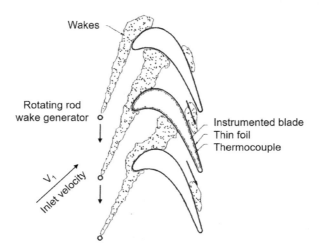

FIGURE 15.16 Conceptual view of the effect of the unsteady wake on the blade model.

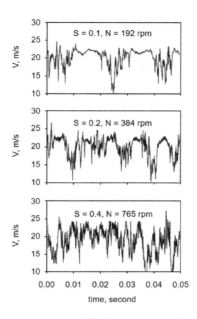

FIGURE 15.17   Typical instantaneous velocity profiles caused by the unsteady wake.

From the results of the instantaneous velocity, phase-averaged mean velocity, and turbulence intensity profiles, it is expected that the unsteady wake, with
higher Strouhal numbers, will produce a larger impact on the downstream blade.
Figure 15.18 shows the unsteady, passing wake promotes earlier and broader boundary layer transition and causes much higher heat transfer coefficients on the suction
surface, whereas the passing wake also significantly enhances heat transfer coefficients on the pressure surface and leading edge region ($X/C = 0$), where X/C is the
ratio of surface measured from airfoil leading edge to airfoil chord length. The above
calculated wake-induced intermittency values, Equation (15.25), were employed to
predict transitional region heat transfer coefficients and compared well with experimental measured results. However, upstream of transition, the theory underpredicted
heat-transfer levels. It implies that laminar boundary has already been disturbed by
unsteady wake before transition. Note that Nusselt numbers at the begining of transition and at the end of transition were based on experimentally measured values as
shown in Figure 15.18.

Example of Turbulence Spot-Induced Transition:

The above-mentioned normal turbulent spot-induced intermittency values,
Equations (15.15) and (15.16), were employed to predict transitional region heat
transfer coefficients and compared well with measured results at high free stream
turbulence experiments [9]. They studied the influence of mainstream turbulence on surface heat transfer coefficients of a gas turbine blade model (similar
to Figure 15.16). A five-blade linear cascade in a low-speed wind tunnel facility
was used in the experiments. The mainstream Reynolds numbers were 100,000,
200,000, and 300,000 based on the cascade inlet velocity and blade chord length.

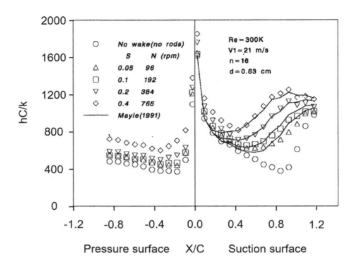

FIGURE 15.18 Unsteady wake effect on turbine blade surface heat transfer.

The grid-generated turbulence intensities at the cascade inlet were varied between 2.8% and 17%. A hot-wire anemometer system measured turbulence intensities, mean- and time-dependent velocities at the cascade inlet, outlet, and several locations. A thin-foil thermocouple instrumented blade determined the surface heat transfer coefficients. The results showed that the mainstream turbulence promoted earlier and longer boundary layer transition, caused higher heat transfer coefficients on the suction surface, the leading edge region ($X/C = 0$), and significantly enhances the heat transfer coefficient on the pressure surface. The onset of transition on the suction surface boundary layer moved forward with increased mainstream turbulence intensity and Reynolds number (similar to Figure 15.18). However, upstream of transition, the theory underpredicted heat-transfer levels. It implies that laminar boundary has already been disturbed by high free stream turbulence before transition. The heat transfer coefficient augmentations and peak values on the suction and pressure surfaces were affected by the mainstream turbulence and Reynolds number.

Example of Turbulence Spot and Unsteady Wake-Induced Transition:

Additionally, the above-mentioned normal turbulent spot-induced and wake-induced intermittency values, Equation (15.24), were also employed to predict transitional region heat transfer coefficients and compared well with measured results of combined unsteady wake and high free stream turbulence experiments [10]. They defined a mean turbulence intensity ($\overline{T}u$) for defining the turbulence generated by the combination of both free-stream turbulence and unsteady wakes. With increasing mean turbulence intensity ($\overline{T}u$) through wake strengh or turbulence intensity level, the transition location on the suction surface moved closer to the leading edge ($X/C = 0$). The results show a good match in the region downstream of onset of transition (similar to Figure 15.18). However, upstream of transition, the theory underpredicted heat-transfer levels. It implies that laminar boundary has already been disturbed by high free strean turbulence and unsteady wake before transition.

The above examples show the ability of Mayle's transition theory to predict time-averaged heat-transfer levels for turbine blades under free stream turbulence, unsteady wake, or the combination effects. In addition, one of the most important things is to know where the laminar boundary layer begins to transition Re_{xt} for a given flow Reynolds number Re_x. As mentioned above, either Equations (15.9), (15.10), or (15.17) can be used for onset critical Reynolds number prediction for transition. Note that several important parameters affect the transition location and the transition distance such as free stream turbulence intensity, length scale, pressure gradient, surface curvature, surface roughness, film cooling, etc. From literature, high free stream turbulence, smaller length scale, unsteadiness, flow deceleration, concave surface, film cooling, and surface roughness promote earlier laminar boundary layer transition with shorter transition distance. Refer to Mayle [5] or Han et al. [11] Chapter 2 for the details.

15.3.4 SURFACE ROUGHNESS-INDUCED TRANSITION MODEL

For gas turbines, surface roughness is of concern due to initial manufacturing finish and high temperature combustion deposits after many hours of operation. Combustion deposits may make the turbine airfoil surface rough after many hours of service, and this roughness could be detrimental to the life of the turbine due to increased heat-transfer levels that are much higher than smooth surface conditions. In real engines, the roughness size varied over the surface from about 2–10 μm depending upon the location of airfoil suction or pressure surface. Figure 15.19 shows the momentum thickness Reynolds number at the onset of transition as a function of the roughness parameter and free-stream turbulence intensity. It implies the laminar boundary layer is easier to transition to turbulent boundary layer with increasing roughness size for a given flow condition. That means, transition momentum thickness Reynolds number $Re_{\theta t}$ decreases with larger roiughness height K_s for a giving boundary layer flow momentum thickness θ_t. This can be correlated as shown in Equation (15.26).

$$Re_{\theta t} = 100 + 0.43 \exp\left[7 - 0.77\left(\frac{K_s}{\theta_t} \right) \right] \qquad (15.26)$$

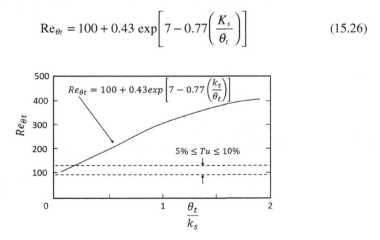

FIGURE 15.19 The momentum thickness Reynolds number at the onset of transition as a function of the roughness parameter and free-stream turbulence intensity.

For example, $\dfrac{\theta_t}{K_s} = 2$, $\mathrm{Re}_{\theta t} = 400$; $\dfrac{\theta_t}{K_s} = 1$, $\mathrm{Re}_{\theta t} = 300$; $\dfrac{\theta_t}{K_s} = 0.5$, $\mathrm{Re}_{\theta t} = 200$.

It is important to note that $\mathrm{Re}_{\theta t}$ is around 100–120 for high turbulence levels between 5% and 10%. The surface rougness effect is diminished under high free-stream turbulence condition. Refer to Mayle [5] for the details.

Example of Roughness and Turbulence-Induced Transition:

Figure 15.20 presents the heat-transfer coefficients for the combined effect of free-stream turbulence and the surface roughness on an airfoil [12]. Three experimental cases are shown: The first case is for a smooth surface with Tu = 5.5%; the second case is for a smooth surface with Tu = 10%; the third case is for the rough surface at Tu = 10%. Comparing the first two cases for the smooth surface, it is evident that the heat-transfer coefficients are significantly enhanced on the pressure surface due to increased turbulence. The suction surface results show that the transition location has moved upstream due to increased free-stream turbulence from $s/L = 1.0$ to $s/L = 0.25$, where s/L is the ratio of surface measured from airfoil leading edge to airfoil chord length. This is the typical results as earlier discussed. For comparing the smooth surface to the rough surface at Tu = 10%, the pressure surface is unaffected by the surface roughness at this high turbulence. The already enhanced heat-transfer coefficients due to high free-stream turbulence are unaffected by the surface roughness. However, the effect on suction surface is significant. The transition location does not seem to be affected by the rough surface. But the length of transition is greatly reduced by addition of surface roughness. A combination of surface roughness with high free-stream turbulence causes the boundary layer to undergo transition more rapidly than for the high free-stream turbulence case only, as clearly shown in Figure 15.20.

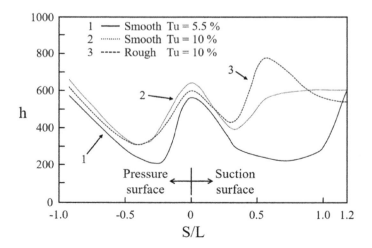

FIGURE 15.20 Effect of surface roughness and free-stream turbulence on airfoil heat transfer coefficient distributions.

15.3.5 FILM COOLING-INDUCED TRANSITION

For turbine blade film cooling, relatively cool air is injected from the inside of the blade to the outside surface, which forms a protective layer between the blade surface and hot mainstream. Film cooling can be modeled as coolant jet in crossflow at a given inclined angle as sketched in Figure 15.21. Film cooling performance depends primarily on the coolant-to-hot-mainstream pressure ratio (P_c/P_t), temperature ratio (T_c/T_g), film cooling hole location (leading edge, trailing edge, pressure and suction sides, end-wall, and blade tip), and geometry (hole size, spacing, shape, angle from the surface, and number of rows) under representative engine flow conditions such as Reynolds number, Mach number, combustion-generated high free-stream turbulence, stator-rotor unsteady wake flow, etc. The coolant-to-mainstream pressure ratio is related to the coolant-to-mainstream mass flux ratio (blowing ratio), while the coolant-to-mainstream temperature ratio is related to the coolant-to-mainstream density ratio. In a typical gas turbine airfoil, the P_c/P_t ratios vary from 1.02 to 1.10 with the corresponding blowing ratios vary approximately from 0.5 to 2.0. Whereas the T_c/T_g values vary from 0.5 to 0.85, the corresponding density ratios vary approximately from 2.0 to 1.5 [13].

In general, the higher the pressure ratio, the better the film-cooling protection (i.e., reduced heat transfer rate to the airfoil) at a given temperature ratio, while the lower the temperature ratio, the better the film-cooling protection at a given pressure ratio. However, a too high-pressure ratio (i.e., blowing too much coolant) may reduce the film-cooling protection because of jet penetration into the mainstream (jet lift off from the surface). Data from numerous available studies in the open literature suggest a blowing ratio near unity is optimum, with severe penalties at either side. The best film cooling design is to reduce the heat transfer rate to the airfoils using a minimum amount of cooling air from compressors.

It is well known that film cooling is to protect turbine airfoil from high-temperature combustion gases; however, airfoil surface heat transfer coefficients can be increased due to cooling jet interaction with mainstream flow. In addition, film cooling may cause early boundary layer transition due to mixing between cooling jet and mainstream. The following is an outline of the relationship between heat transfer

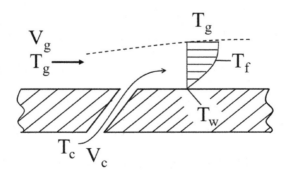

FIGURE 15.21 Film cooling model.

coefficient enhancement and film coooling effectiveness protection under film injection condition. Refer to Han et al. [11] Chapter 3 for the details.

From Figure 15.21, heat flux without film injection q_0'', with film injection q'', and heat flux ratio can be written as:

$$q_0'' = h_0 \left(T_g - T_w \right)$$

$$q'' = h \left(T_f - T_w \right) \tag{15.27}$$

$$\frac{q''}{q_0''} = \frac{h}{h_0} \cdot \frac{T_f - T_w}{T_g - T_w} < 1$$

where

$$\frac{h}{h_0} > 1$$

$$\frac{T_f - T_w}{T_g - T_w} < 1$$

$$\eta = \text{film cooling effectiveness} = \frac{T_g - T_f}{T_g - T_c} = 0 \sim 1$$

Therefore

$$\frac{q''}{q_0''} = \frac{h}{h_0} \left[1 - \eta \left(\frac{T_g - T_c}{T_g - T_w} \right) \right] = \frac{h}{h_0} \left[1 - \frac{\eta}{\phi} \right] \tag{15.28}$$

where

$$\phi = \frac{T_g - T_w}{T_g - T_c} = \text{overall cooling effectiveness}$$

In turbine application, $\phi \cong \dfrac{1600°C - 1000°C}{1600°C - 600°C} \cong 0.6$

T_g = combustion gas temperature

T_c = coolant air temperature

T_f = film temperature (mixing temperature of combustion gas and coolant air)

T_w = blade surface temperature

h_0 = heat transfer coefficient without film cooling

h = heat transfer coefficient with film cooling

M = blowing ratio, coolant-to-mainstream mass flux ratio, $\left(\rho V \right)_c / \left(\rho V \right)_g$

From above equations, one would like to have a film cooling design with higher film cooling effectiveness η and lower heat transfer coefficient augmentation $\dfrac{h}{h_0}$ due to jet interaction with mainstream. Therefore, heat flux ratio $\dfrac{q''}{q_0''}$ with film cooling will be reduced as shown in Equation (15.28). For example, if $\eta = 0.3$, $\dfrac{h}{h_0} = 1.2$, $\phi = 0.6$, then $\dfrac{q''}{q_0''} = 0.6$. It means heat flux with film cooling will be reduced 40% as compared to that without film cooling. This is why the turbine airfoils can be protected by proper film cooling design. However, one of negative impacts of film cooling is to promote early boundary layer transition which will potentially increase heat transfer coefficient to the turbine airfoil.

Example of Film Cooling-Induced Transition:

The effect of unsteady wake on surface heat transfer coefficients of a gas turbine blade was experimentally simulated using a spoked wheel type wake generator [8]. The experiments were performed with a five airfoil, linear cascade in a low-speed wind tunnel facility. The cascade inlet Reynolds number based on the blade chord varied from 100,000 to 300,000. The same unsteady wake facility shown in Figure 15.16 was employed for the film cooling study. Figure 15.22 presents a two-dimensional view of the fillm-cooled turbine-blade model. There is one cavity used to supply coolant to the row of film holes on the suction side of thurbine blade model.

Figure 15.23 presents the three types of hole geometries studied: cylindrical hole, fan-shaped hole, and laidback fan-shaped hole.These three hole geometries are typical for turbine airfoil film cooling designs. It is expected that the expanded holes would signifcantly improve thermal protection of the surface downstream of the ejection location as compared to the cylindrical hole. One row of nine holes of each shape located near the blade suction-side gill-hole region ($X/SL = 0.12$) has been employed, where X/SL is the ratio of surface measured from airfoil leading edge to airfoil chord length. The diameter of the cylindrical hole d and diameter of the cylindrical inlet section of the expanded hole is also d. For all geometries the inclination angle is 40°, the pitch-to-diameter ratio is $P/d = 5.3$, and the length-to-diameter ratio is 7.9. The lateral expansion angle of both expanded holes is 7.24°. The exit forward expansion angle of the laidback fan-shaped hole is 25°. For the fan-shaped and laidback fan-shaped hole, the calculation of the blowing ratio was based on the inlet cross-sectional area of these holes, that is, the same as the cylindrical hole. That means the same blowing ratio provides the same amount of coolant ejected under

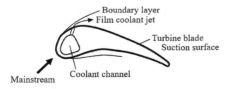

FIGURE 15.22 Cross-sectional view of film-cooled blade model.

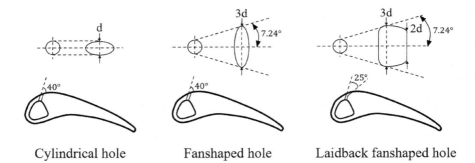

FIGURE 15.23   Film cooling hole geometries.

the same mainstream condition. Thus, the blowing ratio of the shaped holes can be directly compared to those of the cylindrical hole, which makes it more convenient to evaluate the effect of the hole exit shape on heat transfer coefficient augmentation and boundary layer transition [14].

Figure 15.24 presents the effect of hole shape on spanwise-averaged Nusselt-number distributions for cases with a high blowing ratio of $M = 1.2$. For the cases under steady flow condition, the Strouhal number $S = 0$, the spanwise averaged Nusselt number is higher than that of no film-hole case for all three kinds of holes. Film injection through cylindrical holes produces the highest spanwise-averaged Nusselt numbers in the region immediately downstream of the film injection location; this is due to the strongest jet interaction with mainstream for the cylindrical hole film cooling as compared with the reduced interaction for both shaped hole film cooling. Whereas fan-shaped and laidback fan-shaped hole injection have higher spanwise-averaged Nusselt numbers at the downstream of the blade surface. The correspoding boundary layer transition location is moved up from $X/SL = 0.7$ for no film cooling case to $X/SL = 0.5$ for cylindrical hole injection and to $X/SL = 0.3$

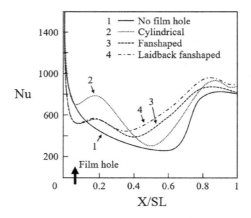

FIGURE 15.24   Effect of hole shape on Nusselt number distributions for steady flow ($S = 0$) at blowing ratio $M = 1.2$.

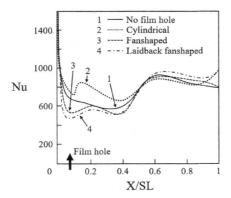

FIGURE 15.25 Effect of hole shape on Nusselt number distributions for unsteady flow ($S = 0.1$) at blowing ratio $M = 1.2$.

for both shaped hole injection. It implies that, for steady flow, boundary transition is promoted by the film coolant injection.

With the addition of unsteady wake, the Strouhal Number $S = 0.1$, as shown in Figure 15.25, the hole effect is intensified at the film injection region of the blade. In the region immediately downstream of film injection, $X/SL > 0.1$, the spanwise-averaged Nusselt numbers for fan-shaped and laidback fan-shaped hole injection are much lower than that of cylindrical hole injection. Note that Nusselt numbers for both shaped hole film injection are lower than the no film injection case. From literature, it shows that both shaped holes provide higher film cooling effectiveness than the cylindrical hole. This is why both shaped holes have been used for turbine airfoil film cooling designs. The differences of the spanwise-averaged Nusselt numbers for different hole injection are reduced at the downstream of the blade surface. However, the boundary-layer transition location is almost the same, at $X/SL = 0.4$, for the three types of film hole and the no film-hole case when there is an unsteady wake effect. It implies that boundary layer transition is dominated by the unsteady wake. This is because the film hole injection is located at $X/SL = 0.1$, the jet-mainstream interaction effect is diminished at $X/SL > 0.4$ where transition is dominated by unsteady wake [14].

15.4 HEAT TRANSFER CORRELATION FOR LAMINAR, TRANSITIONAL, AND TURBULENT FLOW

A correlation for laminar, transitional, and turbulent flow heat transfer in flat-plate boundary layers was published by Lienhard [15]. He developed different approximations through a detailed consideration of multiple data sets for $0.7 \leq Pr \leq 2.57$; $4,000 \leq Re_x \leq 4,300,000$, and varying levels of freestream turbulence up to 5% for smooth, sharp-edged plates at zero pressure gradient. The result was in good agreement with the available measurements and applied over the full range of Reynolds number for either a uniform wall temperature (UWT) or a uniform heat flux (UHF) boundary condition. The correlation should be matched to the estimated transition condition of any particular flow.

The following is a brief outline of transition region correlation; refer to Lienhard [15] for the details. Data supporting those equations span $0.7 \leq \mathrm{Pr} \leq 2.57$ and $4,000 \leq \mathrm{Re}_x \leq 4,300,000$ with free stream turbulence levels up to 5%. Correlation applies to smooth, sharp-edged, flat plates with zero streamwise pressure gradient at either UWT or UHF.

Laminar region:

$$\mathrm{Nu}_L = 0.332\,\mathrm{Re}_x^{1/2}\mathrm{Pr}^{1/3} \quad \mathrm{UWT}$$

$$\mathrm{Nu}_L = 0.453\,\mathrm{Re}_x^{1/2}\mathrm{Pr}^{1/3} \quad \mathrm{UHF} \qquad (15.29)$$

With an unheated starting length of x_0 (UWT or UHF), use

$$\mathrm{Nu}_L \cdot \left[1 - \left(x_0/x\right)^{3/4}\right]^{-1/3}$$

Transition region:

$$\mathrm{Nu}_t = \mathrm{Nu}_L \cdot \left(\mathrm{Re}_x/\mathrm{Re}_{xt}\right)^c \qquad (15.30)$$

where

$$c = 0.9922\log_{10}\mathrm{Re}_{xt} - 3.013 \quad \text{for} \quad \mathrm{Re}_{xt} < 5 \times 10^5 \qquad (15.31)$$

Turbulent region (for UWT and UHF):

$$\mathrm{Nu}_T = \frac{\mathrm{Re}_x\mathrm{Pr}\left(C_f/2\right)}{1 + 12.7\left(\mathrm{Pr}^{2/3} - 1\right)\sqrt{C_f/2}} \qquad (15.32)$$

where

$$C_f = \frac{0.455}{\left[\ln\left(0.06\mathrm{Re}_x\right)\right]^2}$$

$$\mathrm{Nu}_T = 0.0296\,\mathrm{Re}_x^{0.8}\,\mathrm{Pr}^{0.6} \quad \text{for gases only}$$

Combining the equations:

$$\mathrm{Nu}_x = \left[\mathrm{Nu}_L^5 + \left(\mathrm{Nu}_t^{-10} + \mathrm{Nu}_T^{-10}\right)^{-1/2}\right]^{1/5} \qquad (15.33)$$

Therefore, transitional heat transfer coefficients can be predicted at a given Re_x, if Re_{xt} value as well as fully laminar heat transfer coefficients Nu_L are provided. These Re_{xt} and Nu_L values are obtained from experiments and directly correlated to free stream turbulence intensity levels and surface roughness conditions. The important thing is to obtain the critical Reynolds number Re_{xt} and its fully laminar heat transfer Nu_L, and then transitional region heat transfer Nu_t can be obtained.

Note that the c values in transition region as shown in Equation (15.31) serve the same purpose as the intermittency γ values in transition region shown in Equations (15.15) and (15.16). Both c and γ values shown in these equations depend on free stream turbulence intensity and onset of transition location Re_{xt} which needed to be pre-determined from experiments.

REMARKS

Based on Tollmien–Schlichting wave instability theory, laminar boundary can begin transition into turbulent boundary layer at certain critical Reynolds number. This type of gradual transition named normal or natural transition. The critical Reynolds number for normal transition decreases with increasing free stream turbulence level. The normal transition theory can also be obtained by Emmons turbulent spot production concept in which laminar boundary layer can quickly bypass Tollmien–Schlichting wave and transition into turbulent boundary. This type of abrupt transition named bypass transition. The intermittency can be used to correlate with turbulent spot production rate and then with free stream turbulence intensity level. This intermittency can also be extended to correlate with unsteady wake induced bypass transition. Transition region heat transfer can be predicted at a given intermittency value if critical Reynolds number and fully laminar heat transfer are provided for the cases of turbulent spot induced transition, unsteady wake induced transition, roughness, film cooling, or a combination of them. However, one of the most important needed information is the critical Reynolds number for transition initiation that must be pre-determined for a given flow conditions. Note that critical Reynolds number for the onset transition can be based on x-distance, boundary layer thickness, displacement thickness, or momentum thickness, these laminar boundary layer parameters are correlated to each other at a given flow conditions.

REFERENCES

1. H. Schlichting, *Boundary-Layer Theory*, Sixth Edition, McGraw-Hill Book Company, New York, 1968.
2. H.W. Emmons, "The laminar-turbulent transition in boundary layer: Part I," *Journal of Aeronautical Science*, Vol. 18, pp. 490–498, 1951.
3. M.F. Blair, "Influence of free-stream turbulence on turbulent boundary layer heat transfer and mean profile development, Part I: Experimental data. Part II: Analysis of results," *ASME Journal of Heat Transfer*, Vol. 105, pp. 33–47, 1983.
4. M.F. Blair, "Influence of free-stream turbulence on boundary layer transition in favorable pressure gradients," *ASME Journal of Engineering for Power*, Vol. 104, pp. 743–750, 1982.
5. R.E. Mayle, "The role of laminar-turbulent transition in gas turbine engines," *ASME Journal of Turbomachinery*, Vol. 113, pp. 509–537, 1991.
6. R.E. Mayle, and K. Dullenkopf, "A theory for wake-induced transition," *ASME Journal of Turbomachinery*, Vol. 112, pp. 188–195, 1990.

7. K. Dullenkopf, A. Schulz, and S. Wittig, "The effect of incident wake conditions on the mean heat transfer of an airfoil," *ASME Journal of Turbomachinery*, Vol. 113, pp. 412–418, 1991.

8. J.C. Han, L. Zhang, and S. Ou, "Influence of unsteady wake on heat transfer coefficients from a gas turbine blade," *ASME Journal of Heat Transfer*, Vol. 115, pp. 904–911, 1993.

9. L. Zhang and J.C. Han, "Influence of mainstream turbulence on heat transfer coefficients from a gas turbine blade," *ASME Journal of Heat Transfer*, Vol. 116, pp. 896–903, 1994.

10. L. Zhang and J.C. Han, "Combined effect of free-stream turbulence and unsteady wake on heat transfer coefficients from a gas turbine blade," *ASME Journal of Heat Transfer*, Vol. 117, pp. 296–302, 1995.

11. J.C. Han, S. Dutta, and S. Ekkad, *Gas Turbine Heat Transfer and Cooling Technology*, Taylor & Francis Group, New York, 2000.

12. A. Hoffs, U. Drost, and A. Boles, "Heat Transfer Measurements on a Turbine Airfoil at Various Reynolds Numbers and Turbulence Intensities Including Effects of Surface Roughness," ASME Paper 96-GT-169, 1996.

13. J.C. Han, "Advanced cooling in gas turbines 2016 max Jakob memorial award paper," *ASME Journal of Heat Transfer*, Vol. 140(11), pp. 1–20, 2018.

14. S. Teng, J.C. Han, and P.E. Poinsatte, "Effect of film-hole shape on turbine-blade heat transfer coefficient distribution," *AIAA Journal of Thermophysics and Heat Transfer*, Vol. 15(3), pp. 249–256, 2001.

15. J.H. Lienhard V, "Heat transfer in flat-plate boundary layers: A correlation for laminar, transitional, and turbulent flow," *ASME Journal of Heat Transfer*, Vol. 142, pp. 1–14, 2020. Article No. 061805.

16 Turbulent Flow Heat Transfer Enhancement

16.1 HEAT TRANSFER ENHANCEMENT METHODS

A well-known method to increase the heat transfer from a surface is to roughen the surface either randomly with a sand grain or by use of regular geometric roughness elements on the surface. For example, Figure 16.1a shows the cross-section of circular tube with sand grain roughness. However, the increase in heat transfer is accompanied by an increase in the resistance to fluid flow. Many investigators have studied this problem in an attempt to develop accurate predictions of the behavior of a given roughness geometry and to define a geometry which gives the best heat-transfer performance for a given flow friction. Another well-known method to enhance the heat transfer from a surface is to roughen the surface by the use of artificially repeated ribs on the surface. For example, Figure 16.1b shows the cross-section of circular tube with rib roughness. However, the increase in heat transfer is accompanied by an increase in the pressure drop of the fluid flow. Many investigations have been directed toward developing predictive correlations for a given rib geometry and establishing a geometry which gives the best heat transfer performance for a given pumping power.

For the results of roughened surfaces to be most useful, general correlations are necessary for both the friction factor and the heat-transfer coefficient which cover a wide range of parameters. In fully developed turbulent flow, theoretical approaches to the problem of momentum and heat transfer in smooth and rough tubes have been available for many years. These approaches are based on similarity considerations. Two surfaces are said to have geometrically similar roughness if the geometry of their roughness is the same in all aspects except for a scale factor. For example, sand grain roughness is a geometrically similar roughness, and repeated rib roughness can be treated as the geometrically similar roughness. Friction similarity law and heat transfer similarity law have been developed to correlate friction factor and heat transfer coefficient data for sand grain roughness as well as for repeated rib roughness over a wide range of geometric parameters and flow Reynolds numbers and Prandtl numbers.

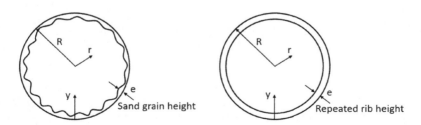

FIGURE 16.1 (a) Circular tube with sand grain roughness, (b) with repeated rib roughness.

DOI: 10.1201/9781003164487-16

16.1.1 Sand Grain Roughness

An early study of the effect of roughness on friction and the velocity distribution was reported by Nikuradse in 1933. He conducted a series of experiments with pipes roughened by sand grains. One of the first studies of heat transfer in rough tubes was conducted by Cope in 1941 [1]. They derived semi-empirical correlations for the friction factors and the heat transfer coefficients from the law of the wall similarity for flow over rough surfaces. The similarity law concept was further developed by Nikuradse in 1950 [2], who applied it successfully to correlate the friction data for fully developed turbulent flow in tubes with sand grain roughness. Based on a heat-momentum transfer analogy, Dipprey and Sabersky in 1963 [3] successfully developed the heat transfer similarity law for fully developed turbulent flow in tubes with close-packed sand-grain roughness, which is complementary to Nikuradse's friction similarity law. Therefore, the friction factor can be predicted for flow in a tube with a given sand-grain roughness height-to-diameter ratio (e/D) and Reynolds number, and the heat transfer coefficient can be predicted for flow in a tube with a given sand-grain roughness height to diameter ratio (e/D), Reynolds number, and Prandtl number.

16.1.2 Repeated Rib Roughness

A number of friction and heat transfer measurements were reported for repeated-rib roughness in tube flow such as Sams in 1952 [4]. In the nuclear reactor area, considerable data were reported for repeated-rib roughness in an annular flow geometry in which the inner annular surface is rough and the outer surface is smooth such as Wilkie in 1966 [5]. To simulate the geometry of fuel bundles in advanced gas-cooled nuclear reactors, White and Wilkie in 1970 [6] studied the heat transfer and pressure loss characteristics of a rib roughened surface with different helix angles. They found the ribs with helix angles performed better than the ribs with perpendicular angles. Webb et al. in 1971 [7] performed experiments on a tube with repeated ribs as sketch in Figure 16.2. They covered a wide range of rib height to hydraulic diameter ratio ($0.01 < e/D < 0.04$) and pitch to height ratio ($p/e = 10, 20, 40$), and the repeated ribs were aligned normal to the main stream direction (transverse or perpendicular ribs).

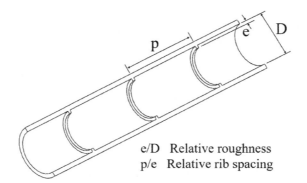

e/D Relative roughness
p/e Relative rib spacing

FIGURE 16.2 Characteristic dimensions of repeated-rib roughness.

Han et al. in 1978 [8] investigated turbulent flow heat transfer in parallel-plate channel with repeated ribs as sketch in Figure 16.3. They reported the effects of rib height-to-hydraulic diameter ratio ($0.05 < e/D < 0.10$), pitch-to-height ratio ($p/e = 5$, 10, 15), rib shape ($\phi = 40°$ to $90°$), and flow angle of attack ($\alpha = 90°$ to $20°$) on friction factor and heat-transfer coefficients. They found the ribs with flow angle of attack (angled ribs, $\alpha = 45°$) performed better than the ribs normal to the flow (transverse or perpendicular ribs, $\alpha = 90°$). Gee and Webb in 1980 [9] confirmed the tube flow with repeated ribs having helix angles performed better than that with normal angles.

Analytical methods for predicting the friction factors and the heat transfer coefficients for turbulent flow over rib roughened surfaces are not available because of the complex flow, such as separation, reattachment, and recirculation, created by periodic rib roughness elements as shown in Figure 16.4. Therefore, heat transfer designers still depend on the semi-empirical correlations over a wide range of rib geometry for the friction and heat transfer calculations.

The above-mentioned law of the wall similarity for flow over sand grain roughness surfaces could be applied to flow over repeated rib roughness surfaces. With repeated-rib roughness, for a given flow attack angle, rib shape, and pitch-to-height ratio, tests with a different height-to-hydraulic diameter ratio (e/D) represent geometrically similar roughness. However, when the values of P/e, flow attack angle α or rib shape are varied, the surfaces are not geometrically similar. Surfaces which are not geometrically similar required modifications to the roughness and heat transfer functions found by similarity considerations. The friction and heat transfer similarity laws were extended to correlate the friction and heat transfer data for turbulent flow

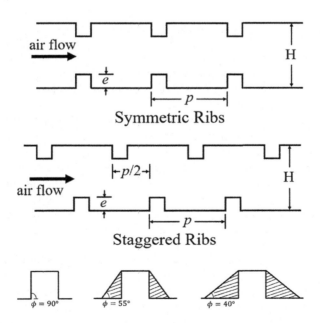

FIGURE 16.3 (a) Symmetric and staggered ribs; (b) model of rib shapes.

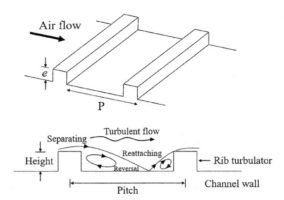

FIGURE 16.4 Schematic of flow separation-reattachment and rib orientations in heat transfer coefficient enhancement.

in tubes with repeated rib roughness by Webb et al. in 1971 [7], for flow between parallel plates by Han et al. in 1978 [8], for flow in tubes by Gee and Webb in 1980 [9], and for flow in annuli by Dalle Donne and Meyer in 1977 [10].

16.1.3 RECTANGULAR CHANNELS WITH REPEATED RIB ROUGHNESS

In the gas turbine blade cooling area, Han et al. in 1985 [11] investigated turbulent flow heat transfer in a square channel with repeated ribs cast only on two opposite walls. They found the square duct with two opposite walls having angled ribs performed better than that with normal ribs. To simulate the geometry of cooling passages in advanced gas turbine engines, Han and Park in 1988 [12] investigated turbulent flow heat transfer in the rectangular channels with five aspect ratios (AR = 4:1, 2:1, 1:1, 1:2, 1:4) having two opposite walls cast with repeated ribs, as sketch in Figure 16.5, which are closely modeled the realistic turbine blade cooling designs. They further confirmed the rectangular channel with two opposite walls having angled ribs performed better than that with normal ribs.

Fully developed turbulent heat transfer and friction in tubes, annulus, between parallel plates, or in rectangular channels with repeated-rib roughness have been studied extensively. Based on those previous studies, the effects of rib height-to-equivalent diameter ratio e/D, rib pitch-to-height ratio P/e, rib shape (rib cross section), and rib angle of attack α on the turbulent flow heat transfer coefficient and friction factor over a wide range of Reynolds number are well established. With repeated-rib roughness, for a given flow attack angle, rib shape and pitch-to-height ratio, tests with a different height-to-hydraulic diameter ratio e/D represent geometrically similar roughness. However, when the values of P/e, flow attack angle α or rib shape ϕ are varied, the surfaces are not geometrically similar. Surfaces which are not geometrically similar required modifications to the roughness function and heat transfer function found by similarity considerations. The friction and heat transfer similarity laws were extended to correlate the friction and heat transfer data for turbulent flow in square channels by Han et al. in 1985 [11], for flow in rectangular channels by Han and Park in 1988 [12] and by Han in 1984 and 1988 [13,14].

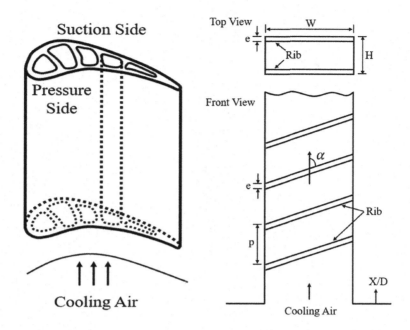

FIGURE 16.5   (a) Sketch of an internally cooled turbine airfoil; (b) rectangular channel with a pair of opposite ribbed walls.

16.2   FRICTION AND HEAT TRANSFER SIMILARITY LAWS FOR FLOW IN CIRCULAR TUBES

16.2.1   FLOW IN CIRCULAR TUBES WITH SAND GRAIN ROUGHNESS

For the results of roughened surfaces to be most useful, general correlations are necessary for both the friction factor and the heat-transfer coefficient which cover a wide range of parameters. In fully developed turbulent flow, theoretical approaches to the problem of heat and momentum transfer in smooth and rough tubes have been available for many years. These approaches are based on similarity considerations. Two surfaces are said to have geometrically similar roughness if the geometry of their roughness is the same in all aspects except for a scale factor. For example, sand grain roughness as shown in Figure 16.1a is a geometrically similar roughness.

16.2.1.1   Friction Similarity Law

Refer to Section 10.4.1, the friction factor for turbulent flow in a smooth tube can be predicted at a given Reynolds number. Perform average velocity across the tube from the law of wall for velocity profile (refer to Section 10.2), then the friction factor can be obtained. The procedure is outlined below.

For tube flow with smooth surface, the law of wall for velocity profile is:

$$u^+ = 2.5 \ln y^+ + 5.5 \qquad (16.1)$$

$$\frac{u}{u^*} = 2.5\ln\left(\frac{yu^*}{\nu}\right) + 5.5$$

$$\frac{\overline{u}}{u^*} = \frac{1}{\pi R^2}\int_0^R\left[2.5\ln\left(y\cdot\frac{u^*}{\nu}\right) + 5.5\right]2\pi r\,dr$$

$$= \frac{2.5}{\frac{1}{2}R^2}\int_0^R\ln\left(y\cdot\frac{u^*}{\nu}\right)r\,dr + 5.5$$

Let $y = R - r$, $dy = -dr$, as defined in Figure 10.7, and RHS first integration becomes

$$\int_0^R\ln\left(y\cdot\frac{u^*}{\nu}\right)r\,dr$$

$$= \int_0^R\ln\left(y\cdot\frac{u^*}{\nu}\right)(R-y)\,(-dy)$$

$$= \int_R^0 R\ln\left(y\cdot\frac{u^*}{\nu}\right)(-dy) + \int_R^0 y\ln\left(y\cdot\frac{u^*}{\nu}\right)dy$$

$$= -\frac{\nu}{u^*}\int_R^0 R\ln\left(y\cdot\frac{u^*}{\nu}\right)d\left(\frac{yu^*}{\nu}\right)$$

$$+ \frac{\nu^2}{u^{*2}}\int_R^0\left(\frac{yu^*}{\nu}\right)\ln\left(y\cdot\frac{u^*}{\nu}\right)d\left(\frac{yu^*}{\nu}\right)$$

Let $\frac{yu^*}{\nu} = x$, perform integration by part:

$$\int_0^x\ln x\,dx = \left[x\ln x - x\right]_0^x$$

$$\int_0^x x\ln x\,dx = \left[\frac{1}{2}x^2\ln x - \frac{1}{4}x^2\right]_0^x$$

The RHS terms become

$$= \left[-\frac{vR}{u^*}\left(\frac{yu^*}{v}\right)\ln\left(\frac{yu^*}{v}\right) + \frac{yu^*}{v} \right]_R^0 + \frac{v^2}{u^{*2}}\left[\frac{1}{2}\left(\frac{yu^*}{v}\right)^2 \ln\left(\frac{yu^*}{v}\right) - \frac{1}{4}\left(\frac{yu^*}{v}\right)^2 \right]_R^0$$

$$= \frac{vR}{u^*}\left[\left(R\frac{u^*}{v}\right)\ln\left(R\frac{u^*}{v}\right) - R\frac{u^*}{v} \right] + \frac{v^2}{u^{*2}}\left[-\frac{1}{2}\left(R\frac{u^*}{v}\right)^2 \ln\left(R\frac{u^*}{v}\right) + \frac{1}{4}\left(R\frac{u^*}{v}\right)^2 \right]$$

$$= R^2 \ln\left(R\frac{u^*}{v}\right) - R^2 - \frac{1}{2}R^2 \ln\left(R\frac{u^*}{v}\right) + \frac{1}{4}R^2$$

Therefore

$$\frac{\bar{u}}{u^*} = \frac{2.5}{\frac{1}{2}R^2}\left[\frac{1}{2}R^2 \ln\left(\frac{Ru^*}{v}\right) - \frac{3}{4}R^2 \right] + 5.5$$

$$= 2.5\ln\left(\frac{Ru^*}{v}\right) - 3.75 + 5.5$$

$$= 2.5\ln\left(\frac{Ru^*}{v}\right) + 1.75$$

where referring to the definition in Section 10.2,

$$\frac{\bar{u}}{u^*} = \frac{\bar{u}}{\sqrt{\dfrac{\tau_w}{\rho}}} = \frac{\bar{u}}{\sqrt{\dfrac{f\dfrac{1}{2}\rho\bar{u}^2}{\rho}}} = \frac{\bar{u}}{\bar{u}\sqrt{f/2}} = \sqrt{\frac{2}{f}} \qquad (16.2)$$

Thus

$$\sqrt{\frac{2}{f}} = 2.5\ln\left(\frac{D}{2}\frac{\bar{u}\sqrt{f/2}}{v}\right) + 1.75$$

$$\sqrt{\frac{2}{f}} = 2.5\ln\left(\frac{\mathrm{Re}_D}{2}\sqrt{f/2}\right) + 1.75 \qquad (16.3)$$

The above friction factor equation is the same as Karman-Prandtl friction equation for smooth tube (see Section 10.4.1).

For tube flow with sand grain roughness as defined in Figure 16.1a:

For sand grain roughness, the dimensionless velocity u^+ is expected to decrease by the roughness function C, this is because the friction factor increases with roughness as expected. The following similar law of wall for velocity profile can be assumed (see Section 10.4.1):

$$u^+ = 2.5\ln y^+ + 5.5 - C\left(\frac{eu^*}{v}\right) \tag{16.4}$$

Follow the above-mentioned integration procedures, the friction factor can be obtained

$$\sqrt{\frac{2}{f}} = 2.5\ln\left(\frac{\mathrm{Re}_D}{2}\sqrt{f/2}\right) + 1.75 - C\left(\frac{eu^*}{v}\right) \tag{16.5}$$

However, the term $C\left(\dfrac{eu^*}{v}\right)$ still needs to be determined. Equation (16.4) can be re-arranged as

$$u^+ = 2.5\ln\left(\frac{yu^*}{v}\right) + 5.5 - C\left(\frac{eu^*}{v}\right)$$

$$= 2.5\ln\left(\frac{y}{e}\frac{eu^*}{v}\right) + 5.5 - C\left(\frac{eu^*}{v}\right)$$

$$= 2.5\ln\left(\frac{y}{e}\right) + 2.5\ln\left(\frac{eu^*}{v}\right) + 5.5 - C\left(\frac{eu^*}{v}\right)$$

$$u^+ = 2.5\ln\left(\frac{y}{e}\right) + R\left(e^+\right) \tag{16.6}$$

where

$$R\left(e^+\right) = 2.5\ln\left(e^+\right) + 5.5 - C\left(e^+\right)$$

$$= \text{Roughness function, depending on } e^+$$

$$e^+ = \frac{eu^*}{v} = \text{Roughness Reynolds number, depending on roughness height } e$$

Follow the same above-mentioned integration procedure to obtain $\bar{u}$ and f for sand grain roughness. Start from Equation (16.6):

$$u^+ = 2.5\ln\left(\frac{y}{e}\right) + R(e^+)$$

$$= 2.5\ln\left(\frac{y}{R}\frac{R}{e}\right) + R(e^+)$$

$$\frac{u}{u^*} = 2.5\ln\left(\frac{y}{R}\right) + 2.5\ln\left(\frac{R}{e}\right) + R(e^+)$$

$$\frac{\bar{u}}{u^*} = \frac{1}{\pi R^2}\int_0^R\left[2.5\ln\left(\frac{y}{R}\right) + 2.5\ln\left(\frac{R}{e}\right) + R(e^+)\right]2\pi r\ dr$$

$$= \frac{2.5}{\frac{1}{2}R^2}\int_0^R\ln\left(\frac{y}{R}\right)r\ dr + 2.5\ln\left(\frac{R}{e}\right) + R(e^+)$$

where RHS first term can be obtained as previous procedure with integration by part:

$$\int_0^R\ln\left(\frac{y}{R}\right)r\ dr$$

$$= \int_0^R(\ln y - \ln R)r\ dr$$

$$= \int_0^R r\ln y\,dr - \int_0^R r\ln R\,dr$$

$$= \int_R^0(R-y)\ln y(-dy) - \ln R\left[\frac{1}{2}r^2\right]_0^R$$

$$= -R\int_R^0\ln y\,dy + \int_R^0 y\ln y\,dy - \frac{1}{2}R^2\ln R$$

$$= -R\left[y\ln y - y\right]_R^0 + \left[\frac{1}{2}y^2\ln y - \frac{1}{4}y^2\right]_R^0 - \frac{1}{2}R^2\ln R$$

$$= -R[0 - R\ln R + R] + \left[0 - \frac{1}{2}R^2\ln R + \frac{1}{4}R^2\right] - \frac{1}{2}R^2\ln R$$

$$= R^2\ln R - R^2 - \frac{1}{2}R^2\ln R + \frac{1}{4}R^2 - \frac{1}{2}R^2\ln R$$

$$= -\frac{3}{4}R^2$$

Insert back into the above integration, therefore,

$$\frac{\overline{u}}{u^*} = \frac{2.5}{\frac{1}{2}R^2}\left[-\frac{3}{4}R^2\right] + 2.5\ln\left(\frac{R}{e}\right) + R\left(e^+\right)$$

(16.7)

$$= -3.75 + 2.5\ln\left(\frac{R}{e}\right) + R\left(e^+\right)$$

As noted, $\dfrac{\overline{u}}{u^*} = \sqrt{\dfrac{2}{f}}$

Thus, Friction Similarity Law for sand grain roughness is

$$\sqrt{\frac{2}{f}} = -3.75 + 2.5\ln\left(\frac{R}{e}\right) + R\left(e^+\right)$$

$$R\left(e^+\right) = \sqrt{\frac{2}{f}} + 2.5\ln\left(\frac{e}{R}\right) + 3.75$$

$$R\left(e^+\right) = \sqrt{\frac{2}{f}} + 2.5\ln\left(\frac{2e}{D}\right) + 3.75$$

(16.8)

$$e^+ = \frac{eu^*}{v} = \frac{e}{D}\mathrm{Re}_D\sqrt{\frac{f}{2}}$$

(16.9)

The above equation is the so-called friction similarity law. The roughness function $R\left(e^+\right)$ is a general function determined empirically for each type of geometrically similar roughness. The roughness function for Nikuradse's [2] close-packed sand-grain roughness has the constant value 8.48 when e^+ is greater than 70; i.e., in completely rough regime, $R\left(e^+\right)$ does not depend on e^+. However, it is expected to be different for different geometrical roughness configurations.

$$R\left(e^+\right) = 8.48$$

(16.10)

$$\text{for } e^+ = \frac{e}{D}\mathrm{Re}_D\sqrt{\frac{f}{2}} > 70 = \text{completely rough region}$$

Since $R\left(e^+\right) = 8.48$, f can be predicted from the above friction similarity law equation (16.8) for a given e/D. For completely rough region, e^+ is independent of Re_D. For example, if $e/D = 0.01$, f can be calculated from Equations (16.8) and (16.10), $f = 0.0095$.

16.2.1.2 Heat Transfer Similarity Law

Based on heat and momentum transfer analogy, refer to Section 10.3, the law of wall for temperature profile in circular tube was obtained. The procedure is outlined below:

For smooth tube, referring to Chapter 10, the law of wall for velocity profile from Prandtl mixing length theory is shown below

$$\frac{\tau_w}{\rho}\left(1 - \frac{y}{R}\right) = (\nu + \varepsilon_m)\frac{du}{dy}$$

$$u^+ = 2.5\ln y^+ + 5.5 \tag{16.1}$$

Based on heat and momentum transfer analogy, the law of wall for temperature profile can be shown as

$$q_w''\left(1 - \frac{y}{R}\right) = -(\alpha + \varepsilon_H)\frac{dT}{dy}$$

$$\frac{\varepsilon_m}{\nu} = \frac{\varepsilon_H}{\nu}$$

$$T^+ = \int_0^{y^+} \frac{1 - \left(y^+/R^+\right)}{\dfrac{1}{Pr} + \dfrac{\varepsilon_H}{\nu}}\, dy^+$$

$$T^+ = 2.5\ln y^+ + T^+\left(\text{at } y^+ = 30\right) - 2.5\ln(30) \tag{16.11}$$

For tube flow with sand grain roughness, the above analysis has shown the law of wall similarity for velocity as

$$u^+ = 2.5\ln\left(\frac{yu^*}{\nu}\right) + 5.5 - C\left(\frac{eu^*}{\nu}\right)$$

$$= 2.5\ln\left(\frac{y}{e}\right) + 2.5\ln\left(\frac{eu^*}{\nu}\right) + 5.5 - C\left(\frac{eu^*}{\nu}\right)$$

$$= 2.5\ln\left(\frac{y}{e}\right) + R\left(e^+\right) \tag{16.6}$$

The law of wall similarity for temperature can also be written as

$$T^+ = 2.5\ln\left(\frac{yu^*}{v}\right) + T^+\left(at\ y^+ = 30\right) - 2.5\ln(30)$$

$$= 2.5\ln\left(\frac{y}{e}\right) + 2.5\ln\left(\frac{eu^*}{v}\right) + T^+\left(at\ y^+ = 30\right) - 2.5\ln(30) - C\left(\frac{eu^*}{v}\right)$$

$$= 2.5\ln\left(\frac{y}{e}\right) + G\left(e^+, Pr\right) \tag{16.12}$$

where

$$T^+\left(at\ y^+ = 30\right) = T^+\left(at\ y^+ = 5\right) + 5\ln\left(1 + \frac{Pr\cdot y^+}{5} - Pr\right)$$

$$= 5Pr + 5\ln(1 + 6Pr - Pr)$$

$$= 5Pr + 5\ln(1 + 5Pr)$$

$$= \text{function of } Pr$$

$$G\left(e^+, Pr\right) = 2.5\ln\left(\frac{eu^*}{v}\right) + T^+\left(at\ y^+ = 30\right) - 2.5\ln(30) - C\left(\frac{eu^*}{v}\right)$$

$$= \text{Heat transfer function, depending on } e^+ \text{ and } Pr$$

Follow the same procedures as outlined above for obtaining average velocity, the average temperature across the tube can be determined by integrating Equation (16.12) as following,

$$\overline{T}^+ = \frac{1}{\pi R^2}\int_0^R \left[2.5\ln\left(\frac{y}{e}\right) + G\left(e^+, Pr\right)\right]2\pi r\ dr$$

$$= \frac{1}{\frac{1}{2}R^2}\int_0^R 2.5\ln\left(\frac{y}{e}\right)rdr + G\left(e^+, Pr\right)$$

$$= \frac{1}{\frac{1}{2}R^2}\int_0^R 2.5\left[\ln\left(\frac{y}{R}\right) + \ln\left(\frac{R}{e}\right)\right]rdr + G\left(e^+, Pr\right)$$

$$= \frac{2.5}{\frac{1}{2}R^2}\int_0^R \ln\left(\frac{y}{R}\right)rdr + 2.5\ln\left(\frac{R}{e}\right) + G\left(e^+, Pr\right)$$

Note that, from above-mentioned integration by part,

$$\int_0^R \ln\left(\frac{y}{R}\right) r\, dr = -\frac{3}{4} R^2$$

Thus, the dimensionless average temperature across the tube becomes

$$\bar{T}^+ = \frac{2.5}{\frac{1}{2}R^2}\left(-\frac{3}{4}R^2\right) + 2.5\ln\left(\frac{R}{e}\right) + G\left(e^+, \mathrm{Pr}\right)$$

$$= -3.75 + 2.5\ln\left(\frac{R}{e}\right) + G\left(e^+, \mathrm{Pr}\right) \tag{16.13}$$

From Section 10.3, the dimensionless temperature is defined as

$$T^+ = \frac{T_w - T}{\dfrac{q_w''}{\rho C_p u^*}}$$

Therefore, the average temperature across the tube can also be obtained from the above dimensionless temperature by integration as follows:

$$\bar{T}^+ = \frac{1}{\pi R^2} \int_0^R \frac{T_w - T}{\dfrac{q_w''}{\rho C_p u^*}} \cdot 2\pi r\, dr$$

$$= \frac{1}{\dfrac{q_w''}{\rho C_p u^*}} \left(\frac{1}{\pi R^2} \int_0^R T_w \cdot 2\pi r\, dr - \frac{1}{\pi R^2} \int_0^R T \cdot 2\pi r\, dr \right)$$

$$= \frac{1}{\dfrac{q_w''}{\rho C_p u^*}} (T_w - T_b)$$

$$= \frac{\rho C_p (T_w - T_b) u^*}{q_w''}$$

where

$$T_b = \text{bulk mean average temperature}$$

$$= \frac{1}{\pi R^2} \int_0^R T \cdot 2\pi r\, dr$$

And from definition of Stanton number St,

$$St = \frac{q_w''}{\rho C_p \bar{u}(T_w - T_b)}$$

$$= \frac{q_w''}{\rho C_p (T_w - T_b)} \frac{1}{\bar{u}}$$

$$= \frac{u^*}{\bar{T}^+} \cdot \frac{1}{\bar{u}}$$

$$= \frac{\bar{u}\sqrt{f/2}}{\bar{T}^+} \cdot \frac{1}{\bar{u}}$$

Thus, the average dimensionless temperature across the tube can be related to Stanton number and friction factor as

$$\bar{T}^+ = \frac{\sqrt{f/2}}{St} \tag{16.14}$$

Combing with $\bar{u}^+$ from Equation (16.7),

$$\bar{u}^+ = -3.75 + 2.5\ln\left(\frac{R}{e}\right) + R(e^+) \tag{16.7}$$

Equation (16.13) becomes

$$\bar{T}^+ = \bar{u}^+ - R(e^+) + G(e^+, \mathrm{Pr})$$

Note $\bar{u}^+ = \dfrac{\bar{u}}{u^*} = \sqrt{2/f}$ and combine with Equation (16.14); therefore, Equation (16.13) becomes

$$\frac{\sqrt{f/2}}{St} = \sqrt{2/f} - R(e^+) + G(e^+, \mathrm{Pr})$$

or

$$G(e^+, \mathrm{Pr}) = R(e^+) + \frac{\sqrt{f/2}}{St} - \sqrt{2/f}$$

$$= R(e^+) + \frac{\dfrac{f}{2St} - 1}{\sqrt{f/2}} \tag{16.15}$$

The above equation is called the heat transfer similarity law. The $G(e^+, \text{Pr})$ is called heat transfer function; it depends on roughness Reynolds number e^+ and Prandtl number Pr.

Heat transfer function $G(e^+, \text{Pr})$ is related to roughness function $R(e^+)$, Stanton number St is related to friction factor f. From Dipprey and Sabersky [3] for close-packed sand-grain roughness, experimentally determined $G(e^+, \text{Pr})$ as

$$G(e^+, \text{Pr}) = 5.19(e^+)^{0.2} \text{Pr}^{0.44} \tag{16.16}$$

For $0.0024 < e/D < 0.049$, $1.2 < \text{Pr} < 5.94$

For example, if $e^+ = 100$, $e/D = 0.01$, $R = 8.48$ from Equation (16.10), $f = 0.0095$ from Equation (16.8) and Equation (16.10), $\text{Pr} = 0.72$, $G = 11.14$ from Equation (16.16), Stanton number can be calculated from Equation (16.15) as St $= 0.00401$.

Summary for tube flow with close-packed sand grain roughness

$$R(e^+) = \sqrt{\frac{2}{f}} + 2.5\ln\left(\frac{2e}{D}\right) + 3.75 \tag{16.8}$$

$$e^+ = \frac{e}{D}\text{Re}_D \sqrt{f/2} \tag{16.9}$$

$$R(e^+) = 8.48, \quad \text{for } e^+ > 70 \text{ completely rough region} \tag{16.10}$$

$$G(e^+, \text{Pr}) = R(e^+) + \frac{\dfrac{f}{2\text{St}} - 1}{\sqrt{f/2}} \tag{16.15}$$

$$G(e^+, \text{Pr}) = 5.19(e^+)^{0.2} \text{Pr}^{0.44} \tag{16.16}$$

For a given Re_D and e/D, f can be obtained from Equations (16.10) and (16.8), then e^+ can be calculated from Equation (16.9), then $G(e^+, \text{Pr})$ can be obtained from Equation (16.16) for a given Prandtl number, and finally St can be estimated from Equation (16.15). For example, if $\text{Re}_D = 100,000$, $e/D = 0.02$, you can calculate f and e^+. If $\text{Pr} = 0.72$, you can calculate $G(e^+, \text{Pr})$ and then calculate St.

16.2.2 FLOW IN CIRCULAR TUBES WITH REPEATED RIB ROUGHNESS

For tube flows with the repeated-rib roughness with a different height-to-hydraulic diameter ratio represent geometrically similar roughness, as shown in Figure 16.2. However, when the values of P/e are varied, the surfaces are not geometrically similar. Surfaces which are not geometrically similar required modifications to the roughness and heat transfer functions found by similarity considerations. Based on the friction similarity law, Webb et al. [7] found a successful friction correlation for turbulent tube flow with repeated-rib roughness by accounting the geometrically non-similar roughness parameter, the p/e ratio, as shown in Figure 16.6. That was

$$R\left(e^{+}\right) = 0.95\left(P/e\right)^{0.53}, \quad \text{for } e^{+} > 35 \tag{16.17}$$

The friction factor can be found by combining Equations (16.8) and (16.17). Based on the heat transfer similarity law, Webb et al. [7] found a successful heat transfer similarity relationship for turbulent flow with repeated-rib roughness by accounting the geometrically nonsimilar geometry, the p/e ratio, as shown in Figure 16.7. That was

$$G\left(e^{+}, \Pr\right) = 4.5\left(e^{+}\right)^{0.28} \Pr^{0.57} \tag{16.18}$$

For $10 \leq P/e \leq 40, \quad 0.01 \leq e/D \leq 0.04, \quad 0.72 \leq \Pr \leq 37.6$

Webb et al.'s experimental data indicated that the effect of pitch to height ratio on $G\left(e^{+}, \Pr\right)$ is small for P/e ratios between 10 and 40, so P/e does not appear explicitly in the $G\left(e^{+}, \Pr\right)$ equation. The Stanton number can be found by combining Equations (16.8). (16.9), (16.15), (16.17), and (16.18).

For a given Re_{D}, p/e, and $e/D, f$ can be obtained from Equations (16.8) and (16.17), then e^{+} can be calculated from Equation (16.9), then $G\left(e^{+}, \Pr\right)$ can be obtained from Equation (16.18) for a given Prandtl number, and finally St can be estimated from Equation (16.15). For example, if $\mathrm{Re}_{D} = 100,000$, $e/D = 0.02$, $p/e = 10$, you can calculate f and e^{+}. If $\Pr = 0.72$, you can calculate $G\left(e^{+}, \Pr\right)$ and then calculate St.

Gee and Webb [9] reported the heat transfer and friction characteristics for turbulent air flow in circular tubes with three helix angles ($\alpha = 30°$, $49°$ and $70°$) all having a rib pitch-to-height ratio $P/e = 15$ and height-to-diameter ratio $e/D = 0.01$. Both Stanton number and friction factor decrease with a decrease in the helix angle. But the friction factor decreases faster than the Stanton number. The data are correlated in a form to permit performance prediction with any relative roughness size (e/D).

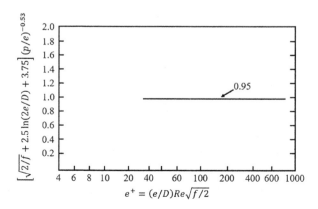

FIGURE 16.6   Final friction correlation for repeated-rib tubes.

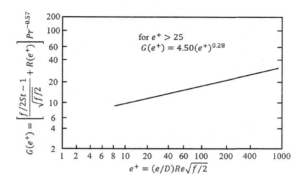

FIGURE 16.7 Final beat transfer correlation including the Prandtl number dependency.

From the previous derived friction similarity law by considering the geometrically nonsimilar roughness parameters of helix angle (α), roughness function $R\left(e^{+}\right)$ shown in Equation (16.8) is replaced by

$$R\left(e^{+}\right)\left[\left(\alpha/50\right)^{0.16}\right] \tag{16.19}$$

For $e/D = 0.01$, $P/e = 15$, $Re_D = 6{,}000$ to $65{,}000$, $e^{+}= 6$ to 50
Pr $= 0.72$, $\alpha = 30°$ to $70°$

From the previous derived heat transfer similarity law by accounting the geometrically nonsimilar roughness parameters of helix angle (α), heat transfer function $G\left(e^{+},\mathrm{Pr}\right)$ shown in Equation (16.15) is replaced by

$$G\left(e^{+},\mathrm{Pr}\right)\left[\alpha/50°\right]^{j} \tag{16.20}$$

where $j = 0.37$ for $\alpha < 50°$, $j = -0.16$ for $\alpha > 50°$

From thermal performance comparison, the preferred helix angle α is approximately $49°$, and the best roughness Reynolds number $e^{+} = 20$. They expected Equations (16.19) and (16.20) could be applied for other e/D ratios, based on the

similarity consideration as $e^{+} = \dfrac{e}{D}\,\mathrm{Re}_D\,\sqrt{f/2}$ shown in Equation (16.9).

16.2.3 Flow between Parallel Plates with Repeated Rib Roughness

With repeated-rib roughness, for a given flow attack angle, rib shape and pitch-to-height ratio, tests with a different height-to-hydraulic diameter ratio represent geometrically similar roughness. However, when the values of P/e, flow attack angle, or rib cross-section are varied, the surfaces are not geometrically similar. Surfaces which are not geometrically similar required modifications to the roughness and heat transfer functions found by similarity considerations.

Han et al. [8] extended the similarity concept to correlate the friction data for turbulent flow between parallel plates with repeated-rib roughness by considering the geometrically nonsimilar roughness parameters of p/e, rib shape (ϕ), and flow attack angle (α), as sketched in Figure 16.3. From the previous derived friction similarity law, the correlation of roughness function versus roughness Reynolds number is shown in Figure 16.8 for flow between parallel plates with geometrically nonsimilar repeated rib roughness parameters.

$$R(e^+) = 4.9(e^+/35)^m / \left[(\phi/90°)^{0.35} (10/P/e)^n (\alpha/45)^{0.57} \right] \tag{16.21}$$

For $5 \le P/e \le 20$, $0.032 \le e/D \le 0.102$, $\mathrm{Re}_D = 3{,}000$ to $30{,}000$,

$\mathrm{Pr} = 0.72$, $\phi = 40°$ to $90°$, $\alpha = 20°$ to $90°$

where

$$e^+ = \frac{e}{D_h}\mathrm{Re}_D \sqrt{f/2}$$

$$\mathrm{Re}_D = \frac{\bar{u}D_h}{\nu}$$

D_h = hydraulic diameter
$m = -0.4$ if $e^+ < 35$
$m = 0$ if $e^+ \ge 35$
$n = -0.13$ if $P/e < 10$
$n = 0.53\,(\alpha/90)^{0.71}$ if $P/e \ge 10$

The friction factor can be found by combining Equations (16.8) and (16.21). Equation (16.21) is the general correlation for the friction factor measured for 11 different dissimilar rib-type geometries and 16 different geometries in all. When the roughness Reynolds number is 35 or larger the flow is in the completely rough regime. The form of the equation changes at P/e of 10. This roughly corresponds to the rib spacing necessary for the boundary layer to re-attach to the plate and regrow along the plate.

Based the heat transfer similarity law, Han et al. [8] extended to correlate the heat transfer data for turbulent air flow between parallel plates with repeated-rib roughness

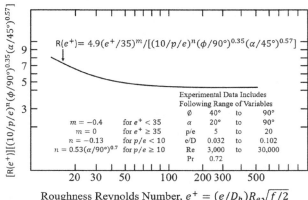

FIGURE 16.8 Final friction correlation.

by accounting the geometrically nonsimilar roughness parameters of P/e, rib shape, and flow attack angle. The final heat-transfer correlation shown in Figure 16.9 is where

$$G(e^+) = 10(e^+/35)^i / [\alpha/45°]^j \tag{16.22}$$

where
$i = 0$ if $e^+ < 35$
$i = 0.28$ if $e^+ \geq 35$
$j = 0.5$ if $\alpha < 45°$
$j = -0.45$ if $\alpha \geq 45°$

For the correlation St is based on the total heat transfer area. Note that the experiments have shown that $G(e^+)$ is not a function of ϕ or P/e. The Stanton number can be found by combining Equations (16.8), (16.9), (16.15), (16.21), and (16.22).

The above correlation is applied for air flow only. To approximately account the Prandtl number effect, $G(e^+)$ in Equation (16.20) may be replaced by $G(e^+) (0.72/Pr)^{0.57}$. This is taken from Webb's experimental data where the Prandtl number was varied from 0.72 to 37.6.

For a given Re_D, p/e, e/D, ϕ, and α, f can be obtained from Equations (16.8) and (16.21), then e^+ can be calculated from Equation (16.9), then $G(e^+, Pr)$ can be obtained from Equation (16.22) for a given Prandtl number, and finally St can be estimated from Equation (16.15). For example, if $Re_D = 50,000$, $e/D = 0.05$, $p/e = 10$, $\phi = 90°$, and $\alpha = 45°$, you can calculate f and e^+. If $Pr = 0.72$, you can calculate $G(e^+, Pr)$ and then calculate St.

Most investigations based the heat-transfer coefficient on the projected surface area rather than the total area. When St is based on the projected area, Equation (16.22) becomes

$$G(e^+) = 8(e^+/35)^i / [\alpha/45°]^j \tag{16.23}$$

where i and j have the same values as in Equation (16.22).

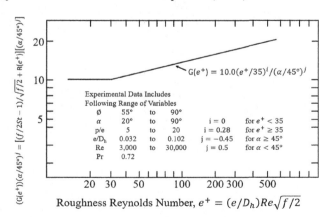

FIGURE 16.9 Final heat-transfer correlation with Stanton number based on total area.

16.3 FRICTION AND HEAT TRANSFER SIMILARITY LAWS FOR FLOW IN RECTANGULAR CHANNELS

16.3.1 FLOW IN A RECTANGULAR CHANNEL WITH FOUR-SIDED, RIB-ROUGHENED WALLS

16.3.1.1 Friction Similarity Law

For rectangular channel flow with smooth surface, assume the law of the wall for velocity profile in tube flows can also be applied to rectangular channel flows as

$$u^+ = 2.5 \ln y^+ + 5.5$$

Follow the same procedure outlined before for circular tube, the average velocity across the rectangular channel as defined in Figure 16.10 can be determined by integration as follows:

$$
\frac{\bar{u}}{u^*} = \frac{1}{A_c} \int \left[2.5 \ln \left(\frac{y u^*}{v} \right) + 5.5 \right] dA_c
$$

$$
= \frac{1}{\frac{1}{2} a \cdot b} \int_0^{\frac{a}{2}} \left[2.5 \ln \left(\frac{y u^*}{v} \right) + 5.5 \right] \cdot b \cdot dy
$$

$$
= \frac{1}{\frac{1}{2} a} \int_0^{\frac{a}{2}} 2.5 \ln \left(\frac{y u^*}{v} \right) \cdot \frac{v}{u^*} \cdot d \left(y \frac{u^*}{v} \right) + 5.5
$$

$$
= \frac{1}{\frac{1}{2} a} \cdot 2.5 \cdot \frac{v}{u^*} \cdot \left[\frac{y u^*}{v} \cdot \ln \left(\frac{y u^*}{v} \right) - \frac{y u^*}{v} \right]_0^{\frac{a}{2}} + 5.5
$$

$$
= \frac{1}{\frac{1}{2} a} \cdot 2.5 \cdot \frac{v}{u^*} \cdot \left(\frac{\frac{a}{2} u^*}{v} \cdot \ln \left(\frac{\frac{a}{2} u^*}{v} \right) - \frac{\frac{a}{2} u^*}{v} \right) + 5.5
$$

$$
= 2.5 \ln \left(\frac{\frac{a}{2} u^*}{v} \right) - 2.5 + 5.5
$$

Note $\dfrac{\bar{u}}{u^*} = \sqrt{2/f}$, thus

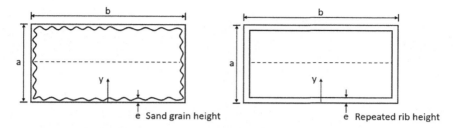

FIGURE 16.10 Rectangular channel with (a) sand grain roughness; (b) repeated rib roughness.

$$\sqrt{2/f} = 2.5\ln\left(\frac{\frac{a}{2}u^*}{v}\right) + 3$$

For the rectangular channel with channel height a (narrow side) and channel width b (wider side), Reynolds number is defined based on channel hydraulic diameter as

$$\mathrm{Re}_D = \frac{\bar{u}D_h}{v}$$

$$D_h = \frac{4ab}{2(a+b)} = \frac{2ab}{a+b}$$

$$\frac{a}{2} = D_h \cdot \frac{a+b}{4b} = \frac{D_h}{2} \cdot \frac{a+b}{2b}$$

Therefore, the above equation can be re-arranged as

$$\sqrt{2/f} = 2.5\ln\left(\frac{D_h}{2}\frac{u^*}{v} \cdot \frac{a+b}{2b}\right) + 3$$

$$= 2.5\ln\left(\frac{D_h}{2}\frac{\bar{u}\sqrt{f/2}}{v} \cdot \frac{a+b}{2b}\right) + 3$$

$$= 2.5\ln\left(\frac{\mathrm{Re}_D}{2}\sqrt{\frac{f}{2}}\right) + 3 + 2.5\ln\left(\frac{a+b}{2b}\right) \qquad (16.24)$$

For rectangular channel with sand grain roughness as defined in Figure 16.10a, assume the law of wall for velocity profile is similar to tube flow and can be shown as

$$u^+ = 2.5\ln\left(\frac{yu^*}{v}\right) + 5.5 - C\left(\frac{eu^*}{v}\right)$$

$$= 2.5\ln\left(\frac{y}{e}\frac{eu^*}{v}\right) + 5.5 - C\left(\frac{eu^*}{v}\right)$$

$$= 2.5\ln\left(\frac{y}{e}\right) + 2.5\ln\left(\frac{eu^*}{v}\right) + 5.5 - C\left(\frac{eu^*}{v}\right)$$

$$= 2.5\ln\left(\frac{y}{e}\right) + R\left(e^+\right)$$

where

$$R\left(e^+\right) = 2.5\ln\left(\frac{eu^*}{v}\right) + 5.5 - C\left(\frac{eu^*}{v}\right)$$

Follow the same procedure as before, the average velocity across the rectangular channel with sand grain roughness can be obtained as following:

$$u^+ = 2.5\ln\left(\frac{y}{\frac{a}{2}}\frac{\frac{a}{2}}{e}\right) + R\left(e^+\right)$$

$$\frac{\bar{u}}{u^*} = \frac{1}{\frac{1}{2}ab}\int_0^{\frac{a}{2}}\left(2.5\ln y - 2.5\ln\left(\frac{a}{2}\right)\right)b\,dy + 2.5\ln\left(\frac{\frac{a}{2}}{e}\right) + R\left(e^+\right)$$

The first RHS term:

$$\frac{1}{\frac{1}{2}a}\left[2.5(y\ln y - y) - 2.5\ln\left(\frac{a}{2}\right)\right]_0^{\frac{a}{2}}$$

$$= \frac{1}{\frac{1}{2}a}\left[2.5\left(\frac{a}{2}\ln\frac{a}{2} - \frac{a}{2}\right) - 2.5\ln\left(\frac{a}{2}\right)\frac{a}{2}\right]$$

$$= -2.5$$

Thus

$$\frac{\bar{u}}{u^*} = \sqrt{\frac{2}{f}} = -2.5 + 2.5\ln\left(\frac{\frac{a}{2}}{e}\right) + R\left(e^+\right)$$

or

$$R(e^+) = \sqrt{\frac{2}{f}} + 2.5\ln\left(\frac{e}{\frac{1}{2}a}\right) + 2.5$$

where

$$\frac{e}{\frac{1}{2}a} = \frac{e}{D_h}\frac{4b}{(a+b)}$$

$$\ln\left(\frac{e}{\frac{1}{2}a}\right) = \ln\left(\frac{e}{D_h}\frac{4b}{(a+b)}\right) = \ln\left(\frac{e}{D_h}\right) + \ln\left(\frac{4b}{(a+b)}\right)$$

Therefore, friction similarity law for rectangular channels with sand grain roughness is

$$R(e^+) = \sqrt{\frac{2}{f}} + 2.5\ln\left(\frac{e}{D_h}\right) + 2.5\ln\left(\frac{4b}{(a+b)}\right) + 2.5$$

or

$$R(e^+) = \sqrt{\frac{2}{f}} + 2.5\ln\left(\frac{2e}{D_h}\right) + 2.5\ln\left(\frac{2b}{(a+b)}\right) + 2.5 \qquad (16.25)$$

where

$$e^+ = \frac{e}{D_h}\mathrm{Re}_D\sqrt{f/2} \qquad (16.26)$$

$$D_h = \frac{2ab}{(a+b)}$$

a = narrow-side length of rectangular channel
b = wider-side length of rectangular channel
$R(e^+)$ = roughness function to be determined experimentally for rectangular

channels with sand grain roughness $\dfrac{e}{D_h}$.

The above-derived friction similarity law in rectangular channels with four-sided sand grain roughness can be applied for rectangular channels with four-sided repeated rib roughness, but the experimentally determined value of $R(e^+)$ with repeated rib roughness $\left(p/e, \dfrac{e}{D_h} \right)$ is different from that with sand grain roughness $\left(\dfrac{e}{D_h} \right)$.

16.3.1.2 Heat Transfer Similarity Law

Based on the above-mentioned heat and momentum transfer analogy:

$$u^+ = 2.5\ln\left(\frac{y}{e}\right) + R(e^+)$$

$$T^+ = 2.5\ln\left(\frac{y}{e}\right) + G(e^+, \mathrm{Pr})$$

From above integration of $\bar{u}^+$, similarly, the integration of average dimensionless temperature across the rectangular channel $\bar{T}^+$ can be obtained as

$$\bar{T}^+ = \frac{1}{\frac{1}{2}ab} \int_0^{\frac{1}{2}a} \left[2.5\ln\left(\frac{y \quad \frac{1}{2}a}{\frac{1}{2}a \quad e}\right) + G(e^+, \mathrm{Pr}) \right] b \; dy$$

$$\bar{T}^+ = -2.5 + 2.5\ln\left(\frac{\frac{1}{2}a}{e}\right) + G(e^+, \mathrm{Pr})$$

Since

$$\bar{u}^+ = -2.5 + 2.5\ln\left(\frac{\frac{1}{2}a}{e}\right) + R(e^+)$$

Thus

$$\bar{T}^+ = \bar{u}^+ - R(e^+) + G(e^+, \mathrm{Pr})$$

Since $\bar{T}^+ = \dfrac{\sqrt{f/2}}{\mathrm{St}}$, and $\bar{u}^+ = \sqrt{2/f}$, therefore,

$$\frac{\sqrt{f/2}}{St} = \sqrt{2/f} - R\left(e^+\right) + G\left(e^+, \mathrm{Pr}\right)$$

or

$$G\left(e^+, \mathrm{Pr}\right) = R\left(e^+\right) + \frac{\sqrt{f/2}}{St} - \sqrt{2/f}$$

From friction similarity law for sand grain roughness

$$R\left(e^+\right) = \sqrt{2/f} + 2.5\ln\left(\frac{2e}{D_h}\right) + 2.5\ln\left(\frac{2b}{a+b}\right) + 2.5 \qquad (16.25)$$

Therefore, heat transfer similarity law for sand grain roughness becomes

$$G\left(e^+, \mathrm{Pr}\right) = \frac{\sqrt{f/2}}{St} + 2.5\ln\left(\frac{2e}{D_h}\right) + 2.5\ln\left(\frac{2b}{a+b}\right) + 2.5 \qquad (16.27)$$

where

$G\left(e^+, \mathrm{Pr}\right)$ = heat transfer function, to be experimentally determined for rectangular channels with four-sided sand grain roughness.

The above-derived heat transfer similarity law can be applied to rectangular channels with four-sided repeated rib roughness $\left(p/e, \frac{e}{D_h}\right)$, as sketched in Figure 16.10b, but $G\left(e^+, \mathrm{Pr}\right)$ value will be different.

A family of repeated-rib roughness is defined as geometrically similar if the rib pitch, rib shape, and rib angle of attack are not varied. In order to correlate the friction and heat transfer data for geometrically nonsimilar repeated-rib roughness, the above-derived roughness and heat transfer functions may be expressed by

$$R = \phi_1\left(e^+, P/e, \alpha, \text{shape}, \dots\right)$$

$$G = \phi_2\left(e^+, \mathrm{Pr}, P/e, \alpha, \text{shape}, \dots\right)$$

16.3.2 Flow in Rectangular Channels with Two Opposite-Sided, Rib-Roughened Walls

In some applications, such as gas turbine airfoil cooling design, periodic rib roughness is cast onto two opposite walls of internal cooling passages to enhance the heat transfer to the cooling air. The sketch of an internally cooled turbine airfoil is shown in Figure 16.5. The internal-cooling passages can be approximately modeled as rectangular channels with a pair of opposite rib-roughened walls, as shown in Figure 16.5. The channel aspect ratio (W/H) was varied from 1 to 2 and to 4. In each channel, the rib turbulators were placed on the opposite walls with all parallel with an angle-of-attack α of 90°, 60°, 45°, or 30°. The rib height-to-hydraulic diameter ratio (e/D) was varied from 0.047 to 0.078, and the rib pitch-to-height ratio (P/e) was varied from 10 to 20 for Reynolds numbers from 10,000 to 60,000.

Based on the correlation for four-sided smooth channels and the similarity law for four-sided ribbed channels, a general prediction model for the average friction factor and average heat transfer coefficients was developed for flow in rectangular channels with two smooth and two opposite rib-roughened walls. Refer to Han [13] for the details.

The friction factor for fully developed turbulent flow in a four-sided smooth channel, as shown in Figure 16.11, can be defined by

$$f_s = \frac{\tau_s}{\frac{1}{2}\rho \bar{V}_s^2}$$

Similarly, the friction factor for fully developed turbulent flow in a four-sided ribbed channel, as seen in Figure 16.11, may be defined as

$$f_r = \frac{\tau_r}{\frac{1}{2}\rho \bar{V}_r^2}$$

This analysis concerns fully developed turbulent flow in a rectangular channel with two opposite ribbed walls, as shown in Figure 16.11. The average friction factor for this case may be expressed by the following equation.

$$\bar{f} = \frac{\bar{\tau}}{\frac{1}{2}\rho \bar{V}^2}$$

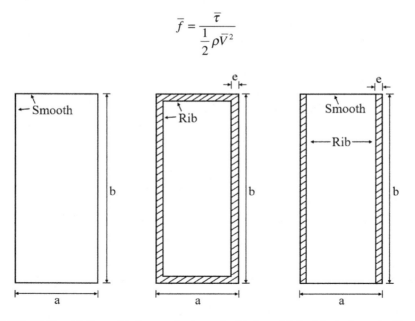

FIGURE 16.11 (a) Four-sided smooth channel; (b) four-sided ribbed channel; (c) two-smooth and two opposite-ribbed channel.

The average wall shear stress $\bar{\tau}$ can be related to the shear stresses produced by the two smooth and the two ribbed walls by

$$\bar{\tau}(2a+2b)L = (2a\tau_{2s} + 2b\tau_{2r})L$$

In this analysis, it is assumed that the total shear force in a channel with two smooth and two ribbed walls can be approximated by combining the shear force from two smooth walls in a four-sided smooth channel with the shear force from two ribbed walls in a four-sided ribbed channel, i.e.,

$$(2a\tau_{2s} + 2b\tau_{2r})L \cong (2a\tau_s + 2b\tau_r)L$$

It is noted that τ_{2s} is not necessarily equivalent to τ_s and that τ_{2r} is not necessarily equivalent to τ_r. Most likely τ_{2s} would be slightly higher than τ_s due to the adjacent ribbed walls, while τ_{2r} would be slightly lower than τ_r because of its adjacent smooth walls. However, since $\tau_{2s} << \tau_{2r}$ and a < b for most of the two-opposite rib-roughened channels, the above assumption should be reasonable. Combining above friction factor and shear stress equations and assuming that the fluid dynamic energy in each channel is about the same, i.e., $\frac{1}{2}\rho\bar{V}_s^2 = \frac{1}{2}\rho\bar{V}_r^2$, thus, the average friction factor $\bar{f}$ can be found from

$$\bar{f} = \frac{af_s + bf_r}{a+b} \tag{16.28}$$

Equation (16.26) gives the average friction factor as a weighted average of the four-sided smooth channel friction factor f_s and the four-sided ribbed channel friction factor f_r. These friction factors are weighted by the smooth wall width a and the ribbed wall height b, respectively. The friction factor in a four-sided smooth rectangular channel as well as in a four-sided rib-roughened rectangular channel can be referred to the above-mentioned friction similarity law.

The prediction method for heat transfer will be very similar to that for friction factor as described earlier if the heat and momentum transfer analogy is applied. For fully developed turbulent flow in four-sided smooth channels and four-sided ribbed channels, the corresponding Stanton numbers can be defined as

$$St_s = \frac{q_s''}{GC_p\left(\bar{T}_w - T_b\right)_s}$$

$$St_r = \frac{q_r''}{GC_p\left(\bar{T}_w - T_b\right)_r}$$

The average Stanton number for fully developed turbulent flow in a rectangular duct with two smooth and two opposite ribbed walls may be expressed as

$$\overline{St} = \frac{q''}{GC_p\left(\bar{T}_w - T_b\right)}$$

where

$$\overline{q}''(2a + 2b)L = \left(2aq''_{2s} + 2bq''_{2r}\right)L$$

If the analogous assumption for heat transfer follows the same form as was made for friction factor, then

$$\left(2aq''_{2s} + 2bq''_{2r}\right)L \cong \left(2aq''_s + 2bq''_r\right)L$$

By combining above equations and assuming that the temperature gradient in each case is about the same, i.e., $\left(\overline{T}_w - T_b\right) = \left(\overline{T}_w - T_b\right)_s = \left(\overline{T}_w - T_b\right)_r$, the average Stanton number can be written as

$$\overline{\mathrm{St}} = \frac{a\mathrm{St}_s + b\mathrm{St}_r}{a + b} \tag{16.29}$$

Equation (16.29) gives the average Stanton number as a weighted average of the four-sided smooth channel Stanton number St_s and the four-sided ribbed channel Stanton number St_r. These Stanton numbers are weighted by the smooth wall width a and the ribbed wall height b, respectively. The Stanton number in a four-sided smooth rectangular channel as well as in a four-sided rib-roughened rectangular channel can be referred to the above-mentioned heat transfer similarity law.

Summary: Friction and Heat Transfer Similarity Laws for Flow in a Rectangular Duct with Two Opposite-sided, Rib-roughened Walls

Based on the above-mentioned analysis, there appear to have two approaches to obtain the roughness functions R and G. The first approach is experimentally correlating R and G from Equations (16.25) and (16.27) by measuring f and St for flow in rectangular channels (given b/a) with four-sided ribbed walls (given e/D_h, P/e, …). The second approach is correlating R and G from Equations (16.28) and (16.29) by measuring $\overline{f}$ and St_r for flow in rectangular channels (given b/a) with a pair of opposite ribbed walls (given e/D_h, P/e, …).

Most of available experimental data are based on the second approach to correlate R and G by measuring $\overline{f}$ and St_r for flow in rectangular channels (given b/a) with two opposite rib-roughened walls (given e/D_h, P/e, …). This is because the results can be directly applied for the design of turbine airfoil cooling passages. From the above analysis, the average friction factor for flow in rectangular channels, by letting $a = H$ and $b = W$, can be rewritten as

$$\overline{f} = \left(Hf_s + Wf\right)/(H + W)$$

$$f = \overline{f} + (H/W)\left(\overline{f} - f_s\right)$$

By inserting the above expression of f into Equation (16.25), the roughness function R can be written as

$$R(e^{+}) = \left\{2/\left[\bar{f} + (H/W)(\bar{f} - f_{s})\right]\right\}^{1/2} + 2.5\ln\left\{(2e/D)[2W/(W + H)]\right\} + 2.5$$

$$(16.30)$$

where f_s can be approximately calculated from the Blasius equation for smooth circular tubes as

$$f_s = f(FD) = 0.046\mathrm{Re}^{-0.2} \qquad (16.31)$$

Equation (16.30) implies that the friction roughness function R can be experimentally determined (or correlated) by measuring the average friction factor $(\bar{f})$ for fully developed turbulent flow in rectangular channels (given W/H) with two opposite, geometrically similar, ribbed walls (given e/D), and by combining with the f_s calculated from Equation (16.31). Refer to Han [14] for the details.

The correlation of the friction roughness function R shown in Figure 16.12 can be obtained as

$$R = 3.2\left[(P/e)/10\right]^{0.35} \qquad (16.32)$$

For $e^{+} \geq 50,\ 10 \leq P/e \leq 20,\ 0.021 \leq e/D \leq 0.078,$
　$1 \leq H \leq 4,\ \ \mathrm{Re}_D = 8{,}000$ to $80{,}000,\ \ \mathrm{Pr} = 0.72$

Note that R in Equation (16.32) is independent of e^{+}. This implies that the average friction factor is almost independent of Reynolds number (i.e., in the fully rough

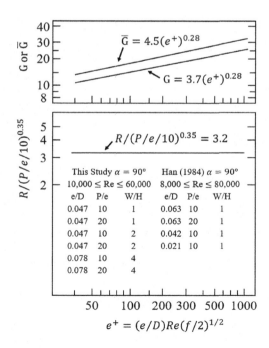

FIGURE 16.12　Friction and heat transfer correlations.

region with $e^+ \geq 50$). After R is determined experimentally from Equation (16.32), the average friction factor $\left(\overline{f}\right)$ can be predicted by combining Equations (16.30), (16.31), and (16.32) for a given rib geometry (e/D, P/e), channel aspect ratio (W/H), and Reynolds number Re. And then the friction factor with four-sided ribbed walls f can also be calculated from $\overline{f}$ and f_s.

It is assumed that the ribbed-side-wall Stanton number has a constant value for either two opposite-sided or four-sided ribbed channels (i.e., St = St$_r$). Similarly, combining with Equation (16.25), the heat transfer function G shown in Equation (16.27) can then be determined by

$$G\left(e^+,\mathrm{Pr}\right) = R\left(e^+\right) + \left\{\left[\overline{f} + (H+W)\left(\overline{f} - f_s\right)\right] / (2\mathrm{St}_r) - 1\right\} /$$

$$\left\{\left[\overline{f} + (H+W)\left(\overline{f} - f_s\right)\right] / 2\right\}^{1/2} \tag{16.33}$$

According to the heat transfer similarity law derived in Equation (16.33), the measured Stanton number on the ribbed-side-wall St$_r$ (assuming St$_r$ = St), the average friction factor $\overline{f}$, the channel aspect ratio W/H, and R could be experimentally correlated with the heat transfer roughness function $G\left(e^+,\mathrm{Pr}\right)$. The effect of the P/e ratio on the $G\left(e^+,\mathrm{Pr}\right)$ is negligible. A plot of G versus e^+ is also shown in Figure 16.12 as

$$G = 3.7\left(e^+\right)^{0.28} \quad \text{for} \quad e^+ \geq 50 \tag{16.34}$$

After G is correlated experimentally from Equation (16.34), the ribbed-side-wall Stanton number$\left(\mathrm{St}_r\right)$ can be predicted by combining Equations (16.30), (16.31), (16.32), (16.33), and (16.34) for a given e/D, P/e, W/H, and Re.

Since both the smooth-side-wall Stanton number $\left(\mathrm{St}_s\right)$ and the ribbed-side-wall Stanton number $\left(\mathrm{St}_r\right)$ were measured, the average Stanton number $\left(\overline{\mathrm{St}}\right)$ between the smooth side and the ribbed side walls was obtained. Assuming that Equation (16.30) can be used to correlate the average heat transfer data by replacing G and St$_r$ by $\overline{G}$ and $\overline{\mathrm{St}}$, the correlation of the $\overline{G}$ shown in Figure 16.12 can be expressed by

$$\overline{G} = 4.5\left(e^+\right)^{0.28} \quad \text{for} \quad e^+ \geq 50 \tag{16.35}$$

If G, $\overline{G}$, R, and $\overline{f}$ are known for a given e/D, P/e, W/H, and Re, the ribbed-side-wall Stanton number St$_r$ and the average Stanton number $\overline{\mathrm{St}}$ can be predicted, respectively, from Equation (16.33), and from the same Equation (16.33) by replacing G and St$_r$ by $\overline{G}$ and $\overline{\mathrm{St}}$. After determining St$_r$ and $\overline{\mathrm{St}}$ from above correlations, the smooth-side-wall Stanton number St$_s$ can be found by

$$\mathrm{St}_s = \overline{\mathrm{St}} + (W/H)\left(\overline{\mathrm{St}} - \mathrm{St}_r\right) \tag{16.36}$$

The roughness functions, $R\left(e^+\right)$ in Equation (16.28) and heat transfer function $G\left(e^+,\mathrm{Pr}\right)$ in Equation (16.33), can also be extended to correlate the effects of geometrically nonsimilar parameters such as rib spacing P/e, rib angle α, rib shape. Refer to Han and Park [12] for the details. The correlation of the friction roughness function R including the rib angle α effect shown in Figure 16.13 can be written as

$$R / \left[(P/e/10)^{0.35} (W/H)^m \right] = 12.31 - 27.07(\alpha/90°) + 17.86(\alpha/90°)^2 \qquad (16.37)$$

For $e^+ \geq 50$, $10 \leq P/e \leq 20$, $0.047 \leq e/D \leq 0.078$, $30° \leq \alpha \leq 90°$,

   $1 \leq H \leq 4$, $Re_D = 10{,}000$ to $60{,}000$, $Pr = 0.72$
   where

   $m = 0$,   for $\alpha = 90°$

   $m = 0.35$,   for $\alpha \langle 90°$,   if $W/H \rangle 2$,   let $W/H = 2$

Note that R in Equation (16.37) is independent of e^+. This implies that the average friction factor is almost independent of Reynolds number (i.e., in the fully rough region with $e^+ \geq 50$). After R is determined experimentally from Equation (16.37), the average friction factor $\left(\bar{f} \right)$ can be predicted by combining Equations (16.30), (16.31), and (16.37) for a given rib geometry $(e/D, P/e)$, rib angle α, channel aspect ratio (W/H), and Reynolds number.

   The correlation of the heat transfer function G shown in Figure 16.14 can be written as

$$G = 2.24(W/H)^{0.1} \left(e^+ \right)^{0.35} (\alpha/90°)^m (P/e/10)^n \qquad (16.38)$$

for a square channel $(W/H = 1)$: $m = 0.35$, $n = 0.1$
   for rectangular channels $(W/H = 2, 4)$: $m = 0$, $n = 0$.

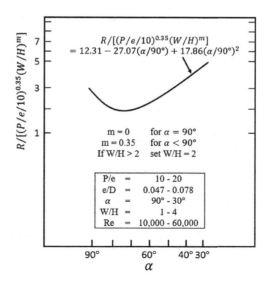

FIGURE 16.13   Final friction factor correlation.

After G is correlated experimentally from Equation (16.38), the ribbed-side-wall Stanton number (St_r) can be predicted by combining Equations (16.30), (16.31), (16.33), (16.37), and (16.38) for a given e/D, P/e, W/H, and Re. Assuming that Equation (16.33) can be used to correlate the average heat transfer data by replacing G and St_r by $\overline{G}$ and $\overline{St}$, the correlation of the $\overline{G}$ shown in Figure 16.14 can be expressed by

$$\overline{G} = 1.2G \qquad (16.39)$$

After determining St_r and $\overline{St}$ from above correlations, the smooth-side-wall Stanton number St_s can be found by

$$St_s = \overline{St} + (W/H)\left(\overline{St} - St_r\right) \qquad (16.40)$$

16.4 HEAT TRANSFER ENHANCEMENT FOR TURBINE BLADE INTERNAL COOLING APPLICATIONS

Typical Rotor Blade Cooling System

One of important heat transfer enhancement applications is for advanced cooling system designs of high temperature gas turbine blades. Gas turbine cooling technology is complex and varies between engine manufacturers. The cross-section of the cooling channels changes along the cord length of the blade due to the blade profile.

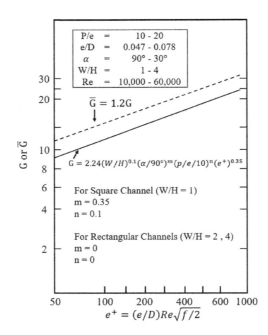

FIGURE 16.14 Final heat transfer correlation.

The internal cooling channels near the blade leading edge have been modeled as narrow rectangular channels with aspect ratios of AR = 1:4 and 1:2. Toward the trailing edge, the channels have wider aspect ratios of AR = 2:1 and 4:1. AR = aspect ratio = W/H = (rib-side-wall width) divided by (smooth-side wall height). Figure 16.15 shows the typical cooling technology with three major internal cooling zones in a turbine rotor blade with strategic film cooling in the leading edge, trailing edge, pressure and suction surfaces, and blade tip region. Jet impingement cooling can be used mostly in the leading-edge region of the rotor blade, due to structural constraints on the rotor blade under high-speed rotation and high centrifugal loads. Typically, a blade mid-chord region uses serpentine coolant passages with rib turbulators on the inner walls of the rotor blades, while the blade trailing-edge region uses short pins due to space limitation and structural integration. It is important to determine the associated coolant passage pressure losses for a given internal cooling design. This can help in designing an efficient cooling system and prevent local overheating on the rotor blade. Refer to Han et al. [15] for the details.

16.4.1 Heat Transfer Enhancement in Rectangular Channels with Rib Turbulators

In the early 1970s cooling designs of gas turbine blades, repeated transverse ribs (orthogonal ribs or α = 90-degree ribs, oriented 90-degree to the coolant flow) are cast on two opposite walls of internal cooling channels to enhance heat transfer.

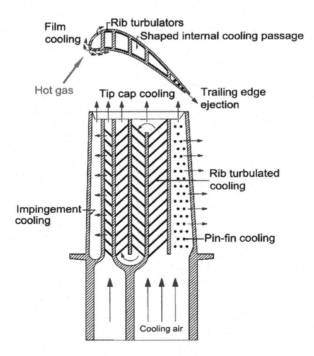

FIGURE 16.15 Schematic of a modern cooled turbine blade.

Internal cooling channel heat transfer enhancement mechanism is due to the flow separation-reattachment tripped by repeated transverse ribs (turbulence promoters, rib turbulators, trip strips). These flow separations re-attach the boundary layer to the heat transfer surface thus increasing the heat transfer coefficient. Repeated transverse ribs mostly disturb only the near-wall flow for heat transfer enhancement consequently the pressure drop penalty by repeated transverse ribs is affordable for turbine blade internal cooling designs. In general, repeated transverse ribs ($\alpha = 90°$) used for early turbine cooling designs are nearly square in cross-section with a typical relative rib height of 5% of channel hydraulic diameter ($e/D = 0.05$), and a rib spacing-to-height ratio of 10 ($p/e = 10$).

16.4.1.1   Angled Ribs and Heat Transfer Correlation

In 1980s, Han and his co-workers [11–14] systematically investigated heat transfer enhancement in rectangular channels with two opposite repeated-rib walls with a typical relative rib height of 5%–10% channel hydraulic diameter ($e/D = 0.05$–0.1), and a rib spacing-to-height ratio (p/e) of 5–20. They proposed that the repeated angled ribs (α oriented 30°, 45°, or 60° to the coolant flow), as sketched in Figures 16.5 and 16.15, performed better than the earlier repeated transverse ribs over a wide range of rectangular channels (channel aspect ratio AR varied from 4/1, 2/1, 1/1, ½, and ¼) applicable for gas turbine blade cooling designs. They found that the ribbed side heat transfer enhancements are about three times and the pressure drop penalties are about 4–8 times the values for 45 and 60-degree ribs compared to a smooth channel for Reynolds numbers between 15,000 and 60,000. The pressure drop penalties are only 2–4 times for the 45- and 60-degree angled ribs with the same level of heat transfer enhancement for the narrow aspect ratio channels (AR = ½ and ¼). However, for the same level of heat transfer enhancement in a broad aspect ratio channel (AR = 4/1), the pressure drop penalties are as high as 8–16 times the friction factor in a smooth channel for angled ribs. They concluded that the narrow aspect ratio channel performs better than a broad aspect ratio channel with angled ribs. They developed heat transfer and friction correlations for a wide range of rib turbulators geometry (such as rib height, angle, shape, and spacing), rectangular cooling channel aspect ratio, and Reynolds number. These correlations have been widely cited by turbine cooling researchers as baseline data comparison and used by turbine cooling designers for advanced heat transfer analysis and optimal cooling design prediction for newly developed turbine blades. All major turbine manufacturers utilize this new angled-rib cooling concept to replace earlier transverse-rib configurations for improving their turbine blade internal cooling designs. Refer to Han et al. [15] Chapter 4 for the details.

16.4.1.2   High Performance V-Shaped and Delta-Shaped Ribs

In 1990s Han and his co-workers [15] further proposed the high-performance 3-dimensional turbulence promoters, such as V-shaped ribs, and delta-shaped vortex generators for advanced turbine cooling system designs, as sketched in Figure 16.16. They found that the ribbed side heat transfer augmentations for 45- and 60-degree.

V-shaped ribs are higher than the previous-mentioned 45- and 60-degree-angled ribs for Reynolds numbers between 15,000 and 80,000. This is because the V-shaped ribs induce four-cell vortices as compared to the two-cell vortices induced by the angled ribs as sketched in Figure 16.16. And the ribbed side heat transfer enhancements for 45- and 60-degree broken angled ribs or broken V-shaped ribs are much higher than the corresponding 45- and 60-degree nonbroken angled ribs or nonbroken V-shaped ribs. However, the corresponding friction factor enhancements are comparable with each other for broken and nonbroken rib configurations. These promising V-shaped or delta-shaped ribs have been considered to be integrated into the new generation turbine blade cooling designs.

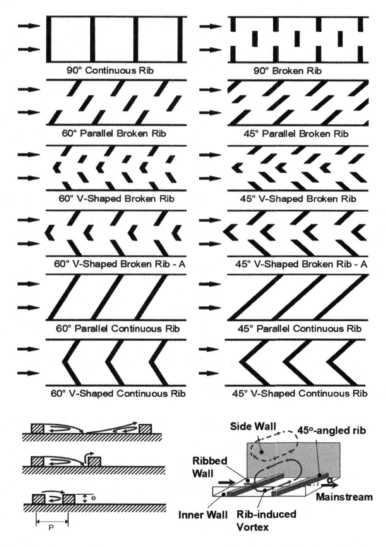

FIGURE 16.16 Upper: high-performance ribs for turbine blade internal cooling; lower: concept of flow separation-reattachment from ribs and a pair of vortices induced by angled ribs.

In 2000s Han and his co-workers [16] performed systematic experiments to measure heat transfer and pressure losses in a square channel with 45-degree round and sharp-edged ribs ($e/D = 0.1$ to 0.18, $p/e = 5$ to 15) at a wide range of Reynolds numbers ranging from 30,000 to very high flows of $Re = 400,000$. These high Reynolds are typical of land-based gas turbines. With round-edged ribs, the friction was lower, resulting in a smaller pressure drop. The heat transfer coefficients for the round-edged ribs, on the other hand, were similar to sharp-edged ribs. Figure 16.17 shows heat transfer correlation for a square channel with 45° round-edged ribs at high roughness Reynolds numbers.

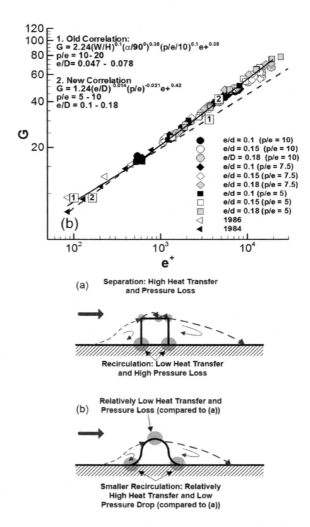

FIGURE 16.17 Heat transfer correlation for a square channel with 45° round edged ribs at high roughness Reynolds numbers.

16.4.2 Heat Transfer Enhancement with Rib Turbulators in Rotating Channels

16.4.2.1 Rotor Blade Internal Cooling

Rotational effects on the coolant passage heat transfer were not fully recognized until the late 1980s and early 1990s. Both Coriolis and rotating buoyancy forces can alter the flow and temperature profiles in the serpentine coolant passages and affect their surface heat transfer coefficient distributions. Since the direction of the Coriolis force is dependent on the direction of rotation and cooling flow, the Coriolis force has a different direction in the multipassage serpentine channels. The rotation direction remains the same for the two-passage channels, for example, but the direction of cooling flow gets reversed from the first channel to the second after the 180-degree turn. Therefore, the direction of the Coriolis force is opposite in these two channels, and the resultant rotation-induced secondary-flow (a pair of vortices) is different. Figure 16.18 shows the conceptual view of combined effects of Coriolis and rotational buoyancy on flow-temperature distribution [15,16]. For radial outward flow in the first channel, the Coriolis force shifts the core flow toward the trailing wall. If both trailing and leading walls are symmetrically heated, then faster-moving coolant near the trailing wall would be cooler than the slow-moving coolant near the leading wall. Rotational buoyancy is caused by

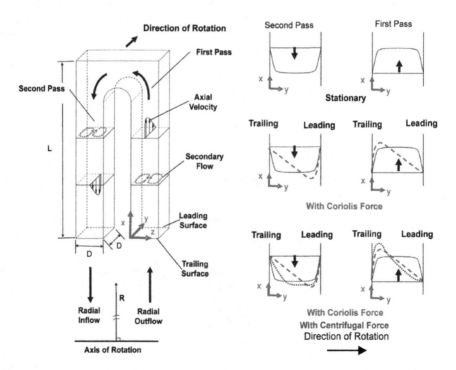

FIGURE 16.18 Conceptual view of coolant flow distribution through a two-pass rotating channel.

a strong centrifugal force that pushes cooler heavier fluid away from the center of rotation. In the first channel, rotational buoyancy affects the flow in a fashion similar to the Coriolis force and causes a further increase in the flow near the trailing wall of the first channel (to enhance heat transfer on the trailing wall), whereas the Coriolis force favors the leading side of the second channel. The rotational buoyancy in the second channel tries to make the flow distribution more uniform in the channel. Figure 16.19 shows the predicted secondary-flow (rotation induced vortices) and temperature distribution in a rotating two-pass rectangular channel by Al-Qahtani et al. [17]. It clearly shows that cooler fluid is shifted toward the trailing wall in the first channel radially outward flow and toward the leading wall in the second channel radially inward flow.

16.4.2.2 Heat Transfer Correlation with Rotation Number and Buoyancy Parameter

In addition to Reynolds number and Prandtl number, rotation number and buoyancy number are important parameters in determining flow and heat transfer in rotating channels. The rotation number (Ro) has been widely accepted to establish the strength of rotation by considering the relative strength of the Coriolis force compared to the bulk inertial force. The buoyancy number (Bo) is used to include the effects of density variation (centrifugal effects) and is defined as the ratio of the Grashof number to the square of the Reynolds number; both of which are based on the channel hydraulic diameter. Typical rotation numbers for aircraft engines are near 0.25 with Reynolds numbers in the range of 30,000. Increasing the range of the rotation number and buoyancy number is very important since gas turbine engineers can utilize these parameters in their analysis of heat transfer under rotating conditions.

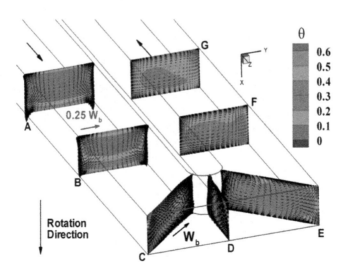

FIGURE 16.19 Secondary flow vectors and dimensionless temperature contours in a rotating two-pass rectangular smooth channel.

16.4.2.3 Summary of Rotation parameters

For turbulent air flow in rectangular channels, Nusselt number can be predicted by a given Reynolds number. In addition to Reynolds number, Rotation number and Buoyancy number are important parameters under rotation conditions. The following is a brief summary of these parameters.

$$\text{Re}_D = \frac{\rho V D_h}{\mu} = \frac{V D_h}{\nu}$$

$$= 10{,}000 \text{ to } 60{,}000 \text{ for aircraft turbines}$$

$$\text{Nu}_0 = \frac{h D_h}{k} = 0.023\,\text{Re}_D^{0.8}\text{Pr}^{0.4} \text{ for smooth channel heat transfer correlation}$$

$$\text{Ro} = \text{Rotation number}$$

$$= \frac{\text{Coriolis force}}{\text{Inertia force}}$$

$$= \frac{\Omega D_h}{V} = 0.2 \sim 0.5$$

$$\Omega = \text{RPM rotational speed}$$

$$D_h = \text{cooling channel hydraulic diameter}$$

If without rotation, Grashof number for natural convection is due to temperature gradient and gravity:

$$\text{Gr} = \frac{\text{buoyancy force}}{\text{viscous force}}$$

$$= \frac{(T_w - T_\infty)\beta g L^3}{\nu^2}$$

$$= \text{Grash of number}$$

where $\beta = \dfrac{1}{T}$ for air.

For rotation, replace g by $R\Omega^2$, L by D_h, T_∞ by T_c, T by T_c, thus, rotation buoyancy number is

$$\text{Bo} = \frac{\text{Gr}}{\text{Re}^2} = \frac{(T_w - T_c)\cdot \dfrac{1}{T_c}\cdot R\Omega^2 \cdot D_h^{\,3}}{\nu^2 \cdot \left(\dfrac{V D_h}{\nu}\right)^2}$$

$$= \text{Rotation Buoyancy number}$$

Re-arrange rotation buoyancy number as

$$\text{Bo} = \frac{\Delta\rho}{\rho} \cdot \frac{R\Omega}{V} \cdot \frac{D_h\Omega}{V}$$

$$= \frac{\Delta\rho}{\rho} \cdot \left(\frac{D_h\Omega}{v}\right)^2 \cdot \frac{R}{D_h}$$

$$= DR \cdot \text{Ro}^2 \cdot \frac{R}{D_h}$$

$$= 0.1 \sim 3.0$$

where

$$\text{DR} = \frac{T_w - T_c}{T_c} = \frac{\rho_c - \rho_w}{\rho_c} = \frac{\Delta\rho}{\rho}$$

ρ_c = coolant density based on coolant temperature

ρ_w = coolant density based on channel wall temperature

DR = coolant density ratio = $0.2 \sim 0.5$

R = distance between center of rotation and cooling channel (rotation arm length)

D_h = channel hydraulic diameter

$\dfrac{R}{D_h}$ = ratio of rotation arm length and channel hydraulic diameter

$\quad$ = $30 \sim 70$ for small to large turbines

$\text{Bo} \cong 0.2$ for aircraft turbine

$\cong 1 \sim 3$ for large land-based turbines

In general, Nusselt number depends on Reynolds number, Rotation number, and Buoyancy number:

$$\text{Nu} \sim \text{Ro} \text{ for a given } \text{Re}_D$$

$$\text{Nu} \sim \text{Bo} \text{ for a given } \text{Re}_D$$

$$\text{Nu} \sim DR \cdot \text{Ro}^2 \cdot \frac{R}{D_h} \text{ for a given } \text{Re}_D$$

16.4.2.4 Rotational Effect on Coolant Channel Heat Transfer

There are only limited papers available in the open literature that study the rotational effect on rotor coolant passage heat transfer (for example, several pioneer papers published by researchers reviewed and cited in Chapter 5 of Han et al. [15]). In 1990s Wagner et al. [18] and Johnson et al. [19] from the United Technologies Research Center studied heat transfer in rotating multipass coolant passages with square cross-section and smooth walls as well as ribbed walls at uniform wall temperature conditions. For the square channel with smooth wall, results show that the heat transfer can enhance 2–3 times on the trailing surface (at Bo = 1) and reduce up to 50% on the leading surface (at Bo = 0.2) for the first-pass radial outward flow passage as sketch in Figures 16.20 and 16.21; however, the reverse is true for the second-pass radial inward flow passage due to the flow direction change as sketch in Figures 16.22 and 16.23. Without considering the rotational effect, the coolant passage would be over-cooled on one surface while over-heated on the opposite surface. Results also show that the heat transfer difference between leading and trailing surfaces is greater in the first-pass than that in the second-pass due to the centrifugal buoyancy opposite to the flow direction, as sketched in Figure 16.18. They also studied heat transfer in rotating

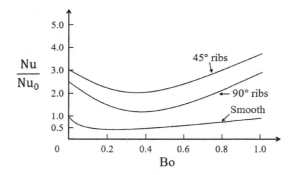

FIGURE 16.20 Nusselt number ratio versus buoyancy number on the leading surface of first-pass radially outward flow.

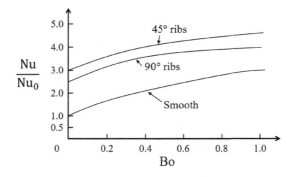

FIGURE 16.21 Nusselt number ratio versus buoyancy number on the trailing surface of first-pass radially outward flow.

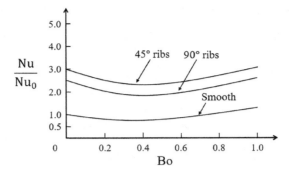

FIGURE 16.22 Nusselt number ratio versus buoyancy number on the trailing surface of second-pass radially inward flow.

multipass coolant passages with a square cross-section with two-opposite 90-degree and 45-degree rib roughened walls at uniform wall temperature conditions. Results show that rotation and buoyancy, in general, have less effect on the rib roughened coolant passage than on the smooth-wall coolant passage. This is because the heat transfer enhancement in the ribbed passages is already up to 2.5 (for 90-degree ribs at Bo = 0) ~ 3.0 (for 45-degree ribs at Bo = 0) times higher than in the smooth passages (at Bo = 0); therefore, the rotational effect is still important but with a reduced percentage. Results also show that, like a nonrotating channel, the 45-degree skewed/angled ribs perform better than 90-degree transverse ribs and subsequently better than the smooth channel.

16.4.2.5 Channel Aspect Ratio and Orientation Effect

Most of the above-mentioned studies dealt with square channels. However, a rectangular, triangular, or wedge passage is required in order to maintain the integrity of the gas turbine blade internal cooling design. The rectangular channel changes the effective rotation-induced secondary flow pattern (a pair of vortices) from that of a square duct as sketched in Figure 16.24. For this reason, one cannot simply apply

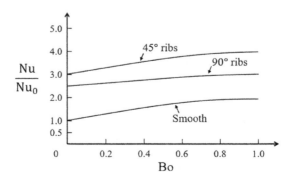

FIGURE 16.23 Nusselt number ratio versus buoyancy number on the leading surface of second-pass radially inward flow.

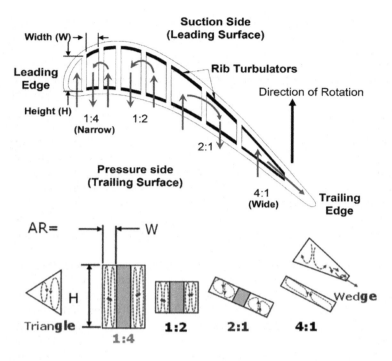

FIGURE 16.24 Typical turbine blade internal cooling channel with rotation-induced vortices.

the results in a square channel to that of a rectangular channel. In 2000s, Han and his co-workers [15] studied heat transfer in a two-pass rectangular rotating channel (AR = 2:1) with 45-degree angled ribbed walls. They found that the effect of rotation on the two-pass rectangular channel is very similar to that on the two-pass square channel but the leading surface heat transfer coefficient does not vary much with the rotation compared with the square channel case. The 4:1 ribbed channel was employed to examine the overall thermal performance (not only the heat transfer augmentation but also the pressure drop penalty). They found that the overall thermal performance of the discrete V-shaped and discrete W-shaped ribs are better than the V-shaped and W-shaped ribs and subsequently better than the standard angled ribs for both stationary and rotation cases. They also reported the rotation effect increased heat transfer on the trailing wall but decreased the heat transfer on the leading wall in the first pass of both the 1:2 and 1:4 channels. In the second pass, however, the difference of the heat transfer between the leading and trailing walls was dramatically reduced under the rotation condition when compared to the first pass. It was suggested that the 180-degree turn induced vortices dominate the rotation induced vortices for the low aspect ratio channels. They also concluded that the orientation of rectangular channel relative to direction of rotation would affect rotation-induced vortices size/strength and hence change heat transfer results on leading and trailing surfaces. Refer to Han and his co-workers [15,16] for the details.

16.4.3 Heat Transfer Enhancement with Impingement Cooling and Pin-Fin Cooling

Leading-Edge Impingement Cooling: Jet impingement cooling is most suitable for the leading edge of the blade where the thermal load is highest and a thicker cross-section of this portion of the blade can suitably accommodate impingement cooling as shown in Figures 16.15. There are many studies focused on the effects of jet-hole size and distribution, jet-to-target surface distance, spent-air cross flow, cooling channel cross-section, and the target surface shape on the heat transfer coefficient distribution. Recent studies have considered the combined effects of leading-edge inner surface with rib or pin roughness coupled with jet impingement for further heat transfer enhancement. Many gas turbine blade impingement cooling papers reviewed and documented in Chapter 4 of Han et al. [15] and [20].

Rotational Effect on Impingement Cooling: In general, it has been reported that the rotation decreases the impingement heat transfer on both leading and trailing surfaces with more effect on the trailing side (up to 20% heat transfer reduction). Overall, the effectiveness of the jet impingement is reduced under rotating conditions due to deflection from the target surface by centrifugal forces. Refer to Han et al. [15] Chapter 5 for the details.

Trailing-Edge Pin Fins Cooling: Pin-fins are mostly used in the narrow trailing edge of a turbine blade where impingement and ribbed channels cannot be accommodated due to manufacturing constraint as shown in Figures 16.15. Heat transfer in turbine pin-fin cooling arrays combines the cylinder heat transfer and end-wall heat transfer. Due to the turbulence enhancement caused by pins (wakes and horseshoe vortex), heat transfer from end-walls is higher than smooth wall cases. Long pins can increase the effective heat transfer area and perform better than short pins. There have been many investigations that studied the effects of pin array (inline or staggered), pin size (length-to-diameter ratio = 0.5–4), pin distribution (streamwise- and spanwise-to-diameter ratio = 2–4), pin shape (with and without a fillet at the base of the cylindrical pin; oblong, cube, and diamond shaped pins as well as the stepped diameter cylindrical pins), partial length pins, flow convergence and turning, and with trailing edge coolant extraction on the heat transfer coefficient and friction factor distributions in pin-fin cooling channels. Many gas turbine blade pin-fins cooling papers reviewed and documented in Chapter 4 of Han et al. [15] and [20].

Rotational Effect on Pin Fins Cooling: The effect of rotation on heat transfer in narrow rectangular channels (AR = 4:1–8:1) with typical pin-fin array used in turbine blade trailing edge design and oriented at 150-deg with respect to the plane of rotation. Results show that turbulent heat transfer in a stationary pin-fin channel can be enhanced up to 3.8 times that of a smooth channel; rotation enhances the heat transferred from the pin-fin channels up to 1.5 times that of the stationary pin-fin channels. Most importantly, for narrow rectangular or wedge-shaped pin-fin channels oriented at 135-deg with respect to the plane of rotation, heat transfer enhancement on both the leading and trailing surfaces increases with rotation. This provides positive information for the cooling designers. Refer to Han et al. [15] Chapter 5 for the details.

REMARKS

Based on law of wall for velocity and temperature profiles, roughness, and heat transfer functions have been developed to correlate friction and heat transfer data for tube flows with similar sand grain roughness geometry. The friction and heat transfer similarity laws have been extended for tube flows with repeated-rib roughness over a wide range of similar and nonsimilar geometric parameters. Turbulent flow heat transfer and friction in tubes, annulus, between parallel plates, or in rectangular channels with artificially repeated-rib roughness have been studied extensively. The effects of rib height-to-equivalent diameter ratio e/D, rib pitch-to-height ratio P/e, rib shape (rib cross section), and rib angle of attack α, on friction and heat transfer over a wide range of Reynolds number have been correlated with the roughness and heat transfer functions. These correlations can be applied for aerospace, solar collector, nuclear fuel element, electronic cooling, and general heat exchanger applications. One of important heat transfer enhancement applications is for advanced cooling system designs of high-temperature gas turbine blades. Both friction similarity law and heat transfer similarity law have been successfully applied to correlate friction factor and Stanton number for turbulent air flow in rectangular channels with two opposite-sided rib-roughened walls. The effects of Coriolis and rotation buoyancy forces on heat transfer enhancement in rectangular channels with turbulence promoters have also been investigated. More studies are needed for rotating blade-shaped coolant channels (realistic cooling passage geometry, shape, and orientation) with high-performance rib-turbulators. To further augment heat transfer, the compound cooling techniques, such as a combination of impinging jets with ribs, pins, dimples, vortex generators, etc., have been explored for advanced cooling systems. In general, the study of higher heat transfer enhancement versus lower pressure drop penalty would continue to identify the best heat transfer performance geometry for turbulent flow heat transfer enhancement.

REFERENCES

1. W. F. Cope, "The friction and heat-transfer coefficients of rough pipes," *Proceedings of the Institution of Mechanical Engineers*, Vol. 145, pp. 99–105, 1941.
2. J. Nikuradse, "Laws for Flow in Rough Pipes," *NACA* TM 1292, 1950.
3. D. F. Dipprey and R. H. Sabersky, "Heat and momentum transfer in smooth and rough tubes at various Prandtl Number," *International Journal of Heat and Mass Transfer*, Vol. 6, pp. 329–353, 1963.
4. E.W. Sams, "Experimental Investigation of Average Heat Transfer and Friction Coefficients for Air Flowing Circular Pipes Having Square-Thread Type Roughness," *NACA* RM-E 52-D17, 1952.
5. D. Wilkie, "Forced Convection Heat Transfer from Surfaces Roughened by Transverse Ribs," In *Proceedings of the 3rd International Heat Transfer Conference*, Vol. 1. AIChE, New York, 1966.
6. L. White and D. Wilkie, "The heat transfer and pressure loss characteristics of some multi-start ribbed surfaces," In *Augmentation of Convection Heat and Mass Transfer*, Edited by A. E. Bergles and R. L. Webb, ASME, New York, 1970.
7. R.L. Webb, E.R.G. Eckert, and R.J. Goldstein, "Heat transfer and friction in tubes with repeated-rib roughness," *International Journal of Heat and Mass Transfer*, Vol. 14(4), pp. 601–617, 1971.

8. J.C. Han, L.R. Glicksman, and W.M. Rohsenow, "An investigation of heat transfer and friction for rib-roughened surfaces," *International Journal of Heat and Mass Transfer*, Vol. 21, pp. 1143–1156, 1978.

9. D.L. Gee and R.L. Webb, "Forced convection heat transfer in helically rib-roughened tubes," *International Journal of Heat and Mass Transfer*, Vol. 23, pp. 1127–1136, 1980.

10. M.D. Dalle Donne and L. Meyer, "Turbulent convective heat transfer from rough surfaces with two-dimensional rectangular ribs," *International Journal of Heat and Mass Transfer*, Vol. 20, pp. 582–620, 1977.

11 J.C. Han, J.S. Park, and C.K. Lei, "Heat transfer enhancement in channels with turbulence promoters," *ASME Journal of Engineering for Gas Turbines and Power*, Vol. 107(1), pp. 628–635, 1985.

12. J.C. Han, and J.S. Park, "Developing heat transfer in rectangular channels with rib turbulators," *International Journal of Heat and Mass Transfer*, Vol. 31(1), pp. 183–195, 1988.

13. J.C. Han, "Heat transfer and friction in channels with two opposite rib-roughened walls," *ASME Journal of Heat Transfer*, Vol. 106(4), pp. 774–781, 1984.

14. J.C., Han, "Heat transfer and friction characteristics in rectangular channels with rib turbulators," *ASME Journal of Heat Transfer*, Vol. 110(2), pp. 321–328, 1988.

15. J.C. Han, S. Dutta, and S. Ekkad, *Gas Turbine Heat Transfer and Cooling Technology*, First edition, CRC Press, Taylor & Francis Group, New York, 2000; Second edition, 2013.

16. J.C. Han, "Advanced cooling in gas turbines 2016 Max Jakob memorial award paper," *ASME Journal of Heat Transfer*, Vol. 140(11), pp. 1–20, 2018.

17. M. Al-Qahtani, Y.J. Jang, H.C. Chen, and J.C. Han, "Prediction of flow and heat transfer in rotating two-pass rectangular channels with 45-degree rib turbulators," *ASME Journal of Turbomachinery*, Vol. 124, pp. 242–250, 2002.

18. J.H. Wagner, B.V. Johnson, R.A. Graziani, and F.C. Yeh, "Heat transfer in rotating serpentine passages with trips normal to the flow," *ASME Journal of Turbomachinery*, Vol. 114, pp. 847–857, 1992.

19. B.V. Johnson, J.H. Wagner, G.D. Steuber, and F.C. Yeh, "Heat transfer in rotating serpentine passages with trips skewed to the flow," *ASME Journal of Turbomachinery*, Vol. 116, pp. 113–123, 1994.

20. L.M. Wright, and J.C. Han, "Heat transfer enhancement for turbine blade internal cooling," *Journal of Enhanced Heat Transfer*, Vol. 21(3–4), pp. 111–140, 2015.

Appendix

Mathematical Relations and Functions

A.1 USEFUL FORMULAS

$$e^x = 1 + \frac{x}{1!} + \frac{x^2}{2!} + \frac{x^3}{3!} + \frac{x^4}{4!} + \cdots$$

$$\sin x = x - \frac{x^3}{3!} + \frac{x^5}{5!} - \frac{x^7}{7!} + \cdots$$

$$\cos x = 1 - \frac{x^2}{2!} + \frac{x^4}{4!} - \frac{x^6}{6!} + \frac{x^8}{8!} - \cdots$$

$$\sinh x = x + \frac{x^3}{3!} + \frac{x^5}{5!} + \frac{x^7}{7!} + \cdots$$

$$\cosh x = 1 + \frac{x^2}{2!} + \frac{x^4}{4!} + \frac{x^6}{6!} + \cdots$$

$$\sinh x = \frac{e^x - e^{-x}}{2}$$

$$\cosh x = \frac{e^x - e^{-x}}{2}$$

$$d\sin x = \cos x\, dx; \quad d\cos x = -\sin x dx$$

$$d\sinh x = \cos x\, dx; \quad d\cosh x = +\sinh x\, dx$$

$$\int \sin x\, dx = -\cos x + c; \quad \int \cos x\, dx = \sin x + c$$

$$\int \sin^2 x\, dx = -\frac{1}{2}\sin x \cos x \frac{1}{2}x + c = -\frac{1}{4}\sin 2x + \frac{1}{2}x + c$$

$$\int \cos^2 x\, dx = \frac{1}{2}\sin x \cos x + \frac{1}{2}x + c = \frac{1}{4}\sin 2x + \frac{1}{2}x + c$$

$$\int \sinh x\, dx = \cosh x + c; \quad \int \cosh x\, dx = \sinh x + c$$

if

$$\int \sinh^2 x \, dx = \frac{1}{2} \sinh x \cosh x - \frac{1}{2} x + c$$

$$\int \cosh^2 x \, dx = \frac{1}{2} \sinh x \cosh x + \frac{1}{2} x + c$$

A.2   HYPERBOLIC FUNCTIONS [1]

x	$\sinh x$	$\cosh x$	$\tanh x$
0.00	0.0000	1.0000	0.00000
0.10	0.1002	1.0050	0.09967
0.20	0.2013	1.0201	0.19738
0.30	0.3045	1.0453	0.29131
0.40	0.4108	1.0811	0.37995
0.50	0.5211	1.1276	0.46212
0.60	0.6367	1.1855	0.53705
0.70	0.7586	1.2552	0.60437
0.80	0.8881	1.3374	0.66404
0.90	1.0265	1.4331	0.71630
1.00	1.1752	1.5431	0.76159
1.10	1.3356	1.6685	0.80050
1.20	1.5095	1.8107	0.83365
1.30	1.6984	1.9709	0.86172
1.40	1.9043	2.1509	0.88535
1.50	2.1293	2.3524	0.90515
1.60	2.3756	2.5775	0.92167
1.70	2.6456	2.8283	0.93541
1.80	2.9422	3.1075	0.94681
1.90	3.2682	3.4177	0.95624
2.00	3.6269	3.7622	0.96403
2.10	4.0219	4.1443	0.97045
2.20	4.4571	4.5679	0.97574
2.30	4.9370	5.0372	0.98010
2.40	5.4662	5.5569	0.98367
2.50	6.0502	6.1323	0.98661
2.60	6.6947	6.7690	0.98903
2.70	7.4063	7.4735	0.99101
2.80	8.1919	8.2527	0.99263
2.90	9.0596	9.1146	0.99396
3.00	10.018	10.068	0.99505
3.50	16.543	16.573	0.99818
4.00	27.290	27.308	0.99933
4.50	45.003	45.014	0.99975
5.00	74.203	74.210	0.99991

(Continued)

x	sinh x	cosh x	tanh x
6.00	201.71	201.72	0.99999
7.00	548.32	548.32	1.00000
8.00	1490.5	1490.5	1.00000
9.00	4051.5	4051.5	1.00000
10.00	11013	11013	1.00000

A.3 BESSEL FUNCTIONS

A.3.1 BESSEL FUNCTIONS AND PROPERTIES [2]

Behaviors of Bessel functions for small arguments:

$$J_0(x) = 1 - \frac{(x/2)^2}{(1!)^2} + \frac{(x/2)^4}{(2!)^2} - \cdots$$

$$J_1(x) = \frac{x}{2} - \frac{(x/2)^3}{1!2!} + \frac{(x/2)^5}{2!3!} - \cdots$$

$$\vdots$$

$$J_\nu(x) = \frac{(x/2)^\nu}{\Gamma(\nu+1)} \left\{ 1 - \frac{(x/2)^2}{1!(\nu+1)} + \frac{(x/2)^4}{2!(\nu+1)(\nu+2)} - \cdots \right\}$$

$$I_0(x) = 1 + \frac{(x/2)^2}{(1!)^2} + \frac{(x/2)^4}{(2!)^2} + \cdots$$

$$I_1(x) = \frac{x}{2} - \frac{(x/2)^3}{1!2!} + \frac{(x/2)^5}{2!3!} - \cdots$$

$$\vdots$$

$$I_\nu(x) = \frac{(x/2)^\nu}{\Gamma(\nu+1)} \left\{ 1 + \frac{(x/2)^2}{1!(\nu+1)} + \frac{(x/2)^4}{2!(\nu+1)(\nu+2)} + \cdots \right\}$$

Behaviors of Bessel functions for large arguments:

$$J_\nu(x) \approx \sqrt{\frac{2}{\pi x}} \cos\left(x - \frac{\pi}{4} - \frac{\nu\pi}{2} \right)$$

$$Y_\nu(x) \approx \sqrt{\frac{2}{\pi x}} \sin\left(x - \frac{\pi}{4} - \frac{\nu\pi}{2} \right)$$

$$I_\nu(x) \approx \frac{e^x}{\sqrt{2\pi x}} \left(1 - \frac{4\nu^2 - 1}{8x} \right)$$

$$K_\nu \approx \sqrt{\frac{\pi}{2x}} e^{-x} \left(1 + \frac{4\nu^2 - 1}{8x} \right)$$

Properties of Bessel functions:

$$\frac{d}{dx}\left[J_0\left(mx\right)\right] = -mJ_1\left(mx\right), \frac{d}{dx}\left[Y_0\left(mx\right)\right] = -mY_1\left(mx\right)$$

$$\frac{d}{dx}\left[I_0\left(mx\right)\right] = -mI_1\left(mx\right), \frac{d}{dx}\left[K_0\left(mx\right)\right] = -mY_1\left(mx\right)$$

A.3.2 BESSEL FUNCTIONS OF THE FIRST KIND [1]

x	$J_0(x)$	$J_1(x)$
0.0	1.0000	0.0000
0.1	0.9975	0.0499
0.2	0.9900	0.0995
0.3	0.9776	0.1483
0.4	0.9604	0.1960
0.5	0.9385	0.2423
0.6	0.9120	0.2867
0.7	0.8812	0.3290
0.8	0.8463	0.3688
0.9	0.8075	0.4059
1.0	0.7652	0.4400
1.1	0.7196	0.4709
1.2	0.6711	0.4983
1.3	0.6201	0.5220
1.4	0.5669	0.5419
1.5	0.5118	0.5579
1.6	0.4554	0.5699
1.7	0.3980	0.5778
1.8	0.3400	0.5815
1.9	0.2818	0.5812
2.0	0.2239	0.5767
2.1	0.1666	0.5683
2.2	0.1104	0.5560
2.3	0.0555	0.5399
2.4	0.0025	0.5202

A.3.3 MODIFIED BESSEL FUNCTIONS OF THE FIRST AND SECOND KINDS [1]

x	$e^{-x}I_0(x)$	$e^{-x}I_1(x)$	$e^{-x}K_0(x)$	$e^{-x}K_1(x)$
0.0	1.0000	0.0000	00	0
0.2	0.8269	0.0823	2.1407	5.8334
0.4	0.6974	0.1368	1.6627	3.2587
0.6	0.5993	0.1722	1.4167	2.3739
0.8	0.5241	0.1945	1.2582	1.9179
1.0	0.4657	0.2079	1.1445	1.6361
1.2	0.4198	0.2152	1.0575	1.4429
1.4	0.3831	0.2185	0.9881	1.3010
1.6	0.3533	0.2190	0.9309	1.1919
1.8	0.3289	0.2177	0.8828	1.1048
2.0	0.3085	0.2153	0.8416	1.0335
2.2	0.2913	0.2121	0.8056	0.9738
2.4	0.2766	0.2085	0.7740	0.9229
2.6	0.2639	0.2046	0.7459	0.8790
2.8	0.2528	0.2007	0.7206	0.8405
3.0	0.2430	0.1968	0.6978	0.8066
3.2	0.2343	0.1930	0.6770	0.7763
3.4	0.2264	0.1892	0.6579	0.7491
3.6	0.2193	0.1856	0.6404	0.7245
3.8	0.2129	0.1821	0.6243	0.7021
4.0	0.2070	0.1787	0.6093	0.6816
4.2	0.2016	0.1755	0.5953	0.6627
4.4	0.1966	0.1724	0.5823	0.6453
4.6	0.1919	0.1695	0.5701	0.6292
4.8	0.1876	0.1667	0.5586	0.6142
5.0	0.1835	0.1640	0.5478	0.6003
5.2	0.1797	0.1614	0.5376	0.5872
5.4	0.1762	0.1589	0.5279	0.5749
5.6	0.1728	0.1565	0.5188	0.5633
5.8	0.1696	0.1542	0.5101	0.5525
6.0	0.1666	0.1520	0.5019	0.5422
6.4	0.1611	0.1479	0.4865	0.5232
6.8	0.1561	0.1441	0.4724	0.5060
7.2	0.1515	0.1405	0.4595	0.4905
7.6	0.1473	0.1372	0.4476	0.4762
8.0	0.1434	0.1341	0.4366	0.4631
8.4	0.1398	0.1312	0.4264	0.4511
8.8	0.1365	0.1285	0.4168	0.4399
9.2	0.1334	0.1260	0.4079	0.4295
9.6	0.1305	0.1235	0.3995	0.4198
10.0	0.1278	0.1213	0.3916	0.4108

A.4 GAUSSIAN ERROR FUNCTION [1]

η	$erf\eta$	η	$erf\eta$	η	$erf\eta$
0.00	0.00000	0.36	0.38933	1.04	0.85865
0.02	0.02256	0.38	0.40901	1.08	0.87333
0.04	0.04511	0.40	0.42839	1.12	0.88679
0.06	0.06762	0.44	0.46622	1.16	0.89910
0.08	0.09008	0.48	0.50275	1.20	0.91031
0.10	0.11246	0.52	0.53790	1.30	0.93401
0.12	0.13476	0.56	0.57162	1.40	0.95228
0.14	0.15695	0.60	0.60386	1.50	0.96611
0.16	0.17901	0.64	0.63459	1.60	0.97635
0.18	0.20094	0.68	0.66378	1.70	0.98379
0.20	0.22270	0.72	0.69143	1.80	0.98909
0.22	0.24430	0.76	0.71754	1.90	0.99279
0.24	0.26570	0.80	0.74210	2.00	0.99532
0.26	0.28690	0.84	0.76514	2.20	0.99814
0.28	0.30788	0.88	0.78669	2.40	0.99931
0.30	0.32863	0.92	0.80677	2.60	0.99976
0.32	0.34913	0.96	0.82542	2.80	0.99992
0.34	0.36936	1.00	0.84270	3.00	0.99998

The Gaussian error function is defined as

$$\text{erf}\,\eta = \frac{2}{\sqrt{\pi}} \int_0^\eta e^{-u^2}\,du$$

The complementary error function is defined as

$$\text{erfc}\,\eta \equiv 1 - \text{erf}\,\eta$$

$$\eta = \frac{x}{\sqrt{4\alpha t}}$$

REFERENCES

1. F. Incropera and D. Dewitt, *Fundamentals of Heat and Mass Transfer*, Fifth Edition, John Wiley & Sons, New York, 2002.
2. V. Arpaci, *Conduction Heat Transfer*, Addison-Wesley Publishing Company, Reading, MA, 1966.

Index

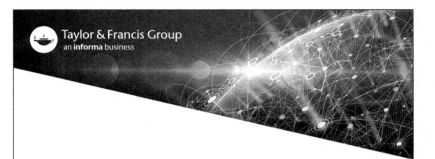

Printed in the United States
by Baker & Taylor Publisher Services